T0180802

Lecture Notes in Computer Science 14365

The series Lecture Notes in Computer Science (LNCS), including its subseries Lecture Notes in Artificial Intelligence (LNAI) and Lecture Notes in Bioinformatics (LNBI), has established itself as a medium for the publication of new developments in computer science and information technology research, teaching, and education.

LNCS enjoys close cooperation with the computer science R & D community, the series counts many renowned academics among its volume editors and paper authors, and collaborates with prestigious societies. Its mission is to serve this international community by providing an invaluable service, mainly focused on the publication of conference and workshop proceedings and postproceedings. LNCS commenced publication in 1973.

Gian Luca Foresti · Andrea Fusiello ·
Edwin Hancock
Editors

Image Analysis and Processing - ICIAP 2023 Workshops

Udine, Italy, September 11–15, 2023
Proceedings, Part I

Editors
Gian Luca Foresti
University of Udine
Udine, Italy

Andrea Fusiello
University of Udine
Udine, Italy

Edwin Hancock
University of York
York, UK

ISSN 0302-9743 ISSN 1611-3349 (electronic)
Lecture Notes in Computer Science
ISBN 978-3-031-51022-9 ISBN 978-3-031-51023-6 (eBook)
https://doi.org/10.1007/978-3-031-51023-6

This Springer imprint is published by the registered company Springer Nature Switzerland AG
The registered company address is: Gewerbestrasse 11, 6330 Cham, Switzerland

Paper in this product is recyclable.

Preface

This volume contains the papers accepted for presentation at the workshops hosted by the 22nd International Conference on Image Analysis and Processing (ICIAP 2023), held in Udine, Italy, from 11 to 15 September 2023. It was co-organised by the Department of Informatics, Mathematics and Physics (DMIF) and the Polytechnic Department of Engineering and Architecture (DPIA) of the University of Udine and sponsored by ST Microelectronics. The conference traditionally covers topics related to theoretical and experimental areas of Computer Vision, Image Processing, Pattern Recognition and Machine Learning, with emphasis on theoretical aspects and applications.

ICIAP 2023 also hosted 15 workshops (including 2 competitions), on topics of great relevance with respect to the state of the art. For the first time, workshops in the same field were collected in two hubs, the Medical Imaging Hub (MIH) and the Digital Humanities Hub (DHH). In addition, an industrial poster session was organised to bring together papers written by scientists working in industry and with a strong focus on application.

In total, 72 workshop papers and 10 industrial poster session papers were accepted for publication. As for the main conference, papers were selected through a double-blind peer review process, considering originality, significance, clarity, soundness, relevance and technical content. Each submission was assigned to at least three reviewers.

This volume contains 10 papers from the industrial poster session, and 31 papers from the following workshops:

- Advances in Gaze Analysis, Visual attention and Eye-gaze modelling (AGAVE)
- Beyond Vision: Physics meets AI (BVPAI)
- Automatic Affect Analysis and Synthesis (3AS)
- International Contest on Fire Detection (ONFIRE)
- Recent Advances in Digital Security: Biometrics and Forensics (BioFor)
- Computer Vision for Environment Monitoring and Preservation (CVEMP)
- Generation of Human Face and Body Behavior (GHB)

The 41 papers accepted for the other workshops are included in the companion volume (LNCS 14366).

AGAVE: Advances in Gaze Analysis, Visual attention and Eye-gaze modelling, organized by Vittorio Cuculo (University of Modena and Reggio Emilia, Italy), Marcella Cornia (University of Modena and Reggio Emilia, Italy), Alessandro D'Amelio (University of Milan, Italy) and Dario Zanca (University of Erlangen-Nuremberg, Germany), provided an overview of recent advances in the field of visual attention, which refers to the human ability to focus on the most relevant parts of images or videos.

BVPAI: Beyond Vision: Physics meets AI, organized by Alessandro Bombini (INFN-Firenze, Italy), Lisa Castelli (INFN-Firenze, Italy) and Stefano dal Pra (INFN-CNAF, Italy), was devoted to nuclear and other physics-based imaging technologies and applications of machine learning and artificial intelligence methods to physics.

3AS: Automatic Affect Analysis and Synthesis, organized by Nadia Bianchi-Berthouze (University College of London, UK), Simone Bianco (University of Milano-Bicocca, Italy), Luigi Celona (University of Milano-Bicocca, Italy), Vittorio Cuculo (University of Modena and Reggio Emilia, Italy), Alessandro D'Amelio (University of Milano, Italy) and Paolo Napoletano (University of Milano-Bicocca, Italy), analyzed the challenging issues related to affective computing, namely how systems should be conceived, validated and compared.

ONFIRE: International Contest on Real-Time Fire Detection on the Edge, organized by Diego Gragnaniello (University of Salerno, Italy), Antonio Greco (University of Salerno, Italy), Carlo Sansone (University of Napoli, Italy) and Bruno Vento (University of Napoli, Italy), was an international competition among methods for real-time fire detection from videos, which were evaluated in terms of accuracy, notification promptness and processing resources.

BioFor: Recent Advances in Digital Security, Biometrics and Forensics, organized by Daniel Riccio (University of Naples Federico II, Italy), Silvio Barra (University of Naples Federico II, Italy) and Javier Lorenzo-Navarro (Universidad de Las Palmas de Gran Canaria, Spain), provided a forum for researchers involved in the fields of security, forensics and biometrics.

CVEMP: Computer Vision for Environment Monitoring and Preservation, organized by Paolo Russo (Sapienza University of Rome, Italy), Fabiana di Ciaccio (University of Florence, Italy) and Diego Marcos (Inria Université Cote d'Azur, France), focused on the usage of image processing, machine learning and computer vision techniques for monitoring and analyzing environmental data.

GHB: Generation of Human Face and Body Behaviour, organized by Mohamed Daoudi (IMT Nord Europe, France) and Stefano Berretti (University of Florence, Italy), provided an overview of recent advances in the context of generating novel sequences of human facial expressions, talking heads or body movements.

The success of ICIAP 2023 was due to the contribution of many people. Special thanks go to the Workshop Chairs (Federica Arrigoni and Lauro Snidaro) and to all the workshop organizers who made such an interesting program possible. We hope that you will find the papers in this volume interesting and informative, and that they will inspire you to further research in the field of image analysis and processing.

September 2023

Gian Luca Foresti
Andrea Fusiello
Edwin Hancock

Organization

General Chairs

Gian Luca Foresti	University of Udine, Italy
Andrea Fusiello	University of Udine, Italy
Edwin Hancock	University of York, UK

Program Chairs

Michael Bronstein	University of Oxford, UK
Barbara Caputo	Politecnico Torino, Italy
Giuseppe Serra	University of Udine, Italy

Steering Committee

Virginio Cantoni	University of Pavia, Italy
Luigi Pietro Cordella	University of Napoli Federico II, Italy
Rita Cucchiara	University of Modena-Reggio Emilia, Italy
Alberto Del Bimbo	University of Firenze, Italy
Marco Ferretti	University of Pavia, Italy
Gian Luca Foresti	University of Udine, Italy
Fabio Roli	University of Cagliari, Italy
Gabriella Sanniti di Baja	ICAR-CNR, Italy

Workshop Chairs

Federica Arrigoni	Politecnico Milano, Italy
Lauro Snidaro	University of Udine, Italy

Tutorial Chairs

Christian Micheloni	University of Udine, Italy
Francesca Odone	University of Genova, Italy

Publications Chairs

Claudio Piciarelli	University of Udine, Italy
Niki Martinel	University of Udine, Italy

Publicity/Social Chairs

Matteo Dunnhofer	University of Udine, Italy
Beatrice Portelli	University of Udine, Italy

Industrial Liaison Chair

Pasqualina Fragneto	STMicroelectronics, Italy

Local Organization Chairs

Eleonora Maset	University of Udine, Italy
Andrea Toma	University of Udine, Italy
Emanuela Colombi	University of Udine, Italy
Alex Falcon	University of Udine, Italy
Andrea Brunello	University of Udine, Italy

Area Chairs

Pattern Recognition

Raffaella Lanzarotti	University of Milano, Italy
Nicola Strisciuglio	University of Twente, The Netherlands

Machine Learning and Deep Learning

Tatiana Tommasi	Politecnico Torino, Italy
Timothy M. Hospedales	University of Edinburgh, UK

3D Computer Vision and Geometry

Luca Magri	Politecnico Milano, Italy
James Pritts	CTU Prague, Czech Republic

Image Analysis: Detection and Recognition

Giacomo Boracchi	Politecnico Milano, Italy
Mårten Sjöström	Mid Sweden University, Sweden

Video Analysis and Understanding

Elisa Ricci	University of Trento, Italy

Shape Representation, Recognition and Analysis

Efstratios Gavves	University of Amsterdam, The Netherlands

Biomedical and Assistive Technology

Marco Leo	CNR, Italy
Zhigang Zhu	City College of New York, USA

Digital Forensics and Biometrics

Alessandro Ortis	University of Catania, Italy
Christian Riess	Friedrich-Alexander University, Germany

Multimedia

Francesco Isgrò	University of Napoli Federico II, Italy
Oliver Schreer	Fraunhofer HHI, Germany

Cultural Heritage

Lorenzo Baraldi	University of Modena-Reggio Emilia, Italy
Christopher Kermorvant	Teklia, France

Robot Vision and Automotive

Alberto Pretto	University of Padova, Italy
Henrik Andreasson	Örebro University, Sweden
Emanuele Rodolà	Sapienza University of Rome, Italy
Zorah Laehner	University of Siegen, Germany

Augmented and Virtual Reality

Andrea Torsello	University of Venezia Ca' Foscari, Italy
Richard Wilson	University of York, UK

Geospatial Analysis

Enrico Magli	Politecnico Torino, Italy
Mozhdeh Shahbazi	University of Calgary, Canada

Computer Vision for UAVs

Danilo Avola	Sapienza University of Rome, Italy
Parameshachari B. D.	Nitte Meenakshi Institute of Technology, India

Brave New Ideas

Marco Cristani	University of Verona, Italy
Hichem Sahbi	Sorbonne University, France

Endorsing Institutions

International Association for Pattern Recognition (IAPR)
Italian Association for Computer Vision, Pattern Recognition and Machine Learning (CVPL)

Contents – Part I

Contents – Part II

Multi-modal Medical Imaging Processing (M3IP)

Federated Learning in Medical Imaging and Vision (FEDMED)

Fine Art Pattern Extraction and Recognition (FAPER)

Pattern Recognition for Cultural Heritage (PatReCH)

Industrial Poster Session

Instance Segmentation Applied to Underground Infrastructures

R. Haenel[1,2](✉), Q. Semler[1], E. Semin[1], S. Tabbone[3], and P. Grussenmeyer[2]

[1] SYSLOR SAS, 57000 Metz, France
{raphael.haenel,quentin.semler,edouard.semin}@syslor.net
[2] ICUBE Laboratory UMR 7357, Photogrammetry and Geomatics Group, University of Strasbourg, CNRS, INSA Strasbourg, 67000 Strasbourg, France
{raphael.haenel,pierre.grussenmeyer}@insa-strasbourg.fr
[3] University of Lorraine, CNRS LORIA UMR7503, Campus Scientifique, 54506 Vandoeuvre-Lès-Nancy, France
antoine.tabbone@loria.fr

Abstract. As underground infrastructures suffer from strikes during the ground excavation procedure resulting from poor subsurface utilities documentation, this paper aims to improve the mapping of such elements by integrating instance segmentation techniques. To perform supervised training of the well-known Mask R-CNN architecture, we created our own dataset based on surveys that we have access to from the SYSLOR company, resulting in around 2600 labelled images. Through several training sessions, performed with K-fold cross validation, we studied the level of contribution of each construction sites selected for the dataset creation. Currently we achieved a very good 38.4 mean average precision (mAP) on three defined classes: sheaths, twisted pipes and smooth pipes.

Keywords: Deep Learning · Instance segmentation · underground infrastructures · dataset creation

1 Introduction

Underground infrastructures refer to all pipes that can be found beneath our feet and contribute to our lives (electricity, gas, public lighting, traffic lights, drinking water, sewerage, telecommunications, internet, etc.). The density of the underground environment increases as urban areas become more congested; pipes cross, intersect and overlap. However, information on the location, twists and turns of these pipes is often inaccurate, incomplete or even missing, resulting in utility strikes when excavating the ground [1].

In order to improve buried pipes mapping, digital methods for capturing information are used, in particular using consumer-grade cameras [23]. Following this methodology, we want to go further by integrating the precise recognition of infrastructures recorded during photo or video acquisition with the support of artificial intelligence and, more specifically, deep learning. This approach is part of the process of automating and improving the mapping of underground infrastructures.

G. L. Foresti et al. (Eds.): ICIAP 2023 Workshops, LNCS 14365, pp. 3–14, 2024.
https://doi.org/10.1007/978-3-031-51023-6_1

In the following sections, we will first see how to apply pipe recognition leading to the establishment of a specific dataset to perform supervised learning of a convolutional neural network (CNN) (Sect. 1). Then the recognition framework will be presented according to well-known metrics (Sect. 2). In Sect. 3, we are going to evaluate these results towards our objectives and also reflect on improvements that can be made.

2 Pipes Recognition

Detection and segmentation are two principles used in deep learning which offer the perception of an object at a given scale: at the scale of the object for one and at the scale of the pixel for the other. In our case, a sparse localization by a bounding box is not sufficient, we need to silhouette the pipes at a pixel level while retrieving a semantic information about the type of infrastructure extracted. We therefore decided to perform instance segmentation in order to recover both semantic and positional information.

2.1 Dataset Creation

In order to perform instance segmentation tasks efficiently, the deep learning model has to be trained according to a certain benchmark that corresponds to the target objects to be detected or extracted. For common context, including street scenes, natural scenes, human recognition and basic objects, there are already several well-known benchmarks [7, 17]. Regarding underground infrastructure and to the best of our knowledge, there is currently no benchmark available. This is because it is a very specific and specialized area that has not yet been studied. In addition, access to these data can be quite restrictive, since, like medical imaging data [15], there are important issues of data confidentiality and security.

In our case, thanks to the SYSLOR company, we have access to several surveys of underground utilities of different sizes, types and configurations, which can be acquired through several sensors [10]. We extracted images from the recorded video of the trenches, filtering out those with significant motion blur. These images were then annotated manually.

2.2 Dataset Description

Given the paradigm of classification purpose described by a three-dimensional graph (Fig. 1) composed by the number of samples needed, the semantic information available and a criterion of data separability and consistency, the aim is to divide the dataset into specific relevant categories in such a way to minimize the intra-class variance or consistency and maximize the inter-class variance or separability.

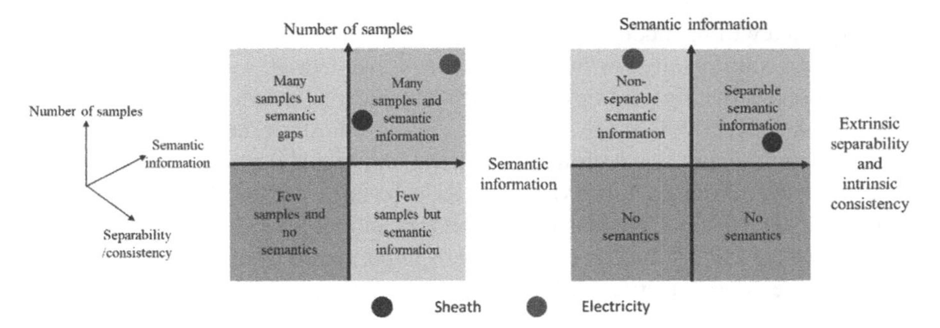

Fig. 1. Classification paradigm

Regarding our task, we are facing two major challenges. The first one is that some underground cables are placed inside a protection element, such as sheaths. Thus, the representation of a type of object can vary drastically in terms of shape and color, as illustrated in Fig. 2. In this case, the grouping of each object within the same class will reduce intra-class consistency, making the classification task more complex. The second one is that we may encounter different levels of complexity in an open trench (Fig. 3), ranging from a single isolated pipe, to two or even several pipes grouped together with potentially transverse elements. Moreover, sometimes a trench presents a specific configuration which is not frequently encountered.

Fig. 2. Overview of electrical utilities components

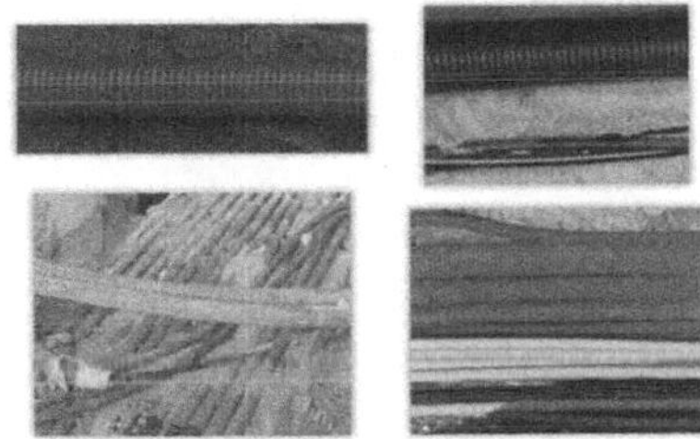

Fig. 3. Different levels of complexity of underground infrastructures

Therefore, to start our study, we decided to use generic classes to cover a large scope of objects. While providing less precise semantic information, we have attempted to maximize the consistency and separability of the classes in order to provide more consistent detection. For example, a class sheath will be preferred to a class electricity, because the striated texture is quite well discernable (Fig. 1). Currently, we have about 2600 labelled images at disposal divided in 41 construction sites and three categories: sheath, twisted pipe (TP) and smooth pipe (SP). Some information on our dataset are provided by Fig. 4. First the number of samples by category (graph A) which accounts for their level of appearance in a trench is presented. The classes are unbalanced because sheaths and smooth pipes encapsulate a greater variety of objects than twisted pipes, which are more specific to one type of object. Graph B shows the number of masks per image distribution that varies according to the site studied and implicitly reflects its level of complexity, a large number of objects reflects a fairly dense site. More specifically, the underground will be much denser in metropolitan areas, which are very urban, than in towns in the countryside. For the first version of this dataset, we can see that we are focusing more specifically on simple configurations with one or two objects in the trench. Finally, graph C describes that the observed pipes are mostly oriented horizontally on images but can take any orientation from vertical to horizontal positions. Trenches are generally surveyed from a point of view parallel to the infrastructure, but this can change depending on the operator responsible for the acquisition. So the point of view of the acquisition on the objects can vary and that it is important to cover as many aspects as possible to have good detection results.

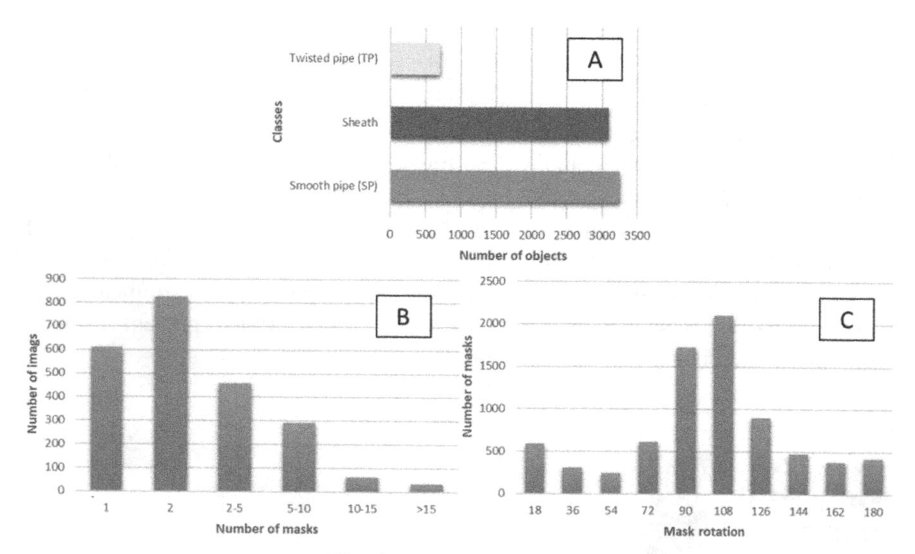

Fig. 4. Dataset description

2.3 Deep Learning Architecture

Instance segmentation methods can be applied differently depending on the order of object positioning and mask generation, resulting in three categories of architecture: single-stage, two-stage and multi-stage [9]. While two-stage and multi-stage methods rely on a certain sequence of dependencies between detection and segmentation, one-stage methods correlate detection and segmentation to achieve high performance in inference speed. Despite considering computation cost, two-stage and multi-stage methods provide more precise mask segmentation [3]. For our specific task, we need a good balance between inference time and segmentation performance, which seems to correspond to a two-stage method.

As we want to implement deep learning techniques to a brand-new category of objects, we want to start by using a simple architecture in order to evaluate the quality of the dataset created for the underground infrastructure detection task. In terms of instance segmentation, Mask R-CNN [11] represents a common state-of-the art two-stage Convolutional Neural Network (CNN) architecture which has proven to be very flexible, easily generalizable to other tasks, simple to train, and particularly effective on use cases very similar to ours [2, 4]. We use the basic architecture to which we add a classification overhead to refine the classification towards real equipment categories (electricity, gas, etc.) (Fig. 5).

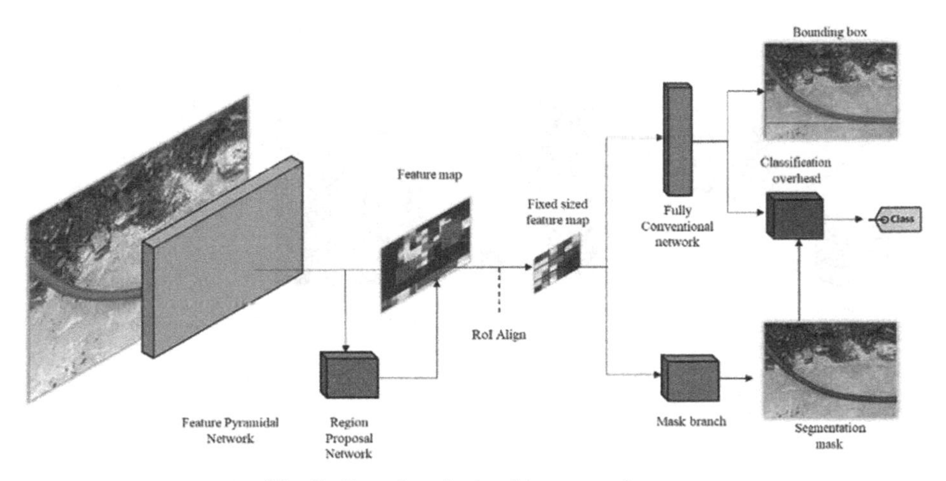

Fig. 5. Deep learning architecture scheme.

2.4 Classification Overhead

As indicated in Sect. 2.2, we have used generic classes to favor the quality of the segmentation rather than precise semantic information. However, when mapping underground infrastructures, the specific type of pipe must be mentioned. This is why, in order to retrieve this information, we have added a classification overlay to the Mask R-CNN labelling branch.

Concerning sheaths, detected objects can be further dissociated according to their color. This feature can be extracted by taking the segmentation polygon resulting from annotation or inference to create a color mask based on the provided image. From there, we can extract an RGB vector by averaging the colors on the color mask. Each annotated object will be transformed into a 3D point resulting in a 3D-colored space with the three-color channels as axis described in Fig. 6. However, classification cannot be performed from there; the classes are not sufficiently discernible because of the relationship between the colors. Therefore, in order to solve the dimensionality reduction problem, we implemented a linear discriminant analysis (LDA) [8] rather than a Principal Component Analysis (PCA) [18] to optimize a data separability criterion by maximizing the difference between the projected means and minimizing the extent of the data projection. This results in a linear combination of features where the projected color space will provide more separable classes.

Finally, the classification procedure consists of a Support Vector Machine (SVM) [13], which is a popular method for classification, regression, and other learning tasks. Trained using the same dataset used to train Mask R-CNN, the SVM produces hyperplanes that best separate the different classes forming the decision boundary so as to maximize the margin between the nearest point of each class. Based on the color feature extracted from the segmentation results, the 3D point will be projected into the final space and the SVM will predict a new class based on its position in the LDA projection space.

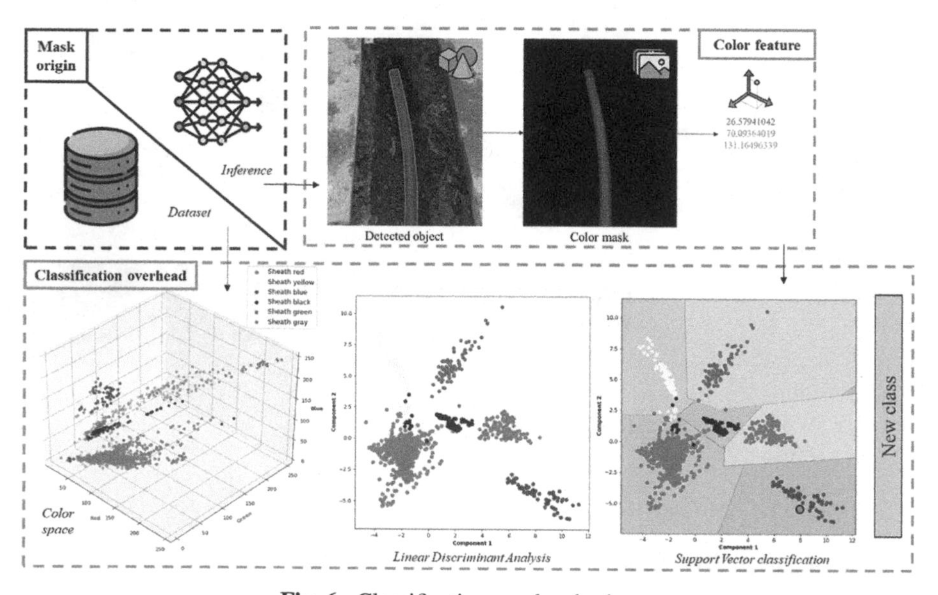

Fig. 6. Classification overhead scheme

3 Experiments

3.1 Training

Depending on the size of a given dataset, training can be performed differently. In our case, we preferred K-fold cross validation with about 2300 images for training (17 sites) and 300 images for validation (24 sites) separated in five groups of data. More than being suited for small dataset, we choose this methodology to further study our dataset. Indeed, it is generally very difficult to measure the contribution of each data and to quantify the information it contains. In the case of unbalanced classes, it is recommended to use stratified cross-validation, i.e., each fold will encapsulate a homogeneous distribution of classes that follows the ratio of the dataset (graph [A] Fig. 4). However, to further assess the degree of generalization that can provide our dataset, the dataset is broken down into folds according to construction sites and not according to images. Therefore, by comparing the different fold compositions in terms of sites with the final metrics, we are going to extract the most prominent sites that can generalize to several trench configurations and perhaps the sites that, in some way, overfit our model and degrade its performance.

During our experiments, we performed several training sessions with a learning rate of 1e-4 for optimal weight approximation while changing the backbone of Mask-RCNN. The idea is to study how the level of features extracted by a given Feature Pyramid Network (FPN) [16] of different depths or types improves the object recognition and segmentation task. We first evaluate multiple variants of the Residual Neural Network (ResNet) [12], basically the most commonly used ResNet50 and ResNet101 denoting the number of layers within the network. Then, we also tested 101-layer ResNeXt [22], which inherits from ResNet, VGG [20] and Inception [21]. This architecture adopts the strategy of exploiting repetitive layer blocks with shortcuts while adding a new dimension based on the split-transform-merge strategy introduced in the Inception architecture. This dimension is represented by cardinality, *i.e.* the size of the set of transformations applied (see Fig. 7). The cardinality illustrates the number of split blocks of the same topology sharing the same parameters.

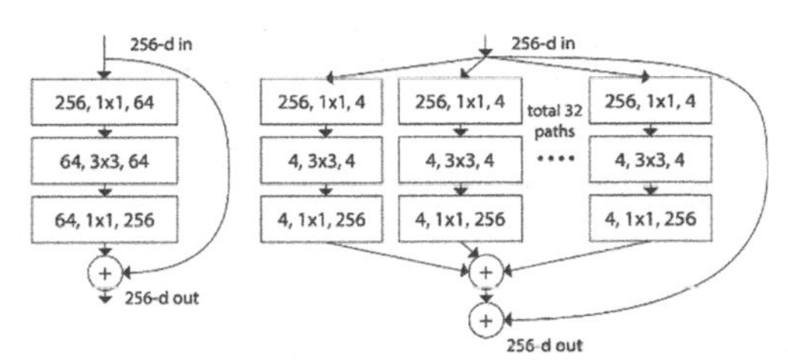

Fig. 7. Difference of structure between ResNet and ResNeXt. **Left:** Convolutional block within ResNet. **Right:** Corresponding block within ResNeXt. A layer is described with the number of channels, filter size and channels

3.2 Instance Segmentation Results

Table 1. Mask R-CNN results (in %) for underground infrastructures extraction

Backbone	AP	AP50	AP75
R-50 FPN	32,7	52.4	34,3
R-101 FPN	35,0	51,8	39,0
X-101 FPN	**38,4**	**56,3**	**40,2**

Table 1 describes the performance of Mask R-CNN with underground infrastructures extraction in terms of average precision (AP) measurements overall and with specific (Intersection over Union) IoU threshold (50 and 75) [19]. By stacking more layers, *i.e.*, increasing the depth of the backbone (50 to 101 layers), the model can be improved as higher-level features encoding more meaningful abstraction information for an image can be extracted. The best results are obtained with the ResNeXt-101 FPN, so for the same network depth, the addition of a dimension provides more benefits. By detailing the AP metric according to the different fold used during training in Table 2, we can further investigate the quality of our dataset. Indeed, we can see that the average accuracy (mAP) has an acceptable deviation across the different folds, which describes good homogeneity between the sites selected in the dataset in terms of the information provided to recognize each of the classes. Concerning the classes, fold 1 suffers from a deficit in the detection of twisted pipes. This means that the most relevant sites for extracting features distinguishing this class were not used for training but for testing as part of the cross-validation procedure (Table 2).

Table 2. Average precision (in %) according to the training fold used with ResNeXt-101

Classes	Fold 1	Fold 2	Fold 3	Fold 4	Fold 5	Mean	Deviation
Sheath	33,0	33,1	34,8	**36,2**	34,8	34,4	1,3
TP	9,9	**45,5**	36,7	30,0	36,7	31,8	13,4
SP	**29,0**	22,8	24,6	24,5	21,0	24,4	3,0
mAP	24,0	**33,8**	32,0	30,3	30,8	30,2	3,7

Overall, the model is efficient but reacts differently to the three classes. Indeed, in the majority of cases, the model is rarely wrong when it predicts an object in the image (more than 70% of precision for the classes), however, depending on the class, the model does not detect certain features completely or at all (between 54.5 and 74.5% of recall).

Sheaths and TP objects stand out with the best results because both type of objects are quite well discretized by their specific characteristics signature. SP results are not as good, mostly because this class encapsulates a wider variety of objects, so the model does not generalize to all objects with the same efficiency, keeping Fig. 1 in mind, the class has a too low consistency so quite often, the model will struggle to detect this type of object effectively, as evidenced by the very low recall values compared to other classes. The Smooth class is also subject to more frequent sources of error than the other two classes. In particular, elements such as shovel handles or forestays used to hold the trench in place can be wrongly detected explaining the precision value compared to sheaths and TP (see Fig. 8 [D, F]).

Table 3. Complete metric computation (in %) for the three retained class

Backbone	Metric	Sheath	TP	SP	Mean
R-50 FPN	Precision	**87,1**	85,1	70,5	80,9
	Recall	68,6	**74,5**	54,5	65,9
	F1 score	73,5	**78,4**	58,9	70,3
	IoU	65,2	**70,1**	51,3	62,2
	AP	34,4	31,8	24,4	30,2
R-101 FPN	Precision	**86,7**	75,4	67,8	76,6
	Recall	**69,3**	64,6	55,9	63,2
	F1 score	**74,1**	68,6	58,4	67,0
	IoU	**66,1**	61,6	50,7	59,4
	AP	**36,9**	32,4	26,7	32,0
X-101 FPN	Precision	**86,1**	81,3	70,8	79,4
	Recall	**73,5**	72,5	58,5	68,2
	F1 score	**77,1**	75,6	61,9	71,5
	IoU	**69,4**	68,5	54,7	64,2
	AP	38,8	**39,0**	26,7	34,8

Figure 8 shows the level of performance achieved by our model in different configurations. First of all, we can observe that the performance of the model will depend on the level of complexity of the trench configuration. For very simple cases, such as single pipe [B, D] or spaced multi-pipes [C], the detection will provide satisfying results. But, the more we converge towards crowded configuration [E, F], the more we will notice a degradation of the detection. This is in accordance with the constitution of the dataset, which focuses on simple cases and is more or less difficult to generalize to complex cases. [A] shows that it is possible that some pipes may still be largely buried during the trench survey, resulting in a lack of visibility and appreciation of the object, which could potentially lead to a lack of detection. Two specific objects in the smooth pipe class are responsible for poorer detection than the other two classes. Firstly, there is a very thin cable (around 1 cm in diameter) [B] which is very rarely detected because the object is confused with the background.

Particular observations can be made from the example [G] in Fig. 8. On the second one, we can see that both groups of pipes (sheath and TP) are considered by the segmentation as one single instance instead of two. This phenomenon reflects a known merging problem encountered with top-down architecture such as Mask R-CNN [14]. Within top-down methods, masks are obtained by segmenting a single instance inside the predicted bounding box. For similar and close instances, mask features are not easily distinguishable from each other, leading to the merging problem where the network segments two similar objects as one instance. Solutions such as the use of object contours [6] or directional information [5, 14] can be used to solve this problem.

What is generally apparent from all these examples is the diversity of configurations that can be encountered. And while our results augur well for the future, the dataset needs to be further completed, as 17 sites do not allow us to generalize, particularly for complex cases.

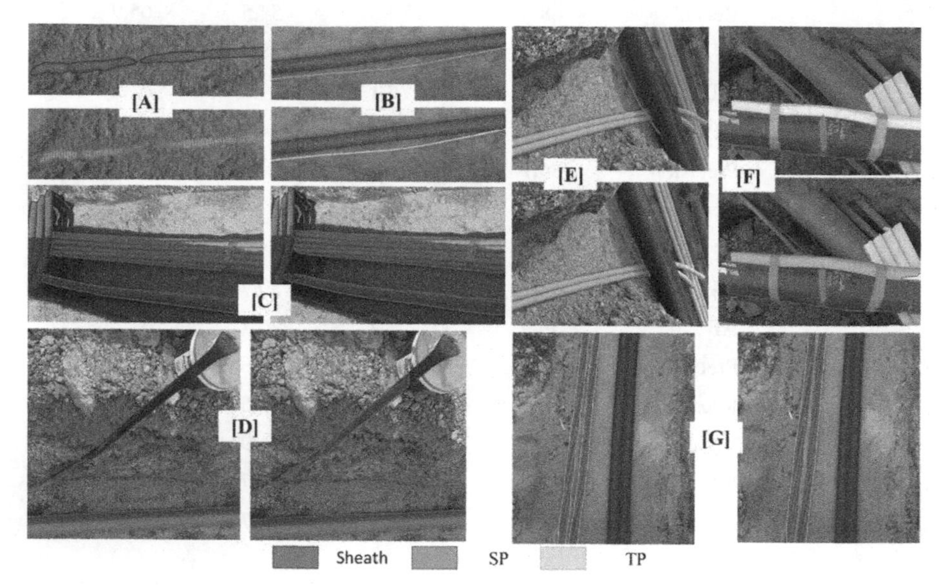

Fig. 8. Examples of segmentation results

3.3 Classification Refinement Results

Sheath	Samples	Precision	Recall	F1-score
Red	64	1.00	1.00	1.00
Yellow	7	1.00	1.00	1.00
Blue	10	1.00	0.40	0.57
Black	10	1.00	1.00	1.00
Green	5	1.00	1.00	1.00
White	8	0.57	1.00	0.73

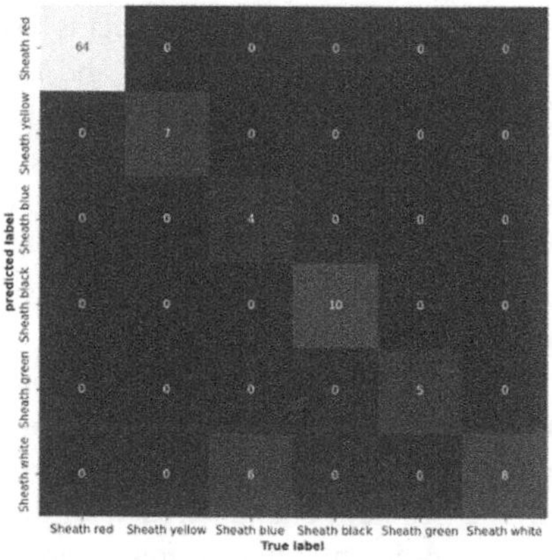

Fig. 9. SVM classification performance

Figure 9 illustrates the classification performance results from the SVM algorithm over a few numbers of samples. Overall, we obtained a very good accuracy of 0.94.

The confusion matrix shows that there is only one prediction error concerning the white sheaths which are confused with blue sheaths. This is due to an illumination effect induced by the sun which can cause bluish reflections on the white sheaths. This problem can be solved by projecting the color space into a three-dimensional component space instead of two within the LDA.

4 Conclusion

This paper stands as the first work aiming to apply deep learning techniques and especially instance segmentation on the extraction of underground infrastructures. Although the pipes are geometrically simple, as they can be represented by a cylinder in perspective view or a rectangle in top view, the environment in which they are located can be quite challenging, with several objects stacked or intersecting each other.

Because there is no available benchmark for this task, we have created our own dataset composed of about 2600 labelled images divided into three classes: sheath, twisted pipe, and smooth pipe. These classes have been selected to discretize objects according to either their texture or geometry by focusing on inter-class separability rather than precise semantic information. This allows to answer favorably the first challenge of variability of the representation of a type of object.

This dataset has then been used to train the Mask R-CNN model with different backbone: ResNet-50, ResNet-101, ResNeXt-101. The last one provides the best results with a 38.4 mAP. Concerning the second challenge related to the variability of the configurations, we efficiently cover simple configurations such as single pipe or spaced multi-pipes. Nevertheless, we have observed that for more complex or novel cases, the detection performance decreases. Therefore, we first need to further improve the dataset by expanding the coverage of possible site configurations and studying image preprocessing. We will then be able to study the possibility of changing the instance segmentation architecture, in particular with an attention module to promote better generalization of the performance of the model.

References

1. Al-Bayati, A.J., Panzer, L.: Reducing damage to underground utilities: lessons learned from damage data and excavators in North Carolina. J. Constr. Eng. Manag. **145**, 04019078 (2019)
2. Barcet, F., Tual, M., Foucher, P., Charbonnier, P.: Using machine learning on depth maps and images for tunnel equipment surveying. Int. Arch. Photogrammetry. Remote Sens. Spat. Inf. Sci. **XLVIII-2-W2–2022**, 1–7 (2022)
3. Bolya, D., Zhou, C., Xiao, F., Lee, Y.J.: YOLACT: Real-time instance segmentation. In: 2019 IEEE/CVF International Conference on Computer Vision (ICCV). IEEE, Seoul, Korea (South), pp. 9156–9165 (2019). https://doi.org/10.1109/ICCV.2019.00925
4. Carreaud, A., Mariani, F., Gressin, A.: Automating the underground cadastral survey: a processing chain proposal. Int. Arch. Photogramm. Remote Sens. Spatial Inf. Sci. **XLIII-B2–2022**, 565–570 (2022). https://doi.org/10.5194/isprs-archives-XLIII-B2-2022-565-2022
5. Chen, L.-C., Hermans, A., Papandreou, G., Schroff, F., Wang, P., Adam, H.: MaskLab: instance segmentation by refining object detection with semantic and direction features. In: 2018 IEEE/CVF Conference on Computer Vision and Pattern Recognition. IEEE, Salt Lake City, UT, pp. 4013–4022 (2018). https://doi.org/10.1109/CVPR.2018.00422

6. Cheng, T., Wang, X., Huang, L., Liu, W.: Boundary-Preserving Mask R-CNN. In: Vedaldi, A., Bischof, H., Brox, T., Frahm, J.-M. (eds.) Computer Vision – ECCV 2020. Lecture Notes in Computer Science, pp. 660–676. Springer International Publishing, Cham (2020)
7. Cordts, M., et al.: The cityscapes dataset for semantic urban scene understanding. In: Presented at the Proceedings of the IEEE Conference on Computer Vision and Pattern Recognition, pp. 3213–3223 (2016)
8. Fisher, R.A.: The use of multiple measurements in taxonomic problems. Ann. Eugen. **7**, 179–188 (1936). https://doi.org/10.1111/j.1469-1809.1936.tb02137.x
9. Gu, W., Bai, S., Kong, L.: A review on 2D instance segmentation based on deep neural networks. Image Vis. Comput. **120**, 104401 (2022)
10. Haenel, R., Semler, Q., Semin, E., Grussenmeyer, P., Tabbone, S.: Evaluation of low-cost depth sensors for outdoor applications. Int. Arch. Photogramm. Remote Sens. Spatial Inf. Sci. **XLVIII-2/W1–2022**, 101–108 (2022). https://doi.org/10.5194/isprs-archives-XLVIII-2-W1-2022-101-2022
11. He, K., Gkioxari, G., Dollar, P., Girshick, R.: Mask R-CNN. In: Presented at the Proceedings of the IEEE International Conference on Computer Vision, pp. 2961–2969 (2017)
12. He, K., Zhang, X., Ren, S., Sun, J.: Deep residual learning for image recognition. In: Presented at the Proceedings of the IEEE Conference on Computer Vision and Pattern Recognition, pp. 770–778 (2016)
13. Hearst, M.A., Dumais, S.T., Osuna, E., Platt, J., Scholkopf, B.: Support vector machines. IEEE Intell. Syst. Appl. **13**, 18–28 (1998)
14. Jena, R., Zhornyak, L., Doiphode, N., Buch, V., Gee, J., Shi, J.: Beyond mAP: Re-evaluating and Improving Performance in Instance Segmentation with Semantic Sorting and Contrastive Flow (2022)
15. Kaissis, G.A., Makowski, M.R., Rückert, D., Braren, R.F.: Secure, privacy-preserving and federated machine learning in medical imaging. Nat. Mach. Intell. **2**, 305–311 (2020)
16. Lin, T.-Y., Dollar, P., Girshick, R., He, K., Hariharan, B., Belongie, S.: Feature pyramid networks for object detection. In: Presented at the Proceedings of the IEEE Conference on Computer Vision and Pattern Recognition, pp. 2117–2125 (2017)
17. Lin, T.-Y., et al.: Microsoft COCO: common objects in context. In: Fleet, D., Pajdla, T., Schiele, B., Tuytelaars, T. (eds.) Computer Vision – ECCV 2014. Lecture Notes in Computer Science, pp. 740–755. Springer International Publishing, Cham (2014)
18. Pearson, K.: LIII. On lines and planes of closest fit to systems of points in space. Lond. Edinb. Dublin Philos. Mag. J. Sci. **2**, 559–572 (1901)
19. Shan, P.: Image segmentation method based on K-mean algorithm. EURASIP J. Image Video Process. **2018**(1), 1–9 (2018). https://doi.org/10.1186/s13640-018-0322-6
20. Simonyan, K., Zisserman, A.: Very Deep Convolutional Networks for Large-Scale Image Recognition (2015). https://doi.org/10.48550/arXiv.1409.1556
21. Szegedy, C., Vanhoucke, V., Ioffe, S., Shlens, J., Wojna, Z.: Rethinking the inception architecture for computer vision. In: Presented at the Proceedings of the IEEE Conference on Computer Vision and Pattern Recognition, pp. 2818–2826 (2016)
22. Xie, S., Girshick, R., Dollar, P., Tu, Z., He, K.: Aggregated residual transformations for deep neural networks. In: Presented at the Proceedings of the IEEE Conference on Computer Vision and Pattern Recognition, pp. 1492–1500 (2017)
23. Yuen, R.Z.M., Boehm, J.: Potential of consumer-grade cameras and photogrammetric guidelines for subsurface utility mapping. ISPRS – Int. Arch. Photogrammetry Remote Sens. Spat. Inf. Sci. **48W1**, pp. 243–250 (2022)

Generating Invariance-Based Adversarial Examples: Bringing Humans Back into the Loop

Florian Merkle[1,2](✉), Mihaela Roxana Sirbu[1], Martin Nocker[1], and Pascal Schöttle[1]

[1] MCI Innsbruck, Innsbruck, Austria
florian.merkle@student.uibk.ac.at, mr.sirbu@mci4me.at, {martin.nocker,pascal.schoettle}@mci.edu
[2] University of Innsbruck, Innsbruck, Austria

Abstract. One of the major challenges in computer vision today is to align human and computer vision. Using an adversarial machine learning perspective, we investigate invariance-based adversarial examples, which highlight differences between computer vision and human perception. We conduct a study with 25 human subjects, collecting eye-gazing data and time-constrained classification performance, in order to study how occlusion-based perturbations impact human and machine performance on a classification task. Subsequently, we propose two adaptive methods to generate invariance-based adversarial examples, one based on occlusion and the other based on second picture patch-insertion. All methods leverage the eye-tracking data obtained from our experiments. Our results suggest that invariance-based adversarial examples are possible even for complex data sets but must be crafted with adequate diligence. Further research in this direction might help better align computer and human vision.

Keywords: Adversarial Machine Learning · Interpretable Machine Learning · HCI

1 Introduction

As computer vision applications become more and more prominent in our daily lives, e.g., in smartphones, digital assistants, (semi-)autonomous cars, or medical diagnostics, so does the comprehensibility of their decisions to humans. Nowadays, the vast majority of computer vision tasks are performed by highly parameterized machine learning models, most prominently deep neural networks (DNNs), convolutional neural networks (CNNs), and transformers. More complex machine learning models can be deployed thanks to advances in computational power and digital storage capacity. Image classification is not the only field where machine learning outperforms humans by certain measures [18]. Thus, one of the biggest challenges is to align machine classification with human classification and make the machine decision explainable to humans [8].

G. L. Foresti et al. (Eds.): ICIAP 2023 Workshops, LNCS 14365, pp. 15–27, 2024.
https://doi.org/10.1007/978-3-031-51023-6_2

The existence of adversarial examples, images that are strategically altered such that machine and human classification do not match, perfectly illustrates the problem of misalignment between human and machine perception. The research area of adversarial machine learning has existed for a long time [2] and is very active to date.

By means of invariance- and sensitivity-based adversarial examples, this study examines the alignment of human and computer vision. We conducted a human-grounded experiment to gather eye-gazing data, which was used to naively generate invariance-based adversarial examples whose effectiveness was consecutively tested on humans. Figure 1 shows some of the images we generated, including the misclassifications by a time-constrained human. From these results, we propose two algorithms to adaptively generate these adversarial examples: one based on occlusion and the other on picture patch-insertion. Our results suggest that, even though creating invariance-based adversarial examples based on complex data sets needs more care than on simple data sets, it is possible to do so.

(a) gas pump misclassified as casette player

(b) parachute misclassified as casette player

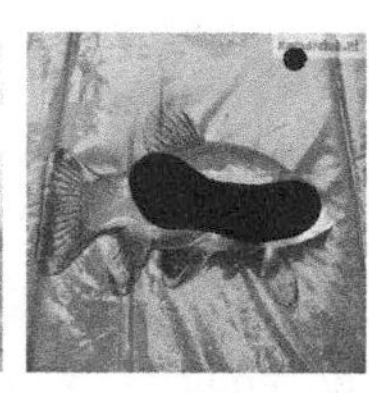

(c) fish misclassified as dog

(d) french horn misclassified as gas pump

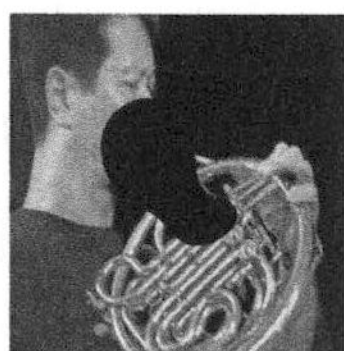

(e) french horn misclassified as dog

Fig. 1. Images that were misclassified by time-constrained humans.

2 Adversarial Examples

Research on adversarial machine learning started in 2004 when it was first explored that spam filters utilizing linear classifiers can be fooled by small changes in the initial email that do not negatively affect the readability of the message but lead to misclassification [7]. In 2013, [21] showed that DNNs are just as prone to adversarial examples as other machine learning algorithms when classifying carefully perturbed input samples.

In a general classification task, we consider data $(x, y) \in \mathbb{R}^d$ where x and y are the samples and their corresponding labels, drawn from an underlying data distribution $\mathcal{D}$. We further consider a classifier C that maps an input to a label prediction $x \mapsto C(x)$. We also use a *label oracle* $\mathcal{O}$, introduced by [23], that maps all inputs to their real class, i.e., $\mathcal{O}(x) = y$ as long as $(x, y) \sim \mathcal{D}$. Thus, we can define an adversarial example as any perturbed image $\tilde{x}$, for which $\mathcal{O}(\tilde{x}) \neq C(\tilde{x})$. Further, adversarial examples are often subject to a constraint regarding the distance between the original image and the adversarial example. This distance is expressed as an L_p norm that shall be less than a certain threshold ϵ, i.e.,

$||x - \tilde{x}|| \leq \epsilon$. Note that the oracle is able to perform perfect classification which means that $(\tilde{x}, y) \sim \mathcal{D}' \neq \mathcal{D}$ indicates a change of the underlying distribution. Thus the semantics change, and therefore perfect classification implies $\mathcal{O}(\tilde{x}) \neq y$.

In the following, we will introduce the two types of adversarial examples that leverage two different failure modes of classifiers.

2.1 Sensitivity-Based Adversarial Examples

The first work on adversarial examples by Szegedy et al. [21] in 2013 referred to sensitivity-based adversarial examples. By calculating the gradients of the model's loss function with respect to the input images, the authors were able to craft a perturbation that was imperceptible to humans but changed the model's prediction to a different class. Using the notation introduced above, sensitivity-based adversarial examples can be defined as having the following properties:

- $C(\tilde{x}) \neq y$, i.e., the model assigns a different class to the perturbed image.
- $\mathcal{O}(\tilde{x}) = y$, i.e., the label oracle assigns the true class to the perturbed image.

The second property is usually proxied by the distance constraint introduced above. If the distance between an original image and its perturbed counterpart is below a certain threshold, humans are expected to assign the same class. There are several works on the deficiency of the use of L_p distance metrics, e.g. [10], but they are still widely used for their simplicity and rigorous evaluation.

Many defenses have been proposed and mostly proven to be ineffective [1,24]. A promising approach is certified robustness in which, within a given L_p-bound ϵ, the model is guaranteed to make the correct prediction by the introduction of noise at both, training and inference time [6]. The intuition is that regardless of how an adversary manipulates the input, as long as the alteration is smaller than ϵ, they will remain inside the bounds and the input will remain correctly categorized. Issues arise as for humans, two distinct bounded L_p disturbances on the same picture may seem quite different. One may be undetectable, while the other may seem to be a whole other picture [23].

2.2 Invariance-Based Adversarial Examples

Naturally, if a specific perturbation also changes an image's semantics, the oracle will assign the image the new label. If these perturbations are created in a way such that the machine classifier does still assign the old label to the image with perturbed semantics, it is an invariance-based adversarial example [23].

As such, invariance-based adversarial examples have the following properties:

- $C(\tilde{x}) = y$, i.e., the model assigns the original label to the perturbed image.
- $\mathcal{O}(\tilde{x}) \neq y$, i.e., the label oracle assigns the new semantically correct label to the perturbed image.

The second property is usually proxied by human labelers as perfect oracles do not always exist. If additionally, the distance between the benign image and the adversarial example is below the respective ϵ-bound that is guaranteed by certified robustness, i.e., $||x - \tilde{x}|| \leq \epsilon$, it is possible to have adversarial examples that fool these certified robust DNNs.

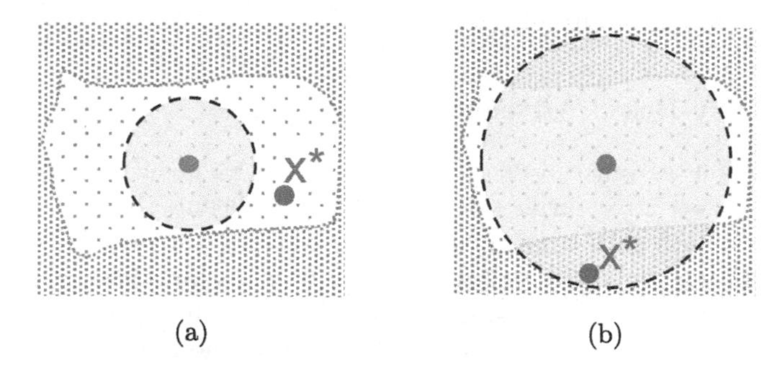

Fig. 2. Both figures (taken from [23]) show the human and the classifier's decision boundary between two classes (red and white). (a) illustrates a sensitivity-based adversarial example, (b) an invariance-based adversarial examples. (Color figure online)

Tramer et al. [23] developed a method for creating invariance-based adversarial instances for MNIST. Their method applies small changes to a sample such that it satisfies the distance constraint but changes its semantics. They utilize an image editor to manipulate pictures at the pixel level manually but also developed an algorithm for crafting these adversarial examples programmatically. They confirm the validity of the adversarial examples by having human subjects classify the altered version. Figure 2 illustrates the two failure modes that arise. The dashed circle illustrates the decision boundary of a (certifiably robust) model. The blue dotted line illustrates the human decision boundary. If a model's decision boundary is too narrow, it allows for sensitivity-based adversarial attacks (Fig. 2a). If the decision-boundary is too wide, invariance-based attacks are possible (Fig. 2b). Note that in the high-dimensional space in which DNNs operate, both failure modes are present for a single model.

3 Human Perception

This section briefly discusses the mechanisms of human perception and highlights similarities and differences between human and computer vision. As the human visual perception system is highly complex we limit ourselves to the areas most important for our work.

Attention is referred to as the act of concentrating on a subset of available stimuli within our focused field of vision [4]. Contrary to widespread assumption, visual attention is not limited to the center of our vision. Human brains control both overt and covert attention, with overt attention being drawn to the high-resolution core field. Covert attention is difficult to quantify because it occurs without eye movement. Thus, both implicit and explicit attention affects our visual activity. We use subtle attention to identify peripheral objects before directing our gaze to that region. The gaze point represents the fundamental eye-tracking data [4]. Fixations in the visual field are responsible for the capacity to direct one's attention to a particular object in the surrounding. [12] explain that two mechanisms regulate fixations: voluntary and involuntary. Voluntary fixation enables humans to intentionally direct their gaze toward an object they choose to look at, while involuntary fixation is the mechanism that maintains the eyes fixed on a particular object once it has been located. It is stated that humans spend approximately 90 % of their viewing time in a fixation state.

Computer vision, specifically the deeper layer of CNNs, resembles many features of human vision [3,26], as depicted in the "cortical object recognition" model [15]. Nevertheless, there are also differences, such as the presentation of images and the human visual system's resilience to picture deterioration. While CNNs receive images in fixed rectangular pixel grids with constant spatial resolution, the human eye has eccentricity-dependent spatial resolution [9], which is high in the center of the visual field and falls off linearly with increasing eccentricity [25]. Additionally, the human visual system is more resilient to picture deterioration, such as contrast loss or additive noise, compared to other visual systems [11].

4 Experiments

Our experiment consists of two phases. In the first phase, we collected eye-gazing data from the participants while solving classification tasks. This data was then used to craft a new set of stimuli in which those parts of the images are occluded where the participants' eyesight is focused. We chose a *MobileNetV3-large* classifier [13] from the *Torchvision* library, pre-trained on ImageNet and finetuned on the ten Imagenette classes until convergence.

4.1 Stimuli

For the original stimuli, we used 100 sample images from the Imagenette data set [14]. Imagenette is a subset of ImageNet that contains ten easy-to-categorize classes (tench, English springer, cassette player, chain saw, church, french horn, garbage truck, gas pump, golf ball, parachute). During the experiments, the class *tench* was changed to *fish*, and the class *English springer* was changed to *dog* to simplify the classes, as we did not anticipate our participants to have knowledge about fish species or dog breeds. Ten images were selected for each class based on human observer classification accuracy. To minimize ambiguity, only images illustrating a single instance of the class object were used, with one exception of an image in the class *dog*. The images were then resized to 640×640 pixels to meet the eye-tracking experiment requirements.

4.2 Human Subjects

This study involved 25 participants, with eleven in the first phase and 14 in the second phase. Participants provided personal information including age, gender, eye health, and vision correction methods. The group of participants consisted of ten men and 15 women, ranging in age from 25 to over 50 years. At the time of the study, all of the participants had normal or corrected-to-normal vision.

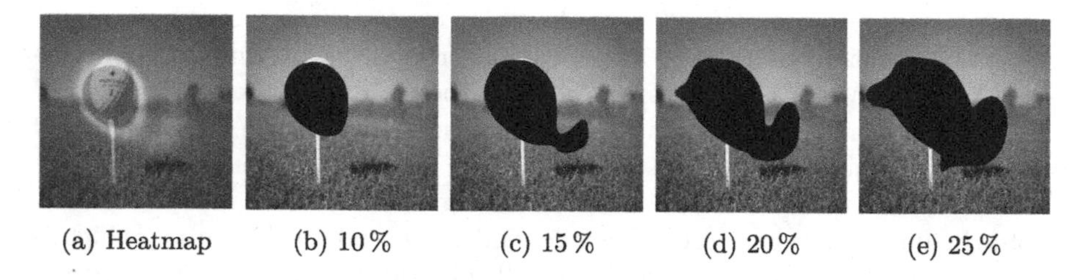

(a) Heatmap (b) 10 % (c) 15 % (d) 20 % (e) 25 %

Fig. 3. (a) Eye-tracking heatmap. (b)-(e) Occluded images for varying occlusion rates. Subcaptions indicate the respective occlusion rate.

4.3 Eye Tracking Experimental Set-Up

The data was collected with a Tobii Pro X2-60 eye-tracker from the year 2020. This device is designed specifically for use in fixation-based research, allowing the collection of gaze data at a rate of 60 frames per second. The device sends out infrared light from the bottom of the screen to generate reflections on the corneas of participants' eyes. A camera on top of the screen captures these reflection patterns and other visual data including the person and its surroundings. Image processing algorithms are used to identify critical features such as the eyes and corneal reflection patterns. These features are used to identify the three-dimensional position of each eyeball and the gaze point on the screen [22].

Participants were seated 50–70 cm from the screen and eye-tracking device. We assessed their ability to perceive the content of the screen, comprehend the language, and operate the computer mouse. Participants were allowed to use single-vision glasses, but multi-power correctives and eye makeup were prohibited. The top-to-bottom gaze angle was restricted to a 36-degree range to optimize tracking [22]. An observer was present in the room during the experiment.

Initially, the device was calibrated with the default calibration settings of the Tobii Studio software, including a red color pointer, gray background, medium speed, and a calibration region that spans the whole screen. Five calibration points are sequentially displayed. The eye-tracker determines the user's eye characteristics and employs these measures with a 3D eye model to calculate gaze data [22]. Upon completion of the calibration process participants were presented with instructions and the demographic data was collected. Subsequently, participants were shown pictures from each class with the corresponding label for five seconds to rule out mistakes due to language difficulties. The test images were then displayed consecutively for one second each. The time constraint is motivated by [9] that found that time-constrained humans might be susceptible to adversarial examples. The first phase used the original stimuli, while the second phase used the partly occluded images. Participants were asked to select the true class from ten classes using a forced-choice questionnaire arranged with the Latin square technique. The output of these experiments includes the participants' classification results and a set of heatmaps indicating eye-gazing data, with one heatmap per participant and per image, where higher values represent a higher focus on specific pixel areas, see Fig. 3a.

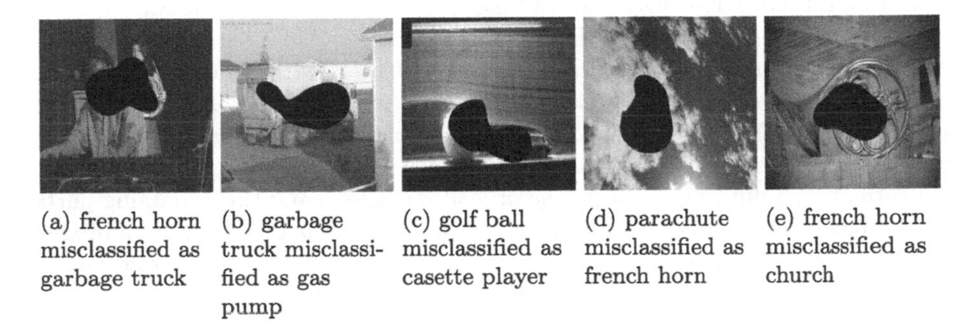

(a) french horn misclassified as garbage truck (b) garbage truck misclassified as gas pump (c) golf ball misclassified as casette player (d) parachute misclassified as french horn (e) french horn misclassified as church

Fig. 4. Images that were misclassified by the classifier. Subcaptions indicate the label assigned by the classifier.

4.4 Creation of Occlusion-Based Adversarial Examples

Eye-gazing data from the first phase is used to create invariance-based adversarial example candidates, which are used as stimuli for the second phase. The heatmaps from the first phase are used to compute binary masks based on a threshold that selects the top heatmap values defined by a certain percentage of all values. A mask's entry is set to one if the corresponding heatmap value exceeds the threshold value, and zero otherwise. The binary masks are then used to occlude the original images. Figure 3 illustrates the same picture with different occlusion rates. An occlusion rate of 10 % was chosen as this value yielded a classification accuracy of 92 %, i.e., the highest result. Higher occlusion rates resulted in accuracy values below 90 %.

Our approach is motivated by the observation that DNNs learn spurious correlations [16,19], i.e., correlation without causation or attributed to a third factor that is not noticeable, e.g., by a human observer. Occluding areas used by humans for classification might still enable a DNN to correctly classify an image, resulting in an invariance-based adversarial example, as defined in Sect. 2.2. Our work is further motivated by [5], which identified sparse pixel-subsets that retained an image's original label using an iterative gradient-based method.

5 Results

As mentioned above, the model's accuracy drops from 100 % to 92 % for our 100-image test set. The model achieved correct classification for all images in the *dog, cassette player, church,* and *gas pump* classes, but misclassified the classes *fish, chainsaw, garbage truck,* and *golf ball* once, and the classes *french horn* and *parachute* twice. Figure 4 displays the images that were misclassified by the model but correctly labeled by all 14 human subjects. As such, these images fulfill our criteria for sensitivity-based adversarial examples: Fooling the classifier, while keeping its semantics.

Over a total of 1 400 classification tasks performed by the 14 human subjects, the accuracy sums up to 98.2 %. Humans performed well in identifying *dogs, cassette players, churches, garbage trucks,* and *golf balls,* with no errors recorded. Few errors were made for the classes *fish* (1), *french horn* (3), *gas pump* (5), and *parachute* (3). However, the class chainsaw had 13 errors, mainly due to one challenging image shown in Fig. 5a. In this image with low lighting, a small occlusion rate of 10 % entirely occludes the object of interest, and the remaining parts of the image suggest a social gathering, leading to misclassification by humans. Only two out of 14 participants correctly classified the image. An overview of the human-assigned labels is shown in Fig. 5a. Similar to the majority of the participants, the classifier assigned the label *french horn.* A similar example is displayed in Fig. 5b, where most participants successfully assigned the correct label, while the model assigned the label *chain saw.* Figure 1 illustrates all images misclassified by at least one participant. Although some classification errors may seem trivial, it should be considered that participants were only given a brief glimpse of the image before being asked to classify it.

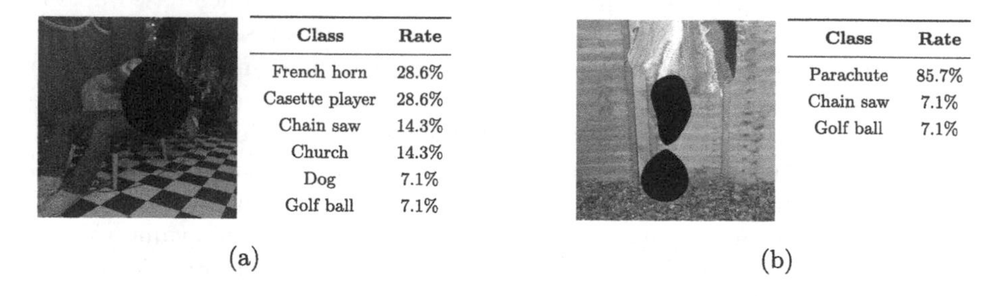

Class	Rate
French horn	28.6%
Casette player	28.6%
Chain saw	14.3%
Church	14.3%
Dog	7.1%
Golf ball	7.1%

(a)

Class	Rate
Parachute	85.7%
Chain saw	7.1%
Golf ball	7.1%

(b)

Fig. 5. Misclassified images by the classifier and time-constrained humans.

Our aim was to generate invariance-based adversarial examples, however, we were only able to find five samples that partially fool participants under time constraints while still being assigned the original label by the classifier. We further found six samples that satisfy the definition of a sensitivity-based adversarial example.

6 Adaptive Approaches

To improve the search for invariance-based adversarial examples, we propose an adjusted approach based on linear search. As some images are more sensitive to occlusion than others regarding correct classification, we employ an adaptive approach. This method maximizes occlusion rates in areas that humans focused their eyesight on during the first experiment while ensuring correct classification by the model. The occlusion rate is increased iteratively in one-percent steps until the model is unable to assign the correct class. The last successful sample is then returned. Figure 6 illustrates this approach.

Inspired by [23] to craft adversarial examples by replacing small objects with other objects, we generate potential invariance-based adversarial examples using a similar approach to our previous adaptive method. However, instead of simply occluding high-interest areas of the image, we cover them with high-interest patches from other images in our test data set. High-interest areas are defined as those pixels with the highest values in the corresponding heatmap obtained in

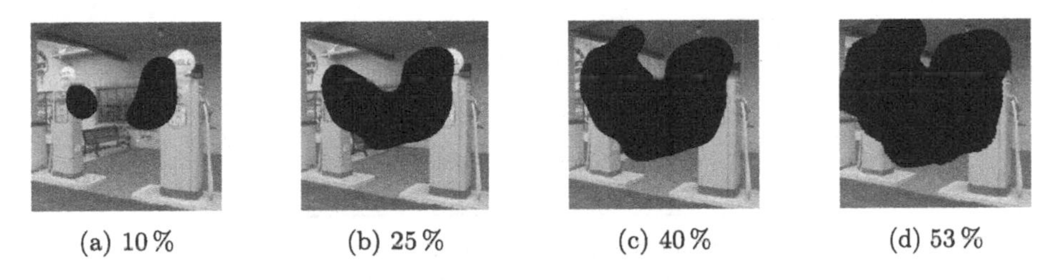

(a) 10 %　(b) 25 %　(c) 40 %　(d) 53 %

Fig. 6. Adaptive occlusion method. The model correctly labels images with an occlusion rate of up to 52 %, e.g., samples 6a–6c. Sample 6d is the first one to be misclassified. The resulting adversarial example has an occlusion rate of 52 %.

Algorithm 1: Generating invariance-based AEs using patch-insertion

Input: x_{base}, x_{target}, $\texttt{heatmap}_{\text{base}}$, $\texttt{heatmap}_{\text{target}}$, label y, model $\texttt{M}$
Result: inv-based AE-candidate $\tilde{x}_{\text{best}}$

1 $\tilde{x} \leftarrow x_{\text{base}}$
2 **do**
3 　$\tilde{x}_{\text{best}} \leftarrow \tilde{x}$
4 　`ratio_occ += 0.01`
5 　**for** case in {base, target} **do**
6 　　$\tau \leftarrow$ `ratio_occ`-quantile of all $\texttt{heatmap}_{\text{case}}$-values
7 　　$\texttt{binary\_mask}_{\text{case}} \leftarrow 1$ where $\texttt{heatmap}_{\text{case}} > \tau$, else 0
8 　　$\texttt{com}_{\text{case}} \leftarrow \texttt{center\_of\_mass}(\texttt{binary\_mask}_{\text{case}})$
9 　**end**
10 　`offset_hor`, `offset_vert` $\leftarrow \texttt{com}_{\text{base}} - \texttt{com}_{\text{target}}$
11 　$x_{\text{target}}^{\text{shifted}} \leftarrow \texttt{pad\_and\_slice}(x_{\text{target}}, \texttt{offset\_hor}, \texttt{offset\_vert})$
12 　$\tilde{x} \leftarrow x_{\text{target}}^{\text{shifted}}$ where $\texttt{binary\_mask}_{\text{base}} = 1$, else x_{base}
13 **while** $\texttt{M}(\tilde{x}) = y$;

the first phase of our experiments. The algorithm is presented in the pseudocode of Algorithm 1. A loop is run until the new candidate adversarial example is misclassified by the model (line 13). Within each iteration, the occlusion ratio is increased (line 4) which defines the heatmap-threshold τ (line 6). Based on that, a binary mask of the high-interest regions in both images is calculated (line 7), followed by the center of mass computation of the masks (line 8). The target image is shifted (line 11) according to the offset between the heatmaps' center of mass (line 10) and updates the candidate adversarial example by copying the shifted target image into the base image's high-interest regions, i.e., where the binary mask of the base image is one (line 12). Figure 7 illustrates the insertion principle, together with selected samples generated with this method.

Both approaches, i.e., occlusion-based and patch-insertion-based, achieve a model accuracy of 100 % as the model's misclassification is the stopping criterion. Human classification performance will be investigated in future work. It is noted, that this approach is not limited to behavioral data but can apply any type of importance map as long as it has an importance score for every pixel.

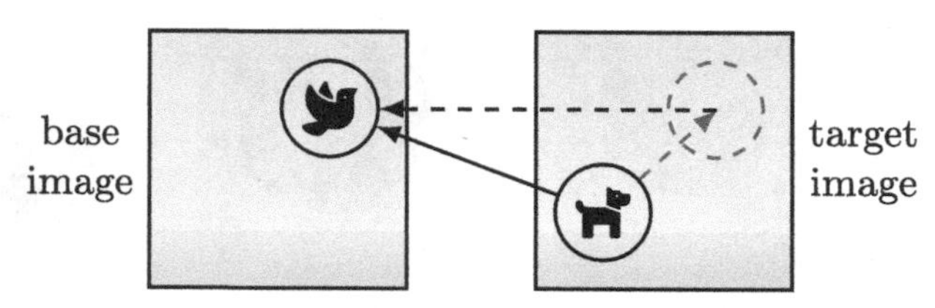

(a) The area of interest, i.e., areas with high heatmap values, of the target image is inserted into the base image's area of interest (black full circles). The target image is shifted by the offset between the heatmaps' center of mass (red dashed arrow).

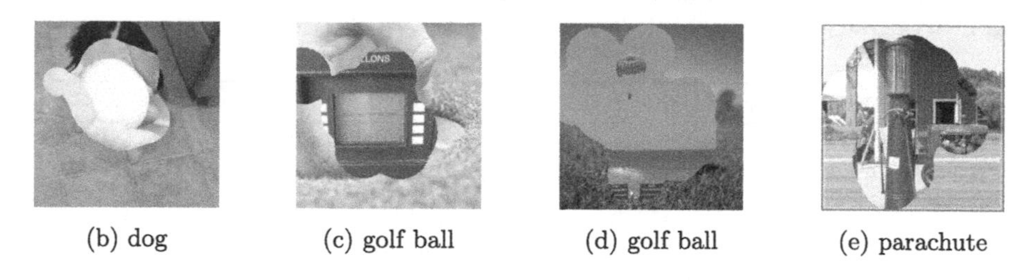

(b) dog (c) golf ball (d) golf ball (e) parachute

Fig. 7. Patch-insertion method. Captions in (b)-(e) denote (correct) model labels.

7 Conclusion

The alignment of human and machine vision remains a significant concern in the development of ever more performant and reliable computer vision systems. In this work, we were able to show that the creation of invariance-based adversarial examples for complex data sets is possible but still poses a challenging task. We show that while in general, the machine classifier and time-constrained humans are quite robust to a moderate occlusion rate of 10 %, there are cases in which both failed to perform the classification task successfully. However, such a naive approach seems to be insufficient to craft invariance-based adversarial attacks reliably. Therefore, we propose an adaptive algorithm to generate invariance-based adversarial example candidates either by occluding the maximum amount of pixels before the machine classification is flipped or by inserting a patch with high user attention from a target image into the area with high user attention of the base image. The resulting images are sure to meet the first requirement for invariance-based adversarial examples, i.e., not changing the machine classification. In order to evaluate the effectiveness of our proposed algorithm, that is the ratio of candidates that meet the second requirement (changing human classification), another round of evaluation with human subjects is planned for future work.

Our work opens up several directions for future research. With our adaptive approaches, we explore the decision boundary of the classifier. It might be worthwhile, yet resource-intensive, to conduct similar approaches with respect to human subjects. Notably, our work and the creation of the eye-gaze-based heat maps have apparent parallels to explainability approaches such as Grad-CAM [20]. Combining the fields of adversarial machine learning, explainability, and human behavior might bring new insights into the relationship between human and computer vision. Further, as the proposed algorithm is not limited to behavioral data, GradCAM-generated heatmaps can be used to create potential invariance-based adversarial examples computationally. Recent research [17] has explored the possibility of hardening neural networks against invariance-based adversarial examples by conducting adversarial training with such samples but was limited to the MNIST dataset due to the non-availability of invariance-based adversarial examples for more complex data sets and the ineffectiveness of the existing algorithm for MNIST data. Our approach might help efforts in this direction to overcome these challenges. Adapting our center of mass-aligned patch-insertion algorithm to insert larger patches and use the target image label might even be a promising data augmentation strategy that makes vision models more invariant to less important image areas.

This work provides essential insights and methods to align human and computer vision further. Our hope is that this will inspire future research to incorporate human beings in their studies, as they are a crucial element in the alignment of computer and human vision but remain often neglected so far.

Acknowledgements. Florian Merkle and Pascal Schöttle are supported by the Austrian Science Fund (FWF) under grant no. I 4057-N31 ("Game Over Eva(sion)"). Martin Nocker is supported under the project "Secure Machine Learning Applications with Homomorphically Encrypted Data" (project no. 886524) by the Federal Ministry for Climate Action, Environment, Energy, Mobility, Innovation and Technology (BMK) of Austria.

References

1. Athalye, A., Carlini, N., Wagner, D.: Obfuscated gradients give a false sense of security: circumventing defenses to adversarial examples. In: International Conference on Machine Learning, pp. 274–283. PMLR (2018)
2. Biggio, B., Roli, F.: Wild patterns: ten years after the rise of adversarial machine learning. Pattern Recogn. **84**, 317–331 (2018)
3. Cadieu, C., et al.: Deep neural networks rival the representation of primate it cortex for core visual object recognition. PLoS Computa. Biol. **10**, e1003963 (2014)
4. Carrasco, M.: Visual attention: the past 25 years. Vision. Res. **51**, 1484–1525 (2011)
5. Carter, B., Jain, S., Mueller, J.W., Gifford, D.: Overinterpretation reveals image classification model pathologies. In: Advances in Neural Information Processing Systems, vol. 34 (2021)
6. Cohen, J., Rosenfeld, E., Kolter, Z.: Certified adversarial robustness via randomized smoothing. In: International Conference on Machine Learning, pp. 1310–1320. PMLR (2019)
7. Dalvi, N., Domingos, P., Sanghai, S., Verma, D.: Adversarial classification. In: Proceedings of the tenth ACM SIGKDD International Conference on Knowledge Discovery and Data Mining, pp. 99–108 (2004)
8. Du, M., Liu, N., Hu, X.: Techniques for interpretable machine learning. Commun. ACM **63**(1), 68–77 (2019)
9. Elsayed, G.F., et al.: Adversarial examples that fool both computer vision and timelimited humans. In: Advances in Neural Information Processing Systems, pp. 3910–3920 (2018)
10. Engstrom, L., Tsipras, D., Schmidt, L., Madry, A.: A rotation and a translation suffice: Fooling CNNs with simple transformations. arXiv:1712.02779 (2017)
11. Geirhos, R., Janssen, D.H.J., Schütt, H.H., Rauber, J., Bethge, M., Wichmann, F.A.: Comparing deep neural networks against humans: object recognition when the signal gets weaker (2018)
12. Guyton, A.C., Hall, J.E.: Textbook of Medical Physiology. Elsevier Inc. (2006)
13. Howard, A., et al.: Searching for mobilenetv3. In: Proceedings of the IEEE/CVF International Conference on Computer Vision, pp. 1314–1324 (2019)
14. Howard, J.: Imagenette (2019). https://github.com/fastai/imagenette/
15. Maximilian, R., Tomaso, P.: Hierarchical models of object recognition in cortex 2. Nat. Neurosci. **2**, 1019–1025 (1999). https://doi.org/10.1038/14819
16. McCoy, R.T., Pavlick, E., Linzen, T.: Right for the wrong reasons: diagnosing syntactic heuristics in natural language inference. In: 57th Annual Meeting of the Association for Computational Linguistics, ACL 2019, pp. 3428–3448. Association for Computational Linguistics (ACL) (2020)
17. Rauter, R., Nocker, M., Merkle, F., Schöttle, P.: On the effect of adversarial training against invariance-based adversarial examples. arXiv preprint arXiv:2302.08257 (2023)
18. Russakovsky, O., et al.: ImageNet large scale visual recognition challenge. Int. J. Comput. Vision **115**(3), 211–252 (2015)
19. Sagawa, S., Koh, P.W., Hashimoto, T.B., Liang, P.: Distributionally robust neural networks for group shifts: on the importance of regularization for worst-case generalization. arXiv preprint arXiv:1911.08731 (2019)
20. Selvaraju, R.R., Cogswell, M., Das, A., Vedantam, R., Parikh, D., Batra, D.: Grad-CAM: visual explanations from deep networks via gradient-based localization. Int. J. Comput. Vision **128**(2), 336–359 (2020). https://doi.org/10.1007/s11263-019-01228-7

21. Szegedy, C., et al.: Intriguing properties of neural networks. In: International Conference on Learning Representations (2014)
22. Tobii Pro AB: Tobii Studio User's Manual. Danderyd, Stockholm (2016). http://www.tobiipro.com/
23. Tramèr, F., Behrmann, J., Carlini, N., Papernot, N., Jacobsen, J.H.: Fundamental tradeoffs between invariance and sensitivity to adversarial perturbations. In: 37th International Conference on Machine Learning 2020, pp. 9503–9513 (2020)
24. Tramer, F., Carlini, N., Brendel, W., Madry, A.: On adaptive attacks to adversarial example defenses. Adv. Neural. Inf. Process. Syst. **33**, 1633–1645 (2020)
25. Van Essen, D., Anderson, C., Felleman, D.: Information processing in the primate visual system: an integrated systems perspective. Science **255**(5043), 419–423 (1992). https://doi.org/10.1126/science.1734518
26. Yamins, D., DiCarlo, J.: Using goal-driven deep learning models to understand sensory cortex. Nat. Neurosci. **19**, 356–365 (2016)

MARS: Mask Attention Refinement with Sequential Quadtree Nodes for Car Damage Instance Segmentation

Teerapong Panboonyuen[1(✉)], Naphat Nithisopa[1], Panin Pienroj[2], Laphonchai Jirachuphun[2], Chaiwasut Watthanasirikrit[1], and Naruepon Pornwiriyakul[1]

[1] MARS Motor AI Recognition Solution, Bangkok, Thailand
teerapong.panboonyuen@gmail.com,
{teerapong,naphat.nit,chaiwasut,naruepon}@marssolution.io
[2] OZT Robotics, Bangkok, Thailand
{panin,laphonchai}@oztrobotics.com

Abstract. Evaluating car damages from misfortune is critical to the car insurance industry. However, the accuracy is still insufficient for real-world applications since the deep learning network is not designed for car damage images as inputs, and its segmented masks are still very coarse. This paper presents **MARS** (**M**ask **A**ttention **R**efinement with **S**equential quadtree nodes) for car damage instance segmentation. Our MARS represents self-attention mechanisms to draw global dependencies between the sequential quadtree nodes layer and quadtree transformer to recalibrate channel weights and predict highly accurate instance masks. Our extensive experiments demonstrate that MARS outperforms state-of-the-art (SOTA) instance segmentation methods on three popular benchmarks such as Mask R-CNN [9], PointRend [13], and Mask Transfiner [12], by a large margin of +1.3 maskAP-based R50-FPN backbone and +2.3 maskAP-based R101-FPN backbone on Thai car-damage dataset. Our demos are available at https://github.com/kaopanboonyuen/MARS.

Keywords: Quadtree Transformer · Mask Transfiner · PointRend · Instance Segmentation

1 Introduction

Evaluation of car damages in Thailand (see Fig. 1) from an accident is a critical aspect of the car insurance business. When a car is involved in an accident, it can result in significant damage to the vehicle, ranging from minor dents and scratches to more severe structural damage. Evaluating the extent of the damage is essential for the insurance company to determine the cost of repairs or replacement. The evaluation process is typically carried out by a trained professional, such as a claims adjuster, who assesses the damage and determines

G. L. Foresti et al. (Eds.): ICIAP 2023 Workshops, LNCS 14365, pp. 28–38, 2024.
https://doi.org/10.1007/978-3-031-51023-6_3

the cost of repairs or replacement. This process helps the insurance company determine the payout amount to the policyholder. The accuracy of the damage assessment is crucial in ensuring a fair payout to the policyholder and avoiding fraudulent claims. The timely and accurate evaluation of car damages is critical to the car insurance business's smooth operation and helps to maintain the trust of policyholders in the insurance industry [11,14,19].

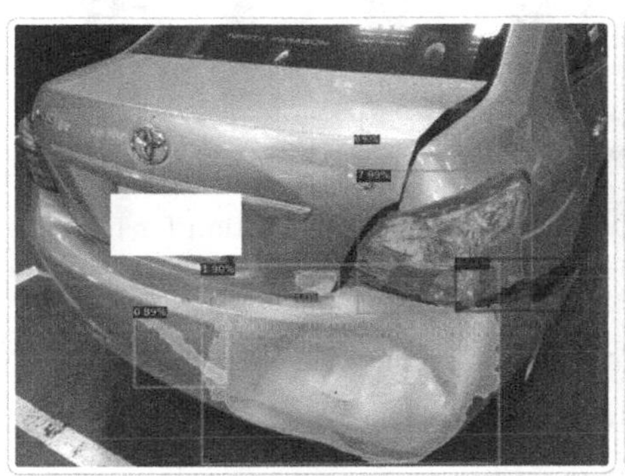
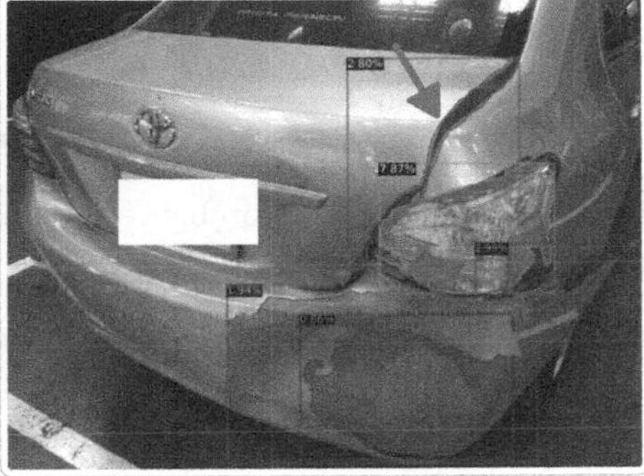

Fig. 1. Instance segmentation with MARS. Our module can be easily incorporated into existing semantic and instance segmentation systems. This paper demonstrates that using MARS instead of the default Sequential Quadtree Nodes in Mask Transfiner [12] leads to much more detailed results.

Recent instance segmentation methods [2–5,9,18,20,21] have been based on object detection pipelines, where objects [12,13] are first localized with bounding boxes and then refined into a segmentation. However, these methods may suffer from false-positive detections and poorly localized bounding boxes that do not cover the entire object. They also do not consider the whole image but independent proposals. In contrast, this paper proposes a new framework, MARS, that models instance masks and considers the image as a whole when making predictions, producing more accurate segmentations. Figure 2 demonstrates that there is still a considerable disparity between the performance of the latest state-of-the-art methods in bounding box and segmentation, particularly for the most recent query-based approaches. This suggests that while detection capability has progressed significantly, the quality of masks has not advanced at the same rate.

Our proposed method is inspired by [12], decomposes and represents the image regions as a quadtree. The central concept of this research is to overcome false-positive detections and imperfectly localized bounding boxes that do not cover the entire object, which is a limitation of detection-based methods. The proposed method considers the entire image while making predictions and aims to address occlusions between different objects. The segmentation maps produced by the proposed method, as shown in Fig. 4, do not require post-processing. We start our end-to-end trained network with a semantic segmen-

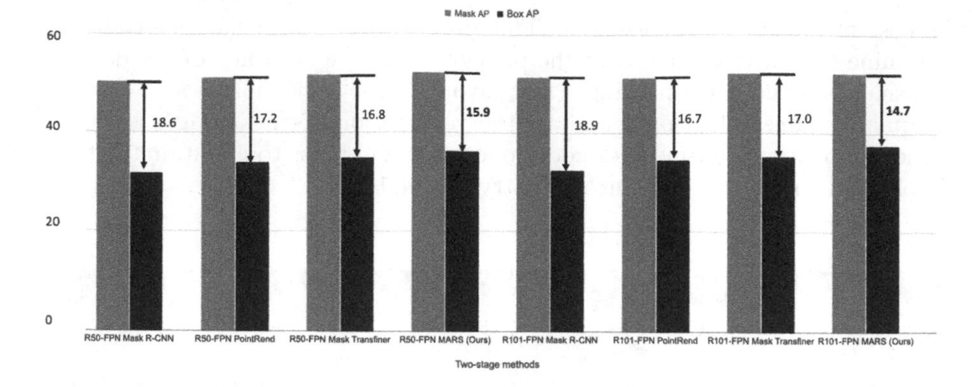

Fig. 2. The performance gap between object detection and segmentation for instance segmentation models on Thai car-damage set using R50-FPN as backbone. Detailed comparisons are in Table 2.

tation module, which generates a variable number of instances for each input image.

The main contributions of this work are two-fold:

- We introduce MARS, a new framework for modeling instance masks that enables our method to generate more precise segmentations. MARS is designed to be straightforward and efficient.
- We propose using self-attention with sequential quadtree nodes in our framework. We modify the relative positional encoding to encode information about spatial distances and relationships between sequential quadtree and quadtree transformer module, which helps capture critical local dependencies and correlations.

Our method, MARS, was extensively evaluated on the Thai car-damage corpus benchmark using quantitative and qualitative analysis. Our results demonstrate that MARS is highly effective in producing accurate segmentations of Thai car-damage imagery, surpassing state-of-the-art methods in high APs family thresholds on the dataset. In addition, our proposed method improves the semantic segmentation task while being trained for the instance segmentation task. Qualitatively, MARS generates sharp object boundaries as shown in Fig. 3 and Fig. 5. Despite the standard intersection-over-union-based metrics being biased towards object-interior pixels and insensitive to boundary improvements, we observed quantitative improvements with MARS. Our method achieved a significant improvement over the robust Mask R-CNN model. Qualitatively, MARS efficiently computes sharp boundaries between objects, as illustrated in Fig. 3.

2 Related Works

Several efforts have been made in the literature to analyze car damage automatically. The instance segmentation field, a complex and challenging area in

Fig. 3. Instance Segmentation on Thai car-damage validation set by a) Mask R-CNN [9], b) PointRend [13], c) Mask Transfiner [12], d) MARS (Ours) using R50-FPN as backbone, where MARS produces significantly more detailed results at high-frequency image regions by replacing Mask R-CNN's default mask head (Zoom in for better view).

machine vision research, has gained significant attention. Its objective is to predict the object class label and the pixel-specific object instance mask, and it locates different types of object instances in various images. Several studies such as [6,7,10,17] have been carried out in this area. This study employs instance segmentation to identify and segment areas of automobile damage in traffic accidents.

Instance Segmentation. Faster R-CNN [8] is extended by Mask R-CNN [9] with the addition of a branch for predicting an object mask in parallel with the existing component for bounding box recognition. Mask R-CNN is simple to train and adds only a small overhead to Faster R-CNN.

PointRend (Point-based Rendering) [13] offers a module that performs point-based segmentation predictions at adaptively selected locations based on an iterative subdivision algorithm. It can be applied flexibly to both instance and semantic segmentation tasks by building on top of existing state-of-the-art models. PointRend is not restricted to instance segmentation and can be expanded to other pixel-level recognition tasks.

Mask Transfiner [12] identifies and decomposes image regions to construct a hierarchical quadtree. They use a multi-scale deep feature pyramid, take sequential input, and execute powerful local and non-local reasoning through the multi-head attention layers. It receives as input sequence the sparsely detected feature points in incoherent image regions across the RoI feature pyramid levels and generates the corresponding segmentation labels. First, it detects and decomposes image regions to build a hierarchical quadtree. Then, all points on the quadtree are transformed into a query sequence for the transformer to forecast the final labels. Unlike previous segmentation methods using convolutions constrained by uniform image grids, it produces high-quality masks with low computation and memory costs.

Car Damage Task. The task of analyzing car damage has been approached by various researchers. Zhang et al. [21] propose an improved Mask R-CNN with ResNet and FPN for feature extraction. Parhizkar et al. [15] introduce a CNN-based approach for accurate vehicle detection and damage localization on the car's exterior surface. Pasupa et al. [16] develop an automatic car part identification system based on Mask R-CNN and GCNet for object detection and semantic segmentation tasks in average weather conditions. Finally, Amirfakhrian et al. [1] present a particle swarm optimization (PSO) algorithm to identify the damaged parts of cars involved in accidents in Iran.

Although previous works have employed mask instance segmentation, there are limitations in its accuracy as some areas with damage may not be visible or segmented correctly. Unlike previous works that use Mask R-CNN for feature representation, our approach uses it to generate masks for instance segmentation. Additionally, our approach involves learning Mask R-CNN specific to each image instead of using global prototypes shared across the car damage dataset.

3 Proposed Method

We next describe Mask Attention Refinement with Sequential Quadtree Nodes (MARS) for instance segmentation. MARS has three components and is illustrated in Fig. 4: (i) node encoder, (ii) sequence encoder, and (iii) pixel decoder like Mask Transfiner [12]. The node encoder first enriches the feature embedding for each incoherent point. The sequence encoder then takes these encoded feature vectors across multiple quadtree levels as input queries. Finally, the pixel decoder predicts their corresponding mask labels. Also, we employ a multi-scale deep feature pyramid. The object detection head then predicts bounding boxes as instance proposals. This component also generates a coarse initial mask prediction at low resolution. Given this input data, we aim to predict highly accurate instance segmentation masks.

Our MARS operates in the detected incoherent regions. Since it operates on feature points on the constructed quadtree, not in a uniform grid, we design a transformer architecture that jointly processes all incoherent nodes in all quadtree levels. Finally, we present the training strategy of MARS along with the implementation details.

MARS Architecture. Figure 4 shows the overall architecture of our MARS. Based on the Mask Transfiner [12], instance segmentation is tackled in a multi-level and coarse-to-fine manner. Instead of using single-level FPN feature for each object, MARS takes as input sequence the sparsely detected feature points in incoherent image regions across the RoI feature pyramid levels, and outputs the corresponding segmentation labels. We interpret the projections of queries, keys, and values in self-attention module as multiple 1×1 convolutions, followed by the computation of attention weights and aggregation of the values before take into sequential quadtree nodes. Then, the sequence transformer encoder of Transfiner jointly processes the encoded nodes from all levels in the quadtree. The transformer thus performs both global spatial and inter-scale reasoning. Each sequence encoder layer has a standard transformer structure, formed by a multi-head self-attention module and a fully connected feed forward network

(FFN). To equip the incoherent points sequence with adequate positive and negative references, we also use all feature points from the coarsest FPN level with small size 7×7. Unlike the standard transformer decoder [4] with deep attention layers, the pixel decoder in MARS is a small two-layer MLP, which decodes the output query for each node in the tree, to predict the final mask labels.

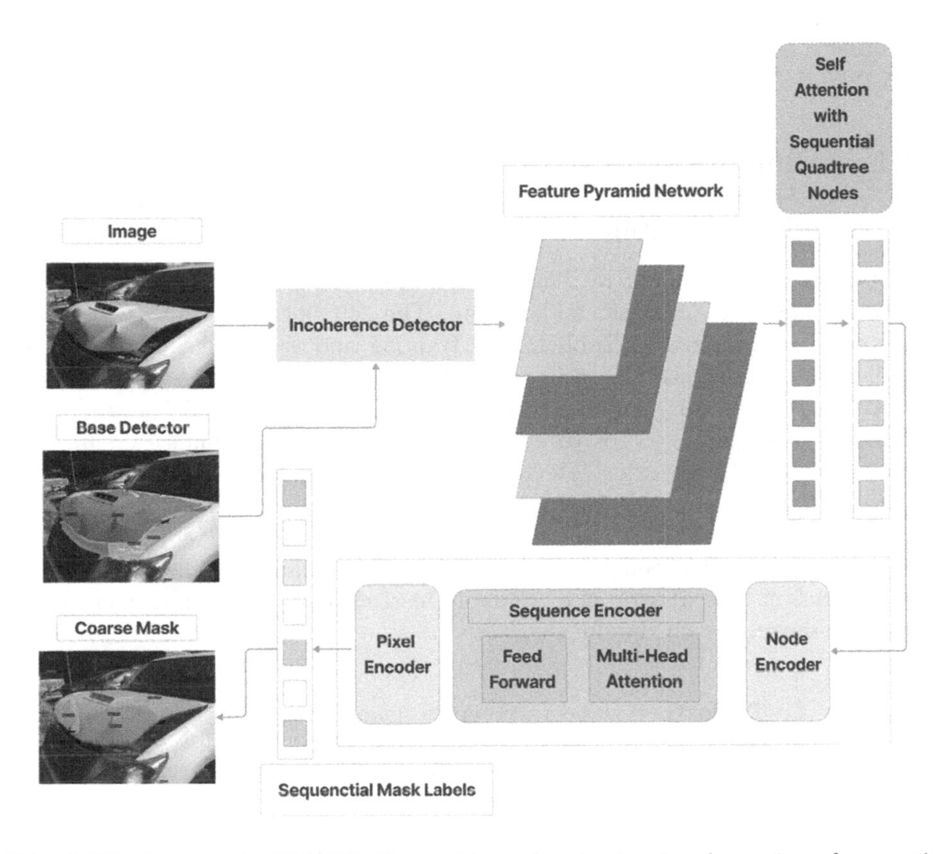

Fig. 4. The framework of MARS. Our end-to-end trained network consists of semantic and instance segmentation modules.

Training and Inference. Based on the constructed quadtree, we develop flexible and adaptive training and inference schemes for MARS, where all detected incoherent nodes across quadtree levels are formed into a sequence for parallel prediction. During inference, MARS follows the quadtree propagation scheme after obtaining the refined labels for incoherent nodes to obtain final object masks. During training, the whole Mask Trans- finer framework can be trained end-to-end. We employ a multi-task loss

$$\mathcal{L} = \lambda_1 \mathcal{L}_{Detect} + \lambda_2 \mathcal{L}_{Coarse} + \lambda_3 \mathcal{L}_{Refine} + \lambda_4 \mathcal{L}_{Inc} \quad (1)$$

Here, $\mathcal{L}_{Refine}$ denotes the refinement with L1 loss between the predicted labels for incoherent nodes and their ground-truth labels. A Binary Cross Entropy loss $\mathcal{L}_{Inc}$ is for detecting incoherent regions. The detection loss $\mathcal{L}_{Detect}$ includes the localization and classification losses from the base detector, e.g., Faster R-CNN or DETR detector. Finally, $\mathcal{L}_{Coarse}$ represents the loss for the initial coarse segmentation prediction used by $\lambda_{\{1,2,3,4\}}$ are hyperparameter weights $\{0.75, 0.75, 0.8, 0.5\}$.

4 Experiments

4.1 Experimental Setup

This paper uses a dataset of Thai car-damage, divided into four categories: Cracked Paint, Dent, Loose, and Scrape. The number of instances for each category is summarized in Table 1. The training and validation data are subjected to data augmentation to improve the models' performance and increase data diversity. The augmented dataset is split into training (60%), validation (20%), and testing (20%) set to evaluate the models' performance on unseen data and real-world scenarios. The data augmentation process addresses several challenges in the dataset, including background variations, unique car damage, unclear or angled body damage, and flipped images. Additionally, cross-class bias was observed between the Scrape, Loose, Dent, and Cracked categories. The experiment's results provide valuable insights into instance segmentation models' performance on the dataset.

Table 1. Object detection performance comparison

Category	Instances
Cracked Paint	273,121
Dent	332,342
Loose	114,345
Scrape	434,237

4.2 Comparison with SOTA

This work compares the performance of the proposed method, MARS, with state-of-the-art instance segmentation methods such as Mask R-CNN, PointRend, and Mask Transfiner on the Thai car-damage dataset. The comparison is based on several metrics, including AP, AP50, AP75, APs, APm, API, and FPS, and the results are presented in Table 2.

The study shows that MARS consistently outperforms Mask R-CNN and PointRend by 4.5 AP and 2.3 AP, respectively, using R50-FPN, and surpasses Mask Transfiner by 2.4 AP using R101-FPN. The superior performance of MARS is not limited to overall AP, but extends to various AP metrics such as APs, APm, APl, AP50, and AP75. For example, MARS significantly improved APs, with a 4.8 AP gain over Mask R-CNN using R50-FPN and a 5.1 AP gain using R101-FPN. Furthermore, MARS outperforms other methods in APm and APl scores, indicating its robustness across different object scales. The higher AP50 and AP75 scores also demonstrate MARS's improved localization and segmentation accuracy compared to other methods.

In addition to its superior performance, MARS is computationally efficient. It almost achieves the highest FPS on the Thai car-damage dataset, possibly because the model was fine-tuned or optimized for this specific dataset. This fine-tuning process could involve adjusting the model's hyperparameters, architecture, or training strategies to improve its efficiency and performance on the Thai car-damage dataset. Task-specific optimizations, such as focusing on learning the dataset's most relevant features, may lead to a more compact model that requires fewer computations during inference, resulting in faster performance.

Table 2. Object detection performance comparison

Method	Backbone	AP	AP50	AP75	APs	APm	APl	FPS
Mask R-CNN [9]	R50-FPN	31.7	50.1	34.7	11.9	29.9	41.3	**8.4**
PointRend [13]	R50-FPN	33.9	51.7	36.4	12.3	31.0	42.2	4.6
Mask Transfiner [12]	R50-FPN	34.9	52.4	37.1	13.8	32.5	45.0	6.7
MARS (Ours)	R50-FPN	**36.2**	**53.0**	**38.9**	**15.8**	**34.6**	**47.3**	6.8
Mask R-CNN [9]	R101-FPN	32.4	51.5	35.1	17.6	33.6	42.0	**8.1**
PointRend [13]	R101-FPN	34.5	52.8	37.0	19.6	35.6	43.7	5.5
Mask Transfiner [12]	R101-FPN	35.1	54.9	37.7	20.9	37.5	44.4	7.1
MARS (Ours)	R101-FPN	**37.5**	**55.7**	**41.2**	**22.7**	**38.7**	**45.1**	7.2

Figure 5 shows qualitative comparisons on the Thai car-damage dataset, where our MARS produces masks with substantially higher precision and quality than previous methods [9,12,13], especially for the challenging regions, such as the dents in the corners of the fender and a black car covered in mud.

4.3 Implementation Details

We utilized advanced hardware and software technologies, including the NVIDIA T4 GPU with 16 GB memory, Intel® Xeon® Scalable (Cascade Lake) 4 vCPUs, 16 GiB RAM, PyTorch (1.13.1) and CUDA (11.7.0), to perform the training and testing tasks for our project. Combining high-performance hardware and an optimized software stack resulted in efficient computations and superior-quality outcomes.

Fig. 5. Example result pairs from SOTA with its standard mask head (columns 3 and 4) vs. with MARS (right image). Note how MARS predicts masks with substantially finer detail around object boundaries.

5 Conclusions

Our new instance segmentation method, MARS (Mask Attention Refinement with Sequential Quadtree Nodes), is introduced in this work. MARS first detects and decomposes image regions to create a hierarchical quadtree. Then, all points on the quadtree are transformed into a self-attention layer before passing through our MARS model to predict car-damage classes. Unlike previous segmentation methods that rely on convolutions with uniform image grids, MARS produces high-quality masks with low computation and memory costs. We demonstrate the effectiveness of MARS on query-based segmentation frameworks and show that it outperforms existing methods on the Thai car-damage dataset. We anticipate the proposed MARS framework being a solid baseline for car-damage instance segmentation tasks.

Acknowledgement. This work was partially supported by Thaivivat Insurance PCL, by providing financial support and expertise in car insurance. We thank Natthakan Phromchino, Darakorn Tisilanon, and Chollathip Thiangdee in MARS (Motor AI Recognition Solution) for annotating the data. Teerapong Panboonyuen and his colleague received no financial support for this article's research, authorship, and/or publication.

References

1. Amirfakhrian, M., Parhizkar, M.: Integration of image segmentation and fuzzy theory to improve the accuracy of damage detection areas in traffic accidents. J. Big Data **8**(1), 1–17 (2021)
2. Arnab, A., Torr, P.H.: Pixelwise instance segmentation with a dynamically instantiated network. In: Proceedings of the IEEE Conference on Computer Vision and Pattern Recognition, pp. 441–450 (2017)
3. Bolya, D., Zhou, C., Xiao, F., Lee, Y.J.: YOLACT: real-time instance segmentation. In: Proceedings of the IEEE/CVF International Conference on Computer Vision, pp. 9157–9166 (2019)
4. Chen, H., Sun, K., Tian, Z., Shen, C., Huang, Y., Yan, Y.: BlendMask: top-down meets bottom-up for instance segmentation. In: Proceedings of the IEEE/CVF Conference on Computer Vision and Pattern Recognition, pp. 8573–8581 (2020)
5. Chen, K., et al.: Hybrid task cascade for instance segmentation. In: Proceedings of the IEEE/CVF Conference on Computer Vision and Pattern Recognition, pp. 4974–4983 (2019)
6. Chen, L.C., Hermans, A., Papandreou, G., Schroff, F., Wang, P., Adam, H.: Mask Lab: instance segmentation by refining object detection with semantic and direction features. In: Proceedings of the IEEE Conference on Computer Vision and Pattern Recognition, pp. 4013–4022 (2018)
7. Deng, D., Liu, H., Li, X., Cai, D.: PixelLink: detecting scene text via instance segmentation. In: Proceedings of the AAAI Conference on Artificial Intelligence. vol. 32 (2018)
8. Girshick, R.: Fast R-CNN. In: Proceedings of the IEEE International Conference on Computer Vision, pp. 1440–1448 (2015)
9. He, K., Gkioxari, G., Dollár, P., Girshick, R.: Mask R-CNN. In: Proceedings of the IEEE International Conference on Computer Vision, pp. 2961–2969 (2017)
10. Iglovikov, V., Seferbekov, S., Buslaev, A., Shvets, A.: TernausNetV2: fully convolutional network for instance segmentation. In: Proceedings of the IEEE Conference on Computer Vision and Pattern Recognition Workshops, pp. 233–237 (2018)
11. Jõeveer, K., Kepp, K.: What drives drivers? Switching, learning, and the impact of claims in car insurance. J. Behav. Exp. Econ. **103**, 101993 (2023)

12. Ke, L., Danelljan, M., Li, X., Tai, Y.W., Tang, C.K., Yu, F.: Mask transfiner for high-quality instance segmentation. In: Proceedings of the IEEE/CVF Conference on Computer Vision and Pattern Recognition, pp. 4412–4421 (2022)
13. Kirillov, A., Wu, Y., He, K., Girshick, R.: PointRend: image segmentation as rendering. In: Proceedings of the IEEE/CVF Conference on Computer Vision and Pattern Recognition, pp. 9799–9808 (2020)
14. Macedo, A.M., Cardoso, C.V., Neto, J.S.M., et al.: Car insurance fraud: the role of vehicle repair workshops. Int. J. Law Crime Justice **65**, 100456 (2021)
15. Parhizkar, M., Amirfakhrian, M.: Car detection and damage segmentation in the real scene using a deep learning approach. Int. J. Intell. Robot. Appl. **6**(2), 231–245 (2022)
16. Pasupa, K., Kittiworapanya, P., Hongngern, N., Woraratpanya, K.: Evaluation of deep learning algorithms for semantic segmentation of car parts. Complex Intell. Syst. **8**(5), 3613–3625 (2022)
17. Tian, Z., Shen, C., Chen, H.: Conditional convolutions for instance segmentation. In: Vedaldi, A., Bischof, H., Brox, T., Frahm, J.-M. (eds.) ECCV 2020. LNCS, vol. 12346, pp. 282–298. Springer, Cham (2020). https://doi.org/10.1007/978-3-030-58452-8_17
18. Wang, X., Zhang, R., Kong, T., Li, L., Shen, C.: SOLOv2: dynamic and fast instance segmentation. Adv. Neural. Inf. Process. Syst. **33**, 17721–17732 (2020)
19. Weisburd, S.: Identifying moral hazard in car insurance contracts. Rev. Econ. Stat. **97**(2), 301–313 (2015)
20. Xie, E., et al.: PolarMask: single shot instance segmentation with polar representation. In: Proceedings of the IEEE/CVF Conference on Computer Vision and Pattern Recognition, pp. 12193–12202 (2020)
21. Zhang, Q., Chang, X., Bian, S.B.: Vehicle-damage-detection segmentation algorithm based on improved mask RCNN. IEEE Access **8**, 6997–7004 (2020)

On-Device Learning with Binary Neural Networks

Lorenzo Vorabbi[1,2](✉), Davide Maltoni[2], and Stefano Santi[1]

[1] Datalogic Labs, 40012 Bologna, Italy
{lorenzo.vorabbi,stefano.santi}@datalogic.com
[2] University of Bologna, DISI, Cesena Campus, 47521 Cesena, Italy
{lorenzo.vorabbi2,davide.maltoni}@unibo.it

Abstract. Existing Continual Learning (CL) solutions only partially address the constraints on power, memory and computation of the deep learning models when deployed on low-power embedded CPUs. In this paper, we propose a CL solution that embraces the recent advancements in CL field and the efficiency of the Binary Neural Networks (BNN), that use 1-bit for weights and activations to efficiently execute deep learning models. We propose a hybrid quantization of CWR* (an effective CL approach) that considers differently forward and backward pass in order to retain more precision during gradient update step and at the same time minimizing the latency overhead. The choice of a binary network as backbone is essential to meet the constraints of low power devices and, to the best of authors' knowledge, this is the first attempt to prove on-device learning with BNN. The experimental validation carried out confirms the validity and the suitability of the proposed method.

Keywords: Binary Neural Networks · On-device Learning · Continual Learning

1 Introduction

Integrating a deep learning model into an embedded system can be a challenging task for two main reasons: the model may not fit into the embedded system memory and, the time efficiency may not satisfy the application requirements. A number of light architectures have been proposed to mitigate these problems (MobileNets [1], EfficientNets [2], NASNets [3]) but they heavily rely on floating point computation which is not always available (or efficient) on tiny devices. Binary Neural Networks (BNN), where a single bit is used to encode weights and activations, emerged as an interesting approach to speed up the model inference relying on packed bitwise operations [4]. However, almost no literature work addresses the problem of training (or tuning) such models on-device, a task which is still more complex than inference because:

G. L. Foresti et al. (Eds.): ICIAP 2023 Workshops, LNCS 14365, pp. 39–50, 2024.
https://doi.org/10.1007/978-3-031-51023-6_4

– quantization is known to affect back propagation and weights update
– popular inference engines (e.g. Tensorflow Lite, pytorch mobile, ecc.) do not support model training

This work proposes on-device learning of BNN to enable continual learning of a pre-trained model. We start from CWR* [5], a simple but effective continual learning approach that limits weight updates to the output head, and designs an ad-hoc quantization approach that preserves most of the accuracy with respect to a floating point implementation. We prove that several state of the art BNN models can be used in conjunction with our approach to achieve good performance on classical continual learning dataset/benchmarks such as CORe50 [6], CIFAR10 [7] and CIFAR100 [7].

2 Related Literature

2.1 Continual Learning

The classical deep learning approach is to train a model on a large batch of data and then freeze it before deployment on edge devices; this does not allow adapting the model to a changing environment where new classes (NC scenario) or new items/variation of known classes (NI scenario) can appear over time. Collecting new data and periodically retraining a model from scratch is not efficient and sometime not possible because of privacy, so the CL approach is to adapt an existing model by using only new data. Unfortunately, this is prone to forgetting old knowledge, and specific techniques are necessary to balance the model stability and plasticity. For a survey of existing CL methods see [8].

In this work we focus on Single Object Recognition task addressing the two CL scenarios of NI and NC; in both cases, the learning phase of the model is usually splitted in *experiences*, each one containing different training samples belonging or not to known classes (this depends on the CL scenario).
CWR* mantains two sets of weights for the output classification layer: cw are the consolidated weights used during inference while tw are the temporary weights that are iteratively updated during back-propagation. cw are initialized to 0 before the first batch and then updated according to Algorithm 1 (for more details see [5]), while tw are reset to 0 before each training mini-batch. CWR*, for each already encountered class (of current training batch), reloads the consolidated weights cw at the beginning of each training batch and, during the consolidation step, adopts a weighted sum based on the number of the training samples encountered in the past batches and those of current batch. The consolidation step has a negligible overhead and can be quantized adopting the same quantization scheme used for CWR* weights. In CWR*, during the first training experience (supposed to be executed offline) all the layers of the model are trained but from the second experience, only the weights of the output classification layer are adjusted during the back-prop stage, to simulate a real case scenario (lines 9–12 of Algorithm 1).

Algorithm 1. CWR* pseudocode: $\overline{\Theta}$ are the class-shared parameters. Both tw and cw refer to the same layer index k of the model.

1: **procedure** CWR*
2: $cw_k = 0$ ▷ k is the index of the classification layer
3: $past = 0$ ▷ number of samples for each class i encountered
4: init $\overline{\Theta}$ random or from pre-trained model
5: **for** each training batch B_j **do** ▷ B_j is the mini-batch of index j
6: expand layer k with neurons for the new classes in B_j never seen before
7: $tw_k[i] = \begin{cases} cw_k[i], & \text{if class } i \text{ in } B_j \\ 0, & \text{otherwise} \end{cases}$
8: train the model with SGD
9: **if** $B_j = B_1$ **then**
10: learn both $\overline{\Theta}$ and tw_k
11: **else**
12: learn tw_k while keeping $\overline{\Theta}$ fixed
13: **for** each class i in B_j **do** ▷ consolidation step
14: $wpast_i = \sqrt{\frac{past_i}{cur_i}}$, where cur_i is the number of patterns of class i in B_j
15: $cw_k[i] = \frac{cw_k[i] \cdot wpast_i + (tw_k[i] - avg(tw_k))}{wpast_i + 1}$
16: $past_i = past_i + cur_i$
17: test the model by using $\overline{\Theta}$ and cw_k

2.2 Binary Neural Networks

Quantization is a technique that yields compact models compared to their floating-point counterparts, by representing the network weights and activations with very low precision. The most extreme quantization is binarization, where data can only have two possible values, namely $-1(0)$ or $+1(1)$. By representing weights and activations using only 1-bit, the resulting memory footprint of the model is dramatically reduced and also the heavy matrix multiplication operations can be replaced with light-weighted bitwise XNOR operations and Bitcount operations. According to [9], that compared the speedups of binary layers w.r.t. the 8-bit quantized and floating point layers, a binary implementation can achieve a lower inference time from 9 to $12\times$ on a low power ARM CPU. Therefore, Binary Neural Networks combine many hardware friendly properties including memory saving, power efficiency and significant acceleration; for some network topologies, BNN can be executed on device without the usage of floating-point operations [10] simplifying the deployment on ASIC or FPGA hardware. For a survey on binary neural networks see [4].

3 On-Device CWR Optimization

3.1 Gradients Computation

In this section we make explicit the weights update in the classification layer; without loss of generality, a neural network $M(\cdot)$ is composed by a sequence of k layers represented as:

$$M(\cdot) = f_{w_k}\left(f_{w_{k-1}}\left(\cdots f_{w_2}\left(f_{w_1}(\cdot)\right)\right)\right) \tag{1}$$

where w_i represents the weights of the ith layer. In CWR* the temporary weights tw_k (lines 10 and 12 of Alg. 1) are updated according to Eqs. 9 and 10, whose quantization is discussed in the next section. Denoting with a_i and a_{i+1}[1] the input and output activations of the ith layer respectively, with $\mathcal{L}$ the loss function, the backpropagation process consists in the computation of two different sets of gradients: $\frac{\partial \mathcal{L}}{\partial a_i}$ and $\frac{\partial \mathcal{L}}{\partial w_i}$.

In CWR* the on-device backpropagation algorithm is limited to the last layer which can be considered a linear layer (with a non-linear activation function) with the following forward formula:

$$a_{k+1} = f_k(o_{k+1}), \quad o_{k+1} = a_k W_k + b_k \tag{2}$$

where a_{k+1} represents the output of the neural network.

Considering a classification task (with M classes) with an unitary batch size, the *Cross-Entropy* loss function is formulated as:

$$\mathcal{H}(y, a_{k+1}) = -\sum_{i=0}^{M-1} y^i log\left(a_{k+1}^i\right) \tag{3}$$

where y^i represents the element of an one-hot encoded vector of ground truth and a_{k+1}^i is the ith output activation sample. Using the softmax as activation for the last layer, reported below:

$$a_{k+1}\left(o_{k+1}^t\right) = \frac{e^{o_{k+1}^t}}{\sum_{j=1}^{M} e^{o_{k+1}^j}} \tag{4}$$

, the gradient formulas for the last classification layer can be expressed using the chain rule:

$$\frac{\partial \mathcal{H}}{\partial W_k} = \frac{\partial \mathcal{H}}{\partial a_{k+1}} \frac{\partial a_{k+1}}{\partial o_{k+1}} \frac{\partial o_{k+1}}{\partial W_k} \tag{5}$$

$$\frac{\partial \mathcal{H}}{\partial b_k} = \frac{\partial \mathcal{H}}{\partial a_{k+1}} \frac{\partial a_{k+1}}{\partial o_{k+1}} \frac{\partial o_{k+1}}{\partial b_k} \tag{6}$$

[1] Note that the output a_{i+1} of level i corresponds to the input of level $i+1$.

The final expression for Eq. 5 using the Eq. 3 as loss function and 4 as non-linear $f_k(\cdot)$ is a well-known result, that can be easily derived:

$$\frac{\partial \mathcal{H}}{\partial W_k} = (a_{k+1} - y)\, a_k \tag{7}$$

$$\frac{\partial \mathcal{H}}{\partial b_k} = (a_{k+1} - y) \tag{8}$$

Using a stochastic gradient descent optimizer with learning rate η, the weights update equation is:

$$W_k^{i+1} = W_k^i - \eta\,(a_{k+1} - y)\, a_k \tag{9}$$

$$b_k^{i+1} = b_k^i - \eta\,(a_{k+1} - y) \tag{10}$$

Therefore in CWR* the temporary weights tw_k (lines 10 and 12 of Alg. 1) are updated according to Eqs. 9 and 10, whose quantization is discussed in the next section.

3.2 Quantization Strategy

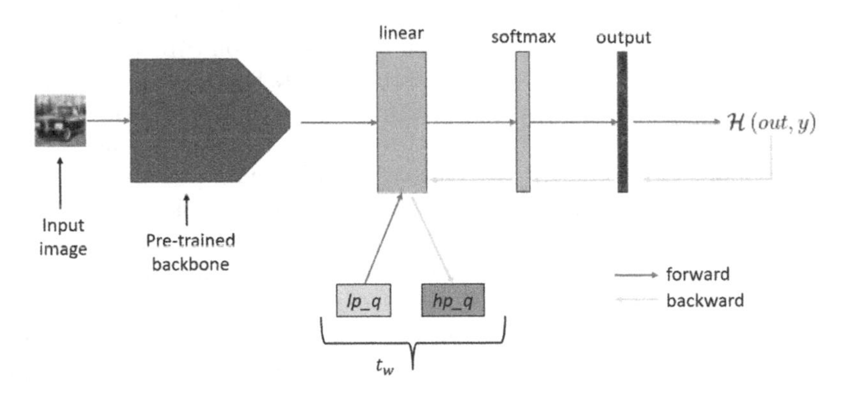

Fig. 1. Double quantization scheme that uses a different quantization level for weights/activations used in forward and backward pass.

Our approach considers two different quantizations: the former uses 1-bit (also called binarization) to represent weights and activations employed by the pre-trained backbone; the latter is used in the last classification layer, to quantize both forward and backward operations. This solution both reduces the latency and simplifies the adaptation of the model on new item/classes encountered.

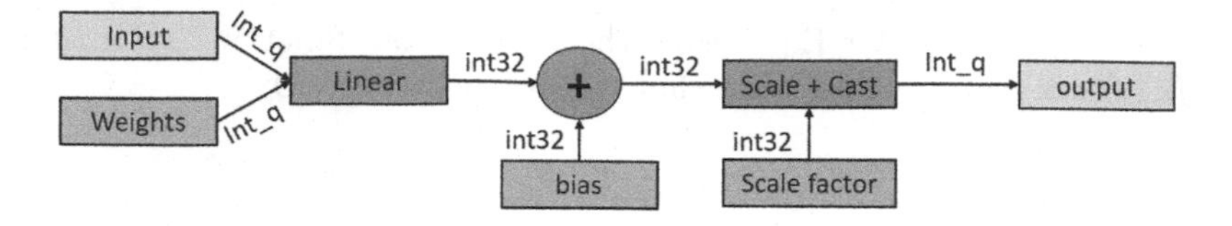

Fig. 2. Quantization scheme adopted using q bits for weights and activations.

In particular, for the last layer quantization we followed the scheme proposed in [11] and implemented in GEMMLOWP library [12]. The quantized output of a 32-bit floating point linear layer, reported in Eq. 2, can be represented as:

$$\overline{o_{k+1}^{int\_q}} = cast\_to\_int\_q\lfloor s_k^{int\_32}\left(\overline{W_k^{int\_q}} a_k^{int\_q} + \overline{b_k^{int\_q}}\right)\rceil \tag{11}$$

The quantization Eq. 11 depends on the number of the quantization q bits used $(8, 16, 32)$, $\bar{\cdot}$ represents the quantized version of a tensor and s^{int_32} is the fixed-point scaling factor having 32-bit precision, as shown in Fig. 2. Similarly to previous works [13,14], we used the straight-through estimator (STE) approach to approximate differentiation through discrete variables; STE represents a simple and hardware-friendly method to deal with the computation of the derivative of discrete variables that are zero almost everywhere.

Based on the results reported in [15–17], the quantization of the gradients in Eqs. 7 and 8 represents the main cause of accuracy degradation during training and therefore we propose to use two separate versions of layer weights W_k, one with low-precision (lp_q) and another with higher precision (hp_q). As shown in Fig. 1, the idea is to use the lp_q version of the weights for the computations that have strict timing deadlines (forward pass), while the hp_q version is adopted during the weight update step (Eqs. 9 and 10), which has typically more relaxed timing constraints (it can be executed also as a background process). Every time a new high-precision copy of weights is computed, a lower version is derived from it and stored.

Gradient quantization inevitably introduces an approximation error that can affect the accuracy of the model; to check the amount of approximation for different quantization levels, for each mini-batch, we compute the Mean Absolute Error (MAE, in percentage) between the floating point gradient and the quantized one for the weight tensor of the CWR* layer (for the dataset CORe50 [6]). The MAE is then accumulated for all training mini-batches of each experience, as shown in Fig. 3. In order to evaluate only the quantization error introduced, both floating-point and quantized gradients are computed starting from the same W_k^i weights (Eq. 9). The plot curves of Figs. 3a and 3b refer respectively to the *quicknet* [9] and *realtobinary* [18] models; it is evident that the quantization error introduced using the lp_q with 8 bits is much larger compared to higher quantization schemes (16/32 bits or floating point) whose gap w.r.t. the floating point implementation is quite low, as pointed out in Sect. 5.

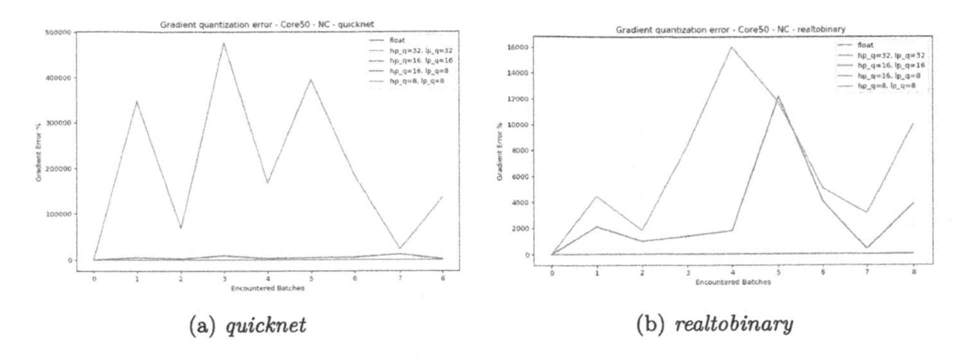

(a) *quicknet* (b) *realtobinary*

Fig. 3. Accumulation of gradient quantization errors (Mean Absolute Error in percentage) between quantized and floating-point versions for each experience. During the first experience the gradient computation is always executed in floating-point.

4 Experiments

We evaluate the proposed approach on three classification datasets: CORe50, CIFAR10 and CIFAR100 with different BNN architectures. The BNN models employed for CORe50 have been pre-trained on ImageNet [19] and taken from Larq repository[2]; instead, the models used for CIFAR10 and CIFAR100 have been pre-trained on Tiny Imagenet[3]. For each dataset, we conducted several tests using a different number of quantization bits with the same training procedure. Our work is targeting a model that could continuously learn and therefore we limited the number of epochs to 10 for the first experience and to 5 for the remaining. The results of Eq. 7 and 9 require the adoption of the Cross Entropy as loss function and the Stochastic Gradient Descent (SGD) as optimizer; the choice of SGD is encouraged as it requires a simple computation with a limited overhead compared to the Adam [20] optimizer. The binarization of weights and activations always happens at training time using an approximation of the gradient (STE introduced in Sect. 3.2 or derived solutions that are model dependent) for *sign* function.

Hereafter we provide some details on the datasets and related CL protocols:

CORe50 [6] It is a dataset specifically designed for Continuous Object Recognition containing a collection of 50 domestic objects belonging to 10 categories. The dataset has been collected in 11 distinct sessions (8 indoor and 3 outdoor) characterized by different backgrounds and lighting. For the continuous learning scenarios (NI, NC) we use the same test set composed ofsessions

[2] https://docs.larq.dev/zoo/api/sota/.
[3] http://cs231n.stanford.edu/tiny-imagenet-200.zip.

#3, #7 and #10. The remaining 8 sessions are split in batches and provided sequentially during training obtaining 9 experiences for NC scenario and 8 for NI. No augmentation procedure has been implemented since the dataset already contains enough variability in terms of rotations, flips and brightness variation. The input RGB image is standardized and rescaled to the size of $128 \times 128 \times 3$.

CIFAR10 and CIFAR100 [7] Due to the lower number of classes, the NC scenario for CIFAR10 contains 5 experiences (adding 2 classes for each experience) while 10 are used for CIFAR100. For both datasets the NI scenario is composed by 10 experiences. Similar to CORe50, the test set does not change over the experiences. The RGB images are scaled to the interval $[-1.0; +1.0]$ and the following data augmentation was used: zero padding of 4 pixels for each size, a random 32×32 crop and a random horizontal flip. No augmentation is used at test time.

On CORe50 dataset, we evaluated the three binary models reported below:

Realtobinary [18] This network proposes a real-to-binary attention matching mechanism that aims to match spatial attention maps computed at the output of the binary and real-valued convolutions. In addition, the authors proposed to use the real-valued activations of the binary network before the binarization of the next layer to compute scaling factors, used to rescale the activations produced after the application of the binary convolution.

Quicknet and QuicknetLarge[9] This network follows the previous works [18,21,22] proposing a sequence of blocks, each one with a different number of binary 3×3 convolutions and residual connections over each layer. Transition blocks between each residual section halve the spatial resolution and increase the filter count. QuicknetLarge employs more blocks and feature maps to increase accuracy.

For CIFAR10 and CIFAR100 datasets, whose input resolution is 32×32, we evaluated the following networks (pre-trained on Tiny Imagenet):

BiRealNet[21] It is a modified version of classical ResNet that proposes to preserve the real activations before the sign function to increase the representational capability of the 1-bit CNN, through a simple shortcut. Bi-RealNet adopts a tight approximation to the derivative of the non-differentiable sign function with respect to activation and a magnitude-aware gradient to update weight parameters. We used the instance of the network that uses *18-layers*[4].

[4] Refer to the following https://github.com/liuzechun/Bi-Real-net repository for all the details.

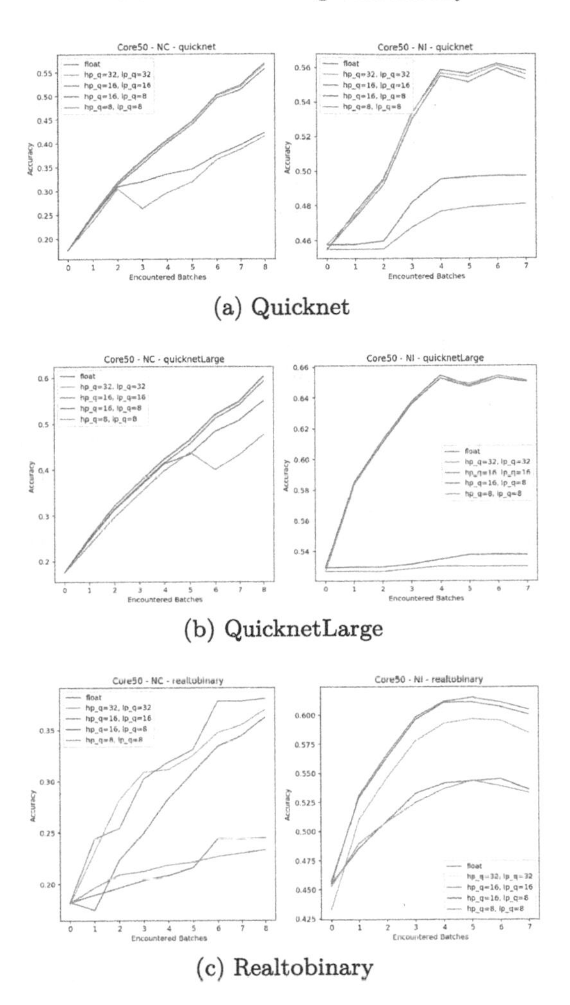

(a) Quicknet

(b) QuicknetLarge

(c) Realtobinary

Fig. 4. CORe50 accuracy results using different quantization methods.

ReactNet[23] To further compress compact networks, this model constructs a baseline based on MobileNetV1 [1] and add shortcut to bypass every 1-bit convolutional layer that has the same number of input and output channels. The 3×3 depth-wise and the 1×1 point-wise convolutional blocks of MobileNet are replaced by the 3×3 and 1×1 vanilla convolutions in parallel with shortcuts in React Net[5]. As for Bi-Real Net, we tested the version of React Net that uses *18-layers*.

Our test were performed on both the NI and NC scenario (discussed in Sect. 2.1). Figure 4, 5 and 6 summarize the experimental results. On CORe50

[5] Refer to the following https://github.com/liuzechun/ReActNet repository for all the details.

dataset (Fig. 4) NC scenario, the quantization scheme *lp_8* gets a consistent accuracy drop over the experiences showing a limited learning capability; instead, the quantizations with *lp_16* and *lp_32* reach the same accuracy level of the floating point model. A similar situation can be observed in the NI scenario with the exception of the QuicknetLarge model where the lower quantization schemes are not able to increase the accuracy of the first experience. For datasets CIFAR10 and CIFAR100 (Fig. 5 and 6) we find similar results for the NI scenario, where the 8-bit quantization scheme limits the learning capability of the model during the experiences. Instead, in the NC scenario, both Bi-Realnet and Reactnet models with *lp_8* quantization, are able to reach an accuracy result closed to the floating-point model. From our analysis it appears that the 8-bit quantization of the gradients limits noticeably the learning ability of a binary model when employed in a continual learning scenario for CWR* method. In order to reach accuracy comparable to a floating point implementation we devise the adoption of at least 16 bits both for *lp* and *hp*; it is worth noting that the computational effort of 16 bits is anyway limited in CWR* because the quantization is confined to the last classification layer.

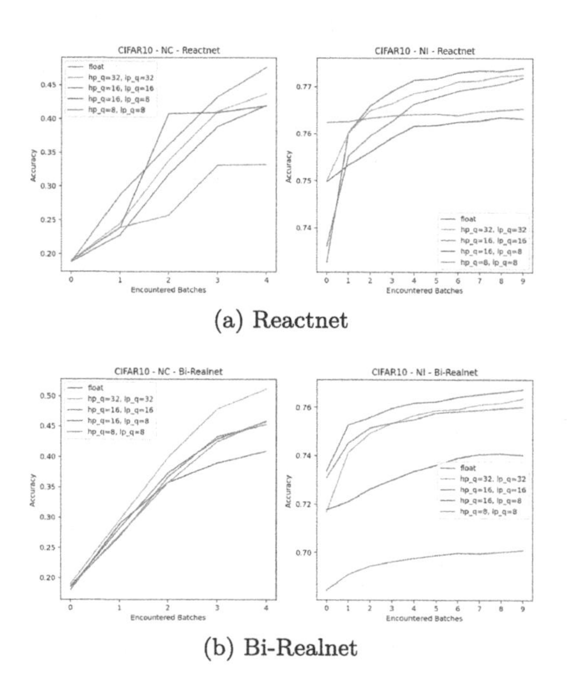

(a) Reactnet

(b) Bi-Realnet

Fig. 5. CIFAR10 accuracy results using different quantization methods.

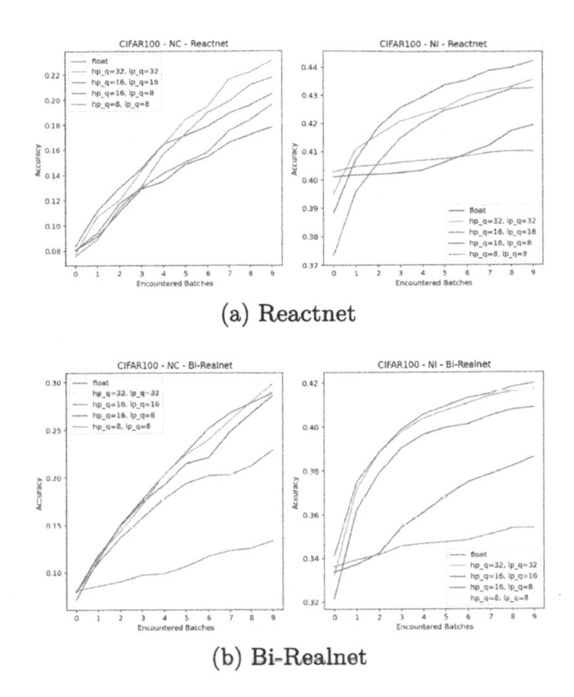

(a) Reactnet

(b) Bi-Realnet

Fig. 6. CIFAR100 accuracy results using different quantization methods.

5 Conclusion

On-device training (or adaptation) can play an essential role in the IoT, enabling the large adoption of deep learning solutions. In this work we focused on implementation of CWR* on edge-devices, relying on binary neural networks as backbone and proposing an ad-hoc quantization scheme. We discovered that 8-bit quantization degrades too much the learning capability of the model, while 16 bits is a good compromise. To the best of authors' knowledge, this work is the first to explore on-device continual learning with binary network; in the future work we intend to explore the application of binary neural networks in combination of CL methods relying on latent replay [24], which is particularly intriguing given the low memory footprint of 1-bit activation.

References

1. Howard, A.G., et al.: Mobilenets: efficient convolutional neural networks for mobile vision applications. arXiv preprint arXiv:1704.04861 (2017)
2. Tan, M., Le, Q.: Efficientnet: rethinking model scaling for convolutional neural networks. In: International Conference on Machine Learning, pp. 6105–6114. PMLR (2019)
3. Cai, H., Zhu, L., Han, S.: Proxylessnas: direct neural architecture search on target task and hardware. arXiv preprint arXiv:1812.00332 (2018)

4. Qin, H., Gong, R., Liu, X., Bai, X., Song, J., Sebe, N.: Binary neural networks: a survey. Pattern Recogn. **105**, 107281 (2020)
5. Lomonaco, V., Maltoni, D., Pellegrini, L.: Rehearsal-free continual learning over small non-iid batches. In: CVPR Workshops, vol. 1, p. 3 (2020)
6. Lomonaco, V., Maltoni, D.: Core50: a new dataset and benchmark for continuous object recognition. In: Conference on Robot Learning, pp. 17–26. PMLR (2017)
7. Krizhevsky, A., Hinton, G., et al.: Learning multiple layers of features from tiny images (2009)
8. De Lange, M., et al.: A continual learning survey: defying forgetting in classification tasks. IEEE Trans. Pattern Anal. Mach. Intell. **44**(7), 3366–3385 (2021)
9. Bannink, T., et al.: Larq compute engine: design, benchmark and deploy state-of-the-art binarized neural networks. Proceedings of Machine Learning and Systems, vol. 3, pp. 680–695 (2021)
10. Vorabbi, L., Maltoni, D., Santi, S.: Optimizing data-flow in binary neural networks (2023). arXiv:2304.00952 [cs.LG]
11. Jacob, B., et al.: Quantization and training of neural networks for efficient integer-arithmetic-only inference. In: Proceedings of the IEEE Conference on Computer Vision and Pattern Recognition, pp. 2704–2713 (2018)
12. Jacob, B., Warden, P., Guney, M.E.: gemmlowp: a small self-contained low-precision gemm library (2017). https://github.com/google/gemmlowp
13. Hubara, I., Courbariaux, M., Soudry, D., El-Yaniv, R., Bengio, Y.: Binarized neural networks. Adv. Neural Inf. Process. Syst. **29** (2016)
14. Mishra, A., Nurvitadhi, E., Cook, J.J., Marr, D.: WRPN: wide reduced-precision networks. arXiv preprint arXiv:1709.01134 (2017)
15. Gupta, S., Agrawal, A., Gopalakrishnan, K., Narayanan, P.: Deep learning with limited numerical precision. In: International Conference on Machine Learning, pp. 1737–1746. PMLR (2015)
16. Das, D., et al.: Mixed precision training of convolutional neural networks using integer operations. arXiv preprint arXiv:1802.00930 (2018)
17. Banner, R., Hubara, I., Hoffer, E., Soudry, D.: Scalable methods for 8-bit training of neural networks. Adv. Neural Inf. Process. Syst. **31** (2018)
18. Martinez, B., Yang, J., Bulat, A., Tzimiropoulos, G.: Training binary neural networks with real-to-binary convolutions. arXiv preprint arXiv:2003.11535 (2020)
19. Russakovsky, O., et al.: ImageNet large scale visual recognition challenge. Int. J. Comput. Vision (IJCV) **115**(3), 211–252 (2015). https://doi.org/10.1007/s11263-015-0816-y
20. Kingma, D.P., Ba, J.: Adam: a method for stochastic optimization. arXiv preprint arXiv:1412.6980 (2014)
21. Liu, Z., Wu, B., Luo, W., Yang, X., Liu, W., Cheng, K.T.: Bi-real net: enhancing the performance of 1-bit cnns with improved representational capability and advanced training algorithm. In: Proceedings of the European Conference on Computer Vision (ECCV), pp. 722–737 (2018)
22. Bethge, J., Yang, H., Bornstein, M., Meinel, C.: Back to simplicity: how to train accurate bnns from scratch? arXiv preprint arXiv:1906.08637 (2019)
23. Liu, Z., Shen, Z., Savvides, M., Cheng, K.-T.: ReActNet: towards precise binary neural network with generalized activation functions. In: Vedaldi, A., Bischof, H., Brox, T., Frahm, J.-M. (eds.) ECCV 2020. LNCS, vol. 12359, pp. 143–159. Springer, Cham (2020). https://doi.org/10.1007/978-3-030-58568-6_9
24. Pellegrini, L., Graffieti, G., Lomonaco, V., Maltoni, D.: Latent replay for real-time continual learning. In: 2020 IEEE/RSJ International Conference on Intelligent Robots and Systems (IROS), pp. 10203–10209. IEEE (2020)

Towards One-Shot PCB Component Detection with YOLO

Gabriele Spadaro[1(✉)], Gaspare Vetrano[2], Barbara Penna[2], Antonio Serena[2], and Attilio Fiandrotti[1]

[1] University of Turin, Turin, Italy
{gabriele.spadaro,attilio.fiandrotti}@unito.it
[2] Optical R&D Lab - SPEA S.p.A. - Automatic Test Equipment, Volpiano, Italy
{gaspare.vetrano,barbara.penna,antonio.serena}@spea.com

Abstract. Consumer electronic devices such as smartphones, TV sets, etc. are designed around printed circuit boards (PCBs) with a large number of surface mounted components. The pick and place machine soldering these components on the PCB may pick the wrong component, may solder the component in the wrong position or fail to solder it at all. Therefore, Automated Optical Inspection (AOI) is essential to detect the above defects even prior to electric tests by comparing populated PCBs with the schematics. In this context, we leverage YOLO, a deep convolutional architecture designed for one-shot object detection, for AOI of PCBs. This architecture enables real-time processing of large images and can be trained end-to-end. In this work we also exploit a modified architecture of YOLOv5 designed to detect small components of which boards are often highly populated. Moreover, we proposed a strategy to transfer weights from the original pre-trained model to this improved one. We report here our experimental setup and some performance measures.

Keywords: YOLO · object detection · PCB · SMD · AOI · defect detection · optical inspection

1 Introduction

Consumer electronic devices such as mobiles, TV sets, etc. are usually designed around one or more Printed Circuit Boards (PCBs) that are populated by automatic *pick and place* machine. Such robots may however fail in soldering a component in the right place, or the joints may be misaligned with respect to the pads, or the component may be installed in the wrong orientation, etc. Therefore, before an electric tests are performed, a visual inspection of the PCB is required to rule out glitches like those described above.[1]

The large amount of components typically populating a modern PCB calls for highly automated, high throughput, optical inspection procedures, so that even components that cannot be easily inspected by electrical probes can be

[1] This work is the result of the collaboration with SPEA, a worldwide leader in electronic component testing solutions that made available the images sued in this work.

G. L. Foresti et al. (Eds.): ICIAP 2023 Workshops, LNCS 14365, pp. 51–61, 2024.
https://doi.org/10.1007/978-3-031-51023-6_5

tested. Traditional PCB optical testing procedures require a camera to be driven over each component according to the PCB schematics, taking a shot of the component, analyze the shot and repeat for each component. The throughput of this workflow (i.e. the number of PCBs processed over time) is limited by i) the momentum of the moving part (the camera or the PCB) during image acquisition ii) the speed of the computer vision algorithm processing the image.

This work proposes a one-shot optical inspection method for component detection in PCBs. We recasted our component detection task into an object detection and classification problem. We rely on a YOLOv5 (You Only Look Once), a deep neural network based one-shot object detector suitable for low latency operations and trainable end-to-end. YOLO requires however large amounts of labeled data, so the first challenge we undertook was creating a sufficiently large training set of annotated whole PCB images from single component crops such as that in Fig. 1b. To obtain a robust component detector, we collected training images from different PCBs under different illumination condition using a SPEA 4080 flying probe tester (Fig. 1a). Next, we improve the YOLOv5 with architectural changes that allow it to perform better at detecting small components while proposing a novel weights transfer method. We experimented with a leave-one-out approach and YOLOv5 shows good detection accuracy especially for very popular classes of components, despite some glitches with small components. The improved YOLOv5 architecture performs significantly better on small components while keeping memory footprint and inference time under control.

(a) SPEA 4080 optical probe

(b) Example of acquired image

Fig. 1. Left: the optical probe of the SPEA 4080 used to acquire the PCB images used in this work. Right: example of the 1278×958 images produced by the optical probe. We annotate the component at the center of the image concatenating the type (e.g., Resistor) and case size (e.g., 6×3 millimeters).

2 Background

Deep convolutional object detection models can be grouped into two classes [17]: two-stage detectors such as the R-CNN family of architectures [4,5,13] and one-stage detectors.

YOLO (You Only Look Once) [10] is a one-stage detector relying on a deep convolutional feature extractor that is able to predict multiple bounding boxes and class probabilities for those boxes directly from full images in one evaluation. Using the whole image to make predictions, YOLO implicitly encodes contextual information about classes as well as their appearance. This allows it to learn generalizable representations of objects. In order to simultaneously predict bounding boxes and class probabilities, YOLO divides the input image in a grid of $S \times S$ cells, where each cell is responsible for the prediction of the object whose center falls within the cell. Each grid cell predicts B bounding boxes, and each bounding box consists of 5 predictions: x, y, w, h, and *confidence*. The coordinates of the box are represented by the predicted values of (x, y) for the center of the box, which is relative to the bounds of the grid cell. On the other hand, the width (w) and height (h), are predicted relative to the whole image. Instead, the confidence is a score described formally in the Eq. 1. This value is intended to reflect how confident the model is about the presence of an object in that box and how accurate it thinks its prediction is. It is easy to see that if no object exists in that cell, this confidence value has to be equal to zero. Otherwise it should be equal to the intersection over union between the ground truth and the predicted box.

$$Pr(Object) * IoU_{pred}^{truth} \quad (1)$$

Each grid cell also predicts C conditional class probabilities $Pr(Class_i|Object)$, where the condition is inherent to the presence of the object in that cell. Dividing the image into a SxS grid and considering B bounding boxes for each grid and C class probabilities, the predictions are encoded as an $S \times S \times (B * 5 + C)$ tensor.

In this work we rely on the fifth revision of YOLO architecture, known as YOLOv5. The code of this model is publicy available on the official Ultralytics repository on GitHub [7] in which YOLOv4 [3] is implemented using PyTorch. The primary distinction of this architecture, as opposed to the one discussed in the previous section, is that it employs anchor boxes instead of predicting bounding box coordinates directly from the image [11]. Anchor boxes are bounding boxes with predetermined dimensions, known beforehand. These boxes are found running a k-means clustering algorithm on training set bounding boxes to automatically find good priors. This is because the network learns how to adjust, and so starting from significant priors makes the learning process easier. Moreover, this architecture consists of a **backbone**, a **neck**, and a **head**. The backbone has the role of extracting relevant features from the input image and it based on a CSPNet [15] called CSPDarkNet-53 followed by a Spatial Pyramid Pooling Layer [6]. The combination of backbone feature layers happens in the neck. This is because the feature layers of the convolutional backbone have to be mixed and taken into account with each other. The head, on the other hand, is where detection happens. In this case we have three detection modules, in order to predict boxes at three different scales using the heads of YOLOv3 [12]. A schema of this architecture can be found in Fig. 3, but ignoring the modules dashed in red.

YOLO has been successfully applied to similar domain in other works. For example in [1,9] the authors adopt this model to locate surface defect on PCBs. In [14], instead, a Transformer-YOLO network detection model is proposed and also in this case applied to locate surface defect. YOLOv3 has been used in [8] to locate components, but not to the level of detail required for our use case.

3 Methodology

This section first describes the procedure we devised to generate the annotated PCB images required to train a YOLOv5. The improved architecture and our proposed weight transfer strategy is then described. At the end we discuss the procedure to train the network.

3.1 Dataset Generation

Towards training YOLO, all objects in each training image must be associated with a bounding box and class identifier. Figure 1b shows an example of the training images we were provided, a few thousand pictures captured from 8 different PCBs. Each image was acquired by the SPEA 4080 floating camera in Fig. 1a. The camera centers on each component according to the CAD schematics and takes a 1278×958 picture of the component and its surroundings. For each image, only the position of the central component in absolute PCB coordinates was however known, i.e. no class label was provided. So, we parsed the CAD schematics extracting for each component the type (resistor, capacitor, etc.) and case size (4×2, 6×3 mm, etc.) from which we inferred the component bounding box. At this point we had images like in Fig. 1b where multiple components are present yet only the central one was annotated. Training YOLO on such images, our preliminary experiments showed, resulted in a large number missed detections (false negatives) at test time, i.e. low recall. Therefore, we developed a workaround for reconstructing whole images of PCBs where each component would be annotated. We point out that each image like in Fig. 1b represents a tile of a jigsaw puzzle of a PCB. We populated an empty canvas with the available images pinned at the position of the central component and reconstructed PCB images like in Fig. 2. The black holes are due to the lack of components in some areas of the PCB that were hence not acquired by the PCB. After solving this jigsaw puzzle, we recomposed the previously generated

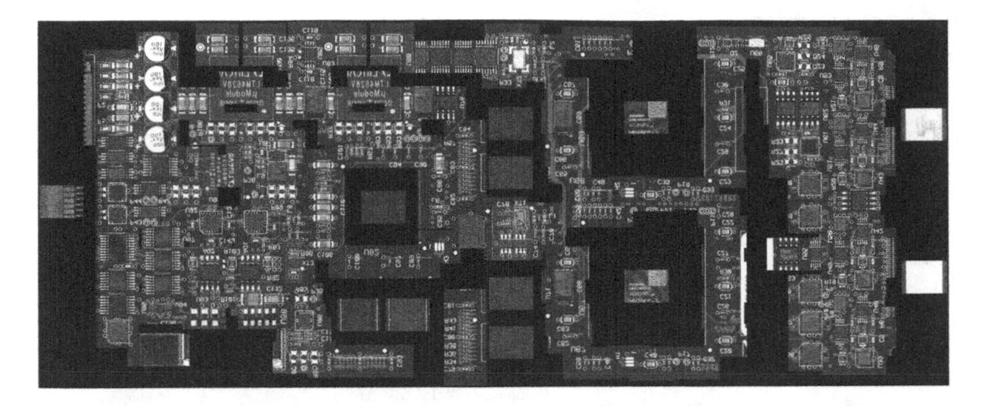

Fig. 2. 60000 × 20000 pixels image of the CPE010 PCB reconstructed and annotated using 354 images like in Fig. 1b. Black areas are due to the lack of components in those areas during the acquisition process.

annotation from each image to the reconstructed image. Following this procedure we successfully reconstructed and annotated a total of 8 PCBs where each component is annotated with a label and a bounding box, as required to train YOLO.

SPEA was eventually able to provide us with the image of two boards acquired as a whole, i.e. these images were not reconstructed from patches as in Fig. 2. With respect to the reconstructed images, there are no visible artifacts at the junction between crops and exposure is more uniform. These boards are identified as MPSDRV608 (10.208 × 18.336) and ZPROMEA50_10 (13.315 × 24.662) in our experiments. With the addition of these two new boards, our dataset consists of 10 different annotated PCBs. Table 1 shows the 36 classes of components considered over these 10 boards. We point out that, give our specific application, the component case size has been incorporated into the class label. Therefore, for a given component (e.g.,resistor) different classes exist based on the case size (e.g. Resistor_0402, Resistor_0603, etc.). It is important to note that all components within the same class have identical dimensions. Some classes created by grouping different components such as Plastic_packaging break this rule however.

3.2 Improving YOLOv5 Architecture for Small Object Detection

It is well known that dealing with small objects is a tough challenge for deep convolutional architectures since repeated convolve-and-pool (or convolve-and-stride) blocks progressively drop spatial details. In [2,16] (and similarly in [8] for YOLOv3), it was proposed to improve YOLOv5 adding an extra detection head relying on the features extracted by the backbone closer to the input, where feature maps preserve fine details. Towards this end, we expand the neck of YOLOv5 to access such feature maps and use them in the new detection head and

Table 1. The PCB component classes considered in this work with number of samples and packgage area over the 10 PCB images we were provided.

Component class	# Samples	μm^2
Resistor_0402	511	756,81
Resistor_0603	967	1.884,59
Resistor_0805	472	3.885,82
Resistor_1206	47	7.076,86
Resistor_1210	2	6.584,29
Resistor_RMINIMELF	3	7.698,42
Resistor Array	92	7.940,38
Resistor_2010	9	18.547,78
Resistor_2512	20	30.649,76
Capacitor_0402	958	794,03
Capacitor_0603	886	1.710,48
Capacitor_0805	404	3.155,46
Capacitor_1206	93	6.296,04
Capacitor_1210	39	13.096,62
Capacitor_Polar_0603	13	3.990,02
Capacitor_Polar_CMKTA	20	8.554,51
Capacitor_Polar_1411P	3	16.262,89
Capacitor_Polar_CMKTB	1	29.971,18

Component class	# Samples	μm^2
Capacitor_Polar_CMKTD	20	48.392,77
Capacitor_Polar_CEVPA8X10	4	69.504,67
Inductor_1210	4	13.988,71
Inductor_IND-XAL4020	4	27.742,08
Inductor_INDIHLP2525CZ01	4	67.996,82
Fuse_0603	8	2.121,57
Fuse_FUSESM	6	21.973,29
Fuse_FUSE-SMDC020	2	24.644,09
Led_0805	56	4.483,81
Led_TEKTONE_LED_1411	4	13.329,49
Connector_CMIMA4VFD_SM	2	59.097,38
Connector_CMIMA6VFD	2	76.665,90
Potentiometer_SMRVAR1	1	33.786,26
Relay_RLPICK-117-1A	52	42.563,62
Switch Array_PULSOMRON	1	56.310,86
Diode_DMELF	2	18.398,49
Cylindrical_diode	71	7.481,30 - 7.538,37
Metallic_packaging	6	23.934,04 - 52.777,34
Plastic_packaging	706	878,41 - 70.537,68

in the subsequent feature aggregations. Figure 3 shows the resulting improved architecture where areas dashed in red refer to the newly added modules.

Unfortunately, the authors of [2] do not detail the initialization of the extra modules nor provide a pretrained network. However, YOLOv5 authors provide a method to transfer weights from any source model to any target model. In detail, it is possible to transfer weights from a source model to a target one under the assumption that the underlying tensor topology and name (i.e., the identifier) is preserved. We hence propose to initialize the blocks in green in Fig. 3a using the weights from the original pretrained model. However, the original model and the improved one have more blocks in common than those transferred with this method. This occurs because the addition of new blocks results in a new numbering of subsequent blocks, and this leads to a change in their names. Taking this into account, we modified the transfer method and transferred the weights of all common blocks between the original and improved models. Moreover, since we used and modified the small version of YOLOv5, the shape of the tensors weights of the added blocks corresponds to the last blocks in the neck of the original YOLOv5 nano (yellow blocks in Fig. 3b). Thus, we could use them in the initialization process of the improved YOLOv5 small model.

3.3 Training Procedure and Evaluation Metrics

YOLOv5 is trained end-to-end minimizing the following multi-part loss function

$$Loss = \lambda_1 L_{cls} + \lambda_2 L_{obj} + \lambda_3 L_{loc} \tag{2}$$

where the three terms represent respectively the Binary Cross Entropy (BCE) loss related to the Classes, the Binary Cross Entropy (BCE) loss related to the

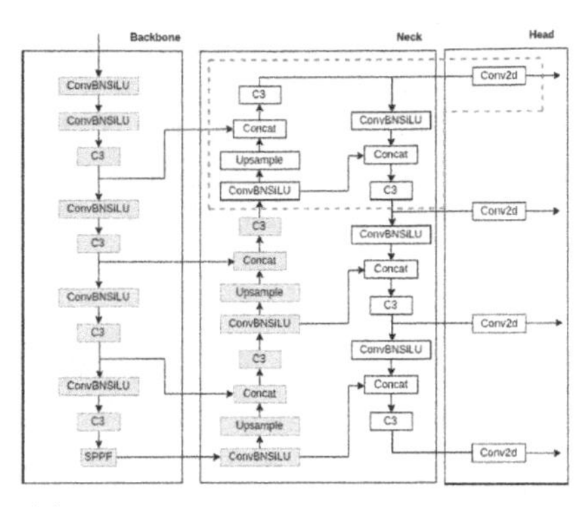

(a) Blocks initialized with pretrained weights using the method proposed by the Ultralytics authors.

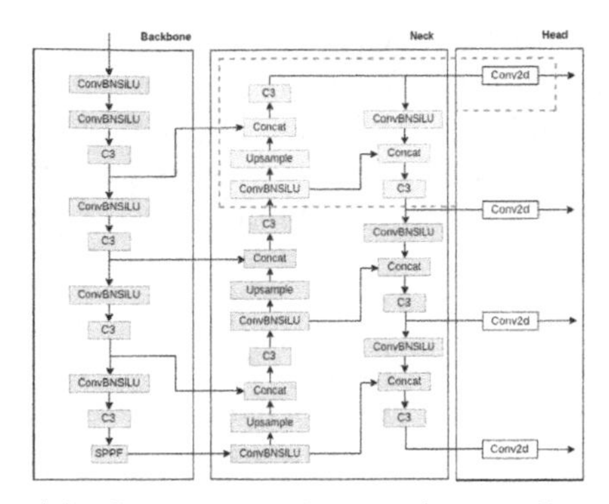

(b) Our proposed transfer weights method. Yellow blocks are initialized using the last neck blocks in the original YOLOv5 nano.

Fig. 3. Comparison of methods for weights transfer. Colored blocks are initialized with weights coming from a pretrained model. Areas dashed in red refer to the newly added modules. (Color figure online)

Objectness, and the Complete-IoU (CIoU) loss for the Location. We trained for 200 epochs 10 different YOLOv5 small models pretrained on the MS COCO dataset. We used the hyper-parameters recommended for YOLO, i.e. a learning rate of 0.01 and a momentum of 0.937. Moreover, we considered 3 warmup epochs having a momentum of 0.8 and a bias learning rate of 0.1. In particular we used the SGD as optimizer and a batch-size of 32 crops over the annotated patches described above.

Due to the large size of the PCB images, we had to crop patches of size 896×896 pixel to fit batches of 32 training images. Eventually, the actual overall training set consisted of 5.490 patches sized 896×896 pixels. Moreover, we augmented our training set of patches with random rotations, flips and a random variation of the brightness since these transformations preserve the size and aspect ratio of the components.

4 Experimental Results

This section reports the results obtained using a leave-one-out approach where one board is left out of the training set and is used for testing. The first section reports the results of the experiments with the original YOLOv5 small, whereas the second section with the improved architecture discussed in Sect. 3.2.

4.1 Experiments with Original YOLOv5

We recall that we trained 10 distinct instances of YOLOv5 small where in turn one PCB was left out of the training set. Table 2 shows the results for each of the 10 PCBs when kept out of the training set. The first 8 rows corresponds to reconstructed PCB images and so we took crops of size 896×896 to avoid including blank areas. The last two rows marked with *, indeed, correspond to the two whole boards we were provided by the end of our experiments. For these boards we were able to test the model by taking larger crops of size 7680×7680 pixels.

Table 2. mAP@0.5 for the board left out of the training set. The first 8 rows refer to 896×896 patches from reconstructed images. The last two rows (*) refer to 7680×7680 patches from whole images.

Board name	All	Resistor		Capacitor		Inference
		0402	0603	0402	0603	(ms)
CPE010	0.734	0.888	0.992	0.617	0.995	11.9
JPAMA30-256K SN	0.908	–	–	–	–	10.9
KDBRLYCMDR3	0.901	–	0.995	–	0.995	11.6
KEXANADUX70V1	0.991	0.978	–	0.995	0.995	10.3
LI122SM-2_CB533	0.818	–	0.838	–	0.985	11.4
SPE010-2	0.994	–	0.995	–	0.991	11.4
ZCPU7Z0	0.806	0.929	0.994	0.824	0.995	10.3
ZPROMEA50_SN	0.796	0.459	0.995	0.624	0.995	12.5
ZPROMEA50_10*	0.941	0.786	0.986	0.989	0.983	265.6
MPSDRV608*	0.796	0.861	0.963	0.879	0.994	266.2
Mean	**0.869**	**0.816**	**0.970**	**0.821**	**0.992**	

Detection accuracy wise, we report the mean average precision (mAP) across all classes. We also report unaggregated results for resistors and capacitors of case size 4×2 since they are the most popular small case components across our set of PCBs. Some component classes were not present in some boards, hence the lack of data in some cells of the table. Moreover, the class distribution at training time varies widely among PCBs, with some classes being in turn under-represented at trainign time depending on the left-out board. For example, CPE010 and ZPROMEA50_SN contribute a lot of small components to the train set, so the performance on these classes drops when they are removed from the train set for being tested. A recurring trend is however the drop in detection accuracy when the component case size drops from 6×3 to 4×2.

Complexity wise, we report the inference time on the test patches. For the larger tested crops of the two whole PCB images, the inference time and the memory footprint are about 265ms and 12GB on a NVIDIA A40 GPU. Such numbers, especially the memory footprint, are not compatible with the hardware constraints SPEA proposed us, although more components can be tested this way.

4.2 Experiments with Improved YOLOv5

As discussed above, the shortcomings of the original YOLOv5 on large crops are large memory footprints and low mAP for small components. Here we tackle the two issues subsampling the images by a four fold factor from 7680×7680 to 1920×1920 and resorting to the improved architecture presented in Sect. 3.2 and comparing the two initialization methods in Fig. 3.

Figures 4 and 5 show the precision-recall curves for the two boards acquired as whole images for the small components classes Resistor_0402 and Capacitor_0402. For both boards, the area under the curves improve for the improved YOLO architecture despite operating at a quarter resolution. In particular, for the ZPROMEA50_10, the best results are obtained by initializing the modified architecture with our method (Fig. 4c). Here we achieved a mAP values of 0.893 for the Resistor_0402, compared to a value of 0.646 obtained with the original model.

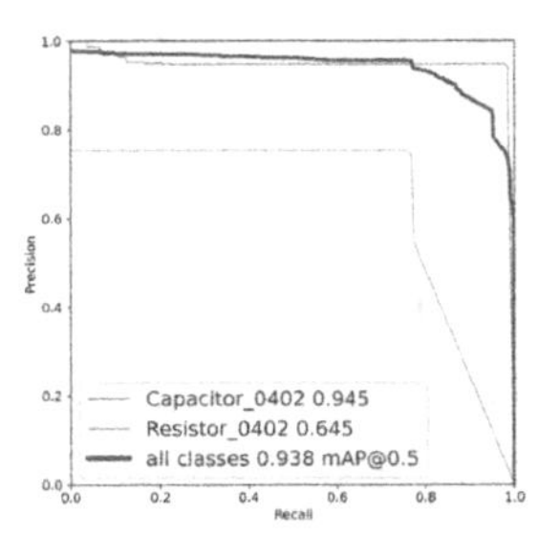

(a) Original YOLOv5s, pretrained on COCO, quarter resolution.

(b) Improved YOLOv5s, default weights transfer, quarter resolution.

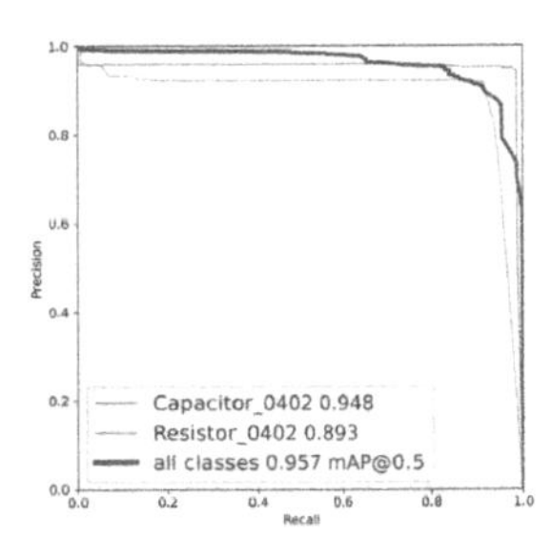

(c) Improved YOLOv5s, proposed weights transfer, quarter resolution.

Fig. 4. Precision-Recall curves for the ZPROMEA50_10 PCB.

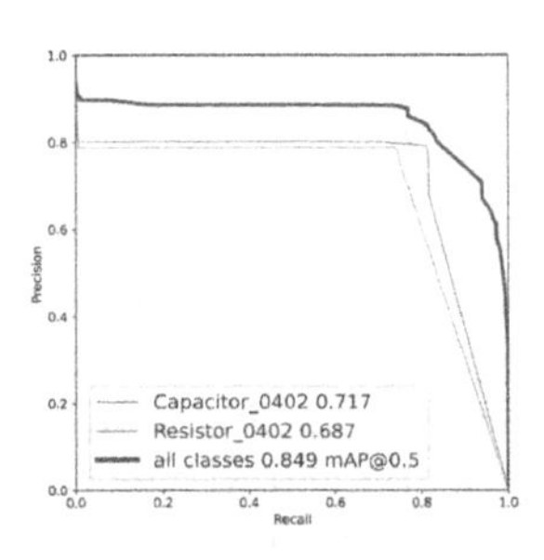

(a) Original YOLOv5s, pretrained on COCO, quarter resolution.

(b) Improved YOLOv5s, default weights transfer, quarter resolution.

(c) Improved YOLOv5s, proposed weights transfer, quarter resolution.

Fig. 5. Precision-Recall curves for the MPSDRV608 PCB.

Table 3 summarizes our results integrating inference time and memory footprint. Here we compared the results of the improved YOLOv5 architecture at quarter resolution with respect to the original architecture both native and quarter resolution. The improved YOLOv5 architectures achieve the best tradeoff between performance, memory footprint and inference time. We point out that the modified YOLOv5 performs about the same on larger components, hence the constantly top mAP across all the classes. Finally, we would like to recall that the improved YOLOv5 was not refined on for fold downscaled images.

Table 3. Comparison between models regarding mAP for small components, memory usage and inference time.

YOLOv5 Type	Down-sample	Image size	Footprint (MiB)	Inference (ms)	mAP@.5 All	Res_0402	Cap_0402
Original	x1	7680x7680	12687	265.6	0.941	0.786	**0.989**
Original	x4	1920x1920	**2661**	**18.0**	0.938	0.646	0.945
Improved	x4	1920x1920	3213	26.5	**0.957**	**0.893**	0.948

(a) Comparison table for the ZPROMEA50_10 PCB.

YOLOv5 Type	Down-sample	Image size	Footprint (MiB)	Inference (ms)	mAP@.5 All	Res_0402	Cap_0402
Original	x1	7680x7680	12687	266.2	0.796	0.861	**0.879**
Original	x4	1920x1920	**2661**	**18.1**	0.849	0.687	0.717
Improved	x4	1920x1920	3215	26.7	**0.866**	**0.866**	0.867

(b) Comparison table for the MPSDRV608 PCB.

5 Conclusion

In this work we explored using a YOLO object detector for automatically detecting PCB components soldering defects. The first challenge we had to undertake was creating a training dataset of completely annotated whole PCB images we recomposed from thousands of unannotated single component images. Our experiments show that performance depends on number of training samples available per class and component size, smaller component yielding poorest results. Moreover, even the small version of YOLOv5 is too memory hungry for performing inferences on entire PCB images. The improved architecture we propose performs well on small components while striking a favorable performance-complexity tradeoff. Our future endeavours includes enlarging our dataset to tens of PCB boards and experimenting on recent YOLO architectures beyond v5.

References

1. Adibhatla, V.A., Chih, H.C., Hsu, C.C., Cheng, J., Abbod, M.F., Shieh, J.S.: Defect detection in printed circuit boards using you-only-look-once convolutional neural networks. Electronics **9**(9), 1547 (2020). https://doi.org/10.3390/electronics9091547, https://www.mdpi.com/2079-9292/9/9/1547
2. Benjumea, A., Teeti, I., Cuzzolin, F., Bradley, A.: YOLO-Z: improving small object detection in yolov5 for autonomous vehicles. CoRR abs/2112.11798 (2021). https://arxiv.org/abs/2112.11798
3. Bochkovskiy, A., Wang, C., Liao, H.M.: YOLOv4: Optimal speed and accuracy of object detection. CoRR abs/2004.10934 (2020). https://arxiv.org/abs/2004.10934
4. Girshick, R.B.: Fast R-CNN. CoRR abs/1504.08083 (2015). https://arxiv.org/abs/1504.08083
5. Girshick, R.B., Donahue, J., Darrell, T., Malik, J.: Rich feature hierarchies for accurate object detection and semantic segmentation. CoRR abs/1311.2524 (2013). https://arxiv.org/abs/1311.2524
6. He, K., Zhang, X., Ren, S., Sun, J.: Spatial pyramid pooling in deep convolutional networks for visual recognition. CoRR abs/1406.4729 (2014). https://arxiv.org/abs/1406.4729
7. Jocher, G.: YOLOv5 by Ultralytics (2020). https://doi.org/10.5281/zenodo.3908559, https://github.com/ultralytics/yolov5
8. Li, J., Gu, J., Huang, Z., Wen, J.: Application research of improved YOLO V3 algorithm in PCB electronic component detection. Appl. Sci. **9**(18), 3750 (2019). https://doi.org/10.3390/app9183750, https://www.mdpi.com/2076-3417/9/18/3750
9. Liao, X., Lv, S., Li, D., Luo, Y., Zhu, Z., Jiang, C.: YOLOv4-MN3 for PCB surface defect detection. Appl. Sci. **11**(24), 11701 (2021). https://doi.org/10.3390/app112411701, https://www.mdpi.com/2076-3417/11/24/11701
10. Redmon, J., Divvala, S.K., Girshick, R.B., Farhadi, A.: You only look once: Unified, real-time object detection. CoRR abs/1506.02640 (2015). https://arxiv.org/abs/1506.02640
11. Redmon, J., Farhadi, A.: YOLO9000: better, faster, stronger. CoRR abs/1612.08242 (2016). https://arxiv.org/abs/1612.08242
12. Redmon, J., Farhadi, A.: YOLOv3: An incremental improvement. CoRR abs/1804.02767 (2018). https://arxiv.org/abs/1804.02767
13. Ren, S., He, K., Girshick, R.B., Sun, J.: Faster R-CNN: towards real-time object detection with region proposal networks. CoRR abs/1506.01497 (2015). https://arxiv.org/abs/1506.01497
14. Tang, J., Liu, S., Zhao, D., Tang, L., Zou, W., Zheng, B.: PCB-YOLO: an improved detection algorithm of PCB surface defects based on YOLOv5. Sustainability **15**(7), 5963 (2023). https://doi.org/10.3390/su15075963, https://www.mdpi.com/2071-1050/15/7/5963
15. Wang, C., Liao, H.M., Yeh, I., Wu, Y., Chen, P., Hsieh, J.: CSPNet: A new backbone that can enhance learning capability of CNN. CoRR abs/1911.11929 (2019). https://arxiv.org/abs/1911.11929
16. Zhan, W., et al.: An improved YOLOv5 real-time detection method for small objects captured by UAV. Soft Comput. **26**, 1–13 (2022). https://doi.org/10.1007/s00500-021-06407-8
17. Zou, Z., Shi, Z., Guo, Y., Ye, J.: Object detection in 20 years: A survey. CoRR abs/1905.05055 (2019). https://arxiv.org/abs/1905.05055

Wildfires Classification: A Comparative Study

Giorgio Cruciata[1(✉)], Liliana Lo Presti[1], Gabriele Ajello[2], Paolo Cicero[2], Giacomo Corvisieri[2], and Marco La Cascia[1]

[1] University of Palermo, 90128 Palermo, Italy
giorgio.cruciata@unipa.it
[2] Italtel S.p.A., 90044 Carini (PA), Italy
gabriele.ajello@italtel.com

Abstract. This paper proposes a comparative study of the performance of deep models for classifying wildfires. First, the paper examines publicly available image datasets for training models to classify fire and smoke. Then, it proposes the Wildfires dataset for the problem of classifying images according to the represented event: fire, smoke and no alarm. The dataset includes images describing real scenarios where images are acquired from a fixed control station far from the place where fire or smoke arises. The paper focuses on convolutional neural networks, residual neural networks, and transformers, and compares model performance on the Wildfires dataset and on a publicly available dataset, FireNet. Based on the experiments conducted in this work, the best accuracy values are achieved by the ResNet-50 network and the Swin-T v2 transformer. On the Wildfires dataset, the latter shows fewer smoke missing and false alarms, and correctly classifies all fire images. Additionally, this paper describes deploying the trained models on an embedded system to develop a fully working prototype for installation in a control station. The experiments show that transformers are not yet suitable for real-time performance when used on embedded systems.

Keywords: Wildfires · deep learning · dataset

1 Introduction

Wildfires are a devastating natural disaster that can have severe consequences for human life and the environment. From 1998–2017, wildfires and volcanic activities have impacted 6.2 million individuals globally, resulting in 2400 deaths due to suffocation, injuries, and burns. Unfortunately, climate changes have led to an increase in the size and frequency of wildfires, exacerbating this already concerning issue [29]. In addition to the loss of human lives, wildfires can also cause extensive damages to wildlife habitats and infrastructures. In order to mitigate the number of wildfire accidents, a large number of methods have been proposed to reduce the damage caused by such accidents.

The traditional approach to detecting wildfires involves manual monitoring by trained personnel, which can be time-consuming and costly. As a result,

G. L. Foresti et al. (Eds.): ICIAP 2023 Workshops, LNCS 14365, pp. 62–73, 2024.
https://doi.org/10.1007/978-3-031-51023-6_6

researchers have been exploring the use of advanced technologies such as remote sensing, image processing, and machine learning [17] to detect and monitor wildfires more efficiently and accurately.

One promising approach to fire classification is the use of Convolutional Neural Networks (CNNs) [16] that can identify areas of high heat, smoke, or flames. Despite such progress, there are still significant challenges to address in order to being able to detect wildfires in real-world scenarios. One issue is the lack of large, high-quality datasets. Existing datasets are limited in size or do not provide enough variation in environmental conditions, making it difficult to train accurate models.

In this paper, we present a comparison of deep learning models to study the accuracy and efficiency of wildfires classification. We implemented CNN- and transformer-based techniques exploiting state-of-the-art models. At the best of our knowledge, no transformer-based technique has been proposed yet for wildfires classification. We also propose a novel dataset built upon several existing datasets, images taken from the internet, and personal shots. Differently than publicly available datasets, our Wildfires dataset is built with a special focus on the classification of fires in wild environments and wooded areas.

We can summarize the contributions of the proposed work as it follows:

- The use of visual transformers for wildfires classification;
- The comparison of CNN- and Visual transformer-based approaches for wildfires classification;
- A novel dataset specifically thought for wildfires classification to be made publicly available[1].

The paper is organized as following. Section 2 provides an overview of the publicly available datasets used to train neural networks. Section 3 reviews techniques based on image processing and on deep learning. Section 4 describes the models we have implemented and our Wildfire dataset. Section 5 presents the experimental results and compares different approaches on the Firenet dataset [15] and on our dataset. Finally, Sect. 6 discusses our findings and describes future research directions.

2 Datasets for Fire and Smoke Analysis

Traditional smoke sensors, which are effective for monitoring enclosed spaces, are not as effective for open spaces, and also do not detect danger until they are reached by steam. Over the years, more efficient and timely systems based on the analysis of images captured by cameras have been developed. However, the development of efficient systems require data. Despite the increasing risk that fire can start, there are not many datasets specifically thought for fire classification and, moreover, most of them are not suitable enough for wildfires classification. For

[1] The dataset is made available upon request via email to Gabriele Ajello (gabriele.ajello@italtel.com).

Table 1. Publicly available classification datasets - Comparison

Dataset	Smoke	Simulated	N. Samples	Image	Video	WF vs F
BoWFire [6]	-	-	226	✓	-	F
FireNet [15]	-	-	2423	✓	-	F
Corsician Fire [28]	✓	-	500	✓	-	WF
Zhang et al. [33]	✓	✓	>20K	✓	-	WF&F
Foggia et al. [9]	-	-	62418/31video	-	✓	WF&F

instance, fires can be considered as dynamic textures [21], namely spatial textures extended to the temporal domain. In this sense, fires are motion patterns characterized by a certain repetitiveness in space and time, changes in illumination, and intrinsic changes to the pattern. The DynTex database [21] consists of over 650 sequences of dynamic texture, including also fires. Fires samples are however limited in this dataset.

Another challenge is that smoke image samples are generally limited in scale and diversity. Thus, in [31,32], a method to systematically produce synthetic smoke images with a wide variation in the smoke shape, background and lighting conditions is adopted for being able to train deep CNN models. Not all datasets annotate the smoke class and, unfortunately, not all works made their data available to the research community. For instance, the work [10] trains a CNN to predict among fire, smoke and negative on a dataset of 5548 images, but the dataset is not publicly available. In [2], it is proposed a compilation of datasets for fire and smoke analysis composed of four datasets: Flickr-FireSmoke, Flickr-Fire, BoWFire, and SmokeBlock. However, the benchmark is not maintained anymore. Among the datasets in the benchmark, we were only able to find the BoWFire dataset [6].

Table 2 details the publicly available datasets we have found online. The column *Smoke* indicates that the class smoke has been included. The column *Simulated* indicates that smoke images are synthetically produced. The last column *Wildfires (WF) vs Fire (F)* shows if the dataset is specifically designed to classify wildfires or generic fire. Note that the dataset [28] only consider fire/smoke images and no negative samples are provided.

Figure 1 shows samples of wildfires and samples taken from the dataset [15]. The latter images depict flames (fire), and are probably not suitable for training networks aiming at classifying wildfires (Table 1).

3 Fire Analysis Techniques

In the early 2000s, the few existing methods for fire classification were mostly based on image processing techniques. In [27], fire is detected frame-by-frame by considering the pixel colors and their changes over time. The motion of the flames was described by a function based on the Wavelet transform. If the colors

Fig. 1. The first row show images of wildfires, specifically selected for their suitability in developing a wildfires classification system. These images are highly relevant to the intended purpose. The second row features images from the FireNet dataset [15], which predominantly depict flames but are not ideal for wildfires classification purposes.

and movement observed in the frames match what is expected from a fire, then the system raises an alarm. The analysis of the frames over time is performed using HMM (Hidden Markov Model). In [11], classification of fire in images follows three main steps: 1) color-based image segmentation; 2) segmentation and filtering of image regions where there is a dynamic texture; 3) background suppression to isolate regions of interest in the image. Fire is detected by analyzing the speed with which the identified regions of interest change color from one frame to another.

In contrast to the above works other papers, including ours, focus on the problem of analyzing a single image, and thus time dimension is not considered. Early works in the domain have used color information to get a pixel-wise classification. The choice of a specific color space was unclear and several authors have attempted to solve the problem in various color spaces. The work in [5] proposes a RGB chromatic segmentation and disorder measurement to detect fire or smoke. Moving regions are segmented, and chromatic features in the RGB space are used to classify the candidate regions into fire and smoke.

The RGB color space has strong dependencies on illumination. If the illumination of image changes, the fire pixel classification can fail. Indeed, in RGB is not possible to separate a pixel's value into intensity and chrominance. For this reason, rather than working on the RGB color space, the work in [3] proposes a generic chrominance model for flame pixel classification in the YCbCr color space. Fire is detected by thresholding the chrominance channels. In [25], segmentation of candidate regions is achieved by K-means clustering. The paper investigates several color spaces and concludes that the blue chrominance channel Cb is the most suitable for fire segmentation with K-means. In [6], RGB images are transformed to gray scale by using a combination of the three color channels that consider the human color perception process in order to bring out the pixels of the flames. Pixel classification is performed through the Bayes and the K-Nearest Neighbor (KNN) classifiers. Texture classification involves the super-pixel extraction and classification. An image is classified as depicting fire if fire is detected by both the classifiers.

With the advent of deep learning, the fire classification problem has mostly moved from pixel-wise classification (fire segmentation) to image-wise classification (deciding whether an image depicts fire or no) or to fire region classification (predicting a bounding-box on the image with potential fire/smoke). Most of the recent methods adopt Convolutional Neural Networks (CNNs). The work [10] is one of the first designing a CNN with this purpose. The model is used to classify three classes: fire, smoke and non-fire. The CNN takes in input RGB images with a size of 64×64 pixels. The network is made of two blocks including two convolutional layers and a MaxPooling layer. The last two layers are instead fully connected. In the first stage, the entire image is classified. Subsequently, the image is divided into smaller 12×12 image patches. Each patch is classified by the network, significantly increasing the accuracy of the proposed approach.

While [33] trains a Faster R-CNN [24] on a simulated dataset to classify smoke, in [30] three different deep models are compared: Faster R-CNN, YOLO v.3 [23], and SSD [18]. Their experiments show that fire can be classified easily especially with Faster R-CNN and SSD, while smoke is harder to identify, with YOLO providing the worst results. In [8], two models are compared: VGG16 [26] and MobileNet [13]. The experimental results show that VGG16 converges faster than MobileNet and achieves higher accuracy values. Similarly, [22] compares VGG16, ResNet50 [12], and DenseNet121 [14]. The results show that the Resnet50 achieves better results. Recently, [4] has tested Faster R-CNN on different datasets while including color-based image processing techniques to improve the classification results.

Attempts to propose novel architectures are found in [15,20]. In [20], a CNNs with 5 convolution layers interleaved with MaxPooling layers, and Dense layers is proposed. A data augmentation technique is employed to increase the network's ability to classify fires. Finally, the whole system has been implemented on a Raspberry PI. The work in [15] proposes a CNN network, FireNet, specifically created for the classification of images into fire and no fire. The network was deployed on a Raspberry Pi 3B.

In addition to low accuracy, existing approaches for early classification of wildfires also suffer from high false-positive rates and missing classification. False positives occur when a system identifies a non-fire event as a fire/smoke. This can lead to unnecessary alarms and drain resources that could be better used for actual emergencies. Missing classification indicates that fire/smoke are not readily classified by the system, determining potential dangers to human lives and wooded areas. With a special focus on accuracy values, false positives and missing classification, in this paper we compare several deep learning-based approaches for the classification of wildfires images.

4 Wildfires Classification

Detecting wildfires in their early stages is challenging due to several factors. Existing approaches for wildfires classification rely on various sources of data, such as satellite imagery, ground-based sensors, and social media.

In this paper, given an image taken from a control station of a wooded area (see Fig. 1, first row), we aim at classifying the image to identify if it depicts fire or smoke. If one of these two classes is detected, the station will raise an alarm. To be useful and usable in real scenarios where embedded systems could be used at the control station, the model must be fast, accurate, and should raise a low number of false alarms.

In this section, we first describe the models we have implemented and compared. Then we provide details about the dataset we assembled in order to test our models. The dataset has been designed for the problem of classifying wildfire images rather than just fire (see Fig. 1).

4.1 Selected Deep Models

Classic CNN-Based Models: Our first CNN model is a *simple CNN* and consists of two main blocks. The first block comprises two convolutional layers (with 16 and 32 3×3 kernels respectively), followed by a 3×3 Max Pooling layer with stride 2. The last block flattens the feature maps and includes two dense layers, each with 100 neurons. The final layer provides the three output classes, with a softmax activation function. In all other layers, we use a LeakyReLU activation function with an alpha parameter of 0.33. The network takes in input RGB images of size 100×100. All pixel values are rescaled in the range $[0, 1]$. In between the two dense layers we use a Dropout layer with a rate of 0.5. The model is optimized by stochastic gradient descent (SGD) with a learning rate of 0.01 and a momentum of 0.9.

We also considered *FireNet*, a CNN specifically designed for the recognition of fire-related images [15]. FireNet consists primarily of a block that repeats three times, comprising a convolutional layer, an average pooling layer, and a dropout layer. The first convolutional layer consists of 16 filters, the second layer consists of 32 filters, and the third layer consists of 64 filters, all with a 3×3 size. The Rectified Linear Unit (ReLU) activation function is employed in all layers, and a dropout rate of 0.5 is used. The remaining part of the network is fully-connected, with three dense layers. The first dense layer contains 256 neurons, followed by a dropout layer with a rate of 0.2. The second dense layer consists of 128 neurons, and the final layer has two neurons, using the Softmax function to generate the output. Input images are scaled to a height and width of 64 pixels. The Adam optimization algorithm is employed, and the activation function used in this network is Adam. In our experiments, we adapted the network such that the final layer comprises three neurons. Indeed, differently than [15], we aim at also identifying smoke in the environment.

Residual Network. We fine-tuned the well-known *Resnet-50* network [12], which was pre-trained on the ImageNet dataset. The input images were resized to 100 pixels and underwent specific pre-processing. The network consists of initial convolutional and MaxPooling layers, followed by three residual blocks. The network architecture includes bottleneck designs for deeper networks. The

  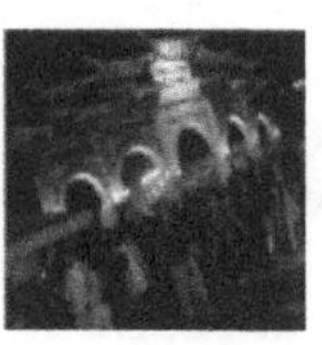

Fig. 2. Negative samples showing similarity to positive ones in colors and shapes.

final layers consist of an average pooled layer and a fully connected layer with 3 neurons. When training the ResNet model, the first 10 epochs were used to train only the dense layers, the remaining 40 epochs are used to train the entire model.

Visual Transformers. In this paper we use two types of Transformers for visual processing: the Vision Transformer (*ViT*) [7] and the Swin Transformer V2 (*Swin-T v2*) [19]. ViT is a model commonly used for image classification networks as well as Swin-T v2, which aims to achieve high performance, particularly during fine-tuning.

In the case of ViT, the input images must have a height and width of 224 pixels, and tokenization is performed by creating patches of size 16×16 pixels. The pixel values are normalized using specific mean and standard deviation values for this network. Patching is employed to divide the original image, and the encoder consists of a normalization layer (Norm), a multi-head attention layer, and a residual block. The output is then normalized again and passed through a Multi-Layer Perceptron (MLP) layer using a Gaussian Error Linear Unit (GELU) function. The optimizer used is AdamW, which implements the Adam algorithm with weight decay.

In the case of Swin-T v2, images are resized to a height and width of 256 pixels, and tokenization is performed by creating patches of size 4×4 pixels. The RGB channels are normalized using appropriate mean and standard deviation values. The output is processed by the Softmax function. Following this level of attention, the encoder structure consists of a normalization layer, a residual block, an MLP, and another normalization layer, concluding with a residual block. The activation function used for the Multi-layer Perceptron is GELU, and the compiler optimizer is AdamW.

4.2 Wildfires Dataset

For training our models, we inspected and tested public and private datasets (the latter were collected at Italtel S.p.A.). As noted in Sect. 2, some datasets, such as Firenet, do not annotate smoke in images, which is an important clue for early fire classification.

We also noticed that a good dataset should have negative samples with visual characteristics similar to those of the images with fire (see Fig. 2) especially in terms of colors and shapes so that the network learns the specific differences between the two situations. Images should mainly concern wooded areas since our goal is the development of a model capable of classifying images from cameras located throughout the territory. Therefore, it must be taken into account that any fire could occur at a significant distance from the place where the camera is located. These two considerations are the inclusion criteria in the dataset.

To construct our Wildfires dataset, a total of 1355 images were collected from various sources, focusing on wildfires and smoke, while discarding all images where just flames are present. In particular, 271 images were downloaded from Google Images, 383 images where selected from the Firenet dataset [15], 76 samples are from the BoWFire dataset [6], 121 images are selected from the dataset in [32], 251 are from [33], 30 images are from the Fire-Detection-Image Dataset [1], and 135 are personal shots. An image belongs to the class smoke if only smoke is visible, instead an image where fire and smoke are visible in wooded area is classified as wildfires.

4.3 Deploying the Model on Embedded Systems

We successfully deployed our trained models on an embedded system, an NVIDIA Jetson Nano equipped with an NVIDIA Tegra X1. To export the models, we employed TensorRT, a SDK developed by NVIDIA specifically for high-performance deep learning inference. TensorRT provides a deep learning inference optimizer for pre-trained models, as well as a runtime for execution. Models were converted into ONNX files and transformed into TensorRT engines. Of course, the performance depended on the number of parameters of the models. On average, without considering the time to load the model parameters and the input, rescaling and preprocessing the input, classifying the test images was performed at around 89 fps with our simple CNN, 59 fps with the FireNet model, 22 fps with the ResNet-50, 5 fps with the ViT transformer, and 0.9 fps with the Swin transformer. Under this point of view, the use of transformers is not convenient when real-time performance are needed. This is an expected result considering the model size.

5 Experimental Results

This section reports the results obtained by our models on the FireNet dataset and on our Wildfires dataset. We note here that, while the FireNet dataset only includes two classes (fire and no fire), our Wildfires dataset includes three classes (fire, smoke, no alarm) and focuses on wooded areas.

When using the FireNet dataset, 2423 images are used for training the models (1124 fire/1299 no fire), while 200 images are used for test purposes (100 fire/100 no fire). On our Wildfires dataset, 1132 images are used for training the models (383 fire, 404 smoke, 345 no alarm), while 223 images are used for test purposes (71 fire, 82 smoke, 70 no alarm).

For each experiment we report: the accuracy value in classification, the number of false alarms, and missing classification. The tested models have been described in Sect. 3 and are the simple CNN (trained from scratch), the FireNet model (trained from scratch or fine-tuned), the fine-tuned ResNet-50, and the fine-tuned visual transformers ViT and Swin-T v2. For all the models, we used data augmentation techniques that apply random rotations to the images in order to improve the parameter learning.

5.1 Results on the FireNet Dataset

Table 2 reports the results achieved on the FireNet test dataset for all the selected models. The simple CNN does not achieve high accuracy values, probably because it is too simple for dealing with the complexity of the problem. As for the FireNet model, we used the pre-trained network by [15], and achieved results similar to that reported in the original paper. Among the CNN-based models, the best performance have been achieved by fine-tuning the ResNet-50 model, especially in terms of false alarms and missing number.

As for the visual transformer, fine-tuning both the models required much more time than training CNNs. While the ViT achieved results inferior to the ones achieved by Firenet, Swin-T v2 reaches an accuracy value comparable to the ResNet model, with only 1 false alarm and 1 missing.

Table 2. Experimental Results on the FireNet dataset

Method	Accuracy	Precision	Recall	F1 Score
Simple CNN (baseline)	78.50%	0.89	0.65	0.75
FireNet [15]	90.00%	0.83	1	0.91
Resnet-50 [12]	**99.50%**	1	0.99	0.99
ViT [7]	85.50%	0.91	0.79	0.84
Swin-T v2 [19]	**99.00%**	0.99	0.99	0.99

5.2 Results on Our Wildfires Dataset

Table 3 reports the results achieved on our Wildfires dataset for all the selected models. In this case, also the FireNet model was retrained, since our dataset includes three classes rather than two. We believe our dataset is more challenging than the FireNet dataset since all models achieve lower performance. Only ViT shows improved performances.

As for the FireNet model, we compared the results achieved when training the model from scratch and when fine-tuning the model on our dataset (considering also data augmentation). In both the cases, we achieve similar accuracy values. However, when fine-tuning the network, we achieve a higher number of missing classification. In particular, when training from scratch, there are 11 false alarms related to fire and 6 related to smoke. Instead, 4 fire and 1 smoke events are missed by the model. When fine-tuning, there are 6 false alarms related to fire, and 5 to smoke. There are also 6 fire and 5 smoke events missing.

Among the CNN-based models, ResNet-50 is again the winner, with only 5 false alarms and 5 missing events. All false alarms refer to the class smoke, while missing classification refers to fire (3 samples) and smoke (2 samples). Swin-T v2 reaches an accuracy value comparable to the ResNet model, but ensures a lower number of missing classification, and slightly inferior false alarms. Interestingly, with the Swin-T model, both false alarms and missing classification refer to samples of the class smoke, which are known to be more difficult to detect.

Table 3. Experimental Results on our Wildfires Dataset (P = Precision, R = Recall)

Method	Accuracy	P Fire	P Smoke	R Fire	R Smoke	F1-Fire	F1-Smoke
Simple CNN (baseline)	79.37%	0.82	**0.96**	0.86	0.93	0.841	0.94
FireNet (from scratch)	85.65%	0.86	0.93	0.94	**0.99**	0.90	**0.96**
FireNet (fine-tuning)	86.10%	0.91	0.94	0.91	0.91	0.91	0.93
Resnet-50 [12]	**94.17%**	**1**	0.94	0.96	0.97	0.98	**0.96**
ViT [7]	90.58%	0.97	0.92	0.94	0.97	0.96	0.95
Swin-T v2 [19]	**93.7%**	**1**	0.95	**1**	0.97	**1**	**0.96**

6 Conclusions and Future Works

The work presented in this paper aims to study the performance of deep learning models for the classification of smoke and fire in images of wooded areas acquired by monitoring stations located remotely from the place where fires can break out. The final goal of the work is to deploy the models developed on embedded systems. After a careful analysis of the existing datasets, it was necessary to build a new dataset because the publicly available datasets do not always include the smoke class, and do not represent realistic scenarios where wildfires starts in wooded areas. This work compares the performance of simple convolutional networks with that of a ResNet-50, and Transformers ViT and Swin-T v2. From the experiments carried out it is possible to conclude that the models that have provided the most accurate results are the Resnet-50 and the Swin-T v2. The models have been tested both on the FireNet dataset, which includes only the fire and no-fire classes, and on our new Wildfires dataset, which also includes the smoke class. Finally, it was studied how each model could be exported and subsequently used on an NVIDIA embedded system in order to obtain a system prototype that can be used in real scenarios. Our experiments also shows that transformers are not suitable yet to be used for real-time applications on embedded systems, despite their performance are very promising. This work deals with the classification of the entire image and, in future work, we will study the possibility to obtain pixel-level segmentation of fires and smoke. A limitation of the current work is that the problem is approached as a single-class classification problem. Images where both fire and smoke are present have been annotated as fire. This can be confusing for the model. Thus, in future work, we will study how to approach the problem as a multi-class classification problem where both fire and smoke can be detected in the images.

References

1. Fire-detection-image-dataset. https://github.com/cair/Fire-Detection-Image-Dataset. Accessed 30 Nov 2010
2. Cazzolato, M.T., et al.: FiSmo: a compilation of datasets from emergency situations for fire and smoke analysis. In: Brazilian Symposium on Databases-SBBD, pp. 213–223. SBC Uberlândia, Brazil (2017)
3. Celik, T., Demirel, H.: Fire detection in video sequences using a generic color model. Fire Saf. J. **44**(2), 147–158 (2009)
4. Chaoxia, C., Shang, W., Zhang, F.: Information-guided flame detection based on Faster R-CNN. IEEE Access **8**, 58923–58932 (2020)
5. Chen, T.H., Wu, P.H., Chiou, Y.C.: An early fire-detection method based on image processing. In: 2004 International Conference on Image Processing, ICIP 2004, vol. 3, pp. 1707–1710. IEEE (2004)
6. Chino, D.Y., Avalhais, L.P., Rodrigues, J.F., Traina, A.J.: BoWFire: detection of fire in still images by integrating pixel color and texture analysis. In: 2015 28th SIBGRAPI Conference on Graphics, Patterns and Images, pp. 95–102. IEEE (2015)
7. Dosovitskiy, A., et al.: An image is worth 16 × 16 words: transformers for image recognition at scale. arXiv preprint arXiv:2010.11929 (2020)
8. Dua, M., Kumar, M., Charan, G.S., Ravi, P.S.: An improved approach for fire detection using deep learning models. In: 2020 International Conference on Industry 4.0 Technology (I4Tech), pp. 171–175. IEEE (2020)
9. Foggia, P., Saggese, A., Vento, M.: Real-time fire detection for video surveillance applications using a combination of experts based on color, shape and motion. IEEE Trans. Circ. Syst. Video Technol. **25**, 1545–1556 (2015)
10. Frizzi, S., Kaabi, R., Bouchouicha, M., Ginoux, J.M., Moreau, E., Fnaiech, F.: Convolutional neural network for video fire and smoke detection. In: 42nd Annual Conference of the IEEE Industrial Electronics Society, IECON 2016, pp. 877–882. IEEE (2016)
11. Gomes, P., Santana, P., Barata, J.: A vision-based approach to fire detection. Int. J. Adv. Rob. Syst. **11**(9), 149 (2014)
12. He, K., Zhang, X., Ren, S., Sun, J.: Deep residual learning for image recognition. In: Proceedings of the IEEE Conference on Computer Vision and Pattern Recognition, pp. 770–778 (2016)
13. Howard, A.G., et al.: MobileNets: efficient convolutional neural networks for mobile vision applications. arXiv preprint arXiv:1704.04861 (2017)
14. Huang, G., Liu, Z., Van Der Maaten, L., Weinberger, K.Q.: Densely connected convolutional networks. In: Proceedings of the IEEE Conference on Computer Vision and Pattern Recognition, pp. 4700–4708 (2017)
15. Jadon, A., Omama, M., Varshney, A., Ansari, M.S., Sharma, R.: FireNet: a specialized lightweight fire & smoke detection model for real-time IoT applications. arXiv preprint arXiv:1905.11922 (2019)
16. Krizhevsky, A., Sutskever, I., Hinton, G.E.: ImageNet classification with deep convolutional neural networks. In: Pereira, F., Burges, C., Bottou, L., Weinberger, K. (eds.) Advances in Neural Information Processing Systems, vol. 25. Curran Associates, Inc. (2012)

17. Lee, W., Kim, S., Lee, Y.T., Lee, H.W., Choi, M.: Deep neural networks for wild fire detection with unmanned aerial vehicle. In: 2017 IEEE International Conference on Consumer Electronics (ICCE), pp. 252–253. IEEE (2017)
18. Liu, W., et al.: SSD: single shot multibox detector. In: Leibe, B., Matas, J., Sebe, N., Welling, M. (eds.) ECCV 2016, Part I 14. LNCS, vol. 9905, pp. 21–37. Springer, Cham (2016). https://doi.org/10.1007/978-3-319-46448-0_2
19. Liu, Z., et al.: Swin Transformer V2: scaling up capacity and resolution. In: Proceedings of the IEEE/CVF Conference on Computer Vision and Pattern Recognition, pp. 12009–12019 (2022)
20. Mohnish, S., Akshay, K., Pavithra, P., Ezhilarasi, S., et al.: Deep learning based forest fire detection and alert system. In: 2022 International Conference on Communication, Computing and Internet of Things (IC3IoT), pp. 1–5. IEEE (2022)
21. Péteri, R., Fazekas, S., Huiskes, M.J.: DynTex: a comprehensive database of dynamic textures. Pattern Recogn. Lett. **31**(12), 1627–1632 (2010)
22. Rahul, M., Saketh, K.S., Sanjeet, A., Naik, N.S.: Early detection of forest fire using deep learning. In: 2020 IEEE Region 10 Conference (TENCON), pp. 1136–1140. IEEE (2020)
23. Redmon, J., Farhadi, A.: YOLO9000: better, faster, stronger. In: Proceedings of the IEEE Conference on Computer Vision and Pattern Recognition, pp. 7263–7271 (2017)
24. Ren, S., He, K., Girshick, R., Sun, J.: Faster R-CNN: towards real-time object detection with region proposal networks. In: Advances in Neural Information Processing Systems, vol. 28 (2015)
25. Rudz, S., Chetehouna, K., Hafiane, A., Laurent, H., Séro-Guillaume, O.: Investigation of a novel image segmentation method dedicated to forest fire applications. Meas. Sci. Technol. **24**(7), 075403 (2013)
26. Simonyan, K., Zisserman, A.: Very deep convolutional networks for large-scale image recognition. arXiv preprint arXiv:1409.1556 (2014)
27. Toreyin, B.U., Cetin, A.E.: Online detection of fire in video. In: 2007 IEEE Conference on Computer Vision and Pattern Recognition, pp. 1–5. IEEE (2007)
28. Toulouse, T., Rossi, L., Campana, A., Celik, T., Akhloufi, M.A.: Computer vision for wildfire research: an evolving image dataset for processing and analysis. Fire Saf. J. **92**, 188–194 (2017)
29. Wallemacq, P., UNISDR, CRED: Economic losses, poverty and disasters 1998–2017, October 2018. https://doi.org/10.13140/RG.2.2.35610.08643
30. Wu, S., Zhang, L.: Using popular object detection methods for real time forest fire detection. In: 2018 11th International Symposium on Computational Intelligence and Design (ISCID), vol. 1, pp. 280–284. IEEE (2018)
31. Xu, G., Zhang, Y., Zhang, Q., Lin, G., Wang, J.: Deep domain adaptation based video smoke detection using synthetic smoke images. Fire Saf. J. **93**, 53–59 (2017)
32. Xu, G., Zhang, Y., Zhang, Q., Lin, G., Wang, J.: Domain adaptation from synthesis to reality in single-model detector for video smoke detection. arXiv preprint arXiv:1709.08142 (2017)
33. Zhang, Q., Lin, G., Zhang, Y., Xu, G., Wang, J.: Wildland forest fire smoke detection based on Faster R-CNN using synthetic smoke images. Procedia Eng. **211**, 441–446 (2018)

A General Purpose Method for Image Collection Summarization and Exploration

Marco Leonardi[2(✉)], Paolo Napoletano[1], Alessandro Rozza[2], and Raimondo Schettini[1]

[1] University of Milano - Bicocca, Milano, Italy
{paolo.napoletano,raimondo.schettini}@unimib.it
[2] lastminute.com, Chiasso, Switzerland
{marco.leonardi,alessandro.rozza}@lastminute.com

Abstract. We propose a flexible framework that can be used to explore large-scale image datasets and summarize photo albums. Our proposed method first groups images based on their semantic content, and then selects the most diverse and aesthetically pleasing images to represent each category. To ensure the selection of high-quality images, we use features extracted from a Convolutional Neural Network to assess their diversity and perceptual properties. The effectiveness of our method is tested using benchmarking datasets and a qualitative study.

Keywords: Summarization · Quality · Aesthetics · CNN

1 Introduction and Background

Today, almost everyone takes photos at any time and for any reason. With the growth of the smartphone industry, anyone can have a high-quality camera in their pocket, and taking a picture is just one click away. Images have become a primary means of communication, and most social networks are now dominated by media content. As the cost of storage decreases, the number of stored photos is increasing, leading to collections of images that are becoming increasingly difficult to explore and enjoy. We are inundated with images: in 2015, on average, 300 million photos were published on Facebook daily [27], while in 2016, an average of 80 million photos were shared every day on Instagram [9].

The challenge of exploring large image collections is also significant in the development of deep learning algorithms for signal or image recognition. Deep neural architectures require large amounts of data for the training process; for example, ImageNet [7] and Microsoft Common Objects in Context [16] each contains millions of images. However, browsing such massive image collections to gain a sense of the data we are working with is nearly impossible.

In advertising and product promotion, summarizing collections of images is fundamental [1,10,22]. For instance, when it comes to the tourism industry, a hotel might have several pictures linked to it, and it's essential to select the most suitable and diverse ones to showcase in order to promote it effectively.

G. L. Foresti et al. (Eds.): ICIAP 2023 Workshops, LNCS 14365, pp. 74–85, 2024.
https://doi.org/10.1007/978-3-031-51023-6_7

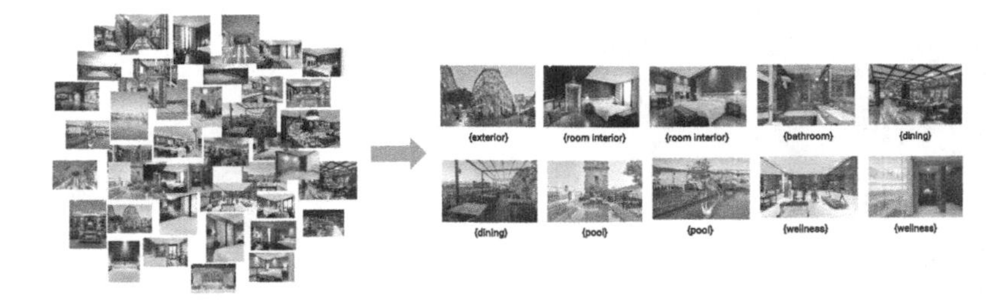

Fig. 1. Example of image summarization of an hotel.

It's crucial to display not only the most aesthetically pleasing pictures but also to portray all the various aspects of a hotel, such as the lobby, the rooms, the facilities, and more. See the example in Figure 1. For the aforementioned reasons, it's clear that the development of approaches able to automatically select the most important and diverse images from a collection of pictures is needed.

A simple yet effective way to automatically summarize a large photo collection is to divide the images into subgroups, for example, by using algorithms such as mean shift [6] or K-means [18], and then selecting one picture for each group. Following this approach, Li *et al.* [14] proposed an automatic organization framework for photo collections based on image content, which exploits a hierarchical clustering technique. In particular their work is focused on general consumer photographs, that typically feature individuals, objects, and landscapes. Consequently the proposed method involves the utilization of human faces.

Similarly Li *et al.* [15] propose an automatic selection system that focuses solely on photos with faces and is based on the aesthetic quality of consumer photos.

One possible way to improve solutions based on clustering techniques is to model the concept of diversity. For example, one can adopt solutions that aim to maximize the contrast of the selected subset, as proposed by Campadelli *et al.* in [3], where they solve the problem of selecting a high-contrast set as a combinatorial optimization problem on graphs.

The problem of summarizing personal photo collections was first formalized by Sinha *et al.* [21]. They defined the problem based on three salient properties that an informative summary should satisfy: *quality*, *diversity*, and *coverage*. In their manuscript, they defined *quality* as the interestingness or attractiveness of the photos in the summary, *diversity* as a measure of its non-redundancy, and *coverage* as a property that reflects how many important concepts are present in both the photolog and the summary. To address this problem, they proposed a summarization framework that optimizes these properties to generate an informative overview.

Other works, such as [4,24,25], focus on photo albums, particularly considering the various events that characterize these albums. In [4], Ceroni et al. propose an expectation-oriented photo selection method that combines various image-related factors, such as image quality, presence of faces, and concept fea-

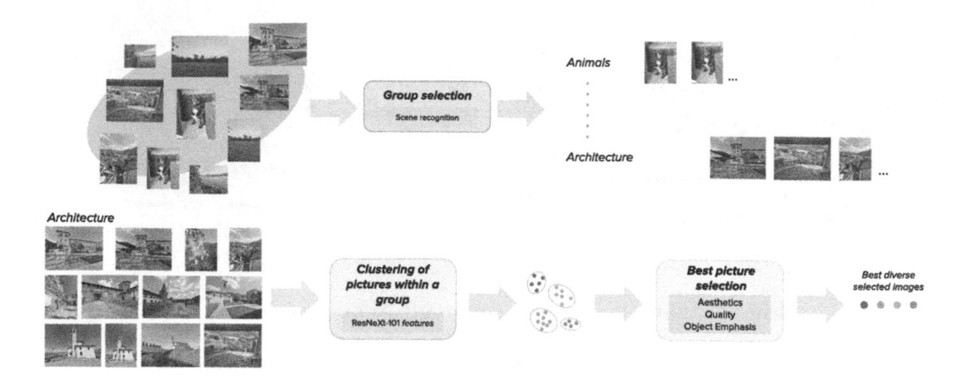

Fig. 2. The overall pipeline of the proposed framework. First, a collection of photos is divided into groups with homogeneous semantic content. Then, for each group, the photos are grouped into k clusters using the K-means algorithm based on features extracted from a ResNeXt-101. Next, for each cluster, the best image is selected based on image aesthetics, quality, and the emphasis on the subject of the photo (*object emphasis*).

tures, with collection-based features, such as album size. Wang et al. in [24] argue that the selection process is influenced by the event type and propose a selection method that takes into consideration the event type and varies its decision accordingly.

However, these methods are limited in their scope as they mainly focused on photo albums, lacking in terms of generality and flexibility.

In this paper, we present a flexible framework that can be used for exploring large-scale image datasets and summarizing photo albums. Our approach is unique in that it focuses on maximizing diversity between selected images while also ensuring high quality and/or aesthetic value.

The proposed method works as follows: images are first grouped based on their semantic content, and representative images are then selected from each group; the final summary consists of all the representative images from each semantic content group.

2 Proposed Method

The proposed method selects the most representative images from a collection of images $\mathcal{H}$ based on several criteria: diversity, quality, and aesthetics. The pipeline, as shown in Fig. 2, consists of three main components: *group selection*, *clustering of pictures within a group*, and *best picture selection*.

The first module (*group selection*) aims to group images based on their semantic content into L groups $\mathcal{G}_1, \cdots, \mathcal{G}_L$. The choice of semantic classes, hence L, is arbitrary and should depend on the task at hand. For example, if we are exploring a dataset, a possible set of semantic classes is the one proposed by Imagenet ($L = 1,000$) [7]. On the other hand, if we want to summarize a photo album, scene categories such as those proposed in the Camera Scene Detection Dataset (CamSDD) ($L = 30$) [20] are more suitable since they cover the most common scenes found in a photo album.

Subsequently, each group of images is processed by the *clustering of pictures within a group* and *best picture selection* modules. In the first stage, the images of a group $\mathcal{G}i$ are divided into k sets of pictures $\mathcal{C}_1, \mathcal{C}_2, ..., \mathcal{C}_k$ based on visual similarity. Where k can be selected arbitrarily, and it determines the quantity of images to be chosen for each semantic class L. Then, the best representative picture of each group $\mathcal{C}_j$ is selected based on three perceptual properties: image quality, image aesthetics, and *object emphasis* (whether the image emphasizes foreground objects). Therefore, each group $\mathcal{G}_i$ is represented by the k most emblematic and diverse images, and the whole summary set $\mathcal{H}$ is made up of $L \times k$ pictures.

2.1 Group Selection

In the first step, a collection of images $\mathcal{H}$ is grouped based on the semantic content into L groups $\mathcal{G}_1, \cdots, \mathcal{G}_L$. Examples of semantic classes include animals, mountains, architecture, and more. Each group is composed of n pictures $\mathcal{G}_l = \{\mathbf{I}_1, \mathbf{I}_2, ..., \mathbf{I}_n\}$ such that $\bigcup_{i=1}^{L} \mathcal{G}_i = \mathcal{H}$ and $\mathcal{G}_i \cap \mathcal{G}_j = \emptyset,\ \forall i, j \in \{1, ..., L\} \mid i \neq j$.

2.2 Clustering of Pictures Within a Group

Subsequently, the proposed method clusters the pictures of a given image group $\mathcal{G}_l$ using the K-Means algorithm [18] into k clusters $\mathcal{C}_1, \mathcal{C}_2, ..., \mathcal{C}_k$ by exploiting the features extracted from the last convolutional layer, after the global average pool, of a pre-trained ResNeXt-101 32×8d [26]. The ResNeXt-101 is pre-trained in a weakly-supervised fashion [19] on 940 million public images with 1.5K hashtags [7], and then fine-tuned on the ImageNet1K dataset to better encode the image content. This phase ensures diversity in the selection process by creating k clusters of visually similar images, such that $\bigcup_{i=1}^{k} \mathcal{C}_i = \mathcal{G}_l$ and $\mathcal{C}_i \cap \mathcal{C}_j = \emptyset,\ \forall i, j \in \{1, ..., k\} \mid i \neq j$.

2.3 Best Picture Selection

Finally, for each cluster $\mathcal{C}_i$ within a given group $\mathcal{G}_l$, the best image is selected based on three perceptual properties: image quality, image aesthetics, and *object emphasis* (i.e., whether the image emphasizes foreground objects). Specifically, given an image $\mathbf{I}$, we compute its perceived aesthetics $a(\mathbf{I})$, visual quality $q(\mathbf{I})$, and grade of *object emphasis* $u(\mathbf{I})$. We define $b(\mathcal{C})$ as the function that returns the best image in a group of images $\mathcal{C}$:

$$b(\mathcal{C}) = \arg\max_{\mathbf{I} \in \mathcal{C}} \frac{a(\mathbf{I}) + q(\mathbf{I}) + u(\mathbf{I})}{3}$$

We can therefore define the best subset $\mathcal{B}$ of images representing $\mathcal{G}$ as:

$$\mathcal{B} = \{b(\mathcal{C}_i)\ \forall i \in \{1, ..., k\}\}$$

3 Datasets

3.1 Automatic Triage for a Photo Series

The Automatic Triage for a Photo Series dataset [5] comprises 15,545 unedited photos with sizes larger than 600 × 800 pixels, organized into 5,953 series. The authors of the dataset collected personal photo albums from participants in a contest-like environment, resulting in over 350 album submissions from 96 contributors. To identify photo series, the authors used SIFT descriptors [17] and color similarity for neighboring images in time. For photos containing human faces, they ensured that each series contained the same group of people. Series with more than 8 photos were split using variant k-means on 116-dimension global features, where the center is a representative picture instead of a mean, as described in [23]. All collected clusters were manually checked to filter out low-quality images or privacy concerns.

After gathering the photo series, a crowd-sourced user study was conducted over Amazon Mechanical Turk (MTurk) [11] to collect human preferences for the best picture in each series. Participants were asked to perform pairwise comparisons of images from the same series using a forced-choice methodology, asking "Imagine you take these two photos and can only keep one. Which one will you choose, and why?" rather than ranking all the photo series to better measure small differences. Participants were also asked to describe why a particular photo was preferred or why the other photo was not preferred.

Figure 3 shows some of the photo series from the Automatic Triage for a Photo Series dataset. To obtain a global ranking for each image series, the authors used the Bradley-Terry model [2], which describes the probability of choosing image I_i over I_j as a sigmoid function of the score difference between the two photos.

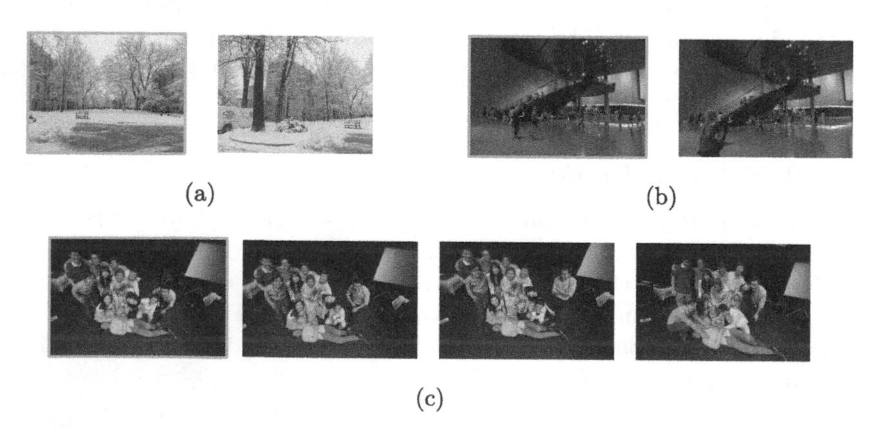

Fig. 3. Examples of photos series from the Automatic Triage for a Photo Series dataset. Each series of images (a, b, c) is highlighted in green the one preferred by the majority of the people.

3.2 Camera Scene Detection Dataset

The Camera Scene Detection Dataset (CamSDD, [20]) is a large-scale dataset containing over 11,000 pictures, each of size 576 × 384 pixels, grouped into 30 of the most important scene categories. The images in the dataset were crawled from Flickr using the same setup as in [8]. After collecting the pictures, images with distorted colors and watermarks, heavily edited pictures, or monochrome images were manually removed.

According to the authors, the dataset was designed to be as diverse as possible, covering different environments and shooting conditions. Therefore, each scene category contains images from various places, viewpoints, and angles. The dataset is also balanced regarding the number of images in each category, with an average of around 350 photos per group.

4 Experimental Setup

4.1 Data Setup

To address the need for a set of images with repeated elements, we grouped the photo series from the Automatic Triage for a Photo Series dataset based on their scenes. We excluded some of the less common scenes from the CamSDD dataset and selected only a subset of the possible scenes. We also combined the *cat* and *dog* scenes into a single master scene called *Animals*. In total, we considered 19 scenes: *Architecture, Backlight, Beach, Blue Sky, Cloudy Sky, Food, Greenery, Group portrait, Indoor, Kids, Mountain, Night shot, Portrait, Snow, Stage concert, Sunset Sunrise, Underwater, Waterfall*, and *Animals*.

To divide the series into 19 scenes, we used a majority vote strategy, where the assigned scene of a series was determined by the scene predicted for the majority of the images in that series. To improve the accuracy of the scene classification, we only considered labels with a probability of 0.7 or higher, resulting in the removal of 55% of the images. Scene dataset contains 19 classes with a total of 5457 images and 2145 series. The average series lenght is 2.53. *Architecture* and *Group portrait* are the most common scenes, and *Waterfall* and *Mountain* being the least common ones.

4.2 Evaluation Metrics

We assessed the performance of the proposed method using three metrics: *Diversity score*, *Selection precision*, and *Average probability*. Given the similarity between images within the same photo series, diversity was measured by calculating the number of unique series in the selection divided by the total number of images selected. The goodness of the selection was evaluated by counting the number of "best" images (i.e., images with the highest probability of selection) among the selected images. The Automatic Triage for a Photo Series dataset was human-annotated, and each image is provided with the probability of choosing one particular image rather than the others. We decided to consider the average of the probability values of the selected images (*Average probability*) in support of the *Selection precision* to measure the quality of the results.

In fact, it could be possible that there is not one single image better than the others, and for this reason it is useful to have also the *Average probability*.

Given n series of not overlapping images $\{\mathcal{S}_1, \mathcal{S}_2, ..., \mathcal{S}_n\}$, from which we have to select k different pictures $\{\mathbf{I}_1, \mathbf{I}_2, ..., \mathbf{I}_k\}$ (where $k \leqslant n$), to define the *Diversity score* we need to introduce the following indicator function:

$$\mathbb{1}(\mathcal{S}_i) := \begin{cases} 1 \ if \ \mathcal{S}_i \cap \{\mathbf{I}_1, \mathbf{I}_2, ..., \mathbf{I}_k\} \neq \emptyset \\ 0 \qquad otherwise \end{cases} \tag{1}$$

Given that the *Diversity score* is $\frac{\sum_{i=1}^{n} \mathbb{1}(\mathcal{S}_i)}{k}$, the *Selection precision* is $\frac{\sum_{i=1}^{k} BEST(\mathbf{I}_i)}{k}$, where:

$$BEST(\mathbf{I}) = \begin{cases} 1 \ if \ p(\mathbf{I}) \ = \max_{\mathbf{I}_i \in \mathcal{S}_i \ | \mathbf{I} \in \mathcal{S}_i} p(\mathbf{I}_i) \\ 0 \ otherwise \end{cases} \tag{2}$$

and $p(\mathbf{I})$ is the probability of the picture $\mathbf{I}$ of being selected.
Finally the *Average probability* can be defined as: $\frac{1}{k} \sum_{i=1}^{k} p(\mathbf{I}_i)$.

It is important to notice that the *Diversity score* ranges between a value of $1/k$ which reflects a low diversity, and an upper bound of 1, indicating the maximum diversity of the set, while *Selection precision* ranges between 0 and 1.

4.3 Experimental Details

The proposed framework was implemented using Python. Perceived image quality was extracted using the method presented in [13], while image aesthetics and *object emphasis* were computed using the system proposed in [12]. To reduce the variability of results, k-means was initialized randomly and executed 10 times to select the best results each time.

The performance of the proposed method was evaluated by averaging over five repetitions.

As the current state-of-the-art solutions are mainly focused on photo albums, we have decide to compute a theoretical range of achievable performances. To do this, we have defined two policies: one for the lower-bound and one for the upper-bound. For the lower-bound, k images are randomly picked from the pool of pictures to summarize. We will reefer to this policy as *Random*. On the other hand, for the upper-bound, the k images are selected by having access to the ground truth, therefore randomly picking the best ones. This will be referred to as the *Oracle* policy.

We evaluated the proposed method for values of k equal to 5, 10, or 20, which are possible numbers of images that a user may want to extract. Notice that, as aforementioned, L is equal to 19.

Table 1. Average results over 5 repetitions in terms of *Diversity score* and *Selection precision*. Due to the time complexity, the policy *High-Contrast Color Sets* is executed a single time.

Diversity score			
Policy name\ k	**5**	**10**	**20**
Random	0.97 ± 0.01	0.95 ± 0.01	0.88 ± 0.01
Oracle	1.0 ± 0.0	1.0 ± 0.0	1.0 ± 0.0
High-Contrast Color Sets	1	0.99	0.98
Proposed method	1.0 ± 0.004	0.98 ± 0.01	0.97 ± 0.01
Selection precision			
Policy name\ k	5	10	20
Random	0.37 ± 0.05	0.41 ± 0.04	0.40 ± 0.02
Oracle	1.0 ± 0.0	1.0 ± 0.0	1.0 ± 0.0
High-Contrast Color Sets	0.45	0.46	0.45
Proposed method	0.58 ± 0.016	0.56 ± 0.01	0.55 ± 0.01
Average probability			
Policy name\ k	5	10	20
Random	0.37 ± 0.03	0.41 ± 0.02	0.43 ± 0.01
Oracle	0.69 ± 0.01	0.70 ± 0.01	0.69 ± 0.01
High-Contrast Color Sets	0.45	0.44	0.45
Proposed method	0.50 ± 0.01	0.50 ± 0.01	0.50 ± 0.01

5 Results

In this section, we compare the proposed method to the *Oracle*, *Random*, and *High-Contrast Color Sets* policies based on the *Diversity score*, the *Selection precision*, and the *Average probability*. Table 1 reports the average performances of the considered policies over five repetitions.

Regarding diversity, as the number of items to be selected increases, the *Diversity score* of the proposed method slightly decreases. In contrast, the gap in terms of *Diversity score* between the proposed method and the *Random* policy becomes noticeable as k increases. This suggests that the proposed method can accurately partition images based on their content. The *High-Contrast Color Sets* policy is also competitive from this perspective.

Regarding *Selection precision*, the *Random* strategy fluctuates around the value of 0.3 with a noticeable variance. In contrast, the proposed method achieves a *Selection precision* of 0.5832 when selecting five images, meaning that three out of five images are the best ones, which decreases to 0.5453 when k is equal to 20. The *High-Contrast Color Sets* policy lies between the *Random* strategy and the proposed method.

The *Average probability* confirms the general behavior highlighted by the *Selection precision*.

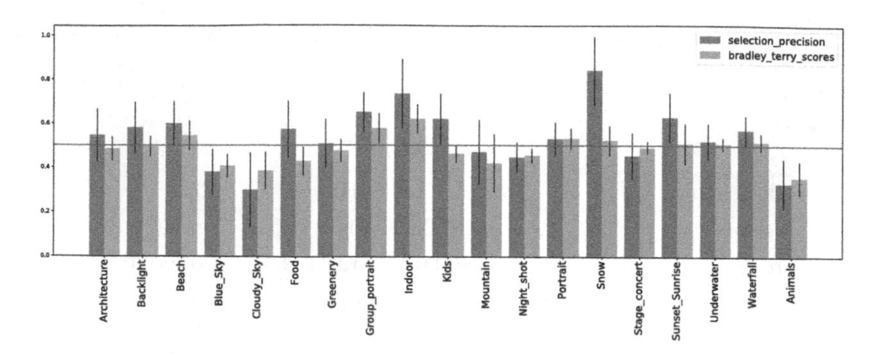

Fig. 4. Per class average *Selection precision* and *Average probability* of the proposed method. The green line indicates the value of 0.5.

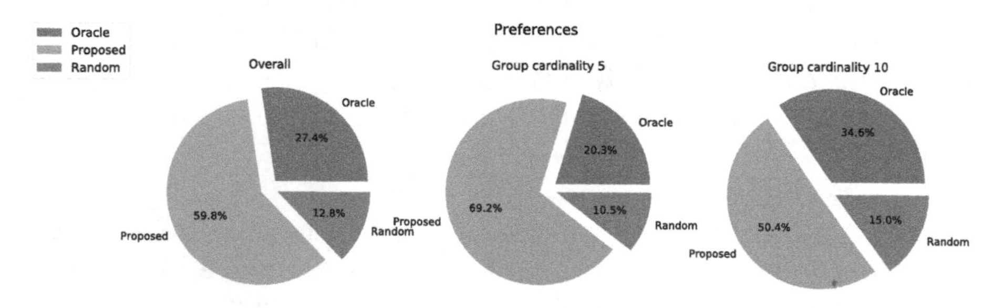

Fig. 5. Preferences distribution over the three considered strategies. The first chart reports the overall decision, while the remaining two are with respect to the group cardinality of 5 and 10 respectively,

In Fig. 4, we present the average performance of each of the considered scene categories in terms of *Selection precision* and *Average probability*. The best-performing category is *Snow*, while the worst one is *Cloudy_sky*. Figure 4 also shows that, except for the *Mountain* scene, scenes with a *Selection precision* of less than 0.5 (e.g., *Blue_sky*, *Cloudy_sky*, *Night_shot*, and *Animals*) do not reflect low values by the *Average probability*, suggesting that the selected photos may not be the worst ones.

5.1 Subjective Results

The ground truth of the Automatic Triage for a Photo Series dataset is closely related to the series of images. Therefore, the results presented so far only reflect the proposed method's ability to select the best images from multiple series, without considering the overall goodness of the selection. To address this, we evaluated the proposed method against the judgments of seven human raters. We asked the raters to select the best image selection from three algorithms: the proposed method, the *Oracle* policy, and the *Random* policy. We evaluated all 19 scenes and two values of k (5 and 10) omitting $k = 20$ due to the large number of

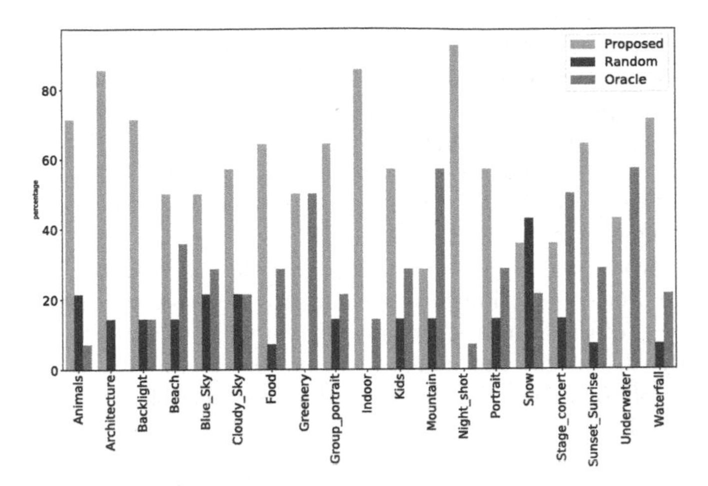

Fig. 6. For each of the 19 considered scenes, and for each of the considered policies, we report the percentage of votes given by the human raters.

images selected. Before the experiments, we instructed the raters to consider the concepts of diversity (i.e., selecting images without repeating subjects), overall aesthetics, and quality of the photos, providing them with visual examples.

We report the percentages of votes obtained by each algorithm in Fig. 5. The first pie chart shows the overall scores, while the other two plots show the scores for group cardinality 5 and 10, respectively. The proposed method was the most selected algorithm, followed by the *Oracle* policy and then the *Random* policy. This trend can be attributed to the fact that the *Oracle* policy selects only the best images, whereas the proposed method aims to select the most beautiful images based on perceptual properties. Our selection strategy appears to perform better for selecting 5 images rather than 10.

Figure 6 shows the percentages of preferences given by the human raters for each of the 19 camera scenes. In almost every situation, the proposed method outperforms the other approaches. It is interesting to compare these results with those in Fig. 4. For scenes with low *Selection precision* (less than 0.5), we found that the *Average probability* is more accurate than the *Selection precision*. Only the *Mountain* scene has a lower *Selection precision* with respect to the *Average probability*, and it is the worst one in Fig. 6. In scenes like *Blue_sky*, *Cloudy_sky*, *Night_shot* and *Animals*, where the low *Selection precision* is not reflected by the *Average probability*, the proposed algorithm is the most selected one, as shown in Fig. 6.

6 Conclusions

In this work, we present a flexible framework that can be used to explore large-scale image datasets and summarize photo albums. Our pipeline maximizes the diversity and quality of the resulting set by first classifying each image according to its camera scene and selecting the best k pictures for each scene group. To achieve this, we use a ResNeXt-101 32×8d to extract features that capture the most diverse pictures, and we employ criteria such as image quality, and aesthetics to select the best images.

Experimental results on the Automatic Triage for a Photo Series dataset demonstrated the effectiveness of our approach, while experiments conducted on human preferences confirmed its capabilities.

In future work, we plan to investigate other perceptual properties to strengthen our image selection process and provide a more accurate and complete description of the images. Additionally, we will explore more sophisticated and recent algorithms for the image diversity block. Finally, we will consider the possibility of collapsing the entire pipeline to select the most beautiful and diverse images all at once.

References

1. booking. Automated image tagging at booking.com (2017)
2. Bradley, R.A., Terry, M.E.: Rank analysis of incomplete block designs: I the method of paired comparisons. Biometrika **39**(3/4), 324–345 (1952)
3. Campadelli, P., Posenato, R., Schettini, R.: An algorithm for the selection of high-contrast color sets. Color Res. App. **24**(2), 132–138 (1999)
4. Ceroni, A., Solachidis, V., Niederée, C., Papadopoulou, O., Kanhabua, N., Mezaris, V.: To keep or not to keep: an expectation-oriented photo selection method for personal photo collections. In: Proceedings of the 5th ACM on International Conference on Multimedia Retrieval, pp. 187–194 (2015)
5. Chang, H., Fisher, Yu., Wang, J., Ashley, D., Finkelstein, A.: Automatic triage for a photo series. ACM Trans. Graphics (TOG) **35**(4), 1–10 (2016)
6. Cheng, Y.: Mean shift, mode seeking, and clustering. IEEE Trans. Pattern Anal. Mach. Intell. **17**(8), 790–799 (1995)
7. Deng, J., Dong, W., Socher, R., Li, L.-J., Li, K., Li, F.-F.: Imagenet: a large-scale hierarchical image database. In: 2009 IEEE Conference on Computer Vision and Pattern Recognition, pp. 248–255. IEEE (2009)
8. Ignatov, A., Kobyshev, N., Timofte, R., Vanhoey, K., Gool, L.V.: Wespe: weakly supervised photo enhancer for digital cameras. In: Proceedings of the IEEE Conference on Computer Vision and Pattern Recognition Workshops, pp. 691–700 (2018)
9. Instagram. Our story: A quick walk through our history as a company (2016)
10. kayak. Hotel image categorization with deep learning (2018)
11. Kittur, A., Chi, E.H., Suh, B.: Crowdsourcing user studies with mechanical Turk. In: Proceedings of the SIGCHI Conference on Human Factors in Computing Systems, pp. 453–456 (2008)
12. Leonardi, M., Napoletano, P., Rozza, A., Schettini, R.: Modeling image aesthetics through aesthetic-related attributes. In: London Imaging Meeting, vol. 2021, Society for Imaging Science and Technology (2021)
13. Leonardi, M., Napoletano, P., Schettini, R., Rozza, A.: No reference, opinion unaware image quality assessment by anomaly detection. Sensors **21**(3), 994 (2021)
14. Li, C.-H., Chiu, C.-Y., Huang, C.-R., Chen, C.-S., Chien, L.-F.: Image content clustering and summarization for photo collections. In: 2006 IEEE International Conference on Multimedia and Expo, pp. 1033–1036. IEEE (2006)

15. Li, C., Loui, A.C., Chen, T.: Towards aesthetics: a photo quality assessment and photo selection system. In: Proceedings of the 18th ACM International Conference on Multimedia, pp. 827–830 (2010)
16. Lin, T.-Y., et al.: Microsoft COCO: common objects in context. In: Fleet, D., Pajdla, T., Schiele, B., Tuytelaars, T. (eds.) ECCV 2014. LNCS, vol. 8693, pp. 740–755. Springer, Cham (2014). https://doi.org/10.1007/978-3-319-10602-1_48
17. Lowe, D.G.: Distinctive image features from scale-invariant keypoints. Int. J. Comput. Vision **60**(2), 91–110 (2004)
18. MacQueen, J., et al.: Some methods for classification and analysis of multivariate observations. In: Proceedings of the Fifth Berkeley Symposium on Mathematical Statistics and Probability, vol. 1, pp. 281–297. Oakland, CA, USA (1967)
19. Mahajan, D., et al.: Exploring the limits of weakly supervised pretraining. In: Ferrari, V., Hebert, M., Sminchisescu, C., Weiss, Y. (eds.) ECCV 2018. LNCS, vol. 11206, pp. 185–201. Springer, Cham (2018). https://doi.org/10.1007/978-3-030-01216-8_12
20. Pouget, A., et al.: Fast and accurate camera scene detection on smartphones. In: Proceedings of the IEEE/CVF Conference on Computer Vision and Pattern Recognition, pp. 2569–2580 (2021)
21. Sinha, P., Mehrotra, S., Jain, R.: Summarization of personal photologs using multidimensional content and context. In: Proceedings of the 1st ACM International Conference on Multimedia Retrieval, pp. 1–8 (2011)
22. Trivago. How we build the image gallery on Trivago (2021)
23. Wang, X.-J., Zhang, L., Liu, C.: Duplicate discovery on 2 billion internet images. In: Proceedings of the IEEE Conference on Computer Vision and Pattern Recognition Workshops, pp. 429–436 (2013)
24. Wang, Y., Lin, Z., Shen, X., Mech, R., Miller, G., Cottrell, G.W.: Event-specific image importance. In proceedings of the IEEE Conference on Computer Vision and Pattern Recognition, pp. 4810–4819 (2016)
25. Wang, Y., Lin, Z., Shen, X., Mech, R., Miller, G., Cottrell, G.W.: Recognizing and curating photo albums via event-specific image importance. arXiv preprint arXiv:1707.05911 (2017)
26. Xie, S., Girshick, R., Dollár, P., Tu, Z., He, K.: Aggregated residual transformations for deep neural networks. In Proceedings of the IEEE Conference on Computer Vision and Pattern Recognition, pp. 1492–1500 (2017)
27. Zephoria: The top 20 valuable Facebook statistics (2015)

Automated Identification of Failure Cases in Organ at Risk Segmentation Using Distance Metrics: A Study on CT Data

Amin Honarmandi Shandiz(✉), Attila Rádics, Rajesh Tamada, Makk Árpád, Karolina Glowacka, Lehel Ferenczi, Sandeep Dutta, and Michael Fanariotis

GE HealthCare, Budapest, Hungary
{Amin.Honarmandishandiz,Attila.Radics,Rajesh.Tamada,Arpad.Makk1, Karolina.Glowacka1,Sandeep.Dutta,Michail.Fanariotis}@ge.com, Lehel.Ferenczi@med.ge.com

Abstract. Automated organ at risk (OAR) segmentation is crucial for radiation therapy planning in CT scans, but the generated contours by automated models can be inaccurate, potentially leading to treatment planning issues. The reasons for these inaccuracies could be varied, such as unclear organ boundaries or inaccurate ground truth due to annotation errors. To improve the model's performance, it is necessary to identify these failure cases during the training process and to correct them with some potential post-processing techniques. However, this process can be time-consuming, as traditionally it requires manual inspection of the predicted output. This paper proposes a method to automatically identify failure cases by setting a threshold for the combination of Dice and Hausdorff distances. This approach reduces the time-consuming task of visually inspecting predicted outputs, allowing for faster identification of failure case candidates. The method was evaluated on 20 cases of six different organs in CT images from clinical expert curated datasets. By setting the thresholds for the Dice and Hausdorff distances, the study was able to differentiate between various states of failure cases and evaluate over 12 cases visually. This thresholding approach could be extended to other organs, leading to faster identification of failure cases and thereby improving the quality of radiation therapy planning.

Keywords: Anomaly Detection · Quality Assurance · Organ at risk segmentation · Failure case identification · Distance metrics · Model performance · Radiation therapy planning · Thresholding

1 Introduction

Organ at risk (OAR) segmentation is an essential step in radiation therapy planning. Accurate delineation of OARs can help to spare these critical structures from high doses of radiation and prevent unwanted side effects [21]. The traditional method for OAR contouring is manual contouring, which is a time-consuming and labor-intensive process. Manual contouring can also be subjective

G. L. Foresti et al. (Eds.): ICIAP 2023 Workshops, LNCS 14365, pp. 86–96, 2024.
https://doi.org/10.1007/978-3-031-51023-6_8

and prone to inter-observer variability, which can affect the accuracy and consistency of the contouring [2–4]. To address these issues, automatic contouring methods have been developed in recent years. These methods use algorithms and machine learning techniques to automatically segment OARs from medical images. Recently, deep learning methods such as Convolutional neural networks have played a significant role in various aspects specially when the data is images [5,9,18–20,22].

According to a study conducted by Fugan et al. [7] an auto-contouring (AC) system was evaluated and found to be in agreement with manual contours within the inter-observer uncertainty level. Additionally, Altman et al. [1] observed that there were no significant dosimetric differences between treatment plans using AC software versus manually contouring. These findings suggest that despite not always being perfect, auto-contours can still be considered clinically acceptable. However, the performance of automatic contouring methods can vary depending on the imaging modality, image quality, and complexity of the OARs being segmented [7]. As such, it is important to evaluate the accuracy and robustness of automatic contouring methods before their implementation in clinical practice.

A conventional approach to enhance the performance of OAR models is to improve the accuracy of the model's predictions. To achieve this, it is necessary to identify cases where the model fails to produce accurate predictions. In the case of OAR models, this requires a visual inspection of the predicted contours to determine if they are closely overlapping the ground truth. Cases where there is unsatisfactory overlap between the predicted and ground truth contours, or where anomalies exist in the predicted images, are selected as failure cases. These cases are then subjected to further analysis to identify the cause of the failure, which involves consultation with experts. However, this approach can be time-consuming, especially when there are many candidate cases to test and no predefined set of candidates to prioritize for failure analysis.

In order to address this issue, our paper proposes a method that utilizes distance metrics to automatically identify candidate cases for failure analysis. By analyzing 20 cases for 6 organs, we determined a threshold that could effectively distinguish these candidates. Instead of manually checking the predicted contours against the ground truth, this approach could save time by automatically identifying potential failure cases based on the HD and Dice metrics [6]. We found that different organs had different values for the proposed threshold, and this method could help decrease the time needed to find failure cases.

2 Methods and Materials

Loss functions play an important role in various tasks. In image segmentation tasks, the role of the loss function is distinguished by measuring the difference between the predicted and ground truth segmentation maps. They guide the training process to optimize model performance. Some common loss functions in common 3D segmentation tasks can be seen in Fig. 1 by Jun Ma et al. [13]. According to the study conducted by Shruti et al. Common loss functions in image segmentation are different and depend on various parameters, although the popular used ones can be cross-entropy loss, dice loss, Hausdorff distance (HD), etc. [11]. In the context of OAR segmentation tasks, the choice of loss functions depends on task-specific requirements such as the nature of the data, the desired characteristics of the segmentation output, and the specific task.

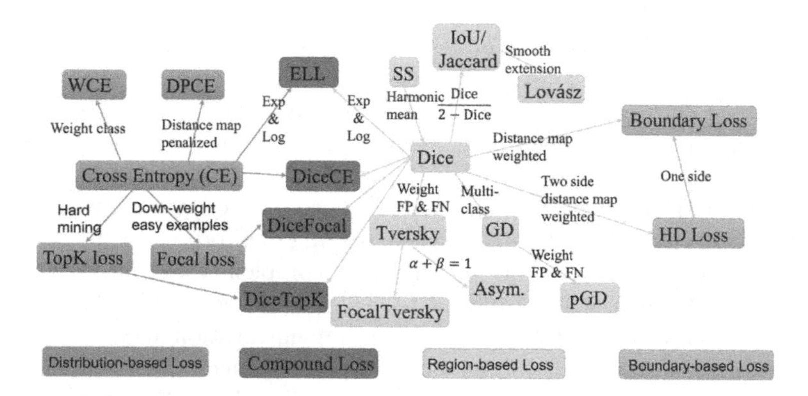

Fig. 1. The Figure shows loss functions for typical 3D segmentation tasks in four different category [8].

2.1 Dice and Hausdorff Distance Metrics

To evaluate the accuracy of the OAR segmentation, we used the Dice and Hausdorff distance metrics. The Dice coefficient measures the similarity between two sets of binary data, while the Hausdorff distance measures the maximum distance between two sets of points. We computed the Dice and Hausdorff distance between the predicted segmentation masks and ground truth masks for each of the six OARs.

The Dice distance, also known as the Dice coefficient, is a widely used metric for evaluating the similarity or overlap between two sets. In the context of medical image analysis, it is often used to assess the accuracy of segmentation algorithms that delineate the boundaries of anatomical structures or regions of interest.

The Dice distance is calculated as the ratio of the intersection of two sets to their combined size, and ranges from 0 to 1, where a value of 1 indicates perfect overlap and a value of 0 indicates no overlap. The formula for Dice distance is:

$$Dice(A, B) = \frac{2\,|A \cap B|}{|A| + |B|} \tag{1}$$

where $\cap$ represents the intersection of the predicted and actual volumes. The Hausdorff distance is another metric used to evaluate the accuracy of segmentation algorithms. It measures the maximum distance between any point on the surface of one volume and the closest point on the surface of the other volume. In other words, it quantifies the "worst-case scenario" of disagreement between two sets. The formula for Hausdorff distance is:

$$Hausdorff(A, B) = \max\{h(a, B), h(b, A)\} \tag{2}$$

where a and b are points on the surfaces of A and B, respectively, and h(a, B) and h(b, A) are the distances from a to the closest point on B, and from b to the closest point on A, respectively.

The Hausdorff distance is often used in conjunction with the Dice distance to provide a more complete picture of the accuracy of a segmentation algorithm. While the Dice distance measures the degree of overlap between two sets, the Hausdorff distance provides information about the extent and location of any discrepancies between the two sets.

2.2 Dataset

We have used two datasets, -i): one containing five organs-at-risk (OARs) that are commonly segmented in radiation therapy planning from an internal dataset and -ii) another one contains one organ from a public dataset for lymph node [15–17]. The internal dataset (acquired by our clinical partners) consists of 20 CT images depicting different areas, scanned using different CT protocols, with slice thickness between 0.5 and 3 mm. These scans were acquired on volunteers, using various multi-slice CT scanners such as (Discovery 750HD, Optima540, etc.). These scans, unlike the ones in the public dataset, are not uniform, and differ in CT image parameter settings (e.g. resolution, pixel spacing).

2.3 Model Architecture

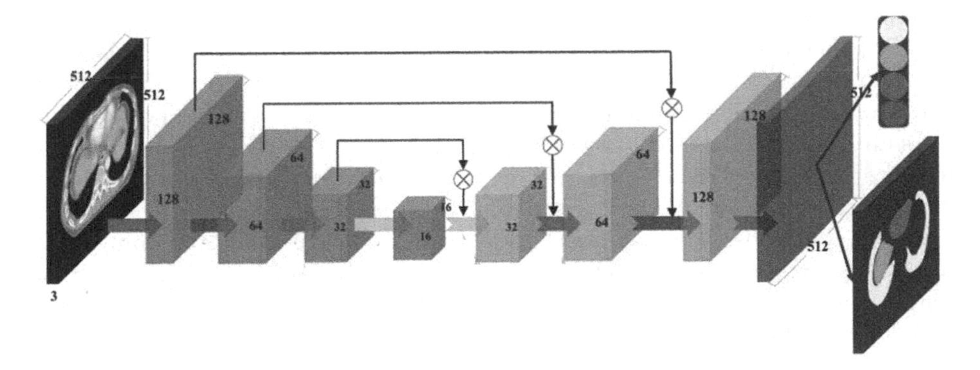

Fig. 2. The figure shows the 3D auto-encoder model for organ at risk segmentation [8].

The Fig. 2 shows a schematic diagram of an auto-encoder model, which is a type of neural network used for tasks such as dimensional reduction and feature extraction. The model consists of two parts: the encoder and the decoder. The encoder takes the input data and compresses it into a lower-dimensional representation or latent space, while the decoder takes this representation and reconstructs the original data. This model can be used for various applications, including image and signal processing.

In our paper, organ contours were created using a novel DL auto segmentation model (DLAS v1.0, GE HealthCare), that the auto encoder structure is like Fig. 2 used in [8]. We employed the encoder part of the model as a feature extractor to obtain a compressed representation of the input images that captures the salient features related to the organs of interest. This compressed representation was then fed into a decoder to generate the organ segmentation contours. This approach allowed us to improve the accuracy and efficiency of the segmentation task compared to manual contouring or other traditional segmentation methods.

2.4 Experimental Setup

Firstly, the contours were produced and reviewed by experts radiologist for five different organs including breasts, Chiasma, Brainstem, Trachea and Femurs. For each organ there are 20 CTs with their ground truth. Then, we converted all of the data to nrrd format and fed them to the 3D auto-encoder model like Fig. 2 with train, evaluation and test partitions respectively 10, 5, 5 which were chosen for the model and batch size of 5.

For training the model we used Dice as loss function. After contours were predicted by the model then two metrics including Hausdorff and Dice [12,14] used to evaluate the model prediction. After visualizing the predicted contours, we compared the results with ground truth to find if there are failure prediction exist in the predicted values. Afterward, we visualize the predicted failed cases versus good cases as presented in Fig. 3.

As it is showed in Fig. 3 for breast organ there are situations where the predicted organs does not overlap with the ground truth or there might be extra lesion predicted outside of the expected organ region. To find these cases we use the characteristics of image distance measures such as Dice and Hausdorff distances. As the Dice's characteristic [6] is suitable to find the overlapped contours and base on characteristics of Hausdorff distance [10] that is considered for distance calculation, we use this metric to find the regions which are predicted far from the ground truth as it is sensitive to the outliers.

We also analysed lymph node data(see Fig. 4) containing abdominal and mediastinal lymph nodes which are publicly available total of 388 mediastinal lymph nodes in CT images of 90 patients and a total of 595 abdominal lymph nodes in 86 patients (for more info see [15–17]).

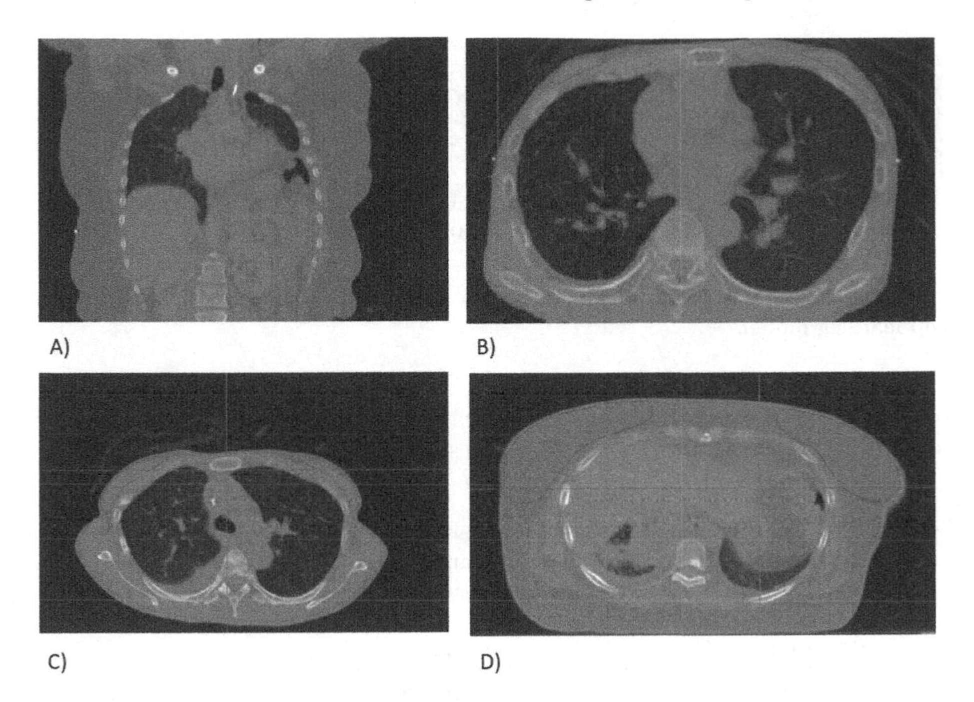

Fig. 3. The figure summarizes four situations expressed in Table 1 for Breasts organ segmentation. The four situations include: A) Shows over segmentation of the model, extra predictions exists in the output beside the region of interest, B) Extra predictions with less overlap with ground truth for the left breast and under segmentation for the right breast organ in some slices, C) good predictions, D) model has good predictions but there are extra predictions (over segmentation) in the left breast. Ground Truth contours are distinguished by red line and predictions are distinguished by green. (Color figure online)

3 Results and Discussion

After calculating Dice and Hausdorff distances for predicted versus ground truth, the purpose was to find the less predicted contours in some slices by low Dice score values. Also, analyzing the HD values could determine if the output is over segmented or not. As you can see in Table 1 with different combination of Dice and Hausdorff values we are able to decide if output contours need to consider for post-processing.

We found that setting a threshold for Dice and Hausdorff distance produced for these six organs could help to distinguish the failure cases, identifying a significant number of candidate cases while keeping the false positive rate low. We also found that in most cases, the identified failures were due to under segmentation between the predicted and ground truth contours, or due to over segmentation by the model.

We compared our approach to a baseline method that relied on visual inspection alone and found that our method could identify candidate cases with higher

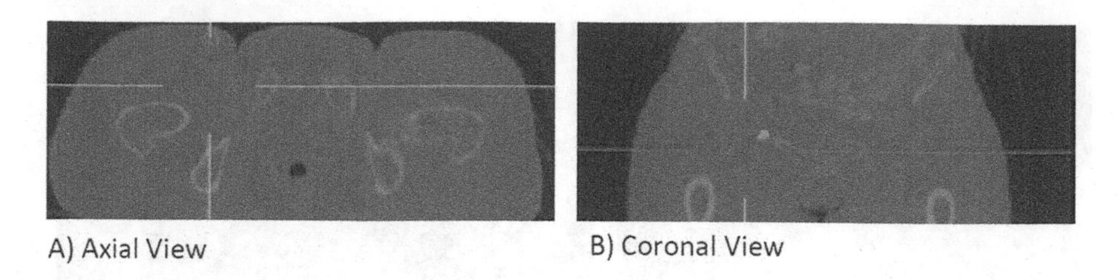

Fig. 4. The figure shows the results of lymph node contouring in CT images using a deep learning model.

speed as it is not needed to visually check them. Our approach is also could be useful in other organs by finding their threshold values.

Table 1. The table summarizes the different situations for anomaly detection in medical imaging based on the Hausdorff distance and Dice similarity coefficient for Breast. The table provides an explanation for each situation based on the predicted results.

Situation	Explanation
HD < 6 and Dice ~ 0.9	Good prediction
$6 <$ HD < 100 and Dice ~ 0.9	Good overlap prediction but there are extra predictions connected and close to GT
Hd > 100 and Dice ~ 0.9	Good prediction but there are extra predictions which are far from respective organ
Hd > 6 and Dice < 0.8	There are extra predictions in slices or far from organ and there is led overlap for some ground truth contours
Hd < 6 and Dice < 0.8	There is no extra predicted contours or lesion but no predicted contours for some GT in some slices

In summary, our research highlights the usefulness of employing a range of distance metrics for detecting inferior or unsatisfactory segmentation predictions in organ-at-risk segmentation, which can enhance the performance of models and facilitate model failure analysis in medical imaging. It should be noted that the appropriate threshold would vary for different organs, and in respect of the size of the organs and contrast and other factors it could change.

In order to evaluate the performance of our proposed approach, First we used the Table 1 values to identify good cases with good overlap and less out of box prediction. Then, we select randomly 12 cases to visually inspect the generated contours. In this step we could see that out of 12 cases which were randomly selected by this approach, the selected ones have less anomaly inside the predicted contours.

As you can see in Table 1 to evaluate the model's performance to detect the anomalies in the predicted contours, we have considered different situations based on the combination values of Hausdorff Distance and Dice coefficient. When Hausdorff distance is less than 6mm and the Dice coefficient is around 0.9, we consider the prediction to be a good one. In the case of Hausdorff between 6 and 100 and the Dice coefficient around 0.9, we have identified the prediction as having a good overlap with the ground truth contours but also having over segmentation because of high value in HD. For Hausdorff distance greater than 6 and the Dice coefficient around 0.9, we acknowledge the prediction as a good one, but with extra predictions located far from the respective organ. Moreover, when Hausdorff is greater than 6 and the Dice coefficient is less than 0.8, we determine that there are over segmentation in slices or far from the organ and there is lack of overlap between ground truth and prediction in some of the slices. Finally, when HD is less than 6 and Dice also is less than 0.8 percent then there are less predicted values for GT and less over segmentation because the HD value is small.

Table 2. Thresholds for Hausdorff distance and Dice score for organ segmentation. The table lists the recommended thresholds for breast and lymph node segmentation based on Hausdorff distance and Dice score metrics. The values are based on expert review of a sample dataset.

Organ	Dice score	Hausdorff Distance
Breast	0.9	6 mm
Lymph Node	0.95	0.5 mm
Femur	0.9	11 mm
Trachea	0.85	19 mm
Chiasma	0.66	3.4 mm
Brainstem	0.8	10 mm

As it is showed in Fig. 5 we visualized different organs and showing their inaccurate prediction that needs to find automatically by our method. We analyzed 20 CTs of 6 organs and we could find threshold values for Dice and HD for these organs as you can see in Table 2 for respective organs. Then we compared 12 cases of each organs which chosen randomly by setting this threshold and compare them visually to find out if this method could distinguish the candidate failure cases.

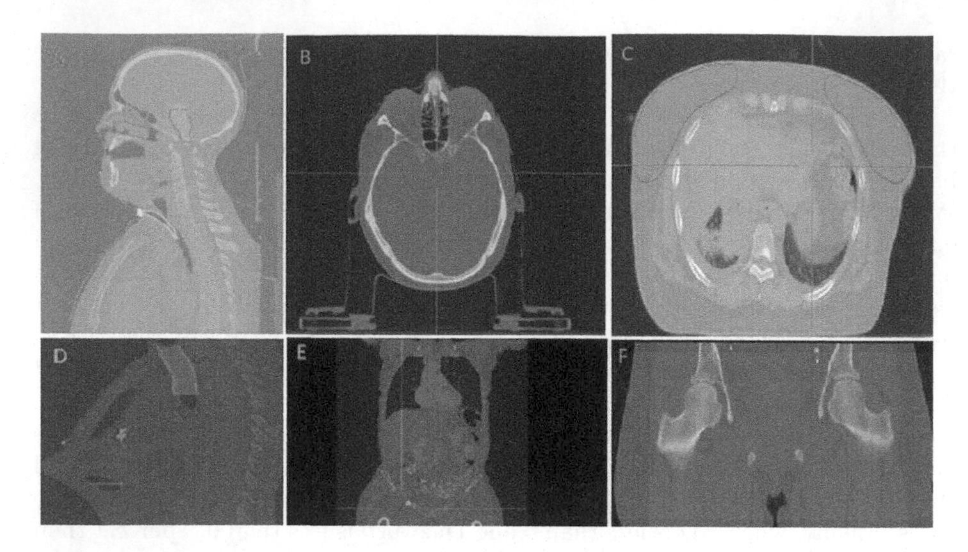

Fig. 5. The figure shows six contoured organs in different boxes, including brainstem, chiasma, breast, trachea, lymph node and femur. These organs are important in radiation therapy planning as they are at risk of damage during treatment. Accurate segmentation of these organs is essential for successful treatment, and auto-contouring models are increasingly being used for this purpose.

4 Conclusion

In conclusion, our paper proposed a method for identifying potential failure cases in organ at risk segmentation using distance metrics, which is a combination of Dice and Hausdorff distances. By analyzing 20 cases of six different organs and visually comparing the ground truth and predicted contours, we were able to determine a threshold for each organ that could potentially decrease the time required for identifying failure cases. Our findings suggest that this approach could be useful for improving the performance of auto contouring models in radiation therapy planning in order to speed up selecting failure cases for post-processing. In future research, we plan to explore the use of this method for anomaly detection after model training and for improving the model performance during clinical decision-making.

Acknowledgments. We would like to thank all the contributors from the Deep Learning Auto Segmentation (DLAS) project for facilitating this study.

References

1. Altman, M.B., et al.: A framework for automated contour quality assurance in radiation therapy including adaptive techniques. Phys. Med. Biol. **60**(13), 5199 (2015)
2. Breunig, J., et al.: A system for continual quality improvement of normal tissue delineation for radiation therapy treatment planning. Int. J. Radiat. Oncol. Biol. Phys. **83**(5), 703–708 (2012)
3. Brouwer, C.L., et al.: 3D Variation in delineation of head and neck organs at risk. Radiat. Oncol. **7**(1), 1–10 (2012)
4. Brouwer, C.L., et al.: CT-based delineation of organs at risk in the head and neck region: DAHANCA, EORTC, GORTEC, HKNPCSG, NCIC CTG, NCRI, NRG Oncology and TROG consensus guidelines. Radiother. Oncol. **117**(1), 83–90 (2015)
5. Csapó, T.G., et al.: Optimizing the ultrasound tongue image representation for residual network-based articulatory-to-acoustic mapping. Sensors **22**(22), 8601 (2022)
6. Dice, L.R.: Measures of the amount of ecologic association between species. Ecology **26**(3), 297–302 (1945)
7. Fung, N.T.C., et al.: Automatic segmentation for adaptive planning in nasopharyngeal carcinoma IMRT: time, geometrical, and dosimetric analysis. Med. Dosim. **45**(1), 60–65 (2020)
8. He, T., et al.: Multi-task learning for the segmentation of organs at risk with label dependence. Med. Image Anal. **61**, 101666 (2020)
9. Honarmandi Shandiz, A., Tóth, L.: Voice activity detection for ultrasound-based silent speech interfaces using convolutional neural networks. In: Ekštein, K., Pártl, F., Konopík, M. (eds.) TSD 2021. LNCS (LNAI), vol. 12848, pp. 499–510. Springer, Cham (2021). https://doi.org/10.1007/978-3-030-83527-9_43
10. Huttenlocher, D.P., Klanderman, G.A., Rucklidge, W.J.: Comparing images using the Hausdorff distance. IEEE Trans. Pattern Anal. Mach. Intell. **15**(9), 850–863 (1993)
11. Jadon, S.: A survey of loss functions for semantic segmentation. In: 2020 IEEE Conference on Computational Intelligence in Bioinformatics and Computational Biology (CIBCB), pp. 1–7. IEEE (2020)
12. Karimi, D., Salcudean, S.E.: Reducing the Hausdorff distance in medical image segmentation with convolutional neural networks. IEEE Trans. Med. Imag. **39**(2), 499–513 (2019)
13. Ma, J., et al.: Loss odyssey in medical image segmentation. Med. Image Anal. **71**, 102035 (2021)
14. Maiseli, B.J.: Hausdorff distance with outliers and noise resilience capabilities. SN Comput. Sci. **2**(5), 358 (2021)
15. Roth, H.R., et al.: A new 2.5D representation for lymph node detection using random sets of deep convolutional neural network observations. In: Golland, P., Hata, N., Barillot, C., Hornegger, J., Howe, R. (eds.) MICCAI 2014. LNCS, vol. 8673, pp. 520–527. Springer, Cham (2014). https://doi.org/10.1007/978-3-319-10404-1_65
16. Seff, A., et al.: 2D view aggregation for lymph node detection using a shallow hierarchy of linear classifiers. In: Golland, P., Hata, N., Barillot, C., Hornegger, J., Howe, R. (eds.) MICCAI 2014, Part I 17. LNCS, vol. 8673, pp. 544–552. Springer, Cham (2014). https://doi.org/10.1007/978-3-319-10404-1_68

17. Seff, A., Lu, L., Barbu, A., Roth, H., Shin, H.-C., Summers, R.M.: Leveraging mid-level semantic boundary cues for automated lymph node detection. In: Navab, N., Hornegger, J., Wells, W.M., Frangi, A.F. (eds.) MICCAI 2015, Part II 18. LNCS, vol. 9350, pp. 53–61. Springer, Cham (2015). https://doi.org/10.1007/978-3-319-24571-3_7
18. Shandiz, A.H., Tóth, L., Gosztolya, G., Markó, A., Csapó, T.G.: Improving neural silent speech interface models by adversarial training. In: Hassanien, A.E., et al. (eds.) AICV 2021. AISC, vol. 1377, pp. 430–440. Springer, Cham (2021). https://doi.org/10.1007/978-3-030-76346-6_39
19. Shandiz, A.H., et al.: Neural speaker embeddings for ultrasound based silent speech interfaces. arXiv preprint arXiv:2106.04552 (2021)
20. Tóth, L., Shandiz, A.H.: 3D convolutional neural networks for ultrasound-based silent speech interfaces. In: Rutkowski, L., Scherer, R., Korytkowski, M., Pedrycz, W., Tadeusiewicz, R., Zurada, J.M. (eds.) ICAISC 2020, Part I 19. LNCS (LNAI), vol. 12415, pp. 159–169. Springer, Cham (2020). https://doi.org/10.1007/978-3-030-61401-0_16
21. Van der Heide, U.A., et al.: Functional MRI for radiotherapy dose painting. Magn. Reson. Imaging **30**(9), 1216–1223 (2012)
22. Yide, Y., Shandiz, A.H., Tóth, L.: Reconstructing speech from real-time articulatory MRI using neural vocoders. In: 2021 29th European Signal Processing Conference (EUSIPCO), pp. 945–949. IEEE (2021)

Digitizer: A Synthetic Dataset for Well-Log Analysis

M. Quamer Nasim[2], Narendra Patwardhan[1,2], Javed Ali[2], Tannistha Maiti[2], Stefano Marrone[1](✉), Tarry Singh[2], and Carlo Sansone[1]

[1] Dipartimento di Ingegneria Elettrica e delle Tecnologie dell'Informazione (DIETI), University of Naples Federico II, Via Claudio 21, 80125 Naples, Italy
{stefano.marrone,carlo.sansone}@unina.it

[2] Deepkapha AI Research., Street Vaart ZZ no 1.d, 9401GE Assen, The Netherlands
{mquamer.nasim,narendra.patwardhan,javed.ali,tannistha.maiti, tarry.singh}@deepkapha.com

Abstract. Raster well-log images are digital representations of paper copies that retain the original analog data gathered during subsurface drilling. Geologists heavily rely on these images to interpret well-log curves and gain insights into the geological formations beneath the surface. However, manually extracting and analyzing data from these images is time-consuming and demanding. To tackle these challenges, researchers increasingly turn to computer vision and machine learning techniques to assist in the analysis process. Nonetheless, developing such approaches, mainly those dependent on machine learning requires a sufficient number of accurately labelled samples for model training and fine-tuning. Unfortunately, this is not a straightforward task, as existing datasets are derived from scanned hand-compiled paper copies, resulting in digital images that suffer from noise and errors. Furthermore, these samples only represent images and not the digital signals of the measured natural phenomena. To overcome these obstacles, we present a new synthetic dataset that includes both images and digital signals of well-logs. This dataset aims to facilitate more effective and accurate analysis techniques, addressing the limitations of current methods. By utilizing this dataset, researchers and practitioners can develop solutions that mitigate the shortcomings of existing methods, ultimately leading to more reliable and precise results in interpreting well-log curves and understanding subsurface geological formations.

Keywords: raster log · digitization · well-log curves · dataset

1 Introduction

Production wells are drilled in economically viable oil or gas fields. During this process, well-logging is conducted to measure different rock properties along the well's depth using drilling tools. These measurements provide valuable information about lithology, porosity, fluid content, and textural variations within the

G. L. Foresti et al. (Eds.): ICIAP 2023 Workshops, LNCS 14365, pp. 97–105, 2024.
https://doi.org/10.1007/978-3-031-51023-6_9

formation. In the past, well-logging data was graphically represented, which had drawbacks like large size, memory requirements, and interference from gridlines. Raster logs, which are scanned copies of paper logs saved as image files, emerged as a cost-effective alternative to preserve well-log data in a future-proof manner [3]. These depth-calibrated raster images not only serve as a digital storage solution but also have the potential to establish a universal computer-readable format for legacy hardcopy data. Despite being often discarded after conversion to vector format, raster imaged well-logs hold significant value for a range of applications beyond resource exploration, such as environmental protection, water management, global change studies, and primary and applied research. By digitizing raster logs using appropriate software, geological subsurface modelling can utilize the digitized data effectively.

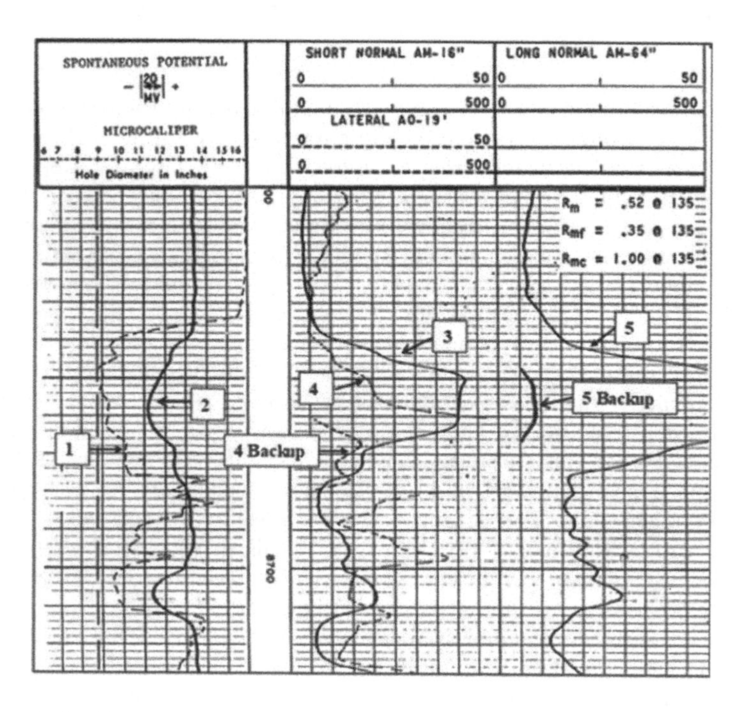

Fig. 1. An illustrative example of a raster well-log. Image taken from [2].

A typical raster log comes with a header, a body, and a footer. The header contains specific information about the well, such as the operating company, well location, and log type. The body is very often based on a standardized log grid, featuring three tracks: The first track is a depth track, while the remaining two tracks are known as "Tracks 2 and 3". The main log section, or graph, displays the depth vertically, while the horizontal scale represents the measurement scale, which can be linear or logarithmic. Inserts are placed throughout the graph to identify each curve, also referred to as traces, readings, or measurements.

Different line types, such as solid, long-dashed, short-dashed, or dotted, are used to differentiate between various measurements. The footer, i.e. the log's final part, includes tool calibrations conducted before and after the log, ensuring accuracy. An example of a raster well-log is reported in Fig. 1.

Dealing with old analog prints presents challenges, particularly in scaling the curves. Thus, a significant part of log analysis involves addressing scaling issues. To address these challenges, researchers are increasingly embracing computer vision and machine learning techniques as valuable tools in the analysis of well-logs. However, implementing these approaches, particularly those relying on machine learning, necessitates a substantial amount of accurately labelled samples to train and fine-tune the models effectively. Unfortunately, obtaining such labelled samples is not a simple task. Indeed, existing datasets are typically generated from scanned paper copies, which introduce noise and errors into the resulting digital images. Moreover, these samples only capture the visual aspects of the well-logs and do not encompass the digital signals representing the actual measured natural phenomena.

To overcome these obstacles, in this work, we introduce a new synthetic dataset encompassing both images and digital signals derived from well-logs. By combining visual representations and the actual measured signals, this synthetic dataset aims to enhance the effectiveness and accuracy of analysis techniques, addressing the limitations of current methods. Researchers and practitioners can leverage this dataset to develop solutions that alleviate the shortcomings associated with existing methods. Ultimately, this advancement will lead to more reliable and precise outcomes in the interpretation of well-log curves and the comprehension of subsurface geological formations.

The rest of the paper is organized as follows: Sect. 1 introduces the motivations supporting this paper; Sect. 2 describes the approach followed to realize the proposed synthetic dataset; Sect. 3 reports some analysis made to estimate the effectiveness of the proposed artificial dataset; finally, Sect. 4 draws some conclusions.

2 Generating Synthetic Well-Log Curves

It is customary in geophysics to employ synthetic curves in research studies as a preliminary step before validating them with actual data. Machine learning techniques, such as random forest, support vector machines (SVM) [1], and deep neural network-based prediction models [4], are utilized to reconstruct well log curves and generate synthetic logs. However, it is worth noting that the generation of these synthetic curves is specifically tailored to target fields and necessitates domain-specific knowledge and expertise. Instead, we intend to design a comprehensive framework capable of generating synthetic well log curves that capture the characteristics of diverse geological formations.

To generate such a synthetic dataset our hypothesis is centred around the notion that curves within a particular track exhibit random overlapping and wrapping factors. To generate synthetic curves, we consider parameters such

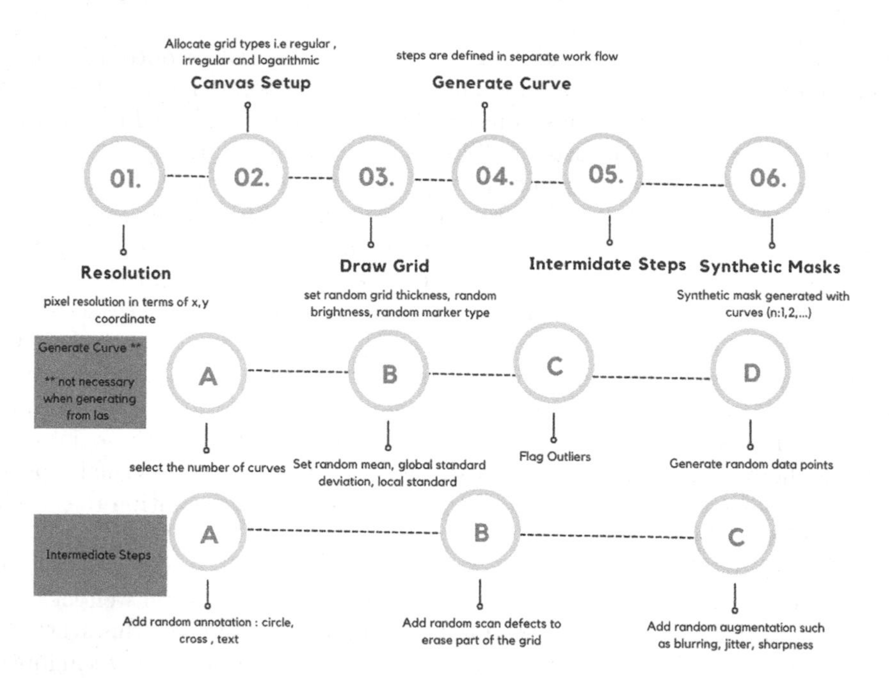

Fig. 2. Pipeline for synthetic dataset generation. It begins with setting the canvas resolution, drawing the grid on the canvas, selecting the number of curves that should be present in the synthetic well-log image and setting the statistics of the curve and generating the random data points. Finally, plot the generated data points on a canvas.

as mean, and standard deviation, also incorporating random noise. The dataset generated in this study comprises two well-log curves per image. The pipeline for generating synthetic curves consists of four main steps (Fig. 2), namely canvas preparation, curve generation, noise and annotation addition, and masks and log generation:

- In the **canvas preparation** step, we establish the resolution of the digital canvas and introduce grids to mimic the erosion present in PDF or image prints. Grids can be linear with a regular scale or logarithmic with a logarithmic scale. To closely resemble raster images, the thickness and brightness of the grid lines are randomly assigned. Additionally, masks are generated without gridlines;
- The **curve generation** phase involves determining the number of curves to be generated for a particular track. These curves exhibit local and global variations. Random mean values, global standard deviation, and local standard deviation are employed to emulate real well-log curves. Furthermore, a parameter defines the number of global variations, allowing for local variation in each segment and ten global variations across the entire curve. Additionally, a flag is utilized to introduce sudden changes in the curve's value, replicating behaviours observed in real well-logs;

- In the third step, we introduce various **synthetic defects** and annotations, resembling those typically observed in real logs due to scan defects. Augmentation techniques such as random blurring, random sharpness, and colour jitter are incorporated to increase the noise within the dataset. Random annotations, including circles, crosses, and random text, are also added to replicate geologists' markings on the logs;
- Finally, in the last step of the pipeline, all the curves that were generated in steps two and three are **merged and added** to the initially empty canvas. This process involves carefully aligning the curves within their respective tracks, ensuring that they are positioned accurately in relation to the depth scale. By combining the curves, we create a comprehensive representation of the synthetic well-log image. The curves' shapes, colours, and line styles are preserved, capturing the inherent characteristics and variations introduced during the curve generation phase. This merging process aims to simulate the appearance of real well-logs and provide a faithful representation of the geological formations and subsurface properties.

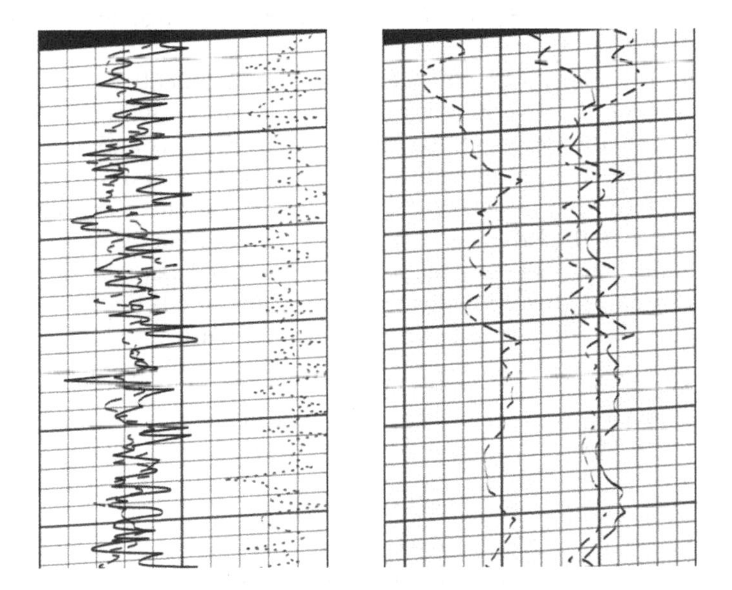

Fig. 3. Comparison between a generated synthetic well-log image (left) and a real one (right).

By successfully combining the generated curves onto the canvas, we achieve a cohesive representation that accurately reflects the synthetic well-logs' characteristics. This step is crucial in the overall process as it enables a holistic view of the synthetic dataset and facilitates the subsequent steps of the analysis pipeline. The resulting merged canvases become a complete and self-contained set of coherent well-log images, encompassing multiple curves within a single visual representation. These images are now ready for further analysis, interpretation, and evaluation. Researchers and practitioners can utilize these synthetic well-log images to develop and refine their analysis techniques, leveraging the generated dataset to train and fine-tune machine learning models or conduct comparative studies against real well-log data. By following the described procedure, we generated a total of 15014 images with two curves and 20000 images with three curves, respectively. Figure 3 reports a comparison between a synthetic and a real well-log image.

3 Dataset Analysis

To assess the effectiveness of the introduced synthetic dataset's effectiveness, we analyzed how well a model for well-log curve extraction performs when trained on a dataset consisting of real well-log curves and tested on our dataset. We used a well-log curve extraction model using state-of-the-art computer vision and machine learning techniques[1]. We trained on a dataset of 10000 images. The images consist of a single track that can either have 2 or 3 well-log curves generated from Log ASCII Standard (LAS) and raster image files obtained from Texas RRC data[2](Fig. 4) .

During the training phase, the model learned to identify and extract the curves by leveraging the diverse range of subtle variations in line styles, colours, and shapes that characterize well-log curves, while also accounting for noise and annotations typically present in real well-log images. After training, we evaluated the model's performance by measuring its accuracy in curve extraction on the test set. We compared (Table 1 the model's extracted curves with the ground truth data, quantifying metrics such as Dice [6] and Tversky [5] losses to assess its overall performance. Additionally, we visually inspected (Fig. 5) the extracted curves to ensure they aligned well with the ground truth.

Table 1. Performance analysis in terms of Dice and Tversky, over the considered test-set, aggregated by considering the MAE, MSE and R2.

Loss	MAE	MSE	R2
Dice	0.0326	0.0037	0.7373
Tversky	0.0305	0.0027	0.8079

Obtained results indicate that the well-log curve extraction model achieved high accuracy and robustness when applied to our synthetic dataset. The model successfully captured the main features and variations of the curves, accurately delineating their shapes and positions within the images. The precision and

[1] The reference to the used approach is hidden to respect the double-blind review stage.

[2] https://www.rrc.state.tx.us/.

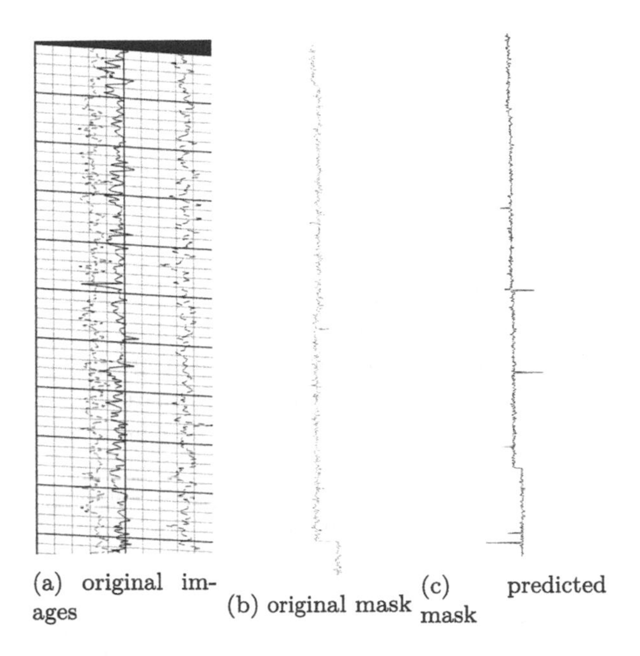

(a) original images (b) original mask (c) predicted mask

Fig. 4. Visual comparison of the segmentation results obtained by the considered machine-learning approach trained on a real dataset and tested on a synthetic test

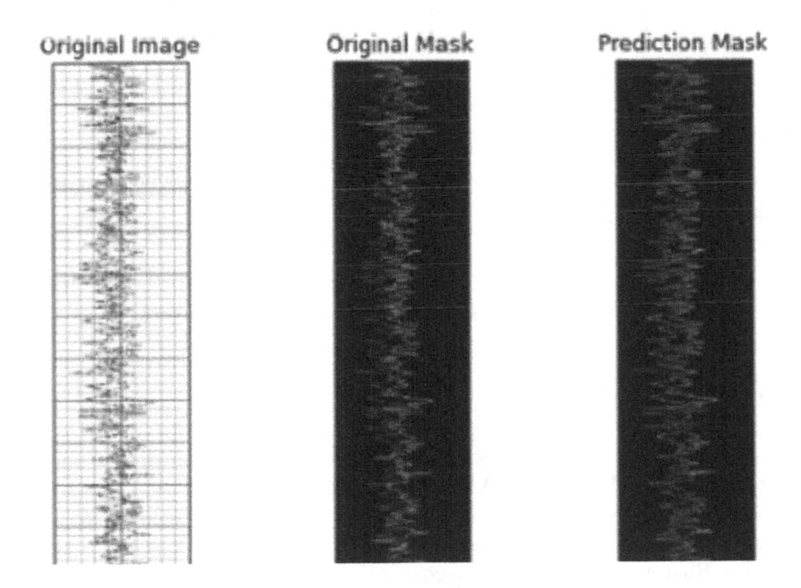

Fig. 5. Visual comparison of the segmentation results obtained by the considered machine-learning approach trained on a real dataset and tested on a synthetic test set.

recall scores were consistently high, indicating a reliable extraction of the well-log curves. This evaluation supports the use of the introduced synthetic dataset, making it a valuable resource for evaluating well-log curve extraction models.

4 Conclusions

In conclusion, this paper introduces a novel approach for generating synthetic well-log curves that is not specific to any targeted oil field. The developed methodology addresses the challenges associated with the availability of labelled data and the limitations of existing methods that rely on scanned hand-compiled paper copies. This approach enables the generation of synthetic curves that capture the complex characteristics of geological formations, including random overlapping and wrapping factors. The proposed pipeline, consisting of canvas preparation, curve generation, noise and annotation addition, and masks and log generation, provides a systematic framework for generating synthetic well-log images. By carefully incorporating parameters such as mean, standard deviation, and random noise, we ensure that the synthetic curves closely resemble real well-log data.

The generated dataset, comprising well-log images with multiple curves, is a valuable resource for training and evaluating analysis techniques in geophysics. Researchers and practitioners can utilize this dataset to develop and refine their algorithms, ultimately leading to more effective and accurate analysis of well-log curves and a better understanding of subsurface geological formations. Future work can involve expanding the dataset to include a broader range of geological scenarios and incorporating additional parameters to enhance the fidelity of the synthetic curves. Additionally, further investigations can explore the application of this approach in other domains of geophysics and its integration with advanced data analysis techniques.

Code base and dataset

The source code and the dataset are available at this url: https://www.earthscan.io/insights/enhancing-well-log-analysis-with-synthetic-datasets.

References

1. Akinnikawe, O., Lyne, S., Roberts, J.: Synthetic well log generation using machine learning techniques. In: SPE/AAPG/SEG Unconventional Resources Technology Conference. OnePetro (2018)
2. Bateman, R.M.: Formation Evaluation with Pre-digital Well Logs. Elsevier (2020)
3. Cisco, S.L.: Raster images offer low-cost well log preservation. Oil Gas J. **94**(51), 40–43 (1996)
4. Kim, S., Kim, K.H., Min, B., Lim, J., Lee, K.: Generation of synthetic density log data using deep learning algorithm at the golden field in Alberta, Canada. Geofluids **2020**, 1–26 (2020)

5. Salehi, S.S.M., Erdogmus, D., Gholipour, A.: Tversky loss function for image segmentation using 3D fully convolutional deep networks. In: Wang, Q., Shi, Y., Suk, H.-I., Suzuki, K. (eds.) MLMI 2017. LNCS, vol. 10541, pp. 379–387. Springer, Cham (2017). https://doi.org/10.1007/978-3-319-67389-9_44
6. Zhao, R., et al.: Rethinking dice loss for medical image segmentation. In: 2020 IEEE International Conference on Data Mining (ICDM), pp. 851–860. IEEE (2020)

CNN-BLSTM Model for Arabic Text Recognition in Unconstrained Captured Identity Documents

Nabil Ghanmi(✉), Amine Belhakimi, and Ahmad-Montaser Awal

AI & ML Center of Excellence, IDNOW, Rennes, France
{nabil.ghanmi,amine.belhakimi,montaser.awal}@idnow.io

Abstract. Optical Character Recognition (OCR) for Arabic text (printed and handwritten) has been widely studied by researchers in the last two decades. Some commercial solutions have emerged with good recognition rates for printed text (on white or uniform backgrounds) or handwritten text with limited vocabulary. In addition to being naturally cursive, the Arabic language comes with additional challenges due to its calligraphy resulting in a variety of fonts and styles. In this work, recent advances in recurrent neural networks are explored for the recognition of Arabic text in identity documents captured in the wild. The unconstrained captures bring additional difficulties as the text has to be first localized before being able to recognize it. Various pre-processing steps are introduced to overcome the difficulties related to the Arabic text itself and also due to the capturing conditions. The presented approach outperforms existing solutions when evaluated using a private dataset and also using the recent MIDV2020 dataset.

Keywords: Arabic Text Recognition · Identity Document · Convolutional Neural Network · Long Short-Term Memory · Connectionsit Temporal Classification · Character Error Rate

1 Introduction

In an economic environment that is increasingly being digitalized, it is becoming essential for companies to rethink their customer journey in order to offer them fast, reliable and ergonomic online services. This particularly concerns the customer onboarding process. For such a purpose, companies have turned to completely digital and remote onboarding as it is a flexible way to acquire new customers without meeting them physically in offices or stores. Typically, the remote access to a service or a product (e.g. telephone or internet contracts, banking, e-gambling, etc.) requires that the customer sends or uploads copies (photos or scans) of his identity documents (or any relevant documents). Then, the data in the document images should be automatically extracted and verified in order to know the customer identity and check the validity of the document as well as the conformity of the extracted data with that provided by the customer

G. L. Foresti et al. (Eds.): ICIAP 2023 Workshops, LNCS 14365, pp. 106–118, 2024.
https://doi.org/10.1007/978-3-031-51023-6_10

in the subscription form. This digital processes was initially enabled by image processing techniques and is now powered by deep learning based approaches.

To make that work properly, a full automatic reading system, from document classification and localization to text reading and verification, is required. Even that this problem has been widely studied and much progress have been made for some alphabet such as latin, reading arabic documents still far from being resolved. It is true that many steps of document reading system are generic and could be directly applied on arabic documents without much difficulty. Nevertheless, the text recognition (aka OCR) is language-dependent and applying, for example, latin OCR to arabic alphabet is not straightforward.

In this paper we focus on arabic text recognition in identity documents. We will present a deep learning based OCR that takes into account the arabic text specifities and the complexity of the identity document background.

2 Related Works

Optical character recognition (OCR) goal is to extract text contained in images (scanned documents or photos taken by a camera) and represent them with a standard encoding like ASCII. OCR systems are usually composed of text detection and recognition steps organized sequentially in most cases. The first step aims at locating text in the input images (scanned documents or photos taken by cameras). Every detected text is then used as an input to the text recognition step.

Most text recognition approaches use a segmentation-based transcription where the input image (containing text blocks) is segmented into text lines, then into words or patches and finally into characters. Many works were focusing on the segmentation task like [8,17] which are based on peaks detection.

Some other methods perform the transcription at word or sentence level. A text line is segmented into words using Hidden Markow Models (HMM) or Artificial Neural Network (ANN) models. The most recent approaches use Recurrent Neural Network (RNN) with Connectionist Temporal Classification (CTC) (also called sequence learning) [11] [15].

In [16], Yousfi et al. presented an Arabic video text recognition system. Features have been extracted from the input images using deep Belief networks and multi layer perceptron. A BLSTM network was used to map the sequence to characters, followed by a CTC output layer. In [4], the authors use a multidimensionnal LSTM and apply a dropout operation on its first layer. Their system was tested on the OpenHaRT (145.000 Arabic handwritten text) with an accuracy of 90.1%.

A combination of CNN and GRU (Gated Recurrent Unit, which is a type of RNN architecture) was used by [14] to recognize Arabic license plates numbers without segmentation. The system achieved an accuracy of 90%. A Similar approach was proposed in [12] where the authors built an end-to-end deep CNN

- RNN model to solve the text-based CAPTCHA images with distortion, rotation and noisy background. This model achieved 99% accuracy but on a fully generated data set.

On the industrial domain, a few OCR systems manage Arabic script recognition such as Sakhr OCR, ABBYY, Nuance, etc. As we are dealing with identity documents, a special focus has been laid on the commercial OCRs dedicated to identity document reading. In that context, we have identified ID reader as an OCR engine that extracts automatically information form several identity documents [5]. It can be applied on 300 dpi images with various fonts, sizes and resolution. Another commercial solution that deals with Arabic identity documents is sky IDentification which is based on deep learning technique.

All these existing OCRs (as well as some others) were evaluated on our private data set of many thousands of fields extracted from various Arabic identity documents. The obtained results were unsatisfactory and did not exceed 43% of accuracy.

Since we focus on deep learning approaches, data sets are a very important aspect for OCR systems. Like for any deep learning application, a good amount of data is required in order to learn the model how to recognize the text in images. Most of the previous works showcase a high accuracy but on private data sets without a fair comparison. This is due to the lack of public data sets and standard bench-marking. In the following, we try to outline the most used Arabic data sets.

Several public data sets are available such as KHATT [6], APTI [13], APTID/MF [2] and PATDB [3]. Even if these data sets present a large variability in terms of capture conditions, image quality, text properties and lexicon, the identity document challenges and characteristics still not considered. In fact, an identity document has a highly textured background, a constrained layout and also a very large lexicon that may contains words not necessarily belonging to the Arabic vocabulary (proper names). Furthermore, this kind of document requires a very accurate recognition as it contains sensitive data.

For our knowledge, few identity document data sets are publicly available. Moreover, they generally contain partial information, or synthetic data. The most known data sets are LRDE IDID [9], BID [7], MIDV [1]. The LRDE IDID contains a few quantity of document samples, which is not useful for significant benchmarking and deep analysis of the processing methods. The BID dataset is composed of synthetic images that are generated by writing text field values on automatically masked document regions, which leads to several imperfections making the data set very different from a real one. As for MIDV data set, even if it contains a lot annotated images, it presents the disadvantage of lack of variability as for each document type, only five different samples are collected under slightly variable conditions. This data set might be useful for document localization, classification or tracking but it is less useful for OCR evaluation as it contains few variable data. Nevertheless, we used a part of this data set in our evaluation.

3 Problem Statement

3.1 Identity Document Reading

Identity documents, such as ID cards, passports, driving licences, resident permits, etc. are a particular type of documents that have a precise design and several security features. They are usually issued by governments to define, prove and verify the holder's identity. The content of these documents is composed of static fields representing the document template and variable fields containing its holder personal data such as his first name, last name, birth date, birth place etc. Extracting these information is of great interest as it allows an easy authentication of the document holder, a robust identity verification and an automated form filling. Nevertheless, this task is very challenging due to the complex background texture that makes text localization and recognition very hard. In fact, the graphical components of the texture (several geometric forms with various styles and colors, complicated patterns, etc.) can be easily confused with textual components and thus disturb the text extraction and recognition system.

3.2 Arabic Text Recognition

Arabic text holds additional difficulties to recognition systems due to its specific style. It is written from right to left and consists of 28 basic letters that connect to each others in order to form words in printed and handwritten text. Thus, Arabic text is always cursive even in its machine-printed form. Most of these letters has 4 different forms according to their positions in the word (beginning, middle, end, or isolated), as shown in Fig. 1.

Fig. 1. Different forms of the letter *'seen'* according to its position: (a) isolated, (b) at the end, (c) in the middle and (d) in the beginning of the word.

Some of these letters do not connect from the left (as for example *'raa'*, *'zaay'*, etc.); and thus they have only two forms (isolated and end). This property produces the piece of Arabic words (PAWs) where one word can be composed in one or more sub-words (PAWs). In addition, many groups of letters share the same body form and are different only by the diacritics. Figure 2 shows 3 examples of this case *'khaa'*, *'haa'* and *'Jiim'* (Fig. 2a), *'raa'* and *'zaay'* (Fig. 2b) and *'baa'* and *'taa'* (Fig. 2c). Other diacritics are used for the vocalization of the text, but they are not essential for the text understandin

4 Proposed Method

The current work was carried out as a part of an automatic system of identity document reading and verification. The full system can be seen as a pipeline composed of several steps. Given, an input image in the wild, the document is firstly localized and classified. Then, the obtained document image is corrected and the text is localized based on a well defined reference layout. Finally, comes the OCR step that takes as input the text bounding boxes extracted by the previous step.

Fig. 2. Example of 3 groups of letters having same structure and different diacritics.

In this section, the proposed OCR for Arabic language is detailed. We will present the model architecture as well as the training parameters. The data pre-processing which is a crucial step for such OCR will also be described.

4.1 Architecture Description

In this work, we have studied two architectures that will be referred, in the following, as basic and optimized respectively. The basic architecture is composed of 3 parts. A feature extraction part that computes the input image features and present them as a reduced space matrix. A sequence modeler part that takes the output of the previous part and outputs an embedding space. The last part is a Connectionist Temporal Classification (CTC) layer that classifies the input into the corresponding letter. More details about this architecture are in Table 1a.

Features Extractor. It is a convolutional neural network composed of three blocks. Each block is composed of a convolution layer with a 3×3 kernel size followed by a max-pool layer with a kernel size of 2×2. The two first max-pool layers have a stride $(2, 2)$ whereas the 3rd max-pool layer has a stride $(2, 1)$ and a padding $(0, 1)$. The number of filters used in the convolution layers within the 3 blocks are respectively 64, 128 then 256. Only the 3rd max-pool layer have a s The input of this feature extractor is $[1 \times 864$ x $48]$ which is an image of the field containing the text. Its output is a matrix with a size of $[256 \times 6$ x $217]$. Even if the extraction part is fairly light weight, the experiments show that the extracted features allows the sequence modeler to easily decode the field image.

Sequence Modeler. We used two layers of bidirectional LSTM each of which contains 240 neurons. The input of this LSTM is the extracted feature vector that is flatten to a vector of size $[256 \times 1302]$. The bidirectional property makes the network able to predict a certain letter based on the sequence from the beginning and from the end. This allows a better reading, exceptionally when the image is of low quality. The output of the sequence modeler is then mapped to the number of alphabets with a linear layer.

CTC. This layer allows the network to output the sequence into a classification soft-max layer. Then a beam search algorithm is performed to transform the output into its corresponding text.

Table 1. (a) Basic and (b) Optimized architecture details.

Layer	Params	Output size
CNN_1	64,3,3	64,48,864
Relu		64,48,864
maxpool_1		64,24,432
CNN_2	128,3,3	128,24,432
Relu_2		128,24,432
maxpool_2		128,12,216
CNN_3	256,3,3	256,12,216
Relu_3		256,12,216
maxpool_3		256,6,217
batch_nor_1		256,6,217
flatten layer		256,1302
blstm x 2	240	240,1302
out_embedding	n_classes	1302,n_classes

(a)

Layer	Params	Output size
CNN_1	64,3,3	64,48,864
Relu		64,48,864
maxpool_1		64,24,432
CNN_2	128,3,3	128,24,432
Relu_2		128,24,432
maxpool_2		128,12,216
CNN_3	256,3,3	256,12,216
Relu_3		256,12,216
maxpool_3		256,6,217
batch_nor_1		256,6,217
map_seq		217,64
blstm x 2	240	240,217
out_embedding	n_classes	217,n_classes

(b)

Optimized Model. In order to make the basic model lighter and faster, the initial architecture is used, while reducing the number of weights by replacing the flatten layer with a mapping linear layer. In the basic model we directly flatten the output of the convolutions to generate a vector of size $[256 \times 1302]$. This flatten layer is replaced in the optimized model by a linear layer, which reduces the size to $[217 \times 64]$. Then, the obtained matrices are reshaped to get the appropriate output vector. Using the same number of neurons in the LSTM, the number of weights is drastically reduced while preserving the accuracy of the models. More specifically, the number of weights for the basic model is around 1.6M while for the optimized model, it is around 800K. The optimized network is then 3 times faster than the basic model (the processing time is reduced from 0.23s to 0.07s). More details about the Optimized model are shown in Table 1b.

4.2 Training Parameters

Both architectures (Basic and Optimized) were trained using the PyTorch framework. The RMSprop optimizer was used with a triangular function to assign the learning rate with values between 0.01 to 0.001. The models were trained on a GTX 1080Ti (12G).

(a) (b)

Fig. 3. (a) Example of field image extracted from an identity document. (b) The horizontal mirroring of field image in (a).

4.3 Data Pre-processing

- **Field image mirroring.** It is worth remembering that a text reading system is generally composed of two main steps: text detection and text recognition. To cope with the complexity of the identity document background and the variability of the length of the text fields, the first step was designed to return bounding box with theoretical maximum width defined in the knowledge base (see Fig. 3a).

  Thus, the extracted images have generally empty regions at their left sides, which may causes some problems for training the OCR models. In fact, during the training, the OCR model tries to align the ground truth with the image starting from left to right which leads to aligning the first characters of the ground truth with the empty region and thus disturbing the training. To cope with this problem the image is horizontally mirrored before being used for the training. In addition to placing the text on the left side of the image, this mirroring reverse the order of the characters in the image which allows a best alignment of these characters to their correspondent in the text ground truth. Figure 3b shows an example of an horizontally mirrored image.
- **Ground truth string reversing.** It is known that Arabic text is right-aligned, i.e. each text field starts at the right side of its dedicated zone. Nevertheless, in some documents, mainly passports, some Arabic fields are left-aligned and they finished at the left side of their dedicated zones (see Fig. 4).

Fig. 4. Left aligned text field.

Unlike the right-aligned fields, we do not need to mirror the image for this kind of fields as the empty area is on its right part. But we need to reverse the ground truth text for the training in order to ensure that each character is correctly aligned with its pixels in the image. In fact, Arabic text is written from right to left and the image is traversed form left to right by the OCR model. Therefore, reversing ground truth is necessary.

5 Experiments

5.1 Dataset

A private data set composed of Algerian, Moroccan and Tunisian identity documents (ID cards and passports) is used for experimenting the proposed model. It is collected from a real production flow and used internally for research purposes. Some statistics on this data set are available in the table 2.

Table 2. Private data set description.

Document class	Nb samples
DZA_ID	6378
DZA_P	14402
MAR_ID	15596
MAR_P	13402
TUN_ID	2632
TUN_P	14924
Total	67334

For each document we extract field images. Each field contain one line of text (First Name, Last Name, ID Number, etc.). The Fig. 5 shows two examples of text fields extracted from a Moroccan identity card.

(a) (b)

Fig. 5. Examples of field images extracted from documents.

The extracted field images are then transformed to gray scale. To solve the problem of variable size on the network input, the images are padded to have size (1800×100) and then resized to (864×48). This resizing make the network lighter and faster during the inference, which is very important in an industrial context.

The data set is split into train, validation and test sets containing respectively 80%,10% and 10% of the total number of samples.

5.2 Evaluation Metrics

- **Character Error Rate (CER).** This metric is based on Levenshtein distance that measures the similarity between two strings. This distance corresponds to the number of deletions, insertions, or substitutions at character-level required to transform the ground truth text (aka reference text) into the OCR output. For each field, the CER is computed using this formula:

$$CER = \frac{D + I + S}{N}$$

where D, I, S correspond to the number of deletions, insertions and substitutions respectively and N corresponds to the total number of the characters in the reference text.
- **Accuracy.** This metric is defined as the number of the text fields that are correctly read by the OCR divided by the total number of text fields.

$$CER = \frac{\text{nb correctly read fields}}{\text{nb all fields}}$$

5.3 Model Training

Data Augmentation. As image labeling is a tedious task, we start training the model using a relatively small data set, then we increased progressively the number of annotated document. The behaviour of the model accuracy in function of the training data size is studied, in order to estimate the optimal data set size (Fig. 6). In addition to the use of manually annotated real data, artificially generated data is used. The artificial images were created using a randomly cropped background from real documents on which a text was written with a random blur and distortion to simulate more realistic data. It is worth noting that the model

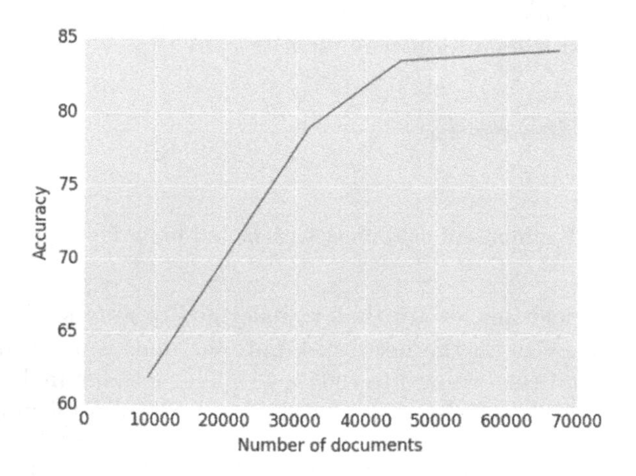

Fig. 6. The effect of the data set size on the accuracy of the model.

generalizes less well when it is trained mostly on generated data than when it is trained using real data. Therefore, we favor using real data, and we serve of artificial data mainly to cope with the unbalanced data problem by adding more samples of the less represented characters in the initial real data set.

Generic vs Specific Model. As we deal with multiple identity document classes (different versions of ID cards, passports, driving licences, etc. from various countries), two different strategies to train the proposed model arise:

- Generic training. This strategy aims to make the model generic, i.e. applicable for several identity document classes, by training it with labelled data extracted from various classes. This is particularly important to reduce the cost of data labelling since a unique data set is required. It is also adapted to deal with the cases of documents for which there is not enough data to use for a specific training.
- Specific training. This strategy consist of training specific model for each document class using labelled data extracted only from this particular class. The main advantage is that such a model will be more accurate since it will encounter less variability in terms of text background, font and styles. Nevertheless, a lot of data from each document class should be labeled, which is tedious and time consuming. Moreover, even if the training is generally straightforward, some specific tuning could be required.

We start our experimentation by trying the generic strategy. During the training, the model achieves an accuracy of about 73% on the validation data set. However, this accuracy has decreased on the test data set, mainly on the document classes that were not present in the training set and only 61% of accuracy is obtained. For the document classes that were present in the training data set, the accuracy on the test set remains comparable to the training accuracy and it varies between 66% and 73% depending on the document class complexity and its frequency in the training set. Based on these results, the generic model is kept for the document classes that do not have enough training data. For other classes, specific models are built.

5.4 Model Evaluation

Once trained, the proposed models are evaluated on two test data sets:

- Private data set: all the documents in this data set belong to classes that were represented in the training data set.
- Public MIDV data set: contrary to the private data set, all the documents in this data set belong to classes that were not seen during the training.

Results on the Private Data Set. Both versions of the proposed model are tested on the private test set and the obtained results are summarized in the Table 3. The optimized model is quite similar in term of performance to the basic model, but it is much faster in inference time. The results show that the generic model achieves comparable accuracy to the specific models for the document classes that have few training samples (TUN_ID and DZA_ID) and the model that presents a more complex background (TUN_PA) than the other classes.

Table 3. Performance of the basic and optimized model on the private test set.

	Basic model		Optim model	
	Acc.(%)	CER. (%)	Acc. (%)	CER. (%)
DZA_ID	**72**	**6.33**	71	7.02
DZA_P	86	4.15	**87**	**3.98**
MAR_ID	**78**	**5.9**	77	6.14
MAR_P	87	3.72	**87**	**3.42**
TUN_ID	**71**	**8.1**	69	8.3
TUN_P	**70**	**7.8**	70	7.9
GENERIC	**69**	**8.3**	68	9.4

These results are compared to those obtained using easyOCR which is a ready-to-use tool based on the work published in [10], as well as two commercial solutions: Nuance and Google OCR. We also evaluate a private OCR developed by a company working on KYC solutions in Arabic countries (we will use the name *"OCR for Arabic KYC"* to designate this OCR). The table 4 shows that the best OCR among those we evaluated achieves an accuracy of about 43%, which proves that reading Arabic text within identity documents is very challenging. The model we proposed is significantly more efficient and exceeds this best OCR by 26 percentage points. This can be explained by the fact that our OCR is specialized on identity documents and has learned how to cope with their specificity, mainly their background complexity.

Table 4. Our OCR outperforms 4 OCRs that we evaluated during our study.

OCR	**Accuracy** (%)
Our OCR	**69**
easyOCR	19.7
Google OCR	20.4
Nuance	5.0
OCR for Arabic KYC	43.1

Results on the Public MIDV Dataset. Two examples of fields extracted from this data set are shown in Fig. 7. As we already said, this data set contains a small number of documents with slightly different angles. In order to have more field images for our test, an augmentation procedure is applied on the existing fields by generating random noise, blur and geometrical distortion. The generic model achieves an accuracy of 61% and 60% using the basic and the optimized version respectively. This low performance can be explained firstly by the fact that these document classes are not seen during the training. Secondly, the quality degradation by adding blur and noise made some cases hard to read.

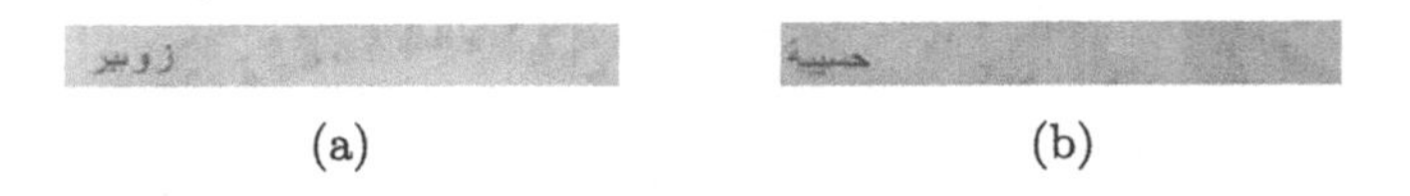

(a) (b)

Fig. 7. Two Examples of field images extracted from MIDV dataset.

6 Conclusion

In this paper, we show that a CNN-BLSTM model, originally designed for Latin text recognition, can be adapted to Arabic script which has challenging characteristics. This adaptation aims to align an input text line image with its ground truth text, during the training of the CNN-BLSTM. Two main operations are proposed, image mirroring and ground truth string reversing, in order to align the image pixels to their correspondent characters. The presented model was experimented on a large private data set and a small public one. The obtained results show that it outperforms several famous commercial OCRs. This can be explained by the fact that our model is dedicated to the identity documents contrary to the other OCRs which are generic and more adapted to full page text that are less challenging. We plan, for the next step, to explore other deep learning models based on transformer architectures. Such models are well adapted for both image understanding and character level text generation, and they have shown very good results on several text recognition problems.

References

1. Bulatovich, B.K., et al. "MIDV-2020: A comprehensive benchmark dataset for identity document analysis". In: 46.2 (2022), pp. 252–270
2. Al-Hashim A.G., Mahmoud, S.A.: Benchmark database and GUI environment for printed Arabic text recognition research. In: WSEAS Trans. Inf. Sci. Appl. 7.4 (2010), pp. 587–597
3. Jaiem, F.K., et al. "Database for Arabic printed text recognition research". In: ICIAP. 2013, pp. 251–259
4. Maalej, R., Tagougui, N., Kherallah, M.:Online Arabic handwriting recognition with dropout applied in deep recurrent neural networks. In: 12th IAPR DAS. IEEE. 2016, pp. 417–421
5. El-Mahallawy, M. S. M.: A Large scale HMM-based omni front-written OCR system for cursive scripts. In: Ph.D. thesis, Cairo University. 2008

6. Mahmoud, S. A., et al.: "KHATT: an open Arabic offline handwritten text database". In: Pattern Recognition 47.3 (2014), pp. 1096–1112
7. Ngoc, M., Fabrizio, J., éraud, T.G.,: Saliency-based detection of identy documents captured by smartphones. In: 13th DAS. 2018, pp. 387–392
8. Ramdan, J., Omar, K., Faidzul. M.: A Novel method to detect segmentation points of Arabic words using peaks and neural network. In: IJASEIT 7.2 (2017), pp. 625–631
9. de Sá Soares, A., das Neves Junior, R. B., Bezerra, B.L.D.:BID Dataset: a challenge dataset for document processing tasks. In: 18th Conference on Graphics, Patterns and Images. SBC. 2020, pp. 143–146
10. Shi, B., Bai, X., Yao. C.: An End-to-End Trainable Neural Network for Image-based Sequence Recognition and Its Application to Scene Text Recognition. In: CoRR abs/1507.05717 (2015). arXiv: 1507.05717
11. Shivakumara, P., et al.:"CNN-RNN based method for license plate recognition". In: CAAI Transactions on Intelligence Technology 3.3 (2018), pp. 169–175
12. Shu, Y., Xu, Y.: End-to-End Captcha Recognition Using Deep CNNRNN Network. In: IEEE 3rd IMCEC. IEEE. 2019, pp. 54–58
13. Slimane, F., et al.: A new Arabic printed text image database and evaluation protocols. In: 2009 10th International Conference on Document Analysis and Recognition. IEEE. 2009, pp. 946–950
14. Suvarnam, B., Ch, V.S.: Combination of CNN-GRU model to recognize characters of a license plate number without segmentation. In: 5th ICACCS. IEEE. 2019, pp. 317–322
15. Yousef, M., Hussain, K.F., Mohammed, U.S.: Accurate, Data-Efficient, unconstrained text recognition with convolutional neural networks. In: CoRR abs/1812.11894 (2018). arXiv: 1812.11894
16. Yousfi, S.: Embedded Arabic text detection and recognition in videos. PhD thesis. Université de Lyon, 2016
17. Zeki, A., Zakaria, M., Liong, C.: The Use of Area-Voronoi diagram in separating Arabic text connected components. In: 3rd ACEA. 2007, pp. 251–288

Advances in Gaze Analysis, Visual attention and Eye-gaze modelling (AGAVE)

Detection and Localization of Changes in Immersive Virtual Reality

Manuela Chessa, Chiara Bassano, and Fabio Solari(✉)

Department of Informatics, Bioengineering, Robotics, and Systems Engineering, University of Genoa, Genoa, Italy
{manuela.chessa,fabio.solari}@unige.it

Abstract. Immersive visualization, i.e. the presentation of stimuli, data, and information with head-mounted displays and virtual reality (VR) techniques, is nowadays common in several application contexts. For effective use of such setups, it is worth studying if the attentional mechanisms are affected (improved or worsened) in any way, or if human performances in detecting changes are similar to what happens in the real world. Here, we focus on assessing the Visual Working Memory (VWM) in VR by using a change localization task, and on developing a computational model to account for experiment outcomes. In the change localization experiment, we have four factors: set size, spatial layout, visual angle, and observation time. The results show that there is a limit of the VWM capacity around 7 ± 2 items, as reported in the literature. The localization precision is affected by visual angle and observation time ($p < 0.0001$), only. The proposed model shows high agreement with the human data ($r > 0.91$ and $p < 0.05$).

Keywords: Immersive Virtual Reality · Visual Attention · Visual Working Memory · Saliency Map · Computational Model

1 Introduction

The growing interest in using immersive visualization devices, e.g., head-mounted displays (HMDs) for Virtual Reality (VR) applications, changed how users interact with the environments, adding new forms of stimuli and data presentation and manipulation. The main difference between standard displays (monitors, smartphones, or tablets) and immersive devices is that in the latter users are surrounded by 360°, visual, 3D, and interactive information. On the one hand, this is similar to what happens in the real world. On the other hand, such information may be manipulated and enriched via software. This is one of the opportunities given by VR applications. However, the implications on human visual perception might be explored. In particular, it is worth analyzing if users can exploit all the potential localizations of visual information, and if the spatial arrangements in immersive VR affect perceptual mechanisms, e.g. the Visual Working Memory (VWM).

G. L. Foresti et al. (Eds.): ICIAP 2023 Workshops, LNCS 14365, pp. 121–132, 2024.
https://doi.org/10.1007/978-3-031-51023-6_11

VWM is an apparatus committed to the active maintenance of visual information for ongoing tasks. To study VWM, in the literature, we find reductionist approaches and real-world or natural behavior paradigms [14]. Reductionist approaches are landmarks in cognition and perception research, although most works are based on 2D images or videos displayed on conventional screens with a visual angle of 30°–40°. VWM is quantified in terms of items: its capacity could be around 4 [4,6] or 7 ± 2 [11,21] items, depending on the task. A common task for assessing the VWM is the *change detection* [14]. The change detection can be studied by using the *one-shot* paradigm [23]: the original and modified images are presented once, separated by a blank screen (or retention mask), and participants have to answer within a certain time. Change Blindness (CB) is the failure to notice an expected or unexpected difference between two very similar images when there is a brief break between them. The real-world paradigm is used by most of the papers in the VR literature, e.g. they consider spatial representation and navigation of a virtual environment, both indoor and outdoor, [12,13,20]. Though studies using the reductionist approach in VR exist, e.g. visual search [15] and cueing [24].

Descriptive models of VWM can be commonly found in the literature, e.g. [2,8,16,17]. Since visual saliency might affect working memory storage capacity, an interest in understanding its neural basis has raised, e.g. [3,5]. Saliency models can also be used to assess the effectiveness of visualizations, e.g. [19,22]. In [28] the authors presented and implemented in a coherent way several methods for the computation of the saliecy map: AIM (Attention by Information Maximization), AWS (Adaptive Whitening Saliency), CAS (Context Aware Saliency), CVS (Covariance-based Saliency), DVA (Dynamic Visual Attention), FES (Fast and Efficient Saliency), GBVS (Graph-Based Visual Saliency), IKN (Itti-Koch-Niebur Saliency Model), IMSIG (Image Signature), LDS (Learning Discriminative Subspaces), QSS (Quaternion-Based Spectral Saliency), RARE2012 (Multi-scale Rarity-based Saliency Model), SSR (Saliency Detection by Self-Resemblance), and SUN (Saliency Using Natural statistics). In [1], the authors developed a paradigm for evaluating the VWM in immersive visualization. Moreover, they developed an image-based computational model based on visual saliency for mimicking the human behavioral data of VWM. Their results confirmed the existence of a limit of VWM capacity of around 7 $\pm$ 2 items, as found in the literature based on the use of 2D videos and images. They also found the visual angle (VA) and the observation time influence performances. Indeed, with VA enlargement as it happens in HMDs, participants need more time to have a complete overview of the presented stimuli. Moreover, their model shows a good agreement with the results of the experiment.

In our paper, similarly to [1], we analyze VWM in immersive VR by using an experiment (replicating the one-shot change detection task [23]), in which participants are required to detect a change in the VR environment, and to indicate the position of the change. Specifically, we consider two 3D layout arrangements: the vertical one, where the objects are at the same distance in front of the observer, as on a wall, and the horizontal one, where the objects' distance with respect to the observer changes, as on a table. We can control the 3D position of the objects and their angular position with respect to the observers. Although, in principle, we can put objects all around the observers, completely exploiting the 3D environment, i.e., all the 360-degree scene around them, we focus on three visual angles (VAs) subtending stimuli presentation space: 40, 80, and 120°. It is worth noting the vertical layout with 40 deg of VA (though the view is 3D) replicates the standard 2D experiment of the literature, thus providing a baseline for the interpretation of our results.

Moreover, we extended the computational model presented in [1] to account for the outcomes of our experiment: specifically, we model both the detection of the change and the error in the pointing task. As input to the model, we use the saliency maps of the views of the virtual environment by showing that the model can mimic human performance.

2 Experiment and Computational Model

2.1 Task

To recreate the one-shot change detection paradigm, we show a memory array of items for an observation time T1 (300 ms, indicated by S, or 900 ms, indicated by L), then a gray canvas fills the full headset field of view (FOV) for an inter stimulus interval (ISI) of 500 ms, and lastly a test array for a period T2 of 5 s, as illustrated in Fig. 1 (right). During the test, the original collection of spheres (i.e. the items) is shown with or without any change, such as adding or removing a sphere, or no change is made (control condition). Once the participant has responded or the answer counter has ended, the test array is removed and a new scene is displayed after a 2-second inter trial interval (ITI).

In the *change localization* task, participants are asked to answer the question *"Where was the change if any?"* in three ways: localizing the position of the changed item through raycasting; pressing the "Equal" button if they have not perceived any change; waiting 5 s, if they have perceived a change but can not indicate where. Because pressing the controllers' trigger button causes a natural oscillation of the wrist, the raycasted position is acquired on release for improved measurement stability.

The virtual environment (VE) is simple to avoid interest prioritization. Items, i.e. the blue spheres, are randomly distributed on the vertical (indicated by V) and horizontal (indicated by H) surfaces (see Fig. 1 left and middle) on an invisible grid composed of $5° \times 5°$ bins. Their radius is 2.5 cm, to occupy around 4° of the VA at a distance of 70 cm, thus fitting the bins without overlapping. The observer is 70 cm from the vertical surface and 55 cm above the table. The far edge of the table is 220 cm from the observer.

We measured VWM capacity at the variation of four different factors: the set size (4, 6, 8, 10, 12 items), the distribution of the objects, i.e. spatial layout (vertical or horizontal) and VA (40, 80 and 120°), and the observation time (300 and 900 ms).

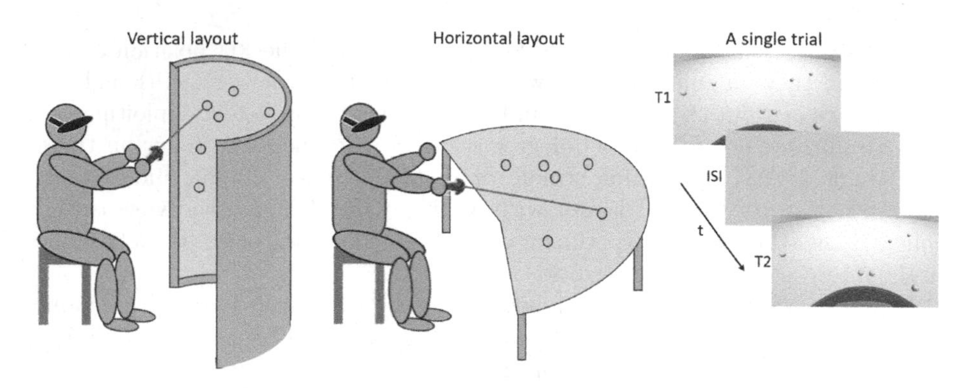

Fig. 1. The setup for the vertical (left) and horizontal (middle) layout. The participant's view of a single trial of the experiment (right).

2.2 Measures

During task execution, we acquire the given answers and the raycasted positions, the spheres' distribution, and the head positions and rotations. We discarded 1.2% of trials, i.e. participants who did not answer on time in the majority of their trials.

Based on the responses for each participant and experimental condition, the hit rate (HR) and false alarm rate (FAR), which are derived from the signal detection theory, are computed [1]. They are common metrics for assessing performance in a one-shot change detection task. FAR is calculated as the number of times participants perceive a change in the control condition (FP) divided by all the trials without any change (FP+TN), HR considers the number of times people accurately report a change (TP) divided by all trials with change (TP+FN).

Finally, we have analyzed the pointing error for each target position, calculated as the Euclidean distance between the center of mass of the raycasted positions and the real target. We have considered only the subset of data with HR exceeding the threshold value of 0.75 HR, in order to remove noisy samples [7,11].

2.3 Apparatus and Procedure

For the experiment, we used an HTC Vive and an Alienware Aurora R5. The virtual environment was developed by using Unity.

The experiment is composed of 1080 trials: the trials were grouped into 12 blocks of 90 trials, and we split the experiment into 3 sessions composed of 4 blocks, with a minimum inter-session break of half a day. The blocks last in a range of 6–13 min. We randomized the experimental runs within subjects and the block presentation order between subjects.

2.4 Participants

Twenty-one participants accomplished the *change localization* experiment (age range 20–36, 25.3 ± 3.7 years), by completing 1080 trials each. They were all students, PhDs, and researchers at the University of Genoa and had to sign an informed consent. They all reported having normal or corrected-to-normal vision and no deficit in stereo vision.

2.5 Computational Model

We aim to model the processing involved in the VWM. Such processing could exploit the information embedded in the saliency map: specifically, in the difference between the saliency map of the first view and the saliency map of the second view of our experiment [1]. If in this difference there is a residual saliency, such a residual saliency might drive the detection of the change in the observed views. It is worth noting that the proposed computational model processes the HMD views, as observed by a participant in our experiment.

We employ all the methods presented in [28] as front end of our model. Although bottom-up saliency methods may be appropriate in simple scenes such as the one we explore here, the model's performance may be influenced by more complex scenarios. In such a case, using top-down information could be advantageous.

However, the change localization might be influenced by the instruction to localize changes, potentially resulting in a visual search task. As reported in the literature, saliency maps are only an important factor in memory tasks and not in search tasks [9,10,26,27]. It is worth noting that our analysis reveals that there are no significant differences between the results of the change detection experiment in [1] and the change localization experiment carried out in this paper. Even though a change localization experiment may result in a visual search task, asking participants to localize the change rather than simply detect it has no effect on their ability to detect a change, since results are not statistically different. Thus, we can consider saliency maps as the front-end of our model.

We have computed HR and FAR as in [1]: the change detection might be measured as the absolute difference between the saliency maps of the two views. The entire saliency map activity may take into consideration how many objects are present in the view, thus making it possible to utilize it to emulate human behavior as the number of objects rises. To account for the uncertainty in human judgments, we also take into consideration the noise that exists in neural processes.

Instead, we have specifically devised how to model the pointing precision, i.e. the distance between the center of mass of the distribution of the raycasted positions, pointed by the participants, and the target position.

To model the error in the pointing task, first, we consider the target position and the salient regions in the difference map $D(x, y) = S_1(x, y) - S_2(x, y)$ (i.e. the difference between the saliency maps of the first and the second view). Since several residual salient regions can be present in $D(x, y)$, we discard the negligible salient regions by thresholding $D(x, y)$ (the value is 0.05 in our simulations). We compute the distance d_{COM} of the center of mass of the residual salient regions with respect to the target position. The smallest distance d_{COM} represents a measure of the modeled pointing precision. However, to take into account the uncertainty for a human subject to judge the position of the center of mass, we consider different components. The difference d_{area} between the area of the target and the area of the selected residual salient region (i.e. the one with the smallest distance d_{COM}). The FAR might play a role in accounting for human error in pointing, thus we consider it in the modeled localization error. Moreover, we should also consider the inherent noise present in the human neural processes and, specifically in this experiment, in the motor side of the pointing task.

The modeled localization error (Loc_M) can be described by

$$\begin{aligned} Loc_M &= c_1 \left(w_1 \, d_{COM} + w_2 \, d_{area} \right) \\ &\quad \max(S_1(x,y) - S_2(x,y))^{\gamma} \left(\sum_x \sum_y S_1(x,y) \right)^{\delta} + n_1, \end{aligned} \tag{1}$$

where w_1 and w_2 are the weights of the linear combination; c_1 is a normalization term to limit the modeled error in the range of human error; γ and δ are static non-linearity; n_1 is the noise from a normal distribution, with a mean of zero and its standard deviation is a fraction of the average localization error. The specific values of the model parameters employed in our simulations are: $w_1 = 1$, $w_2 = 5$, $\gamma = 3$, $\delta = 2$, and $n_1 = 0.4$.

3 Results

3.1 Change Localization Experiment

For each experimental condition, we calculated the average HRs and the FARs. As shown in Fig. 2 the HR tends to decrease as the number of items increases, while the FAR increases. N-ways ANOVA statistical analysis confirms a significant effect of set size over HR ($F(4, 1245) = 33.18$, $p < 0.0001$) and FAR ($F(4, 1245) = 20.53$, $p < 0.0001$). In the first case, all groups marginal means are significantly different from each other ($p < 0.0001$), while in the second case, FARs with 4 and 6 items are similar and statistically different from those referred to a higher number of items. In particular, FAR with 4 items is significantly different from FAR with 8, 10 and 12 ($p < 0.0001$), FAR with 6 items is different from FAR with 10 and 12 ($p < 0.02$) and FAR with 8 items is different from FAR with 12 ($p < 0.02$). The only exceptions are trials with VA 120° and short observation time, where FARs arrive to 0.5 and HRs fast decrease below 0.75. In this task, layouts seem having an influence only on FAR ($F(1, 1248) = 7$, $p < 0.02$). Moreover, while outcomes in the VA 40° and 80° trials are comparable, they are significantly different from trials with VA 120°. Finally, better results are associated to the higher observation time (HR: $F(1, 1248) = 197.9$, $p < 0.0001$;

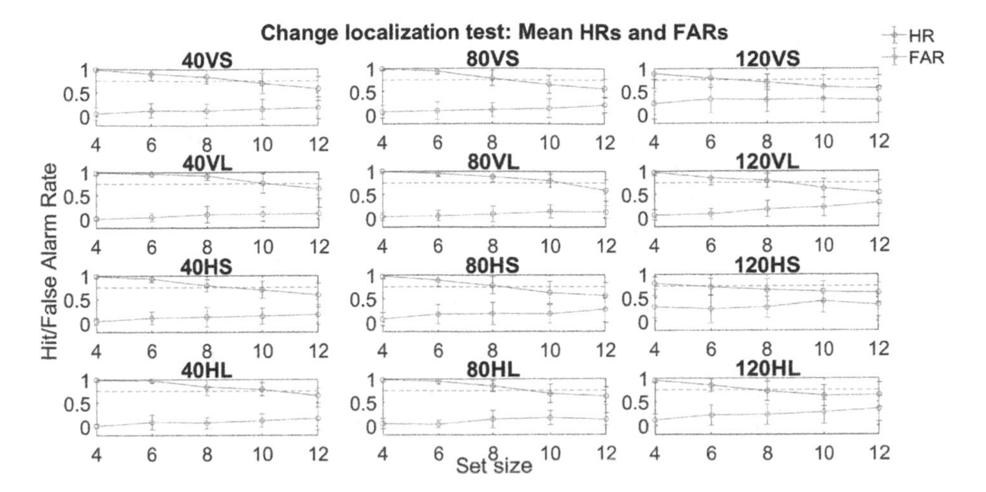

Fig. 2. Mean and standard deviation of HRs and FARs at the variation of the set size in the different experimental conditions, considering the VA (40, 80 or 120°), the layout (V or H) and the observation time (S or L). The dashed line represents the 0.75 threshold.

FAR: $F(1, 1248) = 40.74$, $p < 0.0001$) and smaller VAs (HR: $F(2, 1247) = 38.8$, $p < 0.0001$; FAR: $F(2, 1247) = 85.96$, $p < 0.0001$).

After having identified the subset of data exceeding the 0.75 accuracy threshold, we have computed the per bin HR, in order to understand if the probability of correctly detecting changes was associated with the distance of the modified item from the center of view. Results confirm the ones in [1] for the *change detection* test: position is relevant only in the case of VA 120° and short observation time, where participants could not have a complete overview of the scene, hence the HR decreases in the periphery of the horizontal VA, reaching a value of 0.5, which is the chance level. With longer observation time, however, results become stable in the entire VA. In general, performance is better and with less variability when a longer observation time is provided, decreases with VA enlargement, and is not influenced by the layout or by the distance of the stimulus from the user.

On the subset of data exceeding 0.75 HR, we investigated the effect of layout, VA and observation time over pointing precision. The red dots in Fig. 3 represent the real target and the arrow points to the center of mass of the raycasted positions for that specific target. Missing points are referred to target positions for which we do not have any data. We calculated the Euclidean distance between the center of mass and the target and averaged it across the horizontal and vertical VA, to better highlight the effect of the horizontal or vertical distance from the VA center.

In general, raycasted positions are well clustered around the target and, even in the worst cases participants can recognize the area in which the change was applied. With VA enlargement, the pointing error increases ($F(2, 7493) = 103.58$, $p < 0.0001$) and the greatest inaccuracy is associated to VA 120°. Also observation time ($F(1, 7494) = 14.93$, $p < 0.0001$) and layout ($F(1, 7494) = 752.95$, $p < 0.0001$) seem to play a fundamental role. Pointing error, in fact, decreases when participants are given a longer observation time. Furthermore, in the vertical layout case, we do not find any preferential direction of the pointing error, it is symmetric with respect to the real target but mostly distributed along the vertical axis because participants tended to raycast the target position from below and not frontally. In the horizontal case, instead, we notice an overestimation of depth and the Euclidean error is increasingly higher with distance, even if the actual angular error does not vary significantly. Finally, when participants have to raycast an added object, in general, precision is higher, than when they have to point the location where an item has disappeared ($F(1,7494) = 72.55$, $p < 0.0001$). In this case, in fact, they are more inaccurate since they have to rely on their memory or on allocentric cues.

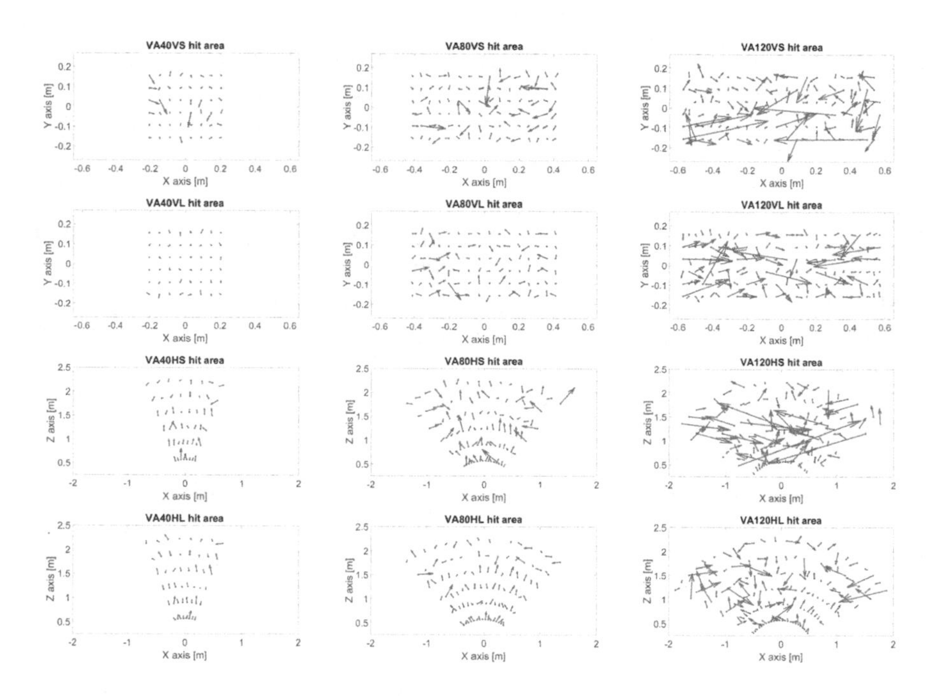

Fig. 3. Mean pointing error (blue arrow) computed considering the center of mass of the distribution of pointed positions around the target (red point) for trials with VA 40° (left), 80° (middle) and 120° (right). V = vertical; H = horizontal; S = 300 ms observation time; L = 900 ms observation time (Color figure online)

3.2 The Proposed Model Accounts for the Patterns of Human Data

Here, we carry out the analysis for the HRs and FARs, and the point precision analysis. For a quantitative evaluation we employ the Pearson correlation score

(r) and the related p-value: this metric is used in assessing agreement with human data [18,25].

It is worth noting that our model processes directly the same visual stimuli of the participants of the change localization experiment. The correlations between modeled HRs and human HRs are similar to the ones in [1] for the change detection experiment. This fact is relevant for the model: indeed, there are no significant differences among the human HRs of change detection and change localization experiments, and the model is able to provide similar agreements with the human data for both experiments. Our model (with the SSR saliency map as front-end) has a high level of agreement with the human data ($r > 0.87$ and $p < 0.05$). On average, also the methods AIM, AWS, DVA, IKN and SUN show a good level of agreement.

On the contrary, with respect to [1], the correlations between modeled FARs and human FARs show a better agreement. The reason could be that the human FARs increase as a function of set size (though it is not statistically significant). In particular, the SSR and, partially, AIM have a high agreement with human data.

Since our experiment aims to assess the localization of the detected change, we compare the modeled pointing precision (see Eq. 1) and human pointing precision. From a quantitative point of view, Table 1 shows the correlations between modeled pointing precision and human pointing precision: our model by using the SSR saliency map has a high level of agreement with the human data ($r > 0.91$ and $p < 0.05$). On average, also the methods AIM, DVA and IMSIG show a good level of agreement.

Figure 4 shows the modeled (blue) and human (red) pointing precision as a function of the number of items for two methods. This allows a visual inspection

Table 1. Correlations r (p-value) between modeled and human pointing precision. The significant agreements with the human data are in bold.

method	VA40H	VA40V	VA80H	VA80V	VA120H	VA120V
AIM	0.90 (0.04)	0.91 (0.03)	0.69 (0.20)	0.96 (0.01)	0.80 (0.11)	0.91 (0.03)
AWS	−0.79 (0.11)	−0.23 (0.71)	0.10 (0.88)	0.05 (0.94)	−0.75(0.15)	−0.24 (0.70)
CVS	−0.30 (0.62)	0.64 (0.24)	0.01 (0.99)	−0.56 (0.32)	−0.60 (0.29)	0.67 (0.22)
DVA	0.27 (0.66)	0.41 (0.50)	0.91 (0.03)	0.85 (0.07)	0.29 (0.64)	0.94 (0.02)
GBVS	−0.41 (0.50)	−0.25 (0.68)	0.56 (0.32)	0.20 (0.75)	0.08 (0.90)	−0.58 (0.31)
IKN	−0.63 (0.26)	−0.58 (0.31)	−0.48 (0.42)	−0.82 (0.09)	−0.54 (0.34)	−0.74 (0.15)
IMSIG	0.19 (0.76)	−0.90 (0.04)	0.14 (0.83)	−0.90 (0.04)	−0.59 (0.29)	0.18 (0.77)
LDS	−0.62 (0.26)	−0.24 (0.69)	−0.25 (0.68)	−0.79 (0.11)	0.07 (0.91)	0.82 (0.09)
QSS	−0.66 (0.23)	−0.83 (0.08)	−0.77 (0.12)	−0.70 (0.19)	−0.47 (0.42)	−0.76 (0.13)
SSR	**0.99 (0.00)**	**0.96 (0.01)**	**0.95 (0.02)**	**0.96 (0.01)**	**0.93 (0.02)**	**0.98 (0.00)**
CAS	−0.14 (0.83)	0.92 (0.03)	0.86 (0.06)	0.92 (0.03)	0.22 (0.72)	0.85 (0.07)
FES	0.93 (0.02)	−0.20 (0.74)	−0.02 (0.98)	−0.88 (0.05)	0.92 (0.03)	−0.00 (1.00)
RARE2012	−0.33 (0.59)	−0.98 (0.00)	−0.47 (0.42)	−0.07 (0.91)	−0.67 (0.22)	−0.05 (0.93)
SUN	0.07 (0.92)	−0.70 (0.19)	0.84 (0.07)	0.90 (0.04)	−0.68 (0.20)	−0.13 (0.84)

of the correlation data of Table 1: there is a high level of agreement between the behavior of modeled and human data by using the SSR saliency map as input to our model.

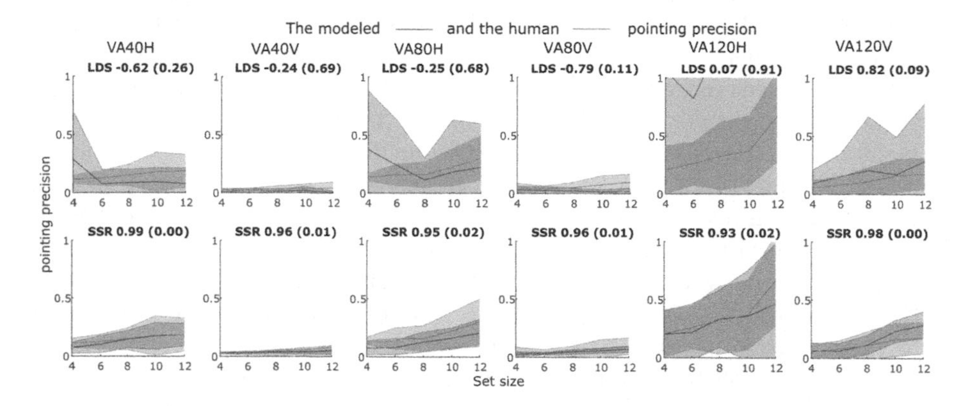

Fig. 4. The modeled (blue) and human (red) pointing precision for two methods as a function of set size. The mean is denoted by the line and the standard deviation by the shaded area. Correlations r (p-value) are reported in the title of each subfigure. (Color figure online)

4 Discussion and Conclusion

In this paper, we analyzed the VWM using a *change localization* experiment. We might assume that *change localization* is associated with a higher cognitive load than a simpler detection task, as the participant is asked to acquire an additional information, i.e. the item position. However, HRs referred to our experiment and the one in [1] are similar and no statistically significant difference has been reported. Moreover, our model correlates very well with the HR and FAR of the *change localization* experiment.

Considering the ability to localize the changes, coherently with HRs, results suggest that localization precision is strongly affected by VA and observation time: pointing error increases with VA enlargement and decreases as participants are given more time to explore the scene. In VA 40° and 80°, once the participant has seen the change, localization is quite precise; however, even in the worst case, the participant is still able to identify at least the part of the scene where the change occurred. In the vertical layout case, there is not a preferential direction for the error nor an influence of height over error magnitude; while in the horizontal case, distances affect results, especially in trials with larger VAs. However, the overestimation of distances is mainly due to the pointing technique used, as the angular error remains stable as the distance increases.

To model the human error in pointing precision, we devised a model (Eq. 1) that takes into account different aspects. We consider a linear combination of the

Euclidean distance between the target and modeled pointing position and the difference between the target's salient area and modeled position. Moreover, we consider the modeled FAR related to the error in the change detection since this can be related to the average error in the pointing task. The results show that the proposed model has a high level of agreement with human data, specifically when it is based on the SSR saliency map. It is worth noting that the noise for modeling the human data of pointing precision is high, i.e. 0.4. This difference with respect to the noise for HR and FAR modeling of change detection in [1] might be due to the motor component of the pointing task. Overall, our results show that the proposed model is able to replicate human data with a good level of agreement.

References

1. Bassano, C., Chessa, M., Solari, F.: Visual working memory in immersive visualization: a change detection experiment and an image-computable model. Virtual Reality **27**, 2493–2507 (2023). https://doi.org/10.1007/s10055-023-00822-y
2. Bays, P.M., Husain, M.: Dynamic shifts of limited working memory resources in human vision. Science **321**(5890), 851–854 (2008)
3. Brunel, N., Wang, X.J.: Effects of neuromodulation in a cortical network model of object working memory dominated by recurrent inhibition. J. Comput. Neurosci. **11**(1), 63–85 (2001)
4. Cohen, M.A., Dennett, D.C., Kanwisher, N.: What is the bandwidth of perceptual experience? Trends Cogn. Sci. **20**(5), 324–335 (2016)
5. Compte, A., Brunel, N., Goldman-Rakic, P.S., Wang, X.J.: Synaptic mechanisms and network dynamics underlying spatial working memory in a cortical network model. Cereb. Cortex **10**(9), 910–923 (2000)
6. Cowan, N.: Metatheory of storage capacity limits. Behav. Brain Sci. **24**(1), 154–176 (2001)
7. Cowan, N., et al.: On the capacity of attention: its estimation and its role in working memory and cognitive aptitudes. Cogn. Psychol. **51**(1), 42–100 (2005)
8. Fougnie, D., Suchow, J.W., Alvarez, G.A.: Variability in the quality of visual working memory. Nat. Commun. **3**(1), 1–8 (2012)
9. Foulsham, T., Underwood, G.: How does the purpose of inspection influence the potency of visual salience in scene perception? Perception **36**(8), 1123–1138 (2007)
10. Foulsham, T., Underwood, G.: What can saliency models predict about eye movements? Spatial and sequential aspects of fixations during encoding and recognition. J. Vis. **8**(2), 6–6 (2008)
11. Franconeri, S.L., Alvarez, G.A., Enns, J.T.: How many locations can be selected at once? J. Exp. Psychol. Hum. Percept. Perform. **33**(5), 1003 (2007)
12. Gras, D., Gyselinck, V., Perrussel, M., Orriols, E., Piolino, P.: The role of working memory components and visuospatial abilities in route learning within a virtual environment. J. Cogn. Psychol. **25**(1), 38–50 (2013)
13. Jaiswal, N., Ray, W., Slobounov, S.: Encoding of visual-spatial information in working memory requires more cerebral efforts than retrieval: evidence from an EEG and virtual reality study. Brain Res. **1347**, 80–89 (2010)
14. Kristjánsson, Á., Draschkow, D.: Keeping it real: looking beyond capacity limits in visual cognition. Attent. Percept. Psychophys. **83**(4), 1375–1390 (2021)

15. Li, C.L., Aivar, M.P., Tong, M.H., Hayhoe, M.M.: Memory shapes visual search strategies in large-scale environments. Sci. Rep. **8**(1), 1–11 (2018)
16. Luck, S.J., Vogel, E.K.: The capacity of visual working memory for features and conjunctions. Nature **390**(6657), 279 (1997)
17. Ma, W.J., Husain, M., Bays, P.M.: Changing concepts of working memory. Nat. Neurosci. **17**(3), 347–356 (2014)
18. Maiello, G., Chessa, M., Bex, P.J., Solari, F.: Near-optimal combination of disparity across a log-polar scaled visual field. PLoS Comput. Biol. **16**(4), e1007699 (2020)
19. Matzen, L.E., Haass, M.J., Divis, K.M., Wang, Z., Wilson, A.T.: Data visualization saliency model: a tool for evaluating abstract data visualizations. IEEE Trans. Visual Comput. Graphics **24**(1), 563–573 (2017)
20. Meilinger, T., Knauff, M., Bülthoff, H.H.: Working memory in wayfinding - a dual task experiment in a virtual city. Cogn. Sci. **32**(4), 755–770 (2008)
21. Miller, G.A.: The magical number seven, plus or minus two: some limits on our capacity for processing information. Psychol. Rev. **63**(2), 81 (1956)
22. Polatsek, P., Waldner, M., Viola, I., Kapec, P., Benesova, W.: Exploring visual attention and saliency modeling for task-based visual analysis. Comput. Graph. **72**, 26–38 (2018)
23. Rensink, R.A.: Change blindness. In: Neurobiology of attention, pp. 76–81. Elsevier (2005)
24. Seinfeld, S., Feuchtner, T., Pinzek, J., Müller, J.: Impact of information placement and user representations in VR on performance and embodiment. IEEE Trans. Visual. Comput. Graph. 1–13 (2020)
25. Sitzmann, V., et al.: Saliency in VR: how do people explore virtual environments? IEEE Trans. Visual Comput. Graphics **24**(4), 1633–1642 (2018)
26. Stirk, J.A., Underwood, G.: Low-level visual saliency does not predict change detection in natural scenes. J. Vis. **7**(10), 3–3 (2007)
27. Underwood, G., Foulsham, T.: Visual saliency and semantic incongruency influence eye movements when inspecting pictures. Q. J. Exp. Psychol. **59**(11), 1931–1949 (2006)
28. Wloka, C., et al.: Smiler: saliency model implementation library for experimental research. arXiv preprint arXiv:1812.08848 (2018)

Pain and Fear in the Eyes: Gaze Dynamics Predicts Social Anxiety from Fear Generalisation

Sabrina Patania[1(✉)], Alessandro D'Amelio[1], Vittorio Cuculo[2], Matteo Limoncini[1], Marco Ghezzi[1], Vincenzo Conversano[1], and Giuseppe Boccignone[1]

[1] PHuSe Lab - Università degli Studi di Milano, Milan, Italy
{sabrina.patania,alessandro.damelio,giuseppe.boccignone}@unimi.it,
{matteo.limoncini,marco.ghezzi3,vincenzo.conversano}@studenti.unimi.it
[2] AImageLab - Università degli Studi di Modena e Reggio Emilia, Modena, Italy
vittorio.cuculo@unimore.it

Abstract. This study presents a systematic approach for analyzing eye movements in the context of fear generalisation and predicting Social Interaction Anxiety Scale (SIAS) scores. Leveraging principles from foraging theory, we introduce a composite Ornstein-Uhlenbeck (O-U) process as a computational model for social anxiety assessment based on eye-tracking data. Through Bayesian analysis, we infer the model parameters and identify a feature set for SIAS score prediction. The results demonstrate the effectiveness of our approach, achieving promising performance using Random Forest (RF) classification. This research offers a novel perspective on gaze analysis for social anxiety assessment and highlights the potential of gaze behaviour as a valuable modality for psychological evaluation.

Keywords: Fear generalisation · Pain · Eye movements · Visual foraging · Anxiety disorders

1 Introduction

If you have been painfully bitten by a vicious dog, you might have acquired fear of all dogs therefrom conceptualised as those harmful barking animals with sharp teeth and claws, four legs and a tail. Under such circumstances, regrettably, you are contending with a fear (over)generalisation problem.

Fear generalisation (FG) describes the phenomenon that learnt fear is not restricted to those exact stimuli with which an aversive experience was originally paired (the specific wicked black beast) but it spreads to perceptually or conceptually similar ones (all black dogs or even all dogs, see Fig. 1) [14].

Clearly, the ability to learn which stimuli in the environment signal threat has an important adaptive advantage (to initiate appropriate defensive responses); a hallmark of human cognition is indeed the adeptness to extract conceptual

G. L. Foresti et al. (Eds.): ICIAP 2023 Workshops, LNCS 14365, pp. 133–144, 2024.
https://doi.org/10.1007/978-3-031-51023-6_12

Fig. 1. Fear learning and generalisation. An aversive episode pairs the perceptual stimulus (conditioned stimulus CS) of a black dog with a painful bite (unconditioned stimulus, US). Learnt fear can subsequently spread over a gradient of harmless stimuli (generalisation stimuli, GSs) more or less similar to the original CS

knowledge from a learning episode [33]. Yet, excessive generalisation may become maladaptive and pathological. Crucially, the overgeneralisation of fear to harmless stimuli or contexts might even turn into a burden to daily life and characteristic of several anxiety disorders [13].

Fear learning and its generalisation effects are usually investigated using the fear-conditioning paradigm [14,16]. Learning is fostered by pairing a perceptual stimulus with an aversive one (e.g., an electric shock) and then the extent of subject's generalisation to stimuli perceptually similar to the original one is assessed via subject's shock expectancy and behavioural/physiological measures (see Sect. 2). To such end, extensive research has been conducted by using geometric shape stimuli (circles, triangles, etc.). Fear learning *in vivo*, however, hardly involves such simple sensory cues; thus, in order to increase the ecological validity of FG studies, recent works [1,29,30] have considered faces as suitable targets. Notably, Reutter and Gamer [29] have provided experimental evidence that the extent of explicit fear generalisation is related to individual patterns of attentional deployment. Precisely, by using visual facial stimuli, eye-tracked participants who dwelled on the distinguishing facial features faster and for longer periods of time were likely to exhibit less fear generalisation. Their analyses though were based on classic measurements (latency of the first fixation, dwell time, etc.); gaze dynamics of viewing behaviour was only indirectly considered.

In this note, based on Reutter and Gamer's publicly available experimental data, we propose a model-based Bayesian analysis of participants' attention deployment (cfr. Sect. 3). Most important, we take a step further by exploiting the inferred model parameters to predict each participant's social anxiety level, by relying on data gathered in the same experiment via the Social Interaction Anxiety Scale (SIAS) questionnaire [29]. Results of our analysis are reported in Sect. 4; altogether, it is shown that gaze dynamics is effective at predicting individual's social anxiety in the fear conditioning context. To the best of our knowledge, the approach we present is novel in this direction.

Beyond the theoretical appeal of the problem, the motivation behind our work stems from the very fact that overgeneralisation of fear behaviours is common in many mental health disorders, including specific phobias, obsessive-compulsive disorder, panic disorder, generalised anxiety disorder, and post-traumatic stress disorder (PTSD) [14]. Further, in the chronic-pain condition, spreading of fear to safe movements may lead to sustained anxiety, dysfunctional avoidance behaviours, and severe disability [?]; indeed, pain-related fear is key to the transition from acute to chronic pain.

All such disorders have a large socio-economic impact, specially after the COVID-19 pandemic outbreak [17]. Steps towards a principled understanding of FG elicited behaviours and their operationalization are thus deemed to be relevant to afford critical insight into its basic mechanisms and to shed light on viable and novel behavioural therapies [5,29].

2 Background and Hypotheses

In this Section we briefly and critically overview the fear conditioning problem. This succint discussion motivates the research hypotheses behind the present work.

Fear Conditioning. The behavioural mechanisms of fear learning are usually investigated using the fear-conditioning protocol in the Pavlovian tradition [26]. In the acquisition phase, one neutral stimulus (CS+), such as a light, a tone or a simple shape, is repeatedly paired with an aversive unconditioned stimulus (US), such as an uncomfortable electric shock, while another neutral stimulus is never paired with the US (CS-). After only a few CS+/US pairings, presentations of the CS+ alone will elicit a conditioned fear response (CR)

FG tests introduce generalisation stimuli (GS), which typically include several stimuli that vary in perceptual or conceptual similarity to the CS+ and are never paired with the US. generalisation of conditioned fear shows that CRs are often elicited by stimuli not associated with the aversive event but which resemble the CS+ along a perceptual or categorical dimension. FG in humans is generally measured by self-report (e.g., ratings of fear or US expectancy) or by gradients of physiological responding (e.g., skin conductance responses, SCRs, fear-potentiated startle responses, FPS, and neuro-functional activation, e.g., BOLD fMRI). Perceptual similarity is acquainted as the most accessible factor to investigate in stimulus generalisation research, as the degree of similarity can be quantified as distance between points along a continuum. Yet, beyond light tones and shapes, it has been argued [12] that real-world fear learning situations are likely to involve complex stimuli (the dog example), rather than simple unidimensional sensory cues, and that humans routinely incorporate prior conceptual knowledge to infer unobserved properties and causal structure of details surrounding an emotional event [3,12]. For instance, category-based fear conditioning has shown that production of defensive responses can be mediated by categorization processes (fear of dogs spreading to other animals or dog-related items and contexts, but see [12] for a discussion). Knowledge about the causal structure of the world might also play an intriguing role [12]. As Rescorla put it [28], Pavlovian conditioning should be best understood as the learning of relations among events so as to allow the organism to represent its environment; in such endeavour, rephrasing Rescorla, the organism acts as a Bayesian information seeker using logical and perceptual relations among events, along with its own prior knowledge, to infer a sophisticated representation of its world. To sum up, FG based on real-world events will necessarily incorporate higher-order processes than what traditionally surmised [12].

Neurobiological Bases. A large amount of work has examined brain, neural, and hormonal mechanisms underlying FG; mechanisms at the lowest level are reviewed in [2]. As to brain regions, earlier work proposed that the hippocampus matches new stimuli to a representation of the original fearful stimulus stored in memory; the degree of similarity then affects activation in downstream regions associated with fear (or safety) in amygdalar, prefrontal, and thalamic brain regions, the amygdala playing a prominent role [21]. However, the idea that the increased amygdala response combined with reduced prefrontal cortex response results in hyperarousal or enhanced affective reactivity is a too simple one [32]. A recent meta-analysis produced a more complex neural working model involving more brain regions [34]. Further, fMRI results by Onat and Buchel [25], obtained using face stimuli, indicate that FG is not passively driven by perception. Rather it is an active process that, besides perceptual similarity, integrates threat identification and ambiguity-based uncertainty to orchestrate a flexible, adaptive fear response. Dysregulation is likely to arise due to deficiency in "top-down" control of emotional response (with a key involvement of the anterior insular cortex) while integrating atypical "bottom-up" detection or appraisal of emotional triggers [15,32]. All together, these results speak for an alternative theoretical framework that is best accomodated within the psychological construction approach where basic ingredients of the mind, namely the large-scale distributed networks within the human brain, interact so that fear dysregulation emerges in all its variety and complexity [4,15,24,32].

Research Hypotheses. The above considerations suggest that, overall, FG is the result of a complex, active process peculiar to the individual and, on the methodological side, the experimental shift to ecological social stimuli such as faces is not an innocent one. Under such circumstances, it is to be expected that, in the endeavour of orchestrating a flexible and adaptive fear response, the vibrant entanglement of threat identification, ambiguity-based uncertainty, perceptual similarity and prior conceptual knowledge, moulds the individual into an active information-seeker. To some extent, Reutter and Gamer experiments [29] and Onat and Buchel analyses [25] substantiate Rescorla's original assumption at the behavioural and neural level, respectively. Onat and Buchel themselves suggest that generalisation is a fundamental ability of humans facing the hurdles in the maze of social context and takes root in foraging animals striving to find food more efficiently [25]. FG, beneath its maladaptive limit, is in the service of the organism to gain insight into the world's events relying upon its own history and skills so to prepare defensive behaviours when encountering situations that were previously experienced as dangerous.

In this perspective, the deployment of visual attention through gaze offers an effectual window on the individual's information-seeking, foraging behaviour (see [6] for an in-depth discussion). Under the visual foraging hypothesis, we assume that a foraging-based analysis of gaze deployment over ecological stimuli is an appropriate approach to characterise the individual's visual information-seeking behaviour within a fear conditioning setting.

More precisely, in such setting we expect that, by inferring model's parameters from participants' eye-tracking data, a compact set of descriptors is provided that can be exploited to weigh latent factors behind the individual's FG process, in particular his/her social anxiety traits.

3 Method

We operationalise and assess our assumptions by considering the Reutter and Gamer FG experiment [29] and their publicly available data, in particular eye-tracking recordings and questionnaires gathered from participants. In what follows for completeness sake, we summarise those aspects of their work relevant for our analyses (Sect. 3.1), leaving to [29] and the accompanying online repository[1] for details.

As to the analyses, a number of works have recently considered eye movements modelling from the foraging perspective (e.g., [7–9,20,22,23]) or the closely related one that exploits sequential decision-making based on drift-diffusion models (e.g. [10,31]). With respect to the analyses needed here such models are overly complex, given their aim of actually simulating gaze shifts, and/or just suitable to cope with simple visual stimuli. Thus, we draw to some extent on the approach proposed in [11] (recapped in Sect. 3.2), which provides a succinct phenomenological model of gaze behaviour adequate to our analyses.

Furthermore, we utilize this gaze model to extract a characterization of the participants, which is subsequently employed for the classification of their Social Interaction Anxiety Scale (SIAS) scores, providing insights into the relationship between gaze behaviour and social anxiety levels.

3.1 Participants and Procedure

The main hypothesis was that participants who deploy more attention toward diagnostic facial features would also show fear generalisation gradients that are curved more strongly, thus indicating less generalisation.

Participants. The participants' sample consisted of 44 individuals (35 female; age $= 25.7 \pm 5.0$ years) that exhibited low levels of social anxiety (SIAS $= 19.7 \pm 8.0$).

[1] https://osf.io/wgqnj/.

Stimuli. Edited facial portrait photographs (642 × 676 pixels resolution) with neutral expression were used as stimuli. Pairs of photographs with the same gender and similar skin color were selected and edited such that either the eyes or both mouth and nose were copied from one face to the other while keeping all other pixels identical. The original and the edited picture differed only in the specified area and were later used as CS+ and CS- This factor, called the diagnostic region, calls for a flexible strategy of facial exploration in order to successfully distinguish between different pairs of stimuli. In total, one male and one female pair was created for each diagnostic region (eyes vs. mouth/nose). Morphs in steps of 20% between CS+ and CS- were generated to test FG.

Procedure. Subjects were seated inside an acoustically shielded cabin and performed a discrimination task to assure that they were able to perceptually distinguish between the stimuli that were sought to be used as CSs. Electrodes for pain application were attached. The intensity of the painful stimulus was then individually adjusted to a moderate level (6 or 7 on an 11-point scale ranging from 0 = no sensation over 4 = minimally painful to a theoretical maximum of 10 = worst pain imaginable). A training task ensured that subjects performed their trial-by-trial shock expectancy rating within the specified time window.

The FG task consisted of three phases. 1) *Habituation*: each of the four selected facial stimuli (one male and one female pair differing in the eye or mouth/nose region, respectively, in counterbalanced fashion) was presented four times without any electrotactile stimulation. 2) *Acquisition*: followed up seamlessly consisting of 32 trials in total. The two stimuli that were assigned to denote the CS+s (i.e., one stimulus of the male and female pair, respectively, in counterbalanced fashion) were reinforced in 75% of the cases. After a short break and recalibration of the eye-tracker, the last phase started. 3) *generalisation*: four intermediate stimuli between the CS+ and the CS- (morphs in steps of 20%) were also presented for each stimulus pair. These GS spanned the generalisation continuum from CS- across GS1 through GS4 to CS+. Each of the ten non-CS+s was presented eight times during the generalisation phase. The two CS+s, however, were presented 16 times with a 50% reinforcement rate in order to reduce extinction. Thus, the generalisation phase consisted of 112 trials. The FG task in total included 160 trials.

Each trial consisted of a face being presented for 6 s. After 4 s, a rating prompt appeared at the bottom of the screen for 2 s. Subjects were asked to indicate the perceived likelihood of an electrotactile stimulation occurring at the end of the trial on a 5-point scale (1 = no shock, 3 = uncertain, 5 = shock certain) but they were not told about different phases or contingencies. The painful stimulation was applied or omitted 5.85 s after stimulus onset.

Eventually, subjects completed the SIAS and a demographic data questionnaire.

Data Recording. Trial-by-trial shock expectancy ratings (via keyboard), heart rate, pupillary responses, electrodermal activity, and eye movements were measured. Eye movements and pupil size were recorded with an EyeLink 1000 Plus system (1000 Hz)

3.2 Proposed Model and Data Analyses

Use the time-varying location vector $\mathbf{x}(t)$ (screen coordinates) to denote the gaze position at time t. Each observed trajectory $\{\mathbf{x}(t),\ t = 0 \dots T\}$ is a realization of a stochastic process $\{X(t),\ t = 0 \dots T\}$, with $X(t) = \mathbf{x}(t)$. A compact description of both local fixational gaze movements, occurring within a selected region of the visual field, and saccadic relocations between regions can be given in terms of a particle randomly wandering but being pulled towards an attractor (the center of a region of interest, RoI) can be formalised as a switching Ornstein-Uhlenbeck (O-U) process. To such end, denote $s_t \in [fix, sac]$, a switching state random variable, indicating whether at time t either a fixational movement within a RoI or a saccadic relocation between RoIs is performed by the observer [10] [8]. Then, the stochastic differential equation (SDE)

$$d\mathbf{x}(t) = \mathbf{B}^{s_t}(\boldsymbol{\mu}^{s_t} - \mathbf{x}(t))dt + \boldsymbol{\Gamma}^{s_t} d\mathbf{W}^{s_t}(t) \tag{1}$$

accounts for the switching O-U process dynamics within and between RoIs sequentially selected by the observer as foci of attention. The term $\mathbf{B}^{s_t}(\boldsymbol{\mu}^{s_t} - \mathbf{x}(t))$ represents the drift towards the attractor point $\boldsymbol{\mu}^{s_t}$, where the 2×2 matrix $\mathbf{B}^{s_t}$ $\mathbf{B}^{s_t} = \begin{bmatrix} B_{ii}^{s_t} & B_{ij}^{s_t} \\ B_{ji}^{s_t} & B_{jj}^{s_t} \end{bmatrix}$ controls the magnitude of the attraction effect; B_{ii} and B_{jj} represent the drift of the process towards the attractor in the i (horizontal) and j (vertical) dimensions, respectively, while the off-diagonal elements $B_{ij} = B_{ji} = \rho_B \sqrt{B_{ii} B_{jj}}$ encode the cross-correlation between drift in both dimensions. The stochastic term $\boldsymbol{\Gamma}^{s_t} d\mathbf{W}^{s_t}(t)$ accounts for diffusion. Akin to $\mathbf{B}^{s_t}$, the 2×2 matrix $\boldsymbol{\Gamma}$ is the control parameter (variances and covariances) of the two driving white noise processes (horizontal and vertical) described by $d\mathbf{W}(t)$. Higher values of variances/covariances generate noisier/more anisotropic gaze trajectories. Given the set of parameters $\boldsymbol{\theta} = \{s_t, \mathbf{B}^{s_t}, \boldsymbol{\Gamma}^{s_t}, \boldsymbol{\mu}^{s_t}\}$, the simulation of a sequence of eye movements $\mathbf{x}(t) \rightarrow \mathbf{x}(t')$, with $t' > t + \delta t$, δt being an arbitrary time step, can be obtained by solving Eq. 1. In generative form, the solution can be written as the conditional sampling of $\mathbf{x}(t')$ given $\mathbf{x}(t)$, i.e., $\mathbf{x}(t') \mid \mathbf{x}(t) \sim P(\mathbf{x}(t') \mid \mathbf{x}(t))$, where the distribution $P(\cdot)$ is the Normal distribution $\mathcal{N}(\cdot)$ (see e.g. [18]):

$$\mathbf{x}(t') \mid \mathbf{x}(t) \sim \mathcal{N}(\boldsymbol{\mu}^{s_t} + e^{-\mathbf{B}^{s_t}\delta t}(\mathbf{x}(t) - \boldsymbol{\mu}^{s_t}), \boldsymbol{\Psi}^{s_t}), \tag{2}$$

where $\boldsymbol{\Psi} = \boldsymbol{D}^{s_t} - e^{-\mathbf{B}^{s_t}\delta t}\boldsymbol{D}^{s_t} e^{-\mathbf{B}^{s_t T}\delta t}$; $\mathbf{B}^{s_t}$ and $\boldsymbol{D} = \frac{\Gamma^2}{2}\mathbf{B}^{-1}$ are 2×2 matrices and the form $e^{-\mathbf{M}}$ denotes the matrix exponential.

The set of model parameters $\boldsymbol{\theta} = \{s_t, \mathbf{B}^{s_t}, \boldsymbol{\Gamma}^{s_t}, \boldsymbol{\mu}^{s_t}\}$ gives a complete description of gaze dynamics and can be inferred as follows. First, the raw eye-tracking data of an individual's gaze trajectory is parsed via the NSLR-HMM algorithm [27], to provide a set of fixational events (saccades are here discarded).

Only the fixations dwelling inside the diagnostic regions, i.e. eyes vs. mouth/nose (cfr Sect. 3.1) are retained. Call $\mathbf{e}^{eye} = [e_1, ..., e_{F_1}]$ the ensemble of F_1 fixations inside the eyes diagnostic area and $\mathbf{e}^{mn} = [e_1, ..., e_{F_2}]$ the group of F_2 fixations inside the mouth/nose diagnostic area. Define $\boldsymbol{\xi} = [\mathbf{e}^{eye} | \mathbf{e}^{mn}]$.

Consider the slice $\mathbf{x}^e = [\mathbf{x}_m, ..., \mathbf{x}_q]$ of the sample $\mathbf{x}(t)$, with $m \geqslant 0$ and $q \leqslant n$; the e index represents a generic fixation $e \in \boldsymbol{\xi}$. The likelihood of the slice, given the parameters $\{\mathbf{B}^e, \boldsymbol{\Gamma}^e\}$ writes $P(\mathbf{x}^e \mid \mathbf{B}^e, \boldsymbol{\Gamma}^e) = \prod_{i=1}^{q-m-1} P(\mathbf{x}^e_{i+1} \mid \mathbf{x}^e_i, \mathbf{B}^e, \boldsymbol{\Gamma}^e)$. Then, the posterior probability of the O-U parameters of the event e given the gaze trajectory slice is recovered via Bayes' theorem

$$P(\mathbf{B}^e, \boldsymbol{\Gamma}^e \mid \mathbf{x}^e) = \frac{P(\mathbf{x}^e \mid \mathbf{B}^e, \boldsymbol{\Gamma}^e) P(\mathbf{B}^e, \boldsymbol{\Gamma}^e)}{P(\mathbf{x}^e)}, \tag{3}$$

where under the mean field approximation $P(\mathbf{B}^e, \boldsymbol{\Gamma}^e) \approx P(\mathbf{B}^e) P(\boldsymbol{\Gamma}^e)$, the LKJ distribution is adopted as the prior for the $\mathbf{B}^e$ and $\boldsymbol{\Gamma}^e$ matrices in order to ensure all positive eigenvalues. Next, the event parameter posterior in Eq. 3 is computed in approximate form via Automatic Differentiation Variational Inference (ADVI) [19] and summarised through its sample average and uncertainty (Highest Density Interval, HDI). The distribution summaries are joined together, thus yielding the vector $\mathbf{v}^e_{(id)}$ for each subject $id \in [1, ..., ID]$, ID being the total number of subjects:

$$\mathbf{v}^e_{(id),k} = [B^{avg,e}_{ii}, B^{hdi,e}_{ii}, B^{avg,e}_{ij}, B^{hdi,e}_{ij}, B^{avg,e}_{jj}, B^{hdi,e}_{jj}, \Gamma^{avg,e}_{ii}, \Gamma^{hdi,e}_{ii}, \Gamma^{avg,e}_{ij}, \Gamma^{hdi,e}_{ij}, \Gamma^{avg,e}_{jj}, \Gamma^{hdi,e}_{jj}]. \tag{4}$$

Eventually, the sequence of events (fixations) - each event e being summarised by the vector $\mathbf{v}^e_{(id),k}$-, characterises the visual behaviour of observer id while scrutinising the stimulus k (image).

Denote:

- $\langle \mathbf{v}^{eye}_{(id),k} \rangle$ and $\langle \mathbf{v}^{mn}_{(id),k} \rangle$ the average fixation feature vector relative to either the eye or mouth/nose diagnostic region associated to the scan path (image) k:

$$\langle \mathbf{v}^{e^{eye}}_{(id),k} \rangle = \frac{1}{F_1} \sum_{a=1}^{F_1} \mathbf{v}^{e^{eye}_a}_{(id),k}, \qquad \langle \mathbf{v}^{e^{mn}}_{(id),k} \rangle = \frac{1}{F_2} \sum_{a=1}^{F_2} \mathbf{v}^{e^{mn}_a}_{(id),k} \tag{5}$$

- $\mathbf{v}_{(id),k}$ the descriptor of scan path k obtained by concatenating the two vectors above:

$$\mathbf{v}_{(id),k} = \left[\langle \mathbf{v}^{eye}_{(id),k} \rangle | \langle \mathbf{v}^{mn}_{(id),k} \rangle \right]; \tag{6}$$

- $\langle \mathbf{v}_{(id)} \rangle$ the summary descriptor of the visual behaviour of observer id, over the set of the K observed stimuli:

$$\langle \mathbf{v}_{(id)} \rangle = \frac{1}{K} \sum_{k=1}^{K} \mathbf{v}_{(id),k}. \tag{7}$$

The categorization of each fixation to its eventual diagnostic region has been carried out utilizing the pre-existing masks within the dataset, which were employed for the purpose of the original study. The account for that was motivated by the potential diagnostic value of these facial regions. By dividing fixations into these two categories, we sought to capture the nuanced dynamics of visual attention within different facial regions and explore their specific contributions to the prediction of social anxiety levels. The extracted parameters and features were derived separately for fixations related to the eyes and fixations related to the region of the nose/mouth, enabling a more fine-grained analysis of the observer's gaze behaviour and its potential relevance for social anxiety prediction. The extracted gaze dynamics parameters were then used as inputs for a Random Forest (RF) classifier to predict the SIAS score. To simplify the prediction task and facilitate the interpretation of results, we transformed the social anxiety prediction problem into a binary classification task. Given that the SIAS scores range from 0 to 80, we decided to use the median score in our dataset, which was found to be 18, as the threshold for binarizing the scores. Individuals with SIAS scores equal to or above the median threshold were considered to have high social anxiety, while those below the threshold were classified as having low social anxiety. The transformation into a binary classification problem allowed us to utilize well-established classification algorithms, such as the Random Forest classifier, to predict social anxiety levels effectively. Prior to selecting the RF classifier, we conducted an evaluation of other classification algorithms, such as the Support Vector Machine (SVM) with a radial basis function kernel and the linear Support Vector Machine classifier (linSVM).

4 Results

We utilized data from a cohort of 43 participants out of the initial pool of 44 participants, as recordings from one participant resulted to be inoperable and had to be excluded from our study.

The evaluation process involved 5-fold cross-validation to ensure robustness and mitigate any potential biases. Performance was assessed using accuracy.

The results of the evaluation revealed that the RF classifier outperformed the other algorithms with a significantly higher accuracy score of 0.73. In comparison, the linSVM and SVM algorithms yielded lower accuracy scores of 0.58 and 0.61 respectively Table 1.

Table 1. Accuracy results of tested algorithms for SIAS classification.

Algorithm	Accuracy (5-fold cv)
linSVM	0.58
SVM	0.61
RF	**0.73**

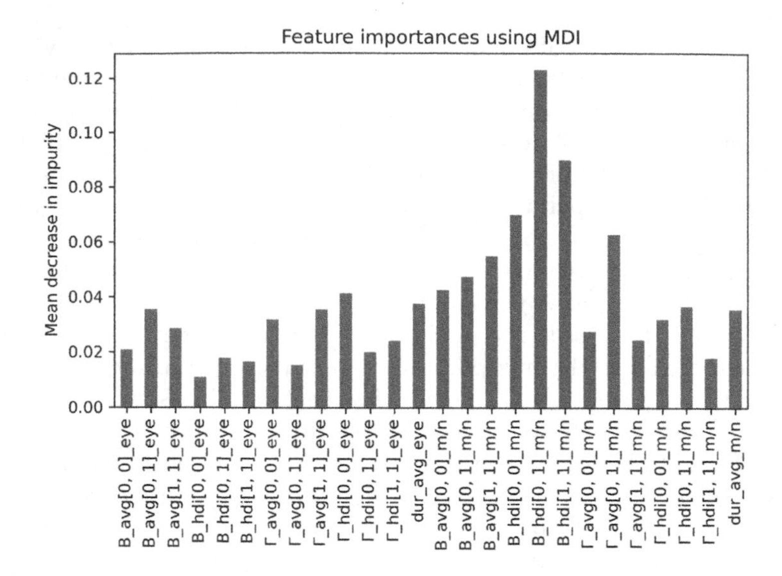

Fig. 2. Results of feature importance estimate, where indexes in brackets indicate the specific element of the matrix.

Additionally, we conducted an analysis to assess the feature importance, computed as the mean of accumulation of the impurity decrease within each tree (Fig. 2). This analysis provided valuable insights into the significance of different gaze dynamics features for the task at hand.

Notably, the results of this analysis revealed that features related to the gaze dynamics of the nose/mouth region hold greater importance in predicting social anxiety levels. Specifically, the feature that exhibited particularly high importance was the standard deviation of the magnitude of fixations drift towards the mean (B matrix).

5 Conclusions

In this paper, we have presented a systematic and principled approach for analyzing eye movements in the context of fear generalisation, specifically focusing on the prediction of Social Interaction Anxiety Scale (SIAS) scores.

In the framework of foraging theory applied to eye movements, we have introduced a composite O-U process to operationalise social anxiety assessment. This phenomenological model captures the exploration-exploitation signature inherent in foraging eye behaviour. By inferring the relevant parameters of the composite O-U model through Bayesian analysis of eye-tracking data, we have identified a feature set that is suitable for predicting SIAS scores.

The results of our study demonstrate the effectiveness of the proposed approach. By utilizing the inferred parameters from the composite O-U model of fixations as features, we have achieved promising performance in predicting SIAS scores using Random Forest (RF) as classification technique.

This research contributes to the body of knowledge by providing a novel perspective on the analysis of gaze for social anxiety assessment. By embracing a model-based approach and leveraging principles from foraging theory, this work can open novel avenues for future research in understanding and utilizing gaze behaviour as a valuable modality for psychological and behavioural assessment.

References

1. Ahrens, L.M., et al.: Fear conditioning and stimulus generalization in patients with social anxiety disorder. J. Anxiety Disord. **44**, 36–46 (2016)
2. Asok, A., Kandel, E.R., Rayman, J.B.: The neurobiology of fear generalization. Front. Behav. Neurosci. **12**, 329 (2019)
3. Barrett, L.F.: The theory of constructed emotion: an active inference account of interoception and categorization. Soc. Cogn. Affect. Neurosci. **12**(1), 1–23 (2017)
4. Barrett, L.F.: Seeing fear: it's all in the eyes? Trends Neurosci. **41**(9), 559–563 (2018)
5. Beckers, T., Hermans, D., Lange, I., Luyten, L., Scheveneels, S., Vervliet, B.: Understanding clinical fear and anxiety through the lens of human fear conditioning. Nat. Rev. Psychol. **2**, 1–13 (2023)
6. Bella-Fernández, M., Suero Suñé, M., Gil-Gómez de Liaño, B.: Foraging behavior in visual search: a review of theoretical and mathematical models in humans and animals. Psychol. Res. **86**(2), 331–349 (2022)
7. Błażejczyk, P., Magdziarz, M.: Stochastic modeling of Lévy-like human eye movements. Chaos (Woodbury, NY) **31**(4), 043129 (2021)
8. Boccignone, G., Cuculo, V., D'Amelio, A., Grossi, G., Lanzarotti, R.: On gaze deployment to audio-visual cues of social interactions. IEEE Access **8**, 161630–161654 (2020)
9. Boccignone, G., Cuculo, V., D'Amelio, A., Grossi, G., Lanzarotti, R.: Give ear to my face: modelling multimodal attention to social interactions. In: Leal-Taixé, L., Roth, S. (eds.) ECCV 2018. LNCS, vol. 11130, pp. 331–345. Springer, Cham (2019). https://doi.org/10.1007/978-3-030-11012-3_27
10. D'Amelio, A., Boccignone, G.: Gazing at social interactions between foraging and decision theory. Front. Neurorobot. **15**, 31 (2021)
11. D'Amelio, A., Patania, S., Bursic, S., Cuculo, V., Boccignone, G.: Using gaze for behavioural biometrics. Sensors **23**(3), 1262 (2023)
12. Dunsmoor, J.E., Murphy, G.L.: Categories, concepts, and conditioning: how humans generalize fear. Trends Cogn. Sci. **19**(2), 73–77 (2015)
13. Dunsmoor, J.E., Paz, R.: Fear generalization and anxiety: behavioral and neural mechanisms. Biol. Psychiat. **78**(5), 336–343 (2015)
14. Dymond, S., Dunsmoor, J.E., Vervliet, B., Roche, B., Hermans, D.: Fear generalization in humans: systematic review and implications for anxiety disorder research. Behav. Ther. **46**(5), 561–582 (2015)
15. Fitzgerald, J.M., DiGangi, J.A., Phan, K.L.: Functional neuroanatomy of emotion and its regulation in PTSD. Harv. Rev. Psychiatry **26**(3), 116 (2018)
16. Ghirlanda, S., Enquist, M.: A century of generalization. Anim. Behav. **66**(1), 15–36 (2003)
17. Kar, N., Kar, B., Kar, S.: Stress and coping during COVID-19 pandemic: result of an online survey. Psychiatry Res. **295**, 113598 (2021)

18. Kloeden, P.E., Platen, E.: Numerical Solution of Stochastic Differential Equations, vol. 23. Springer Science & Business Media, Berlin (2013). https://doi.org/10.1007/978-3-662-12616-5
19. Kucukelbir, A., Tran, D., Ranganath, R., Gelman, A., Blei, D.M.: Automatic differentiation variational inference. J. Mach. Learn. Res. **18**(1), 430–474 (2017)
20. Le, S.T.T., Kristjansson, A., MacInnes, W.J.: Bayesian approximations to the theory of visual attention (TVA) in a foraging task. Q. J. Exp. Psychol. **76**(3), 497–510 (2023)
21. Lissek, S., et al.: Neural substrates of classically conditioned fear-generalization in humans: a parametric FMRI study. Soc. Cogn. Affect. Neurosci. **9**(8), 1134–1142 (2014)
22. Mirza, M.B., Adams, R.A., Friston, K., Parr, T.: Introducing a Bayesian model of selective attention based on active inference. Sci. Rep. **9**(1), 1–22 (2019)
23. Mirza, M.B., Adams, R.A., Mathys, C.D., Friston, K.J.: Scene construction, visual foraging, and active inference. Front. Comput. Neurosci. **10**, 56 (2016), https://www.frontiersin.org/article/10.3389/fncom.2016.00056
24. Mobbs, D., et al.: Viewpoints: approaches to defining and investigating fear. Nat. Neurosci. **22**(8), 1205–1216 (2019)
25. Onat, S., Büchel, C.: The neuronal basis of fear generalization in humans. Nat. Neurosci. **18**(12), 1811–1818 (2015)
26. Pavlov, I.P.: Conditioned Responses. Prentice-Hall Inc., Hoboken (1949)
27. Pekkanen, J., Lappi, O.: A new and general approach to signal denoising and eye movement classification based on segmented linear regression. Sci. Rep. **7**(1), 1–13 (2017)
28. Rescorla, R.A.: Pavlovian conditioning: it's not what you think it is. Am. Psychol. **43**(3), 151 (1988)
29. Reutter, M., Gamer, M.: Individual patterns of visual exploration predict the extent of fear generalization in humans. Emotion (2022). https://doi.org/10.1037/emo0001134
30. Roesmann, K., Wiens, N., Winker, C., Rehbein, M.A., Wessing, I., Junghoefer, M.: Fear generalization of implicit conditioned facial features-behavioral and magnetoencephalographic correlates. Neuroimage **205**, 116302 (2020)
31. Roth, N., Rolfs, M., Hellwich, O., Obermayer, K.: Objects guide human gaze behavior in dynamic real-world scenes. bioRxiv (2023). https://www.biorxiv.org/content/early/2023/03/14/2023.03.14.532608
32. Suvak, M.K., Barrett, L.F.: Considering PTSD from the perspective of brain processes: a psychological construction approach. J. Trauma. Stress **24**(1), 3–24 (2011)
33. Tenenbaum, J.B., Kemp, C., Griffiths, T.L., Goodman, N.D.: How to grow a mind: statistics, structure, and abstraction. Science **331**(6022), 1279–1285 (2011)
34. Webler, R.D., et al.: The neurobiology of human fear generalization: meta-analysis and working neural model. Neurosci. Biobehav. Rev. **128**, 421–436 (2021)

Eye Gaze Analysis Towards an AI System for Dynamic Content Layout

Michael Milliken[1], Andriy Kharechko[1], Ian Kegel[2], Brahim Allan[2](✉), Shuai Zhang[1], and Sally McClean[1]

[1] Ulster University, Belfast, UK
{m.milliken1,a.kharechko,s.zhang,s.mcclean}@ulster.ac.uk
[2] British Telecom, Ipswich, UK
brahim.allan@bt.com

Abstract. As an instance of dynamic content layout in the context of object-based TV broadcasting, graphic insertions within football match replays are presented and discussed. It is offered that an AI system may be an efficient solution to counter occlusion of content. For such a system, aspirational targets are presented, followed by descriptions of an experiment purposed to gather data as to how viewers may experience graphical insertions into football goal replays in comparison to the typically normal football goal replays where no such graphical insertions are made. The experiment takes the form of eye tracking of viewers' visual interactions and subsequent analysis to better understand fixations on and off inserted graphics. A series of metrics is derived, discussed, and the results of which are presented as to how well the experiment may relate to aspirational targets and by how much they may be met at present, as well as potential future experiments based on the current findings.

Keywords: eye tracking · object-based broadcasting · object detection · deep learning · sports

1 Introduction

In traditional (linear) TV broadcasting, every channel is a single scheduled stream of programmes. Modern broadcasting exploits broadcast and IP media streams to deliver content to its consumers with different requirements and use contexts. As the availability of supplementary media content has grown, as well as the proliferation of viewing devices, it has become unnecessary for every user to have an identical experience with the same piece of content, so the need for different versions of the same programme adapted to preferences of its viewers as well as to different types of their devices becomes more and more obvious. The most straightforward solution would be to produce TV content in the highest possible quality and scale it down to address device heterogeneity, but this is often impractical.

Research work supported by Ulster University, British Telecom, and Invest NI.

G. L. Foresti et al. (Eds.): ICIAP 2023 Workshops, LNCS 14365, pp. 145–156, 2024.
https://doi.org/10.1007/978-3-031-51023-6_13

Object-Based Broadcasting (OBB) aims to supply a single version of the content represented by a set of separate media assets (objects) accompanied by metadata (i.e. rules and constraints defined by the production staff) which describes their relationships and associations together with the editorial intent [3]. The objects are independently transmitted to be flattened and rendered at the point of consumption and their exact combination is determined by viewers' preferences, context, platform characteristics and other factors, which creates more personalised yet properly authored overall user experience [2]. In other words, OBB makes the broadcasting system agnostic to data formats and to consumption and production devices, thus enables viewers to have a more interactive and personalised relationship with TV content and tackles issues of heterogeneity of devices, curation of programmes, and potential creation of new types of content and experiences while facilitating efficiency of their production [6].

This paper addresses a basic instance of the problem of dynamic content layout in the context of object-based broadcasting, where media objects (in our case, information graphics) are delivered independently from the video stream and rendered locally within the TV frame in the real time to match user preferences and device characteristics while avoiding occlusions of the regions of interest [1]. We present analysis of eye-tracking results collected in the process of the development of our proof of concept.

This resembles the problem of dynamic captioning (or subtitling), where script locations within non-intrusive regions of the frame are dynamically computed to match speakers' faces, align with their speech and emphasise changes in volume of their voices [4,7,8]. In both cases, displaying media objects (e.g. graphics or captions) closer to the region of interest tackles limitations of the human vision span [10], which becomes more evident when large screens are being watched from relatively close distances, by allowing viewers to simultaneously follow the video feed and auxiliary content with minimal eye strain.

2 AI System

With modern technology it is easy to provide supplemental content, heretofore called graphic or graphics, overlaying or alongside this main content viewed by one or more people. The addition of any graphic however has the very real danger of conflicting with the primary content that is of main concern and interest.

In some situations, the addition of a graphic requires human interactions on the side of whatever entity is providing the content. Either at a production level, broadcaster level, or perhaps even both. Where human interaction is required the throughput of processing these graphic additions is limited to the speed, abilities, experience, and amount of people making the additions. In some cases there may be a discernible inconsistency in placements per person making decisions, unless there are pre-defined plans and patterns in place that are adhered to where possible.

It is key that graphics are positioned consistently and with vigilance so that they are added and removed when appropriate so as to not occlude the primary content on display. For example, the specific motions and movements of a player during an exciting or interesting moment, especially when a goal is being scored. A graphic positioned over the ball or player involved, especially at the wrong time, at the very least, would constitute occlusion of key action. Thus it is easy to say that occlusion would be detrimental to the experience of viewers.

In the interest of consistency and vigilance, an AI system may be an efficient solution to the problem of adding graphics. As discussed, a specific use case is as follows. To provide a graphic for viewers to be positioned on a sequence of multi-angle replays without occluding any key action.

While this serves as an aim for such an AI system, there are constituent targets necessary to approach or meet in order to adequately achieve the aim:

1. Differentiate between different camera angles in a replay sequence.
2. Decide if the clip (or clip sequence) is an appropriate length (or replay speed) to include the graphic.
3. Provide an experience value for the best position for the graphic based on agreed templates (in this case: top-left, bottom-left, top-right, bottom-right).
4. Compare experience values for a series of replays to provide a combined best position to minimise repositioning of graphic across the sequence.
5. Establish a layout value threshold below which it is advisable not to display the graphic.
6. Automate the delay or early removal of the graphic to increase experience values.

Present in these targets is a need to know more about how viewers currently perceive content and graphics in order to make judgements on appropriate durations for graphics and positions that are most visible to viewers. To gather data on these factors, an experiment was necessary, and from such experimental data it was reasoned that ground truths may be established so that a robust AI system may be developed.

3 Experiment

3.1 Overview

The experiment was focused on gathering data to inform general understanding as to non-specific user interactions with TV broadcast football matches by way of short clips in the context of graphics. While the graphics themselves were out of the control of users but added information to the experience. In this case viewers were participants attending a concurrent event, thus were not recruited specifically for the experiment, but drawn from an audience. A total of 18 participants were recruited in this way.

An eye-tracker (Tobii X3-120) was calibrated prior to and used during each participant's viewing of two different clips to record their gaze and fixations while watching each clip in succession on a large screen TV. Tobii Pro Lab was used to display each clip and handle recording of each participant's session. The specific graphical content was not a major consideration, but was very much related to the content being displayed and presented to participants. It was based on a player's career and was comprised of the following: 'Jersey number', 'Player's name', 'Player's photo', 'Team badge', 'Appearances', 'Goals', 'Assists', 'Shots on target', and 'Shots off target' - all information that may be of interest to those who follow the sport. Otherwise the graphic could be comprised of any information desired.

As each participant watched two clips, the first (Clip1) without graphical addition, and the second (Clip2) with graphical addition, this allowed for comparisons to be drawn while not biasing the experience to one clip where the experience would be too familiar and difficult to compare.

3.2 Set-Up

Each participant was situated in front of a large TV. Between the TV and a participant an eye-tracker (Tobii X3-120) was placed, with the appropriate respective angles and distances according to guidelines. Calibration of the eye-tracker was typical, and involved each participant following a circle as it traversed the screen coming to rest briefly at pre-specified points. Participants did not have access to a remote control as the study required only passive watching of content.

3.3 Clips

As mentioned previously, two different clips were used. Each clip came from the same football match [5], but each showed a replay of a different goal more than thirty minutes apart in the football match. So where Clip1 ended, Clip2 did not pick up shortly afterwards maintaining a distinction between the two. Further distinguishing them was the inclusion of a graphic in Clip2. Both were comprised of four different shots, compared one to one they were similar or the same shot type. Within Clip2 the graphics were positioned in either top right or bottom left (just over the score clock), these positions were chosen as the ball was not present there, and existing production guidelines defined outer perimeter space to exclude graphics. For completeness the positions are based on the following templates: Position 1 (top-left), Position 2 (top-right), Position 3 (bottom-left), and Position 4 (bottom-right).

3.4 Data Analysis

The analysis of eye-tracking data involves well-referenced terminology. A fixation is an instance of an eye moving at a relatively low speed with its focus on an Area of Interest (AOI), a defined space within the bounds of stimuli. Such events can be measured as a fixation count for the number of distinct instances which occur within an AOI, or a fixation duration for the length of time before the eye moves at a higher speed or to a point away from the AOI. For any particular AOI, the average or total duration of fixations may also be calculated. The movement of the eye from fixation to fixation is a saccade. Related to these values is the time to first fixation, indicating the time before a fixation first occurs within an AOI. Typically some AOI is defined as data about that part of stimuli to be analysed. Such terminology is applicable to one or more subjects upon which eye tracking is performed, resulting in analysis of samples (individuals) and populations (groups). Considering the population size, the percentage fixated is the percentage of that population that fixated within an AOI. This description of terminology is limited to those used herein and is not exhaustive of all terms to describe data derived from eye tracking.

Following the end of data collection with all participants, the aforementioned Tobii Pro Lab was used for subsequent analysis by way of AOIs and other tools found within the software. In the interest of efficient analysis, given the dynamic content and number of participants, the AOIs were limited and applied to the data collected for each participant equally. The display was split into quadrants, which were themselves split into quadrants so far as to create a total of sixty four equally sized AOIs that covered the whole of the display. The only other AOIs were those specific to the graphics within Clip2 that would collect data specific to them. An example of the AOIs is presented in Fig. 1 which were applied equally to Clip1 and Clip2, with the only consideration being that Clip1 did not include graphics thus only quadrant AOIs are relevant to it.

Fig. 1. Example of AOIs. Criss-crossing quadrants denote display. Lower-left rectangle denotes graphic insert.

AOIs for graphics were active when the graphics were present, thus captured specific details about fixations and gazes within them. For the remainder of the AOIs, they served to capture non-specific data that related to the display and whatever was present within a quadrant at any given time. No individual object was followed throughout, instead the quadrants provided general details about fixations and gazes on whatever was displayed.

4 Results

The data collected through the eye-tracking experiment was analysed and collated as it related to particular targets from Sect. 1. The following metrics were derived and used in the analysis that follows: Time to First Fixation, Total Fixation Duration, Average Fixation Duration, Fixation Counts, Percentage Fixated.

4.1 Graphic Durations

From the fixations within graphic AOIs it was determined that an approximation of appropriate graphic durations could be made. Naturally, analysis was limited to Clip2 as there was no other appropriate clip to compare against. Given that graphics had two potential locations a shot of Clip2, as shown in Fig. 2, but were not present for equal amounts of time, comparisons could be drawn as to how participants interacted with them visually. Comparisons were made across three values relating to the graphics AOIs; Time to First Fixation, Average Fixation Duration, and how many participants fixated/showed interest. The results of which are presented in Tables 1 and 2 for Clip2 Graphic Position 3 and Clip2 Graphic Position 2 respectively, but described and discussed herein.

For a graphic to best serve its purpose there needs to be a maximum number of viewers who see/fixate on it for a suitable amount of time to read or understand it. So derived values should necessarily point to the result that may achieve that best. The number of participants who fixated on graphics were 13 and 16, for Position 3 and Position 2 respectively. Thus to allow for a maximum number of viewers, the time to allow their chance of fixation should be similarly maximised, resulting in 2.01 s and 3.63 s for Position 3 and Position 2 respectively. If a minimum were used, fewer viewers would see it, and even with the average, those who take longer may still miss it entirely. It is not enough to merely give a chance to fixate, but some duration to allow for some chance of viewers understanding what they are looking at. Thus an average is suitable in each case, where the minimum duration leaves little time for this, and the maximum may be akin to distracting from the primary content, resulting in 0.74 s and 1.01 s for Position 3 and Position 2 respectively.

In summation, suggested graphic durations are made by taking the maximum time to allow for first fixations, and the average time of any given fixation, as well as rounding up to the nearest second in the interest of error. These two suggestions for graphic duration are both approximately 3 s and 5 s for Position 3 and Position 2 respectively. As results indicate, 5 s may allow for a greater number of viewers to see and make use of a graphic, even briefly.

Table 1. Select metrics for graphic AOI in Clip2 Position 3

	Minimum	Average	Maximum
Time to First Fixation (secs)	0.49	1.05	2.01
Average Fixation Duration (secs)	0.16	0.74	1.68
Participants Fixated (count)	13	13	13

(a) (b)

Fig. 2. Clip2 Graphic Position 3 (a) and Position 2 (b)

Table 2. Select metrics for graphic AOI in Clip2 Position 2

	Minimum	Average	Maximum
Time to First Fixation (secs)	0.32	1.43	3.63
Average Fixation Duration (secs)	0.12	1.01	3.27
Participants Fixated (count)	16	16	16

4.2 Fixation Durations

While supplemental content may prove useful, it should not greatly take the focus away from the primary content, hence supplemental. Determining how participants fixated on the graphics as opposed to the primary content while the graphics were present was considered important, indicating if a replay should generally be made shorter or longer so that viewers would fixate more on the primary content. An approach to this determination was calculation of how time is distributed over the period when the graphics are present, such as time spent not fixating, fixation duration on primary content, and fixation duration on the graphics.

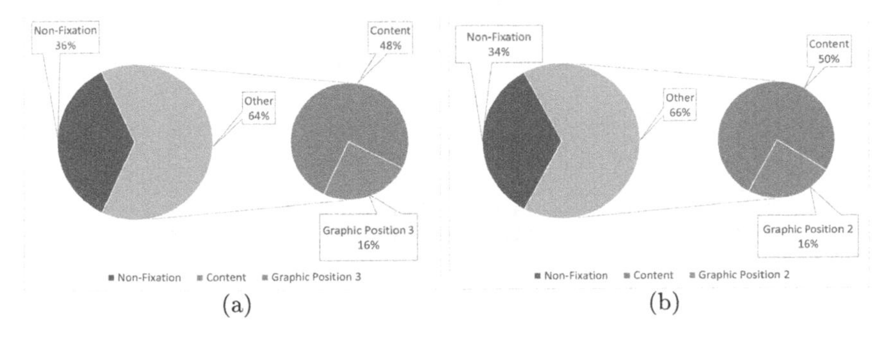

(a) (b)

Fig. 3. Distributions of time during graphic presence for Position 3 in Clip2 (a) and Position 2 in Clip2 (b)

The results of these calculations are presented in Fig. 3 (a) and 3 (b) for Position 3 and Position 2 respectively, both showing that most of the time spent during the periods of the graphics present, the majority is spent fixating, and approximately quarter of that fixation duration is on the graphics, regardless of position.

4.3 Fixation Scores

Positioning of the graphics overlaying the primary content is something to consider, and is especially true when there is danger that they may occlude key events. However, if it is necessary to re-position the graphics too often then viewers are likely to feel frustration with their constant flitting about the screen, which in turn is likely to cause a lack of benefit with little time to understand what is displayed. In the interests of reducing the re-position of the graphics, some analysis is performed on determining the best position from existing templates, and the best position to alleviate the need for re-positioning the graphics.

The measure here was more general than Sects. 4.1 or 4.2, using only the grid spaces as shown in Fig. 1 to collect data, where within each grid space the percentage of participants that fixated within that grid space was converted to a score between 0 and 11 inclusive. 0 being unchanged, 100 equating to 11, with ranges of approximately equal sized steps in between, and pairwise summing the grid spaces to represent each clip overall. Three results are presented for Clip1, Clip2, and a summation of the two in Figs. 4, 5, and 6 respectively, in addition for each Fig. the values are colour-scaled light to dark, low to high.

Within Fig. 4 the rectangle denotes that throughout Clip1 a space of little interest is present, a potentially good position to place the graphics throughout a replay given the very low percentages within, as well as being a large space that would allow for small to medium-sized graphics, although restricted to a portrait orientation. Within Fig. 5 the rectangles in Position 2 and Position 3 denote approximately where the graphics are positioned for some of Clip2's duration. Within Fig. 6 the summation of both Figs. 4 and 6 shows a polygon formed by a threshold of those values below 15, indicating low fixation percentage. The area within the polygon are potentials for minimising repositioning of the graphics throughout replays in general, particularly as given active objects within frame, the focus is central rather than on the perimeter of content.

2	8	7	6	2	0	1	1
3	11	12	9	10	5	3	0
2	13	19	2	18	13	4	1
3	15	14	30	27	22	3	0
0	6	12	33	33	20	1	0
2	12	25	32	22	5	0	0
3	9	20	16	6	0	0	0
4	10	2	0	0	0	0	0

Fig. 4. Clip1 fixation score

0	0	2	2	3	2	10	0
1	4	5	8	7	4	17	4
1	6	8	18	23	12	22	9
0	9	13	27	29	19	15	6
12	19	21	30	17	7	4	0
7	18	16	16	3	2	2	0
5	16	13	5	1	3	1	0
6	9	3	0	0	0	2	0

Fig. 5. Clip2 fixation score

2	8	9	8	5	2	11	1
4	15	17	17	17	9	20	4
3	19	27	44	41	25	26	10
3	24	27	57	56	41	18	6
12	25	33	63	50	27	5	0
9	30	41	48	25	7	2	0
8	25	33	21	7	3	1	0
10	19	5	0	0	0	2	0

Fig. 6. Clip1 and Clip2 fixation score

4.4 Region of Interest Hypothesis

A hypothesis put forward herein states that regions of interest, those regions with majority of fixations, primarily consist of detected objects of interest (i.e. the players and the ball, in the case of football broadcast). This hypothesis evaluated visually by comparing two different heatmaps, one of fixations from gaze data, and one of objects (of types 'person' and 'sports ball') detected using YOLOv3 object detector [11] trained on COCO dataset [9], for each clip. Each heatmap covers the clip from start to end, without specific focus on differing shots/angles.

Figure 7 (a) shows that within Clip1 the regions with majority of fixations fall within detected objects. Figure 7 (b) shows that what occurs for Clip1 also occurs for Clip2, while noting that some regions with majority of fixations also occur within Position 2 and Position 3 where the graphics are found. In each clip those regions of interest with a minority of fixations occur sporadically, some players/officials or the score clock are fixated on. From the experiment described the hypothesis holds true.

5 Discussion

The three results presented in Sects. 4.1, 4.2, and 4.3 relate back to those targets discussed in Sect. 1 as to how the experiment may or may not help to achieve them at present. Only targets 2, 3, and 4 may be discussed reliably in reference to the results. The remaining targets are discussed in reference to limitations within the eye-tracking study.

(a) (b)

Fig. 7. Composite of gaze data and detected object heat maps of Clip1 (a) and Clip2 (b)

5.1 Target 1

As participants look freely upon stimuli, eye tracking may indicate only interest. Across each different shot within both clips there were no clear shapes that would represent different shot types. Further to this, it is possible that given enough people, all of the stimuli would be covered leaving no distinct shapes visible.

5.2 Target 2

Through distribution of time between the primary content and the graphics, it has been shown that generally even a short sequence of a clip is at least an appropriate length for the graphics to be included as they do not draw too much attention from the content. However, to ensure ample time for viewers' interaction with the graphics, a minimum duration greater than 3 s would be appropriate, while allowing for a delay of 5 or more seconds. Although it should be noted given experimental data limited to one type of graphic, the content should be considered for length and average reading speed.

5.3 Targets 3 and 4

Comparisons between positions was limited to Position 2 and Position 3 as those were present in Clip2. Based on approximate location of the graphics in either, Position 2 may be the best position where overall fewer fixations occurred, as it is less likely that any action may be occluded. In terms of a best position to minimise repositioning, summation of both Clip1 and Clip2 showed that Position 4 in either case held few fixations with and without the graphics.

5.4 Target 5

Separation of individual aspects of layout proves difficult when considering eye tracking alone as it indicates where but not what draws fixations, as in Sect. 5.1. While granularity may be applied to what is present for finer details, those details are dependent on an ongoing re-evaluation for size and shape per frame, not to mention possible occlusions. This would at least require a great deal of manual efforts, and at most persistent object detection to accurately fine-tune to objects.

5.5 Target 6

Given the experiment was carried out during an event, limitations were inherent disallowing fine-tuning of graphic durations or a wider range of clips for further comparisons.

6 Conclusions

The experiment's purpose was to explore how gaze patterns of viewers are affected by insertion of an auxiliary graphic (if and how attention is diverted to them) as this would inform the approach towards a general AI system for dynamic content layout towards viewer immersion and a personalised viewing experience. In ways this was successful where as stated in Sect. 5, approximately 5 s is adequate time for the graphics to be present without diverting a majority of attention from content. Thus providing some basis of an AI system in terms of graphic durations.

However, what can be inferred from viewers interacting visually with the inserted graphics may not cover every aspect of a replay, leaving space for further experimental development to more acutely explore targets towards an AI system.

Both targets 1 and 5 are examples of required development as they may not be suitably explored with eye tracking alone. They require some further understanding of replays beyond fixations, necessitating a different approach.

For target 6, the inclusion of a wider variety of not only clips, but some variability in the start, stop, duration of the graphics, and types of the graphics, would help explore that particular target with further eye-tracking experiments.

In short, some aspirational targets of an AI system may be met through eye-tracking experiments, others require a different approach.

Future work would include specific participant recruitment, followed by qualitative data gathering via questionnaires, with potential to consider the cognitive load participants experienced throughout. Intending to better the overall analysis of such experimentation and provide a clearer picture of participants' experiences. Future works would also include exploration of a different methodology to address targets 1 and 5.

References

1. Allan, B., et al.: Towards automatic placement of media objects in a personalised TV experience. Multimedia Syst. **28**(6), 2175–2192 (2022). https://doi.org/10.1007/s00530-022-00974-y
2. Armstrong, M.: Object-based media: a toolkit for building responsive content. In: Proceedings of the 32nd International BCS Human Computer Interaction Conference (HCI). BCS Learning and Development Limited (2018)

3. Armstrong, M., Brooks, M., Churnside, A., Evans, M., Melchior, F., Shotton, M.: Object-based broadcasting - curation, responsiveness and user experience, Tech. Rep., WHP 285, The British Broadcasting Corporation (2014)
4. Brown, A., et al.: Dynamic subtitles: the user experience. In: Proceedings of the ACM International Conference on Interactive Experiences for TV and Online Video, pp. 103–112. TVX 2015, Association for Computing Machinery, New York, NY, USA (2015). https://doi.org/10.1145/2745197.2745204
5. BT Sport: Man City vs Liverpool (1–1, 5–4 on pens) | 2019 Community Shield highlights (2019). https://www.youtube.com/watch?v=k9_tz9bi3rs
6. Evans, M., et al.: Creating object-based experiences in the real world. In: IBC2016. IBC365 (2016)
7. Hong, R., et al.: Video accessibility enhancement for hearing-impaired users. ACM Trans. Multimedia Comput. Commun. Appl. **7S**(1), 1–19 (2011). https://doi.org/10.1145/2037676.2037681
8. Hu, Y., Kautz, J., Yu, Y., Wang, W.: Speaker-following video subtitles. ACM Trans. Multimedia Comput. Commun. Appl. **11**(2), 32 (2015). https://doi.org/10.1145/2632111
9. Lin, T.-Y., et al.: Microsoft COCO: common objects in context. In: Fleet, D., Pajdla, T., Schiele, B., Tuytelaars, T. (eds.) ECCV 2014. LNCS, vol. 8693, pp. 740–755. Springer, Cham (2014). https://doi.org/10.1007/978-3-319-10602-1_48
10. Rayner, K.: The perceptual span and peripheral cues in reading. Cogn. Psychol. **7**(1), 65–81 (1975). https://doi.org/10.1016/0010-0285(75)90005-5
11. Redmon, J., Farhadi, A.: YOLOv3: an incremental improvement. arXiv preprint arXiv:1804.02767 (2018)

Beyond Vision: Physics Meets AI (BVPAI)

A Variational AutoEncoder for Model Independent Searches of New Physics at LHC

Giulia Lavizzari[1,2](✉), Giacomo Boldrini[1,2], Simone Gennai[2], and Pietro Govoni[1,2]

[1] University of Milano-Bicocca, Ed. U2, Piazza della Scienza 3, 20126 Milan, Italy
g.iavittari1@campus.unimib.it
[2] INFN Milano-Bicocca, Ed. U2, Piazza della Scienza 3, 20126 Milan, Italy

Abstract. We present a feasibility study for the use of a generative, probabilistic model, a Variational Autoencoder (VAE), to detect deviations from Standard Model (SM) physics in an electroweak process at the Large Hadron Collider (LHC).

The new physics responsible for the anomalies is described through an Effective Field Theory (EFT) approach: the SM Lagrangian is Taylor-expanded and the higher order terms cause deviations in the kinematic distributions of the observables, and are thus identified by the model as anomalous contributions with respect to SM. Since the training of the model involves almost only SM events, the proposed strategy is largely independent from any assumption on the nature of the new physics signature. To test the proposed strategy we use parton level generations of Vector Boson Scattering (VBS) events at the LHC, assuming an integrated luminosity of 350 fb^{-1}.

Keywords: Variational AutoEncoder · Anomaly Detection · Vector Boson Scattering

1 Introduction

The discovery of the Higgs boson by the ATLAS and CMS collaborations [1,2] in 2012 marked a major milestone in the validation of the Standard Model (SM) of particle physics. The SM theory [3] encapsulates our best understanding of the fundamental structure of matter by describing all the known elementary particles, and three of the four fundamental forces that govern the interactions between them. So far, the SM has been extremely successful in providing accurate experimental predictions and theoretical explanations for many phenomena. Nevertheless, many questions still remain unanswered, indicating the need for an extension of the SM to include new physics contributions: for example the presence of dark matter [4], the relative abundance of matter over antimatter [5] and the so-called hierarchy problem [6]. However, while the CERN Large Hadron Collider (LHC) has collected an unprecedented amount of data, no significant deviations from the SM have been observed so far. Since the number of possible

G. L. Foresti et al. (Eds.): ICIAP 2023 Workshops, LNCS 14365, pp. 159–169, 2024.
https://doi.org/10.1007/978-3-031-51023-6_14

theories Beyond the Standard Model (BSM) and their variations in terms of free parameters is extremely large, and only few of them can be tested, it is possible that so far we have simply chosen the wrong theories to look for.

As a result, many efforts are currently directed towards the development of agnostic analyses with respect to the underlying BSM theory assumed [7–10]. In this framework, Unsupervised Learning represents one of the most popular and powerful choices: the underlying idea is to train a model on known physics and then to use it to detect outliers, i.e. data that follow different patterns from the ones it learnt to recognize during training. In this way the model simply performs an anomaly detection task, without needing any knowledge of a specific new physics signature.

The model we chose to adopt is a Variational AutoEncoder (VAE) [11,12], a generative algorithm that belongs to the family of probabilistic graphical models. The VAE has a similar structure to that of standard AutoEncoders, which operate a dimensionality reduction and then map the lower dimensional representation back to the input space. VAEs, however, feature a regularized latent space to allow the generative task, namely they allow for a well-defined sampling procedure. The VAE makes use of a neural network to encode an input distribution over a multi-dimensional latent space, from which a point is then sampled and decoded through a second network. The model thus learns a probabilistic representation of the data and uses it to generate new samples from the latent space, reconstructing an output as similar as possible to the given input. VAEs are therefore traditionally used mainly for the production of new content, but their generative capabilities can be leveraged also to build a robust anomaly detection strategy. Anomaly detection is the task of identifying unusual or unexpected patterns in a dataset: since VAEs learn to decode generated samples drawn from the same probability distribution of the original dataset, the training is more robust and tolerant to variations in such data than for example standard AutoEncoders. In our case the training sample is a collection of kinematic distributions deriving from a SM process. Once the model is trained, it is asked to reproduce a sample comprising some new physics contributions: since such events follow different patterns than the ones the model learnt during training, they will be reconstructed poorly. By building a metric that accounts for the difference between input and output event by event, one can then isolate the anomalous contributions due to new physics.

To effectively test such a strategy, one needs to have a general but still predictive theory that regroups as many BSM processes as possible. With this in mind, we chose to work within the Standard Model Effective Field Theory (SMEFT) [13] framework, where the SM is seen as the low energy approximation of a more general theory. The low energy footprints of such theory can be parametrized as higher order operators to be added to the SM Lagrangian, which alter the expected kinematic distributions of a given process. SMEFT is therefore a generalized SM extension that can be used to describe small deviations of SM observables. Our aim is to isolate new physics signals while introducing only a minimal dependence on the operators chosen to optimize the analysis.

2 The Physics Use-Case: An Effective Field Theory Interpretation of Vector Boson Scattering

2.1 Same Sign WW Scattering

The process we employed to test the proposed approach is the scattering of vector bosons (VBS) at the LHC. In this event, two partons from the incoming protons radiate vector bosons, which subsequently interact. Such process is an ideal domain for exploring BSM physics, since it can serve as a mean to investigate modifications to the SM electroweak sector. As a matter of fact, VBS is closely linked to the electroweak symmetry breaking mechanism, as in a Higgs-less SM theory its cross-section would diverge at high center-of-mass energies.

Our specific focus was on investigating the set of VBS processes that result in a final state comprising two same-sign W bosons decaying leptonically (Fig. 1). Such process yields a distinct signature in the detector, characterized by two jets with a significant invariant mass in the forward region, two same-sign charged leptons, and missing transverse energy.

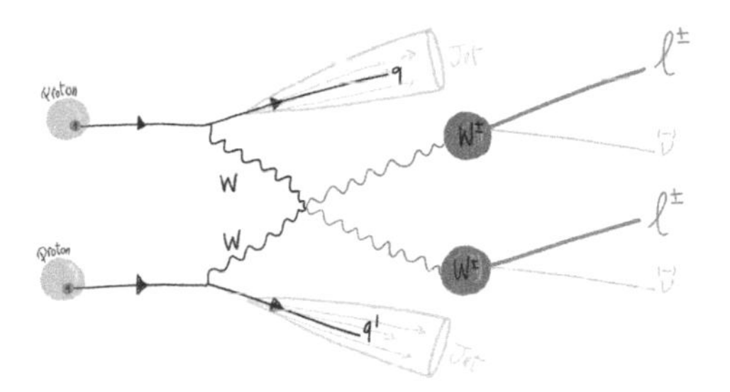

Fig. 1. Topology of same sign WW scattering in the fully leptonic final state.

This study was conducted neglecting any background process that populates the aforementioned final state. We utilized generator-level observables associated with the kinematics of the charged leptons and the final-state partons originating from the initial scattering.

2.2 Modeling the Anomalies: The SM as an Effective Field Theory

In an Effective Field Theory (EFT) interpretation of the SM, a more comprehensive theory is expected to incorporate new matter content at energies well beyond the scale of the LHC. The effects of this theory at low energy can be parametrized by additional terms obtained from the expansion of the SM Lagrangian [14–16]:

$$\mathcal{L}_{EFT} = \mathcal{L}_{SM} + \sum_{i,d>4} \frac{c_i}{\Lambda^{d-4}} \mathcal{O}^{(d_i)} \tag{1}$$

where $\mathcal{O}^{(d_i)}$ denotes a set of dimension d_i operators, c_i represent the Wilson Coefficients quantifying the intensity of these operators, and Λ corresponds to the energy scale of the new physics.

In this study we focus only on dimension 6 operators [17], which represent the first non-zero terms after the SM Lagrangian. Odd-dimensional operators are not taken into account since they would violate the accidental symmetries of the SM. Furthermore, we chose to consider the effect of one single operator at a time. In terms of probability amplitude this translates in the following expression, valid for each operator:

$$A_{BSM} = A_{SM} + \frac{c_\alpha}{\Lambda^2} \cdot \mathcal{A}_{\mathcal{Q}_\alpha} \tag{2}$$

Squaring A_{BSM} yields a quantity proportional to the event probability:

$$|A_{BSM}|^2 = |A_{SM}|^2 + \frac{c_\alpha}{\Lambda^2} \cdot 2Re(A_{SM}A^{\dagger}_{\mathcal{Q}_\alpha}) + \frac{c_\alpha^2}{\Lambda^4} \cdot |\mathcal{A}_{\mathcal{Q}_\alpha}|^2 \tag{3}$$

Here, the first term represents the pure SM contribution, while the following ones introduce a linear (LIN) and quadratic (QUAD) dependence on the EFT amplitudes. These operators collectively modify the kinematic distributions of the given process, as illustrated in Fig. 2. The extent of these modifications depends on the relative weight of the operators.

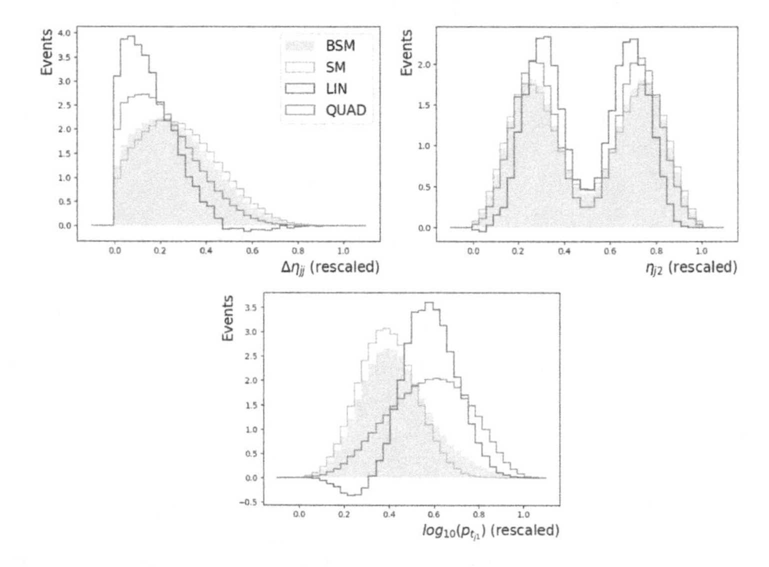

Fig. 2. SM, LIN and QUAD distributions (solid colored lines) for different observables: $\delta\eta_{jj}$ (top left), η_{j2} (top right) and $\log_{10}(p_{T,j1})$ (bottom). The distributions are normalised to an area of 1. The weighted sum of the three contributions gives the overall BSM distribution (filled histogram).

For our analysis, we utilized the SMEFTsim package [18,19] to generate EFT predictions at leading order in the U35 flavour scheme and m_W input scheme for the following operators: $\mathcal{Q}_W$, $\mathcal{Q}_{Hq}^{(1)}$, $\mathcal{Q}_{HW}$, $\mathcal{Q}_{qq}^{(1)}$, $\mathcal{Q}_{qq}^{(1,1)}$, $\mathcal{Q}_{qq}^{(3)}$, and $\mathcal{Q}_{qq}^{(3,1)}$. The parton-level events were generated using MadGraph5_aMC@NLO [23].

3 Variational AutoEncoders

VAEs are probabilistic generative models that allow for building a latent representation of an input, which is then used to reconstruct an output as similar to the input as possible. They were introduced for the first time in 2013 [11] and since then they have found many applications in various fields, mainly for content generation [20] and noise reduction [21]. However, thanks to their generative capabilities, VAEs are also a powerful tool for building anomaly detection strategies that are robust and tolerant to fluctuations in the template data.

3.1 The VAE Architecture

The VAE architecture is organized as shown in Fig. 3: a first deep neural network, known as encoder, maps the distributions of the input into a lower dimensional latent space. The training procedure forces the latent space to be regular, namely to be described by a Gaussian distribution, allowing for more flexibility and expressive power in the model. A second specular network, known as decoder, implements a sort of inverse transformation mapping the latent space back to the input one, thus producing an output that is as similar to the input as possible.

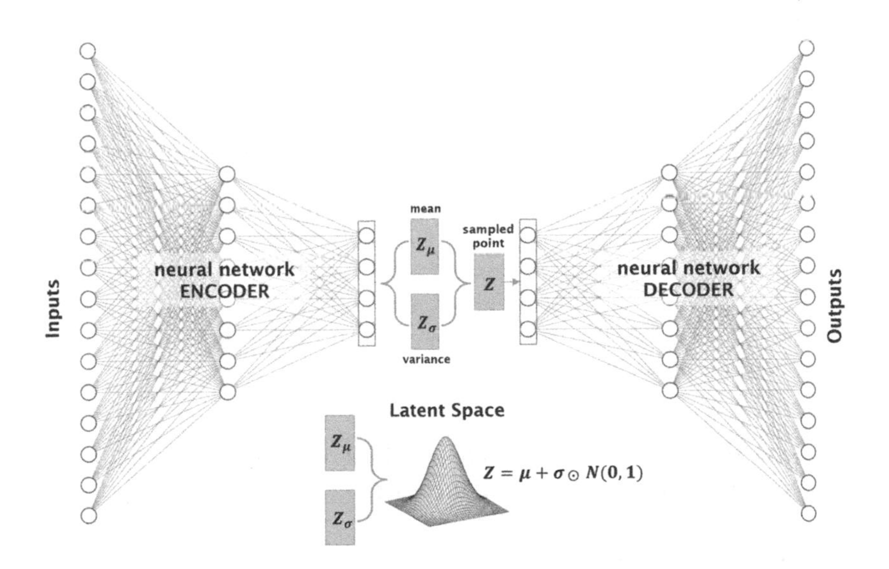

Fig. 3. Schematic representation of the VAE architecture

The model is trained by minimizing a loss function that comprises two terms: the Mean Squared Error (MSE) between input and output, to ensure a good reconstruction of the inputs, and the so-called Kullback-Leibler Divergence (KLD) [22], which accounts for the regularization of the latent space by forcing the latent distributions to be close to gaussians.

3.2 Anomaly Detection with VAEs

Anomaly detection is the task of finding unusual patterns in a dataset: in our case this means identifying events generated by BSM processes, based on the difference in their signature with respect to the known physics.

The VAE model is trained only on SM events, therefore it learns the underlying patterns of known physics. Once trained, the model is evaluated also on BSM events: since the underlying mechanisms are different from the ones the model learnt during training, the output distributions present differences with respect to the input ones (Fig. 4).

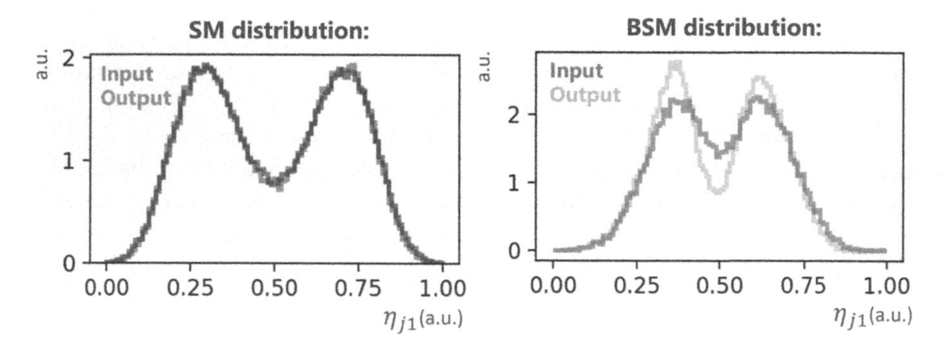

Fig. 4. Comparison between the input and output of the model in the case of SM and BSM distributions for the pseudorapidity of the leading jet.

To assess the quality of the reconstruction of an event we compute the MSE between input and output, averaged on all the observables. By selecting the events for which the MSE is greater than a certain threshold (Fig. 5), we are able to identify an anomaly-enriched region.

4 Embedding a Classification Step in the Training to Optimize for Discrimination

The main limitation of the previously described anomaly detection strategy is that even though our goal is the isolation of EFT events, during the training the model is only optimized to reconstruct a SM sample and the choices that improve the SM reconstruction are not always optimal for discrimination. For example, during our studies we have noticed that a larger latent space produces a better reconstruction for SM events but also decreases the discrimination power

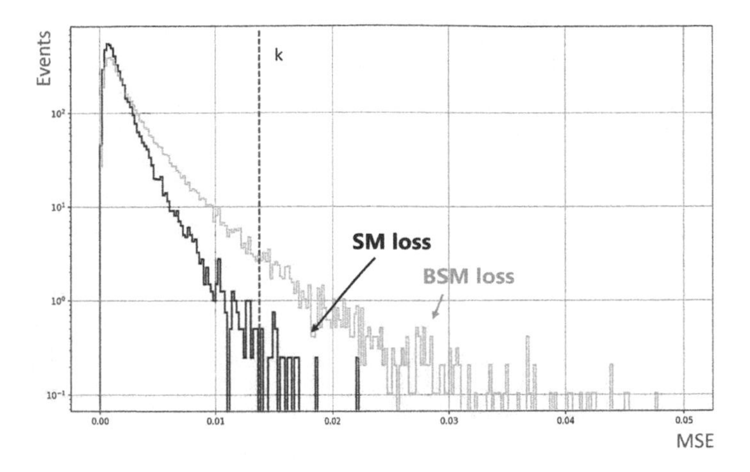

Fig. 5. MSE for SM (blue) and BSM (red) events. Anomalous events lie in the tail of the loss function. (Color figure online)

against EFT perturbations. A possible explanation for this phenomenon is that a larger and therefore more complex latent space allows for the model to better learn the underlying data structure. This results in a better reconstruction of the SM inputs, but also in an improvement in the ability of the model to extrapolate correctly the BSM patterns. Therefore, BSM events result as well better reconstructed and the overall discrimination power of the model is degraded.

In order to overcome such limitation, we introduced a classification step in the training procedure by adding a third component to the model, namely a two-layers neural network that works as a classifier. This allows for fine-tuning the VAE model while performing also a classification task, aiming at enhancing the discrimination power of the model.

This new model is trained both on SM and on a small fraction of EFT events: the VAE part is trained to reconstruct the SM subset via the minimization of the MSE and KLD, then the resulting MSE and KLD losses are given as inputs to the classifier, which is in turn tuned by minimizing a binary cross-entropy. This mechanism allows for embedding the discrimination process within the training, but at the price of gaining an additional dependence on the model used to describe the new physics contribution. To keep the strategy as model independent as possible, we decided to use a single operator during training, and to later evaluate the performances of the model on the other ones as well. The embedding of the auxiliary classifier and this training strategy proved to play a crucial role in enhancing the discrimination power of the model, as shown by the results presented in the following section, while preserving an almost model-independent approach.

Figure 6 illustrates the structure of our model: encoder, decoder and classifier are separate architectures comprising several densely connected layers, later combined into an end-to-end model. Encoder and decoder comprise three layers of dimension 20,10 and 7, respectively, all with a Leaky ReLU activation function. The dimension of the latent space proved to have a strong impact on the performance of the model, and was varied between 3 and 7, yielding the effects described in the previous sections. The classifier comprises two layers, the first having dimension 50 and a linear activation function, the second with dimension 1 and a hard sigmoid activation. The inputs comprise several kinematic and angular distributions encoding the most important features of the physics process considered, namely: the invariant masses of the dilepton and dijet systems (m_{ll} and m_{jj}), the transverse momentum of the two leptons and two jets ($p_{t_{l1}}$, $p_{t_{l2}}$, $p_{t_{j1}}$, $p_{t_{j2}}$), the transverse momentum of the dilepton system ($p_{t_{ll}}$), the missing transverse momentum (MET), the pseudorapidities of leptons and jets (η_{l1}, η_{l2}, η_{j1}, η_{j2}) and the pseudorapidity and azimuthal angle differences between the two jets ($\Delta\eta_{jj}$, $\Delta\Phi_{jj}$).

First we scaled the input distributions between 0 and 1 and we computed the logarithm of the kinematic variables to reduce their dynamic range. We split the SM sample in two subsets comprising 80% and 20% of the total 900000 events, respectively used for training and testing. The whole strategy is implemented through the scikit-learn [24] and TensorFlow [25] libraries.

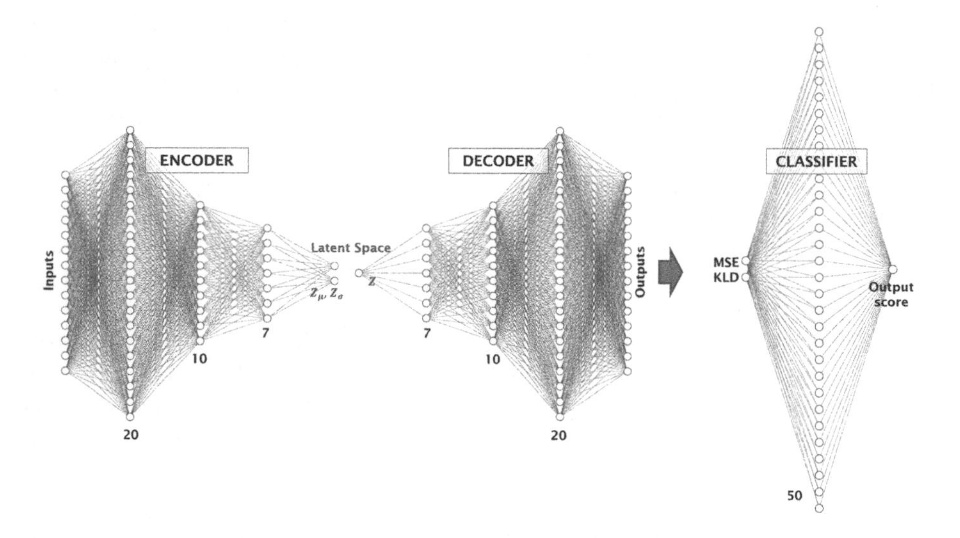

Fig. 6. The VAE model employed. The input variables comprise kinematic and angular distributions of the leptons and the jets.

4.1 Results

To quantify the discrimination power of the VAE and VAE+NN we evaluated the difference between the number of signal and background events, namely the EFT and SM, in the EFT-enriched region isolated by the models. Given our theoretical framework, the amount of events predicted following a BSM prior minus the events predicted considering only the SM contributions equals the amount of events predicted considering the LIN and QUAD operators. Therefore the proxy metric for the significance is the following:

$$\sigma(c_{op}) = \frac{|BSM(c_{op}) - SM|}{\sqrt{SM}} = \frac{|LIN(c_{op}) + QUAD(c_{op}^2)|}{\sqrt{SM}} \tag{4}$$

σ depends on the Wilson coefficients of the operator considered while testing the model. A model is considered sensitive to a given operator if σ is equal or greater than 3 at least for some value of c_{op}.

Our goal is assessing an estimate of the lower detection limits induced by VAE-based models for the various Wilson coefficients. Therefore, for each EFT operator we test the trained model on sets of EFT events obtained considering different values of the Wilson coefficient, ranging between 0.1 and 0.9. For each dataset we computed $\sigma(c_{op})$ and retained the lowest value of c_{op} (if any) for which $\sigma(c_{op}) \geq 3$, listed in Table 1. The smaller the value of c_{op} the model is sensitive to, the better. All the results are obtained considering an integrated luminosity of 350 fb^{-1}, in anticipation of the statistics made available by the LHC Run 3. The results for the VAE+NN refer to a model trained to discriminate against the $\mathcal{Q}_W$ operator: during training c_W is set to 1.

From these results, we can draw several conclusions. Firstly, it is possible to distinguish EFT contributions from known physics using the VAE model. Furthermore, the incorporation of the classifier into the model enhances discrimination capabilities without excessively compromising the generality of the approach. The performance notably improves for $\mathcal{Q}_W$, the operator on which the model was trained, as well as for $\mathcal{Q}_{qq}^{(1)}$ and $\mathcal{Q}_{qq}^{(1,1)}$. The degradation in performance is minimal for $\mathcal{Q}_{qq}^{(3)}$, $\mathcal{Q}_{qq}^{(3,1)}$, and $\mathcal{Q}_{HW}$. Notably, the new model exhibits sensitivity to $\mathcal{Q}_{Hq}^{(1)}$ which was not detected by the simple VAE model, as we are only letting c_{op} vary through a limited range.

Table 1. The value of c_{op} for which $\sigma(c_{op}) = 3$, considering an integrated luminosity of 350 fb^{-1}. The results for the VAE+NN refer to a model trained to reconstruct SM VBS EWK events and to discriminate them from a sample comprising contribution from the $\mathcal{Q}_W$ operator. In the case of c_{Hq}^1, we could not obtain $\sigma = 3$ with the simple VAE, while the VAE+NN proved to be sensitive to such operator.

model	c_W	c_{qq}^1	$c_{qq}^{1,1}$	c_{qq}^3	$c_{qq}^{3,1}$	c_{Hq}^1	c_{HW}
VAE	0.34	0.56	0.29	0.04	0.06	-	0.41
VAE+NN	0.13	0.17	0.18	0.11	0.11	0.61	0.65

5 Conclusions and Future Perspectives

This study demonstrates the feasibility of employing a VAE-like model to isolate an EFT-enriched region. By training the model to recognize known SM physics patterns exclusively, we were able to detect BSM phenomena in a model-independent manner. The effectiveness of this approach was assessed using the scattering of same-sign W bosons, considering an integrated luminosity of 350 fb^{-1}. The obtained results align with the expected outcome of this strategy, which aims to identify a region enriched with potentially interesting events rather than optimizing the signal selection efficiency for specific BSM models.

However, it is important to note that while this study showcases the effectiveness of unsupervised models in detecting EFT perturbations within the SM VBS SSWW signature, a more comprehensive analysis is required. This analysis should incorporate other sources of backgrounds in order to assess the concrete sensitivity reach of the proposed strategy. Further investigations in this direction should specifically address major sources of background for the $2l2\nu + 2j$ final state, such as events with nonprompt or fake leptons, as well as the small QCD-induced production.

References

1. ATLAS Collaboration, Observation of a new particle in the search for the Standard Model Higgs boson with the ATLAS detector at the LHC, Phys. Lett. B 716 (2012) 1–29. arXiv:1207.7214v2
2. CMS Collaboration: Observation of a New Boson at a Mass of 125 GeV with the CMS Experiment at the LHC. Phys. Lett. B **716**, 30–61 (2012). arXiv:1207.7235v2
3. Workman, R.L., et al.: (Particle Data Group), The Review of Particle Physics (2023). Prog. Theor. Exp. Phys. **2022**, 083C01 (2022) and 2023 update
4. Perez Adan, D. (on behalf of the ATLAS and CMS Collaborations): Dark Matter searches at CMS and ATLAS, Rencontres de Moriond 2022: Proceedings of the ElectroWeak Session (2022, La Thuile, Italy). arXiv:2301.10141v1
5. LHCb Collaboration: Measurement of antiproton production from antihyperon decays in pHe collisions at $\sqrt{s_{NN}} = 110$ GeV. Eur. Phys. J. C **83**, 543 (2023). arXiv:2205.09009v2
6. Koren, S.: The Hierarchy Problem: From the Fundamentals to the Frontiers, arXiv e-prints (2020). arXiv:2009.11870v1
7. Collaboration, C.M.S.: MUSiC: a model-unspecific search for new physics in proton-proton collisions at $\sqrt{s} = 13TeV$. Eur. Phys. J. C **81**, 629 (2021). arXiv:2010.02984v2
8. Kasieczka, G., et al.: The LHC Olympics 2020: a community challenge for anomaly detection in high energy physics. Rep. Prog. Phys. **84**, 124201 (2021). arXiv:2101.08320
9. Camaiani, B., et al.: Model independent measurements of Standard Model cross sections with Domain Adaptation. Eur. Phys. J. C **82**, 921 (2022). arXiv:2207.09293v3
10. Krzyzanska, K., Nachman, B.: Simulation-based anomaly detection for multileptons at the LHC. JHEP **2023**, 61 (2023). arXiv:2203.09601v1
11. Kingma, D.P., Welling, M.: Auto-Encoding Variational Bayes, arXiv e-prints (2013). arXiv:1312.6114v11
12. Kingma, D.P., Welling, M.: An Introduction to Variational Autoencoders, Foundations and Trends in Machine Learning: Vol. 12 (2019): No. 4. arXiv:1906.02691v3

13. Ellis, J.: SMEFT Constraints on New Physics Beyond the Standard Model, Contribution to the Proceedings of the BSM-2021 Conference (2021, Zewail City, Egypt), arXiv:2105.14942
14. Buchmuller, W., Wyler, D.: Effective lagrangian analysis of new interactions and flavour conservation. Nucl. Phys. B **268**, 621–653 (1986)
15. Degrande, C., et al.: Effective field theory: a modern approach to anomalous couplings. Annals Phys. **335**, 21 (2013). arXiv:1205.4231v1
16. Brivio, I., Trott, M.: The standard model as an effective field theory. Phys. Rept. **793**, 1–98 (2018). arXiv:1706.08945v3
17. Grzadkowski, B., et al.: Dimension-six terms in the standard model Lagrangian. JHEP **10**, 85 (2010). arXiv:1008.4884v3
18. Brivio, I., et al.: The SMEFTsim package, theory and tools. JHEP **12**, 70 (2017). arXiv:1709.06492v2
19. Brivio, I.: SMEFTsim 3.0 - a practical guide, JHEP 04 (2021) 73. arXiv:2012.11343v3
20. Van Den Oord, A., et al.: Neural Discrete Representation Learning, Proceedings of the 31st International Conference on Neural Information Processing Systems (2017, Long Beach, California, USA), pp. 6309–6318. arXiv:1711.00937v2
21. Im, D.J., et al.: Denoising Criterion for Variational Auto-Encoding Framework. In: Proceedings of the Thirty-First AAAI Conference on Artificial Intelligence (2017, San Francisco, California, USA) 2059–2065. arXiv:1511.06406v2
22. Shlens, J.: Notes on Kullback-Leibler Divergence and Likelihood, arXiv e-prints (2014). arXiv:1404.2000v1
23. Alwall, J., et al.: The automated computation of tree-level and next-to-leading order differential cross sections, and their matching to parton shower simulations. JHEP **07**, 079 (2014). arXiv:1405.0301v2
24. Pedregosa, F., et al.: Scikit-learn: machine learning in python. JMLR **12**, 2825–2830 (2011). arXiv:1201.0490v4
25. Abadi, M., et al.: TensorFlow: large-scale machine learning on heterogeneous distributed systems. arXiv:1603.04467v2

Adaptive Voronoi Binning in Muon Radiography for Detecting Subsurface Cavities

A. Paccagnella[1,2](✉), V. Ciulli[1,2], R. D'Alessandro[1,2], L. Bonechi[2], D. Borselli[2], C. Frosin[1,2], S. Gonzi[1,2], and T. Beni[2]

[1] Dipartimento di Fisica e Astronomia, Università di Firenze, Via G. Sansone 1, 50019 Sesto Fiorentino (Firenze), Italy
andrea.paccagnella@unifi.it

[2] Istituto Nazionale di Fisica Nucleare, Sezione di Firenze, Via B. Rossi 3, 50019 Sesto Fiorentino (Firenze), Italy

Abstract. Muon radiography is an advanced imaging technique that utilizes cosmic muons to visualize the interior of structures and materials, making it highly valuable for subsurface investigations. In this study, we present a measurement conducted using muon radiography at the Temperino mine.

We demonstrate the application of an adaptive binning approach using Voronoi tessellation to enhance image visualization and improve cavity detection.

The results reveal that the adaptive binning technique significantly improves the visibility of regions with cavities.

The combination of muon radiography and adaptive binning through Voronoi tessellation showcases its potential as a powerful tool for subsurface exploration and geological studies, providing a more accurate and reliable approach for cavity detection and characterization.

Keywords: Muon Radiography · Voronoi · imaging

1 Introduction to Muon Radiography

Muon radiography is a technique that utilizes muons from cosmic rays to reconstruct images of otherwise difficult-to-access environments. This technique offers several advantages, including the absence of accelerators to generate the particles interacting with the target under investigation. Moreover, it is a non-invasive technique, both for humans and the object being observed. Additionally, due to the muons' ability to penetrate dense materials over long distances, it is suitable for studying large structures such as mines, hills, and pyramids, as well as highly dense materials like containers of radioactive waste and blast furnaces.

1.1 Cosmic Rays

Cosmic rays are high-energy particles, mainly coming from outside the Solar System. These are generally called primary cosmic rays.

G. L. Foresti et al. (Eds.): ICIAP 2023 Workshops, LNCS 14365, pp. 170–178, 2024.
https://doi.org/10.1007/978-3-031-51023-6_15

When primary cosmic rays enter the Earth's atmosphere they interact with atoms and molecules, especially oxygen and nitrogen, through strong interaction and electromagnetic processes (like pair production). The collision produces a cascade of lighter particles, called secondary cosmic rays, including X-rays and γ, neutrons, mesons (such as pions and kaons), electrons and muons. However, most of the produced particles are unstable particles (i.e. pions and kaons)and tend to decay mostly to muons [1].

The muon belongs to the lepton family and has a mass of 106 MeV/c^2 (about 200 times that of the electron). It is an unstable particle and has a mean life of about 2.2 μs [2]. This lifetime may seem extremely short, but it is sufficient to allow muons to travel long distances in the atmosphere, this is explained via the laws of special relativity. Most muons are created at altitudes of about 15 km by primary cosmic rays and travel with other particles to the Earth in conical showers. They are the most abundant energetic particles arriving at sea level, with a flux of about 1 muon per square centimetre per minute ($\Phi = 1/(\mathrm{cm}^2\mathrm{s})$). The muon flux depends not only on the zenith angles, but also on the muon energy distribution.

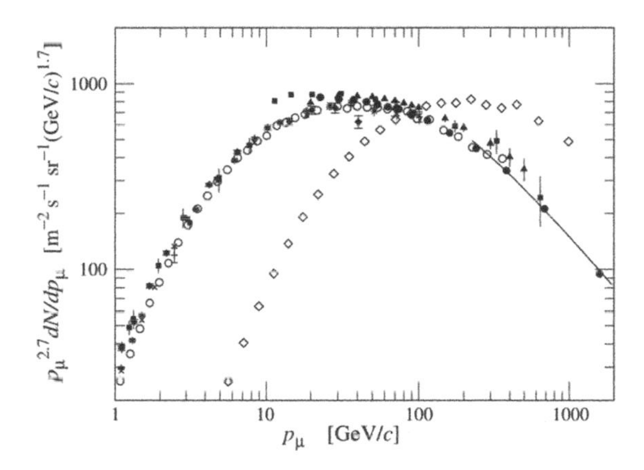

Fig. 1. The spectrum of muons at sea level for $\theta = 75°$ is indicated by the empty diamonds while the other markers refer to the muon spectrum for $\theta = 0°$. The average energy of muons increases as the zenith angle of origin increases. The ordinate is multiplied by $p_\mu^{2.7}$ to compress the spectra. [3]

Figure 1 shows the energy spectrum of cosmic ray muons at sea level for two different zenith angles: for large θ (zenith angle) values, the layer of atmosphere that cosmic rays must cross is greater. Therefore the low-energy muons decay before reaching the ground while the energetic pions have time to decay before interacting, thus increasing the average energy of the muons: this cause a zenith angular muon enhancement [4].

1.2 Muon Radiography

The muon radiography technique exploits the penetration capacity of muons present in cosmic rays to make radiography of very large targets. To date, this is the sole technique that employs cosmic ray muons as a means of investigating relatively dense structures. Since muons are less likely to interact, stop and decay in low density matter than in high density matter, a larger number of muons will travel through the low density regions of target objects in comparison to higher density regions. Muography is a non-invasive technique since it exploits natural radiation present on the entire surface of the earth and we are continuously traversed by these particles.

One of the first muographic measures were those described in [5]. An example of muographic measurement conducted in Italy is [6].

In this study, we intend to present an exemplary measurement conducted using transmission technique at the Temperino mine. Further details about this technique will be provided in the following section.

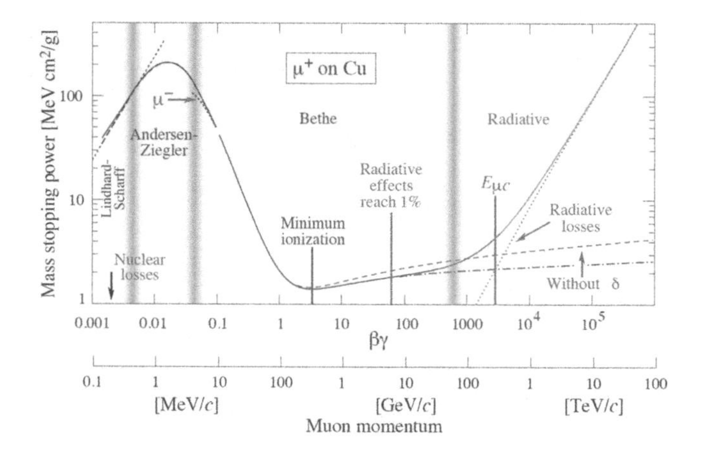

Fig. 2. Stopping power of muons in copper as a function of the kinetic energy [7]. Our energy range of interest is between Minimun Ionizatiom and Radiative Losses.

Transmission Technique. Muon Transmission radiography is based on the energy loss (and the eventual decay) of the muons when they interact with matter, this is shown in the graph in the Fig. 2. A muon with a given momentum $p_{\min}$, before being stopped, will be able to cross on average a certain thickness of material to which we can associate an opacity X. The relationship between muon momentum and average crossed opacity can be found Ref. [8]. Fixed a certain angular direction (θ, φ) and a certain opacity, the expected value of the flux transmitted through the target can be estimated as the integral from $p_{\min}$ to infinity of the differential flux in that direction ($\Phi_{transmitted} = \int_{p_{\min}}^{\infty} d\Phi(\theta, \varphi)\, dp$) [8].

The methodology employed to generate a two-dimensional image of the inner structure of the target consists of three steps. The first step involves observing the number of counts in the presence of the target ($N_{target}(\theta, \varphi)$). The number of muons in each direction depends on the structure of the target, the direction of observation, and the detector's efficiency. The effect of the latter can be reduced by performing the second step, which is the measurement of the free-sky ($N_{free-sky}(\theta, \varphi)$), as it would simplify the relationship between the two measurements by being the same, given that both measurements are conducted with the same detector. This measurement involves detecting the counts in the free-sky with the same detector orientation as the target measurement. By taking the ratio of counts in both configurations, we obtain the measured transmission:

$$T_{measured} = \frac{N_{target}(\theta, \varphi)}{N_{free-sky}(\theta, \varphi)} \cdot \frac{\Delta t_{free-sky}}{\Delta t_{target}}, \tag{1}$$

where Δt_{target} and $\Delta t_{free-sky}$ are the acquisition times in the target configuration and free-sky configuration, respectively.

The third step involves comparing the measured transmission with the expected transmission. The expected transmission can be obtained from simulations disposing of an accurate model that reproduces the differential flux of cosmic rays on the Earth's surface and the known geometry of the target (For more details see [4]). By varying the density of the target ρ in the simulation, a certain number of simulated transmissions $T_{\text{expected}}(\theta, \varphi, \rho)$ are obtained (in this study, we only consider the simulated transmission obtained at a constant density of $2.65\,\text{g/cm}^3$, the standard rock density). The relative transmission is obtained by taking the ratio of the measured transmission to the simulated transmission:

$$T_{relative} = \frac{T_{measured}}{t_{expected}}. \tag{2}$$

Thanks to this last measurement, we obtain a map of relative transmission. Since the transmission is inversely proportional to the density, when the latter is greater than one, it indicates the presence of a region where the density is lower compared to the simulated density. Conversely, when the transmission is less than one, it indicates a denser region.

By further varying the density in the simulation and identifying areas with a relative transmission equal to one, a density map of the studied area can be obtained. For more details and examples of typical plots please refer to [9].

2 The MIMA Detector and Its Application at the Temperino Mine

The muon hodoscope utilized for the reference measurements described in this paper is the MIMA detector (Muon Imaging for Mining and Archaeology). MIMA was designed by the muon radiography group of INFN and the University of Florence [10]. The assembly and calibration of the MIMA detector took place between the end of 2016 and the first half of 2017. It has been employed for muography measurements in various fields, including geology, archaeology, and civil engineering.

The MIMA detector, housed within an aluminium protective cover, has a cubic shape with approximate dimensions of $(50 \times 50 \times 50)$ cm^3. It is mounted on a rotating support that enables altazimuth orientation. The detector (Fig. 3) consists of six tracking planes organized in orthogonal pairs, forming three compact X-Y modules. Each tracking plane features an active surface area of (40×40) cm^2 and comprises 21 plastic scintillators measuring 40 cm in length. These scintillators have a triangular cross-section, with a 4 cm base and a height of 2 cm. The MIMA detector has an angular resolution of 7 milliradians (mrad) and an acceptance angle of approximately 60°.

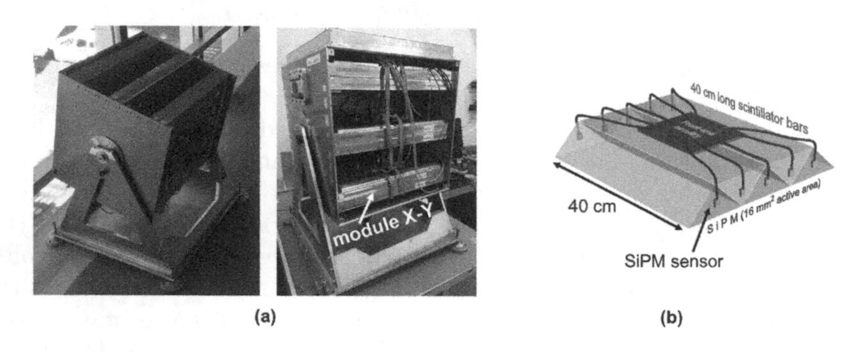

Fig. 3. The MIMA muon tracker. In (a) the detector with the structure of 3 X and Y modules of tracking planes. In (b) the structure of a single plane consisting of triangular scintillator bars (for simplicity only five bars) is shown.

For the following results, we consider the first measurement conducted at the Temperino mine as our main reference. In this particular instance, the MIMA detector was positioned inside the mine with a vertical orientation, namely at $\theta = 0°$ [9]. In Fig. 4, the transmission map obtained without applying any filters can be seen. In this figure, it is possible to distinguish brighter areas from darker areas. The brighter areas, associated with a higher relative transmission, correspond to regions with a lower density than expected, and in this case, they represent cavities. On the other hand, the darker areas, associated with a lower relative transmission, are linked to regions with a higher density than expected. These distinctions in relative transmission help to identify and characterize different features and anomalies within the object under investigation, in this case the Temperino mine, providing valuable information for further analysis and understanding of the studied system.

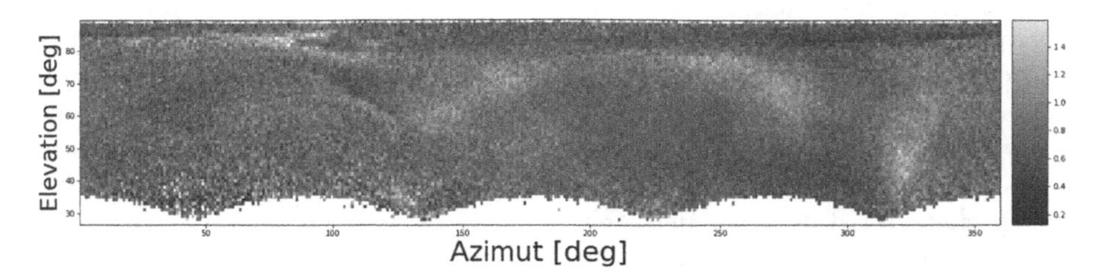

Fig. 4. The relative transmission of measurement 1 conducted at the Temperino mine in altazimuth coordinates, The color bar on the right represents the relative transmission value. [9].

3 Adaptive Tessellation Through Voronoi Binning

When collecting particle counts, as done with the MIMA detector, they are influenced both by the geometry of the detector and the flux variation with the azimuth angle (as describe in Sect. 1.1. In fact, the counts will be higher at the centre of the detector and decrease towards the edges. Adaptive tessellation through Voronoi binning is a technique used for adaptive binning. Voronoi makes each bin of different sizes according to the local signal-to-noise ratio. Then, each data point is assigned to the cell whose centroid is closest to it. By utilizing Voronoi binning, the size and shape of each bin are automatically determined based on the distribution of data points and the local signal-to-noise ratio (S/N).

In this adaptive scheme, regions with a low S/N ratio will result in larger Voronoi cells, accommodating a greater number of data points within each bin. Conversely, regions with a high S/N ratio will yield smaller Voronoi cells, allowing for a finer spatial resolution and more precise representation of the data. This approach ensures that the binning process adapts to the local S/N characteristics of the detector, enhancing the accuracy and reliability of the results obtained.

The procedure for transitioning from one-to-one binning to Voronoi binning is typically carried out through the following steps:

1. Start with the initial one-to-one binning, where each data point is assigned to its bin.
2. Calculate the local S/N ratio for each bin. This can be done by considering the signal and noise levels within each bin or by estimating them based on neighbouring bins, In our case the signal $S = N_i$ is the number of counts while the noise is its square root $N = \sqrt{N_i}$.
3. Determine the desired level of adaptivity for Voronoi binning. This may involve setting thresholds or criteria based on the S/N ratio. In our case we want each time a bin is merged that the new one has a fraction of the S/N equal to 80% of the default one, this in turn was obtained as 2 times the square root of the maximum count ($(S/N)_T = \sqrt{N_{max}}$).
4. Identify bins that meet the adaptivity criteria. These are typically bins with a low S/N ratio, indicating regions where larger bin sizes are required.

5. Apply Voronoi tessellation to the selected bins. Voronoi tessellation calculates the Voronoi cells based on the centroids of the selected bins, generating adaptive bin sizes and shapes.
6. Recalculate the S/N ratio for the newly formed Voronoi bins. This step ensures that the binning remains adaptive and accurately reflects the local S/N characteristics of the detector.
7. Repeat the process iteratively if necessary, considering additional adaptivity criteria or refining the Voronoi binning as needed.

transit individual one-to-one binning to adaptive Voronoi binning, optimizing the bin sizes and adapting to the varying S/N ratios within the detector. For more detailed information consult [11].

By applying Voronoi tessellation to the Fig. 5, it can be observed that the regions with a transmission greater than one (corresponding to the cavities) are significantly more visible and distinct. This can be understood from a visual point of view, where the lighter areas are more marked and wider than the dark ones (these have been manually outlined to make them more evident). The adaptive binning provided by Voronoi tessellation allows for a finer resolution in these regions, resulting in a more accurate representation of the underlying data. We started with the expected counts in the target configuration and chose a signal-to-noise ratio (S/N) threshold equal to twice the square root of the maximum count value. This choice allowed us to transit from an average S/N of 8 to an S/N of 27 in the actual measurement of the target Fig. 6. This enhanced visibility and clarity in the regions of interest can be attributed to the adaptive nature of the binning scheme, which allocates smaller bins to areas with higher transmission values. As a result, the boundaries between different regions become more pronounced, enabling a better characterization and analysis of the data.

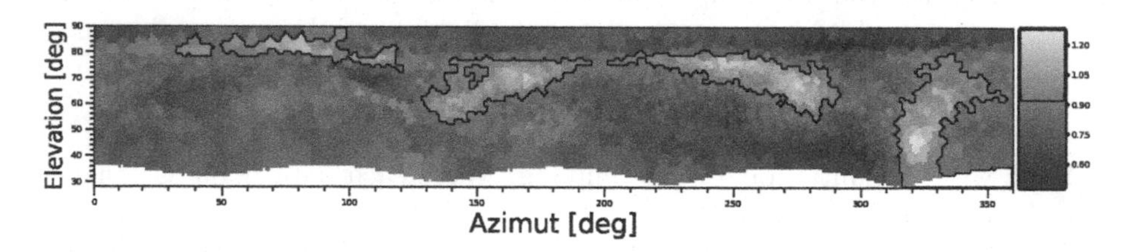

Fig. 5. The relative transmission of the measurement conducted at the Temperino mine after Voronoi tessellation.

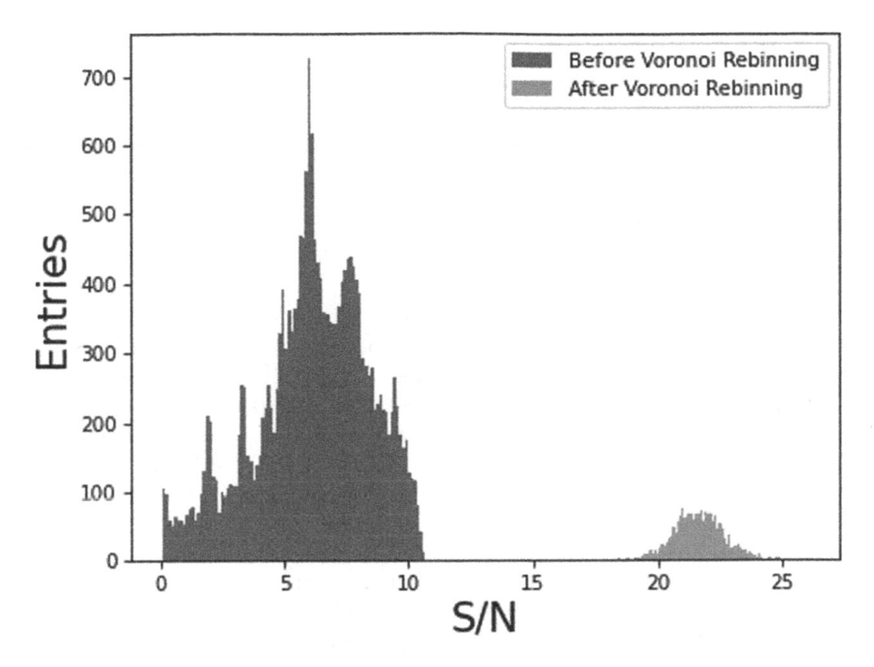

Fig. 6. S/N distribution on counts expected before applying the Voronoi procedure (blue) and after applying it (orange). (Color figure online)

4 Conclusion

In this study, the aim was to introduce muonic radiography and its application in the search for cavities in the Temperino mine, while showcasing how adaptive binning can enhance its performance.

The study demonstrated how applying the Voronoi procedure to the relative transmission image improved visibility, particularly in areas with higher transparency, which corresponded to cavities. The adaptive binning approach allowed for a better representation of data in regions with varying signal-to-noise ratios.

This heightened sensitivity enabled the detection of subtle features and anomalies in the data, leading to a significant improvement in the measured S/N ratio.

Furthermore, the Fig. 5 shows how the use of the voronoi technique can be accompanied by a procedure for automatically identifying the presence of cavities, such as for example a clustering technique or through the supervised training of a neural network.

The enhanced performance of muonic radiography, coupled with adaptive binning, highlights its potential as a powerful tool for studying complex systems and detecting underground structures. The technique's ability to reveal detailed information about the interior of materials and structures makes it a valuable asset in various scientific and engineering applications.

References

1. Grieder, P.K.F.: Chapter 1 - Cosmic ray properties, relations and definitions. In: Grieder, P.K.F. (ed.) Cosmic Rays at Earth, pp. 305–457. Elsevier, Amsterdam (2001). ISBN 978-0-444-50710-5. https://doi.org/10.1016/B978-044450710-5/50003-8, https://www.sciencedirect.com/science/article/abs/pii/B9780444507105500038?via%3Dihub
2. Workman, R.L., et al.: Review of particle physics. In: PTEP 2022, p. 083C01 (2022). https://doi.org/10.1093/ptep/ptac097
3. Grieder, P.K.F.: Chapter 3 - Cosmic rays at sea level, In: Grieder, P.K.F. (ed.) Cosmic Rays at Earth, pp. 305–457. Elsevier, Amsterdam (2001). ISBN 978-0-444-50710-5. https://doi.org/10.1016/B978-044450710-5/50005-1, https://www.sciencedirect.com/science/article/pii/B9780444507105500051
4. Bonechi, L., D'Alessandro, R., Giammanco, A.: Atmospheric muons as an imaging tool. Rev. Phys. **5**, 100038 (2020). ISSN 2405-4283. https://doi.org/10.1016/j.revip.2020.100038, https://www.sciencedirect.com/science/article/pii/S2405428320300010
5. Alvarez, L.W., et al.: Search for hidden chambers in the pyramids. Science **167**(3919), 832–839 (1970). https://api.semanticscholar.org/CorpusID:6195636
6. D'Errico, M., et al.: The MURAVES muon telescope: a low power consumption muon tracker for muon radiography applications. J. Phys. Conf. Ser. **2374**(1), 012190 (2022). https://doi.org/10.1088/1742-6596/2374/1/012190
7. Groom, D.E., Mokhov, N.V., Striganov, S.I.: Muon stopping power and range tables 10 MeV-100 TeV. Atomic Data Nucl. Data Tables **78**(2), 183–356 (2001). ISSN 0092-640X. https://doi.org/10.1006/adnd.2001.0861, https://www.sciencedirect.com/science/article/pii/S0092640X01908617
8. Mokhov, N.V., Groom, D.E., Striganov, S.: Muon stopping power and range tables 10 MeV-100 TeV (2001). https://pdg.lbl.gov/2020/AtomicNuclearProperties/
9. Borselli, D., et al.: Three-dimensional muon imaging of cavities inside the Temperino mine (Italy). Sci. Rep. **12** (2022). https://doi.org/10.1038/s41598-022-26393-7
10. Baccani, G., et al.: The MIMA project. Design, construction and performances of a compact hodoscope for muon radiography applications in the context of archaeology and geophysical prospections. J. Instrum. **13**(11), P11001 (2018). https://doi.org/10.1088/1748-0221/13/11/P11001
11. Cappellari, M., Copin, Y.: Adaptive spatial binning of integral-field spectroscopic data using Voronoi tessellations. Mon. Not. Roy. Astron. Soc. **342**(2), 345–354 (2003). ISSN: 0035-8711. eprint: https://academic.oup.com/mnras/article-pdf/342/2/345/3407404/342-2-345.pdf, https://doi.org/10.1046/j.1365-8711.2003.06541.x

Optimizing Deep Learning Models for Cell Recognition in Fluorescence Microscopy: The Impact of Loss Functions on Performance and Generalization

Luca Clissa[1,2(✉)], Antonio Macaluso[3], and Antonio Zoccoli[1,2]

[1] Department of Physics and Astronomy, University of Bologna, Bologna, Italy
{luca.clissa2,antonio.zoccoli}@unibo.it
[2] National Institute for Nuclear Physics, Bologna, Italy
[3] Agents and Simulated Reality Department, German Research Center for Artificial Intelligence (DFKI), Saarbruecken, Germany
antonio.macaluso@dfki.de

Abstract. In the rapidly evolving domain of fluorescence microscopy, the application of Deep Learning techniques for automatic cell segmentation presents exciting opportunities and challenges. In this work, we investigate the impact of loss functions and evaluation metrics on model performance and generalization in the context of cell recognition.

First, we present extensive experiments with different commonly used loss functions and offer practical insights and guidelines, underscoring how the choice of a loss function can influence model performance.

Second, we conduct a detailed examination of several evaluation metrics with their relative benefits and drawbacks, helping to guide effective model evaluation and comparison in the field.

Third, we discuss how characteristics specific to fluorescence microscopy data impact model generalization. Precisely, we examine how factors such as cell sizes, color irregularities, and textures can potentially affect the performance and adaptability of these models to new data.

Collectively, these insights provide an understanding of the various facets resulting from the application of Deep Learning for automatic cell segmentation, shedding light on best practices, evaluation strategies, and model generalization. Hence, this study can serve as a beneficial resource for researchers and practitioners working on similar applications, fostering further advancements in the field.

Keywords: Loss Function · Generalization · Cell Recognition · Semantic Segmentation · Fluorescence Microscopy

1 Introduction

Fluorescence microscopy is a widely used imaging technique adopted in many life science experiments to observe biological processes at the subcellular scale. This

G. L. Foresti et al. (Eds.): ICIAP 2023 Workshops, LNCS 14365, pp. 179–190, 2024.
https://doi.org/10.1007/978-3-031-51023-6_16

method involves the use of fluorescent compounds, fluorophores, that emit light at a specific wavelength depending on the irradiation they receive. By leveraging the relationship between absorbed and emitted light, specific molecules or biological structures of interest can be observed through fluorescent markers, thus enabling the study of their abundance, activity and interactions under controlled experimental conditions.

For instance, fluorescence microscopy aids research into the study of brain regions responsible for triggering the state of torpor, yielding insights into the neuronal networks linked with this condition [9]. This is particularly relevant since torpor has been associated with radioresistance [3], which in turn could be beneficial for a broad spectrum of medical purposes. Thus, enhancing our understanding of these phenomena may have tremendous impacts, e.g., by mitigating side effects of intensive care, oncology treatments [2] and space travels [3].

Nevertheless, the technical complexity and the manual burden of these analyses often hinder rapid progress in the field. Indeed, these experiments typically rely heavily on semi-automatic techniques that involve multiple steps to acquire and process images correctly. Manual operations like area selection, white balance, calibration and color correction are fundamental in order to identify neurons of interest successfully. As a consequence, this process may be very time-consuming depending on the number of available images. Also, the task becomes tedious when the objects appear in large quantities, thus leading to errors due to fatigue of the operators. Finally, a further challenge is that sometimes structures of interest and image background may look similar, making them hardly distinguishable. When that is the case, the recognition of biological compounds becomes arguable and subjective due to the experimenter's interpretation of such borderline cases, thus leading to an intrinsic degree of arbitrariness.

Given the above issues, the adoption of automated techniques for identifying stained regions presents an opportunity to decrease both processing time and manual effort. Additionally, it offers a means to mitigate errors due to manual operations and fatigue while establishing a systematic "operator effect". In particular, Deep Learning approaches have demonstrated remarkable potential in similar applications where the goal is to detect and/or segment objects [17]. However, performance degrades when dealing with data from domains far from those involved in pre-training, which is typically the case for microscopy images and biological applications [19]. Thus, researchers often need to fine-tune pre-trained models (or train them from scratch) using in-domain data. Despite the availability of large archives of fluorescent microscopy images, the scarcity of corresponding annotations restricts the effectiveness of transfer learning strategies. Also, even when available, public data often present limitations in the number of images, variability of stained structures and annotation formats. These issues pose substantial challenges concerning model generalization and adaptation of open-source approaches. As a result, a common obstacle is the need for a costly annotation phase for obtaining ground-truth labels [27].

This work discusses typical challenges encountered in the analysis of fluorescence microscopy data through Deep Learning. Our contributions include:

- sharing *best practices to enhance model performance*, with a particular focus on the choice of the loss function,
- discussing pros and cons of several *evaluation metrics* and strategies,
- highlighting peculiar data characteristics that may affect *model generalization*.

These arguments are substantiated through comprehensive testing conducted on real microscopy data. Specifically, the **Fluorescent Neuronal Cells v2, yellow dataset** [4,5][1].

2 Related Works

Object segmentation, detection and counting are well-known problems in Computer Vision literature [17]. However, the field of fluorescence microscopy has been predominantly reliant on manual operations for the task of cell recognition. Traditional methods were heavily dependent on human input and expertise for accurate results, often rendering the process laborious and error-prone [6].

To mitigate these issues, several alternatives were proposed in trying to automate this task. Initial attempts relied on hard-coded and adaptive computer vision techniques [7,22]. Deep Learning models have also emerged as viable solutions, driving the field towards more general and reliable automation. Some of these methods tackle directly the problem of counting the objects within an image [28], while others focus first on recognizing cells and then derive the counts as the number of detected objects [12,21].

The more sophisticated strategy consists of framing the problem of cell recognition as a segmentation task. Typically, UNet-like networks are used for this purpose [23] thanks to their effectiveness in many bio-imaging segmentation tasks [29]. These architectures leverage a powerful design involving an encoding block followed by a decoding one. The main idea is to first extract multiple image features, with progressively lower resolution, that help distinguish the various objects present in an image. Then, the decoding branch act as an upsampling path to restore the initial image size and precisely localize the detected objects. Popular variations of the latter architectures are represented by the ResUNet networks family. These integrate residual connections [8] into the UNet model to combat the vanishing gradient problem, which has been shown to improve performance in cell recognition tasks [14,18].

More recently, there has been increasing interest in the potential application of architectures inspired by vision transformers for segmentation tasks, exhibiting promising results also for bio-imaging [10]. This development may signal a new direction for future research in the field.

3 Methods

To assess the impact of different loss functions on model performance, we conduct a series of ablation studies. Loss functions in machine learning serve as

[1] Available at: https://doi.org/10.6092/unibo/amsacta/7347 (in release).

navigational guidance for the learning process, aiming to minimize the difference between the predicted and actual output. Given their nature, different loss functions tend to emphasize different aspects of the learning process [11]. Consequently, the choice of the loss can influence the model's performance, rendering it superior in some examples and inferior in others. Thus, the loss selection has to be tailored to the specific goals and challenges of the analysis.

In the case of fluorescence microscopy, the main difficulties to address for training Deep Learning models are represented by *class imbalance*, *oversampling* and *noisy labels*. Indeed, cell pixels are typically very few compared to background pixels, which requires dedicated strategies to deal with this issue. Also, it is not rare to encounter cell agglomerates in microscopy pictures. When that is the case, on top of detecting where cells are located, the model needs to perform a sharp segmentation so as to clearly separate close-by objects. Finally, another challenge is due to the presence of noisy labels. In fact, the intrinsic variability of the fluorescence phenomenon produces samples where the emissions of the biological structures of interest are not always clearly distinguishable from background noise. Hence, the recognition task presents some borderline cases wherein even human operators have to make a subjective decision. Of course, this affects the accuracy of the annotations one can collect and feed to a training pipeline.

In light of these considerations, we experiment with four alternative loss functions that address one or more of the above issues: weighted Binary Cross Entropy (BCE) [20], Dice loss [26], Focal loss [15], and Focal Tversky loss [1]. The BCE loss computes the cross-entropy between ground-truth and predicted labels. This function is one of the de-facto standards when coming to classification tasks, which makes it also suitable for segmentation (that involves classification at the pixel level). In our study, we adopt a weighted version to tackle class imbalance. Although this mitigates one of the major issues, acting at the pixel level implies that BCE is lacking evaluation at the object level, which is the real focus. This may cause poor segmentation performance, especially on cell boundaries and crowded areas. Another weighted loss is Focal loss. Contrarily to the cross-entropy, Focal loss does not address class imbalance directly. Instead, the weighting schema acts as an oversampling strategy to give more importance to misclassified examples. Hence, this can be used also to mitigate class imbalance since underrepresented classes will tend to be misclassified more. Different from the previous options, Dice loss is designed specifically for segmentation problems. This function measures the overlap between the predicted and target objects, thus promoting both recognition and precise segmentation. As a further advantage, it provides reliable results even in the presence of noisy labels [16]. Nonetheless, small objects typically have a low impact on the total loss. As a consequence, they tend to be neglected when the size of cells is variable, thus hampering the learning process. In response to the latter observation, the Focal Tversky loss was proposed. The main idea is to target segmentation performance directly while also oversampling hard examples, thus bringing together the advantages of Dice and Focal losses. However, this uses a weighting mechanism that is slightly different from the one adopted in the Focal loss.

Table 1. Experiment configurations. We used 3 alternative values as weight for the cell class in BCE: 50, 10 and 200. Similarly, we tested three configurations for the coefficients of combined loss. One balances evenly each term ($\lambda_1 = 0.3, \lambda_2 = 0.3, \lambda_3 = 0.4$), one gives more weight to Dice loss trying to improve performance with overcrowding ($\lambda_1 = 0.2, \lambda_2 = 0.5, \lambda_3 = 0.3$), and the latter replicates the settings adopted in CellViT [10] ($\lambda_1 = 0.5, \lambda_2 = 0.3, \lambda_3 = 0.5$).

	BCE	Dice	Focal	Focal Tversky	Combined	Combined FT
Hyperparameters	w_cell	smooth	gamma	gamma	λ_1, λ_2, λ_3	λ_1, λ_2, λ_3
Values	[50, 100, 200]	1×10^{-6}	2	2	balanced: [0.3, 0.3, 0.4] overcrowd: [0.2, 0.5, 0.3] CellViT: [0.5, 0.3, 0.5]	balanced: [0.3, 0.3, 0.4] overcrowd: [0.2, 0.5, 0.3] CellViT: [0.5, 0.3, 0.5]

Specifically, it leverages the Tversky index, which enables giving more importance to erroneous predictions while over-suppressing the penalization when the example belongs to a class that is globally accurately handled by the model.

In our ablation studies, we start by testing each loss function independently. After that, we also try two combined loss functions obtained as a weighted average of three terms:

$$\text{CombinedLoss} = \lambda_1 \cdot \text{BCE} + \lambda_2 \cdot \text{Dice} + \lambda_3 * \text{Focal} \tag{1}$$

$$\text{CombinedFTLoss} = \lambda_1 \cdot \text{BCE} + \lambda_2 \cdot \text{Dice} + \lambda_3 * \text{Focal Tversky} \tag{2}$$

The idea behind these combinations is to leverage the strengths of individual loss functions to compensate for their respective weaknesses, potentially yielding better segmentation results.

In summary, we put to the test the four losses and the two combined functions in Eqs. (1) and (2). Also, we experiment with three alternative settings for the weights in BCE. Namely, we use three different values for the weight attached to errors on the class *"cell"*, `w_cell`, while *"background"* weight is always equal to 1. Likewise, we test three alternative configurations for Combined and CombinedFT losses ($\lambda_1, \lambda_2, \lambda_3$). Finally, we repeat each of the above configurations with five different initialization seeds, resulting in a total of 60 different runs (see Table 1).

Through this methodological approach, our study provides insight into the effectiveness and suitability of these loss functions for cell recognition in fluorescence microscopy, shedding light on how to best address the typical challenges encountered in this domain.

3.1 Model Training

Each experiment in the ablation studies adopt the same configuration except for the loss function. In practice, we train a network from scratch using the Adam optimizer. The initial learning rate is set based on the "learning rate test" [25]. The training phase continues for 200 epochs with cyclical learning rates [24], and the best model is selected based on the best validation dice coefficient. For all technical details please refer to the GitHub repository[2].

[2] Available at: https://github.com/clissa/fluocells-BVPAI.

Table 2. Test set metrics. The table reports the mean and standard deviation of metrics computed for each loss function configuration across experiments with different seeds.

	F_1 score (IoU)	F_1 score (distance)	MAE	MedAE	MPE (%)
BCE: medium	0.673 ± 0.017	0.827 ± 0.022	2.286 ± 0.245	0.9 ± 0.224	21.498 ± 4.679
BCE: high	0.663 ± 0.033	0.846 ± 0.013	2.583 ± 0.263	1 ± 0	23.281 ± 2.637
BCE: low	0.687 ± 0.017	0.825 ± 0.020	2.151 ± 0.171	0.3 ± 0.447	18.079 ± 3.791
CombinedFT: overcrowd	0.740 ± 0.029	0.848 ± 0.026	1.674 ± 0.161	0 ± 0	9.682 ± 3.221
CombinedFT: balanced	0.744 ± 0.022	0.853 ± 0.022	1.643 ± 0.144	0 ± 0	8.465 ± 1.859
CombinedFT: CellViT	0.728 ± 0.048	0.844 ± 0.030	1.791 ± 0.406	0 ± 0	9.968 ± 3.163
Combined: overcrowd	0.721 ± 0.023	0.837 ± 0.033	1.774 ± 0.132	0 ± 0	10 ± 3.160
Combined: balanced	0.735 ± 0.034	0.845 ± 0.029	1.674 ± 0.166	0 ± 0	9.322 ± 2.074
Combined: CellViT	0.742 ± 0.023	0.849 ± 0.020	1.651 ± 0.188	0 ± 0	9.249 ± 3.039
Dice	0.735 ± 0.020	0.847 ± 0.018	1.700 ± 0.113	0 ± 0	10.621 ± 3.247
FocalTversky	0.781 ± 0.002	0.897 ± 0.003	1.220 ± 1.115	0.1 ± 0.224	6.373 ± 1.607
Focal	0.614 ± 0.027	0.780 ± 0.034	1.517 ± 0.365	0.2 ± 0.4	12.747 ± 3.525

During the assessment, we evaluate the models from the point of view of segmentation, detection and counting. For **segmentation**, we base the association of true and predicted neurons[3] on their overlap in terms of *Intersection-over-Union (IoU)*. In this way, we have a measure of both recognition and segmentation performance. Then, standard metrics such as *precision, recall* and F_1 *score* are discussed as measures of global performance. In the case of **detection**, on the other hand, we adopt a looser matching criterion based on the distance between the target and predicted objects' centers. In this way, we prioritize recognition over precise shape reconstruction. Finally, for **counting** tasks we resort to standard regression metrics, i.e. *Mean Absolute Error (MAE), Median Absolute Error (MedAE)* and *Mean Percentage Error (MPE)*.

4 Results

The results of the experiments are reported grouped by loss function in Table 2. In the following, F_1 *score* (IoU and distance versions) and *MPE* are discussed as the main figures of merit for segmentation, detection and counting performance, respectively, given their relevance in the application under study. MAE and MedAE are also reported to provide a more comprehensive evaluation.

Overall, *Focal Tversky* consistently outperforms other losses by a significant margin with respect to segmentation and detection, showing better metrics and lower variability (see Figs. 1a,1b and 1e). Concerning counting performance, the MPE value for Focal Tversky is the lowest but it shows a much higher standard deviation, which makes it comparable to other losses (see Table 2). Combined losses are the second-best alternatives. No clear difference is observed between

[3] By this we intend the calculation of True Positives (TP), False Positives (FP) and False Negatives (FN).

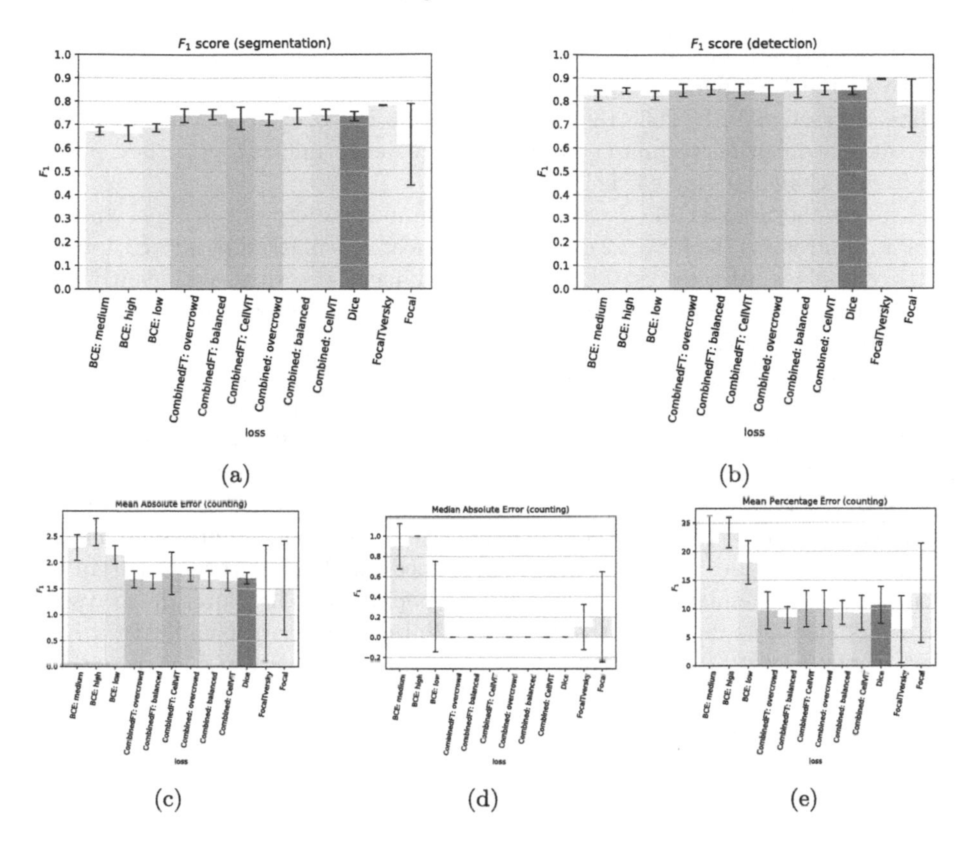

Fig. 1. Test set performance. The figure illustrates the performance metrics relevant to the evaluation in terms of segmentation (Fig. a), detection (Fig. b) and counting (Figs. c to e). The height of the bars indicates the mean value for the corresponding metric across experiments with the same seed, and the error bars are based on the standard deviation.

the versions including Focal loss and the one with Focal Tversky. Within each family, all variants score roughly on par, with a slight advantage in favor of the balanced coefficients configuration. Slightly below there is Dice loss. Focal and BCE experiments follow by a fair margin in terms of segmentation. However, notice how experiments involving Focal loss show high variability. In fact, this loss sometimes produces unstable experiments, that make the model converge towards suboptimal parameters. In terms of detection metrics, BCE achieves competitive results, while Focal loss maintains the gap and a huge variability. On the other hand, the situation is reversed when considering counting performance. This time, the experiments involving BCE losses clearly show the worst performance, even taking into account the uncertainty expressed by the error bars (Fig. 1e). The Focal loss, instead, performs slightly worse than the best model. Notice, also, that the error margin remains high.

To investigate the model's capacity for generalization to new data, we deployed the best-performing model on fluorescence images sourced from an online dataset, S-BSST265 [13]. Although these images are produced through the same acquisition technique (immunofluorescence microscopy), they pose a significant challenge due to their diversity.

Indeed, the S-BSST265 data span nuclei from a variety of cell types and tissues, in contrast to the FNC v2 dataset, which features marked structures located in the cytoplasm of neurons. The S-BSST265 images also vary significantly with respect to multiple parameters including magnification, sample preparation, and signal-to-noise ratio, and they include both normal and diseased cells. As a result, we observe a larger variability in terms of cell sizes compared to FNC v2, as illustrated in the top row of Fig. 2. Moreover, these images display a broad spectrum of appearances. Some have clearly defined boundaries (Figs. 2c and 2d), while others exhibit fuzzier outlines (Fig. 2a). Also, cell pixels show fluctuations in terms of color intensity and present heterogeneous texture and filling patterns, as seen in Fig. 2c. Another challenge arises from the grayscale format of these images. The original model was trained on RGB pictures, wherein the green and blue channels were predominantly populated. Therefore, the grayscale nature of this new dataset introduces an additional layer of complexity to the task. Figure 2 offers a qualitative assessment of model predictions, reporting predicted heatmaps (bottom row) corresponding to selected images (top row) that exhibit peculiar characteristics.

Overall, the model presents a significant level of variability, yielding highly accurate predictions in some examples while behaving poorly in others. In summary, it tends to perform well with images where cells have a size comparable to FNC neurons, demonstrating high confidence, accurate recognition and sharp boundaries (Fig. 2a). Importantly, it seems fairly robust when confronted with smaller cells (Fig. 2b). Nonetheless, the model struggles with bigger objects with irregular texture and color filling. Figure 2c provides an example of this behavior. Upon examination of the predicted heatmaps, the model seems able to identify most objects. However, the recognition is only partial and often limited to parts of their contours. Probably, this is due to the fact that the model was trained to segment the cytoplasm, typically situated around the central nucleus region, which may explain the presence of holes in the object centers. Finally, the model's performance is less satisfactory when tasked with differentiating closely situated objects. As a result, often cell shapes are joined in crowded areas (Fig. 2d).

5 Discussion

In this paper, we explore a use-case of cell segmentation using the Fluocells Neuronal Cells v2, yellow dataset [4]. Specifically, we consider the c-ResUnet network [18] and we focus on the impact of various loss functions on segmentation, detection, and counting performance. Furthermore, we delve into a qualitative assessment of pre-trained models' generalization capabilities when faced with new data having different characteristics.

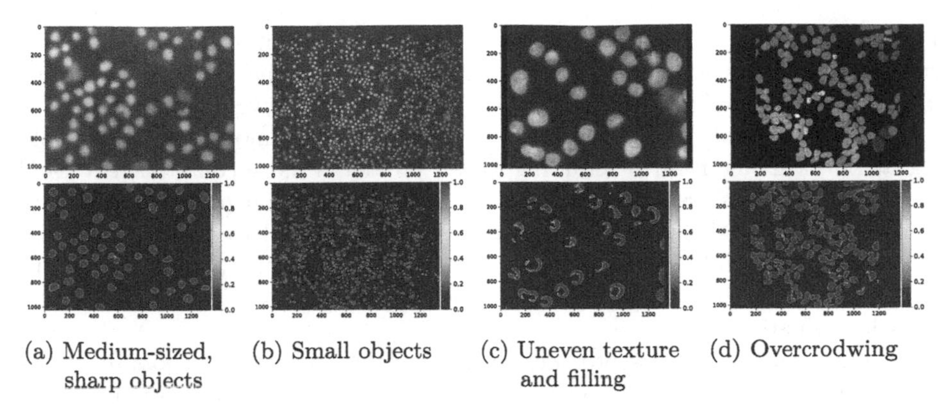

(a) Medium-sized, sharp objects (b) Small objects (c) Uneven texture and filling (d) Overcrodwing

Fig. 2. Generalization performance. Raw predictions of pre-trained c-ResUnet when applied "out of sample" data from `S-BSST265` dataset.

Our findings indicate that the Focal Tversky loss function consistently outperforms others across all tasks, according to the metrics utilized in our study. This superior performance underlines the Focal Tversky loss function as a viable choice for similar segmentation tasks. However, it struggles to separate close-by objects since joining their boundaries only marginally impacts on the loss. For this reason, it needs careful integration with post-processing pipelines tailored to specific data features. For example, this may require adjusting holes filling, small objects removal and watershed hyperparameters based on cell sizes in the target data. Combined losses yield results just slightly below, indicating a promising potential. Interestingly, we observe no apparent performance difference between the version employing the standard Focal loss and the one leveraging the Focal Tversky loss function. Moreover, configurations favoring a more balanced choice of coefficients demonstrate a slight advantage, suggesting that Binary Cross-Entropy (BCE), Dice and focal-based losses nicely balance out their weaknesses. The Dice loss is also fairly competitive with respect to all learning tasks. In contrast, the BCE and Focal loss functions lag behind in terms of segmentation performance. However, they exhibit stronger competitiveness in detection and counting tasks, respectively. Such observations highlight the necessity of careful loss function selection tailored specifically to data features and requirements of future applications.

As for generalization to new data, the model demonstrates significant variability. While certain instances yield highly accurate predicted masks, others are noticeably subpar. We attribute such performance variability to the heterogeneity in cell sizes, irregular colors, and textures. In this respect, data augmentation techniques can be utilized to introduce more diversity into the training data, thereby enhancing the model's robustness to domain shifts.

Concerning overcrowding, we posit that the suboptimal performance arises from the lack of explicit object separation penalization in the Focal Tversky loss. In fact, merging the outlines of proximate cells only has a marginal impact.

Therefore, employing a more comprehensive segmentation loss could present a viable solution. In addition, post-processing operations that are directly tuned to the target data characteristics could significantly improve results.

More in general, the evaluation process described demonstrates how looking at performance from several angles can provide a more comprehensive and thorough model assessment. Additionally, visual inspection plays a key role in identifying insights that numerical metrics are not suited to capture, allowing us to perceive trends, patterns, and outliers in the model's predictions. Thus, a balanced and multifaceted approach is recommended to ensure a robust and holistic evaluation.

In conclusion, our work provides several insights into different loss functions for segmentation, detection, and counting tasks, while simultaneously highlighting key challenges in model generalization. These results may serve as a starting point for model selection and refinement for future research in the field of microscopic fluorescence, thereby informing and enhancing novel studies and applications.

Funding. Research partly funded by PNRR - M4C2 - Investimento 1.3, Partenariato Esteso PE00000013 -"FAIR - Future Artificial Intelligence Research" - Spoke 8 "Pervasive AI", funded by the European Commission under the NextGeneration EU programme.

References

1. Abraham, N., Khan, N.M.: A novel focal Tversky loss function with improved attention U-Net for lesion segmentation. In: 2019 IEEE 16th International Symposium on Biomedical Imaging (ISBI 2019), pp. 683–687 (2019)
2. Bouma, H.R., et al.: Induction of torpor: mimicking natural metabolic suppression for biomedical applications. J. Cell. Physiol. **227**(4), 1285–1290 (2012)
3. Cerri, M., et al.: Hibernation for space travel: impact on radioprotection. Life Sci. Space Res. **11**, 1–9 (2016). https://doi.org/10.1016/j.lssr.2016.09.001. https://www.sciencedirect.com/science/article/pii/S2214552416300542
4. Clissa, L., et al.: Fluorescent neuronal cells v2: multi-task, multi-format annotations for deep learning in microscopy. arXiv preprint arXiv:2307.14243 (2023, under review at Scientific Data)
5. Clissa, L., et al.: Fluorescent neuronal cells. AMS Acta (2023). https://doi.org/10.6092/unibo/amsacta/7347
6. Dentico, D., et al.: C-Fos expression in preoptic nuclei as a marker of sleep rebound in the rat. Eur. J. Neurosci. **30**(4), 651–661 (2009). https://doi.org/10.1111/j.1460-9568.2009.06848.x
7. Faustino, G.M., Gattass, M., Rehen, S., de Lucena, C.J.P.: Automatic embryonic stem cells detection and counting method in fluorescence microscopy images. In: 2009 IEEE International Symposium on Biomedical Imaging: From Nano to Macro, pp. 799–802 (2009). https://doi.org/10.1109/ISBI.2009.5193170
8. He, K., Zhang, X., Ren, S., Sun, J.: Identity mappings in deep residual networks, vol. 9908, pp. 630–645, October 2016. https://doi.org/10.1007/978-3-319-46493-0_38

9. Hitrec, T., et al.: Neural control of fasting-induced torpor in mice. Sci. Rep. **9**(1) (2019). https://doi.org/10.1038/s41598-019-51841-2
10. Hörst, F., et al.: CellViT: vision transformers for precise cell segmentation and classification. arXiv preprint arXiv:2306.15350 (2023)
11. Jadon, S.: A survey of loss functions for semantic segmentation. In: 2020 IEEE Conference on Computational Intelligence in Bioinformatics and Computational Biology (CIBCB), pp. 1–7 (2020)
12. Kraus, O., Ba, J., Frey, B.: Classifying and segmenting microscopy images with deep multiple instance learning. Bioinformatics **32**, i52–i59 (2016). https://doi.org/10.1093/bioinformatics/btw252
13. Kromp, F., et al.: An annotated fluorescence image dataset for training nuclear segmentation methods. Sci. Data **7**(1), 262 (2020)
14. Kumar, P.S., Sakthivel, V., Raju, M., Satya, P.: Brain tumor segmentation of the FLAIR MRI images using novel resUnet. Biomed. Signal Process. Control **82**, 104586 (2023)
15. Lin, T.Y., Goyal, P., Girshick, R., He, K., Dollár, P.: Focal loss for dense object detection. In: Proceedings of the IEEE International Conference on Computer Vision, pp. 2980–2988 (2017)
16. Marcinkiewicz, M., Mrukwa, G.: Quantitative impact of label noise on the quality of segmentation of brain tumors on MRI scans. In: 2019 Federated Conference on Computer Science and Information Systems (FedCSIS), pp. 61–65. IEEE (2019)
17. Minaee, S., Boykov, Y., Porikli, F., Plaza, A., Kehtarnavaz, N., Terzopoulos, D.: Image segmentation using deep learning: a survey. IEEE Trans. Pattern Anal. Mach. Intell. **44**(7), 3523–3542 (2021)
18. Morelli, R., et al.: Automating cell counting in fluorescent microscopy through deep learning with c-ResUnet. Sci. Rep. **11**(1), 22920 (2021). https://doi.org/10.1038/s41598-021-01929-5
19. Ouyang, C., et al.: Causality-inspired single-source domain generalization for medical image segmentation. IEEE Trans. Med. Imaging **42**(4), 1095–1106 (2023). https://doi.org/10.1109/TMI.2022.3224067
20. Pihur, V., Datta, S., Datta, S.: Weighted rank aggregation of cluster validation measures: a Monte Carlo cross-entropy approach. Bioinformatics **23**(13), 1607–1615 (2007)
21. Raza, S.e.A., Cheung, L., Epstein, D., Pelengaris, S., Khan, M., Rajpoot, N.: MIMO-Net: a multi-input multi-output convolutional neural network for cell segmentation in fluorescence microscopy images, pp. 337–340, April 2017. https://doi.org/10.1109/ISBI.2017.7950532
22. Riccio, D., Brancati, N., Frucci, M., Gragnaniello, D.: A new unsupervised approach for segmenting and counting cells in high-throughput microscopy image sets. IEEE J. Biomed. Health Inform. **PP**, 1 (2018). https://doi.org/10.1109/JBHI.2018.2817485
23. Ronneberger, O., Fischer, P., Brox, T.: U-Net: convolutional networks for biomedical image segmentation, vol. 9351, pp. 234–241, October 2015
24. Smith, L.N.: Cyclical learning rates for training neural networks. In: 2017 IEEE Winter Conference on Applications of Computer Vision (WACV), pp. 464–472 (2017)
25. Smith, L.N.: A disciplined approach to neural network hyper-parameters: part 1-learning rate, batch size, momentum, and weight decay. arXiv preprint arXiv:1803.09820 (2018)

26. Sudre, C.H., Li, W., Vercauteren, T., Ourselin, S., Jorge Cardoso, M.: Generalised dice overlap as a deep learning loss function for highly unbalanced segmentations. In: Cardoso, M.J., et al. (eds.) DLMIA/ML-CDS 2017. LNCS, vol. 10553, pp. 240–248. Springer, Cham (2017). https://doi.org/10.1007/978-3-319-67558-9_28
27. Xie, J., Kiefel, M., Sun, M.T., Geiger, A.: Semantic instance annotation of street scenes by 3d to 2d label transfer. In: Proceedings of the IEEE Conference on Computer Vision and Pattern Recognition (CVPR), June 2016
28. Xiong, H., Lu, H., Liu, C., Liu, L., Cao, Z., Shen, C.: From open set to closed set: counting objects by spatial divide-and-conquer. In: Proceedings of the IEEE/CVF International Conference on Computer Vision, pp. 8362–8371 (2019)
29. Zeng, Z., Xie, W., Zhang, Y., Lu, Y.: RIC-Unet: an improved neural network based on Unet for nuclei segmentation in histology images. IEEE Access **7**, 21420–21428 (2019)

A New IBA Imaging System for the Transportable MACHINA Accelerator

Rodrigo Torres[1(✉)], Caroline Czelusniak[2], Lorenzo Giuntini[1,2], Francesca Giambi[3], Mirko Massi[2], Chiara Ruberto[1], Francesco Taccetti[2], Giovanni Anelli[4], Serge Mathot[4], and Alessandra Lombardi[4]

[1] Dipartimento di Fisica e Astronomia, Università degli Studi di Firenze, 50019 Sesto Fiorentino, Italy
rodrigoa.tsaavedra@gmail.com
[2] Istituto Nazionale di Fisica Nucleare, Sezione di Firenze, 50019 Sesto Fiorentino, Italy
[3] Università degli Studi di Firenze, 50121 Firenze, Italy
[4] CERN-European Organization for Nuclear Research, 1211 Geneva 23, Switzerland

Abstract. At LABEC (the INFN ion beam laboratory of nuclear techniques for environment and cultural heritage, located in Florence), a novel transportable accelerator for in-situ ion-beam analysis measurements of cultural heritage materials, MACHINA, has been constructed as part of an international collaboration between the INFN and the CERN. Here we present the most recent developments regarding this project, consisting in the design, construction, and testing of the hardware/software of a data acquisition system for prompt photons of characteristic energies emitted following the interaction between charged particles and matter to construct elemental maps, i.e., grayscale digital images showing the spatial distribution of elements in a material of interest wherein the brightness of each individual pixel correlates to the abundance of a given element.

Keywords: Data acquisition · Imaging · Ion-beam analysis · Particle-induced X-ray emission

Acronyms

CH	Cultural Heritage
DAQ	Data AcQuisition
IBA	Ion Beam Analysis
LABEC	Laboratory of nuclear techniques for Environment and Cultural Heritage
MACHINA	Movable Accelerator for Cultural Heritage In-situ Non-destructive Analysis
PHA	Pulse Height Analysis
PIGE	Particle Induced Gamma-ray Emission
PIXE	Particle Induced X-ray Emission

G. L. Foresti et al. (Eds.): ICIAP 2023 Workshops, LNCS 14365, pp. 191–201, 2024.
https://doi.org/10.1007/978-3-031-51023-6_17

1 Introduction

Ion Beam Analysis (IBA) techniques, consisting in the analysis of secondary radiation and particles produced in the interaction between charged particles and atomic matter (both electrons and nuclei) constitute a family of well established analysis techniques in archaeology and Cultural Heritage (CH) [2,4]. They can provide multi-elemental information of a material down to trace composition levels. Further, IBA techniques can be both non-invasive and non-destructive. For these reasons, IBA techniques have gained a foothold in the studies of CH materials, where the preservation of the materials is essential, and even micro-invasive sampling can be a problem [1,3,10].

IBA techniques utilize MeV ion beams. IBA analyses thus require the use of particle accelerators (such as the Tandem accelerator) that are fixed in place, meaning that these techniques may only be used to analyze materials that can be brought in to the accelerator facility. However, CH materials often cannot be transported as it may be prohibitively expensive, impractical, or damaging to the material (think, for instance, of frescoes or prized paintings).

There is significant demand then for in-situ IBA measurements, i.e. for IBA measurements conducted directly on the premises where the CH material is located. Up until recently, however, no transportable IBA instrumentation existed in the world. Then in 2018, the Movable Accelerator for Cultural Heritage In-situ Non-destructive Analysis (MACHINA) project was born, as a collaboration between the Italian national Institute of Nuclear Physics (INFN) and the CERN (European Center for Nuclear Research), to build the first transportable accelerator in the world for in-situ IBA analyses. MACHINA is a compact accelerator [8] that uses a radio-frequency ion source coupled to two radio-frequency quadrupole cavities [6,7] to provide a 2 MeV proton beam (see Fig. 1). This accelerator will be used by the Opificio delle Pietre Dure (in Florence, Italy) for conservation studies involving the application of the Particle Induced X-ray Emission (PIXE) analysis technique in external beam mode, i.e. with the proton beam extracted from vacuum into the atmosphere.

The first proton beam was extracted from MACHINA during the first quarter of 2022, demonstrating the viability of the instrument as a proton source for IBA analyses. Nevertheless, at the time of beam extraction there was no hardware setup for radiation detection, nor was there a Data AcQuisition (DAQ) system for PIXE. Thus, a new DAQ and imaging system specific for PIXE analysis using the MACHINA hardware was designed, developed, and tested at the Laboratory of nuclear techniques for Environment and Cultural Heritage (LABEC) of the INFN branch in Florence. The following sections will provide more details regarding the designing and testing of this new imaging system.

Fig. 1. The MACHINA accelerator. The ion source is placed at the beginning of the beam line (on the left), the radio frequency quadrupole cavities are the copper structures placed in the middle. At the end of the beam line (on the right) the proton beam is extracted into the atmosphere for IBA analysis. The area footprint of the accelerator is approximately 2.5 m by 1 m, while its weight is about only 500 kg.

2 Methods

In traditional single spot PIXE analyses, a small area of the sample under investigation is irradiated with an ion beam and an energy spectrum of the emitted X-rays is acquired. This analysis provides information about the elemental composition of the sample, but limited only to the small area that was irradiated. Most objects that are of artistic or historical importance tend to have non-uniform structures, even in areas that may seem uniform. This is because there may be small, irregular details present at submillimeter scales. Such non-homogeneous structures become a problem for compositional analyses, since these features may be difficult to identify by visual examination. Therefore, traditional single spot analysis can result in misleading information about the composition of the material under study.

In order to overcome the outlined risk of misleading or ambiguous information, many efforts have been dedicated by the scientific community to develop compositional imaging techniques, using scanning mode acquisition systems. These techniques enable the acquisition of information about both the material composition of the sample and the distribution of detected elements within the scanned area. In other words, the data obtained with imaging techniques may be used to construct elemental distribution maps.

A scanning DAQ system then has been realized for PIXE elemental imaging of macroscopic areas using the external 2 MeV proton beam of MACHINA. Currently, the system can scan continuously areas of up to 60 cm by 20 cm of CH objects of no more than a few kilograms in mass. These limitations exist in part due to the choice of motorized stages used for the scanning system. Nevertheless, a large part of the measurement requests in the CH field may already be fulfilled with this set-up. Further, scans of larger areas and heavier objects will become available in the future with different choices of motorized stages.

The approach chosen for DAQ in scanning mode was to move at constant speed the material of interest in front of the stationary proton beam, acquiring X-ray spectra on-flight for all points contained within a region of interest of the material. To achieve this, first, the detection chain for IBA imaging was set up: the 3D printed hardware was constructed and a suite of libraries and programs were written in C++ for hardware control and management of the multi-detector DAQ procedure specific to the MACHINA hardware. Second, the sample scanning system was set up and synchronized with the DAQ procedure. Third, both the hardware and software of the new imaging system were tested.

The first step was to set up the detection chain for PIXE spectroscopic signals. To do so, a custom support was fabricated using additive manufacturing (3D printing) wherein three silicon detectors were placed. An application specific integrated circuit was used to provide the high-voltage bias to the detectors, read off the temperature sensor inside the detector packaging, and provide live monitoring and regulation of the thermoelectric cooler of the detector (see Fig. 2). The output of each detector preamplifier (a voltage-step signal) was connected to an input stage of a desktop digitizer (CAEN DT5780) that sampled the signal using a 14-bit, 100 MS/s analog-to-digital converter, and applied two algorithms to the digitized samples to provide:

- a self-trigger (a zero-crossing trigger using a bipolar RC-CR2 signal), and
- a trapezoid output signal with height proportional to the energy deposited in the radiation detector.

X-ray spectra may be obtained by performing Pulse Height Analysis (PHA) of this trapezoid output signal, i.e. by binning the trapezoid heights into a histogram. In particular, each individual PHA event (meaning each trigger event) had to be read out one-by-one and assigned to an appropriate position in the region of the sample being analyzed. Thus, a custom library in C++ was developed, following the design precepts of the LABEC laboratory (open source-code, modularity, and accessible user experience), to handle an arbitrary number of digitizer boards. For each board, the library would handle:

- the communication bus between the readout CPU and the board,
- programming of the digital algorithm parameters,
- data readout, and
- hardware synchronization of the acquisition start/stop among different boards through transistor-transistor logic signals.

Fig. 2. Custom support for the MACHINA detection set up. In the image, three silicon detectors are shown. Two of these detectors (on the left- and right-hand side) will be used for PIXE measurements, while the third will be used for indirect charge normalization. A brass cone is visible at the bottom-center, which is the brass housing of the X-ray tube used to test the DAQ system described in this work. In the final MACHINA set-up, this custom support will be placed on top of the last accelerator station, and the beam extraction nozzle will be placed where the X-ray tube brass housing is shown in the picture.

The second step then was to set up a positioning and scanning system to probe an entire region of interest of a sample with the proton beam, and to correlate (for any given time interval) the observed PHA events with the position of the sample in front of the beam. To achieve this, two precision linear stages were used.

The X-Y motorized stages (see Fig. 3) allow for scanning a sample in front of the stationary beam. The area of the sample to be studied is divided into a rectangular lattice using horizontal (Δx) and vertical (Δy) intervals. The center of the cells formed by the (Δx, Δy) intervals define the pixel coordinates of the elemental image. The analysis area is then scanned by moving the sample in front of the stationary beam. The sample is first translated horizontally at a constant speed, followed by a single vertical Δy step, followed by a horizontal translation again, and so on until the scan is done.

X-ray spectra are acquired in-flight and assigned to each lattice cell. The cell coordinates and spectra are saved in a multidimensional array, called datacube. The first two indices of this array are a position matrix and define the pixels of the image. The third index of the array is instead a vector of counts per spectral bin (i.e. an energy spectrum).

Fig. 3. The MACHINA scanning set-up developed. The red crown supporting the semiconductor detectors is visible in the center. This crown will be mounted above the end station of the MACHINA beam line. The scanning apparatus is visible on the right. The sample is affixed to the front face of the carbon-fiber panel (the black rectangle on the right), and is thus not visible in the image. (Color figure online)

Thus, each element of the position matrix (i.e. each pixel of the image) is associated to an energy spectrum. To obtain an elemental image, a region of interest of the spectrum is chosen (such as the photopeak corresponding to a particular element). The area of this region of interest may be determined by integrating the counts found therein. This operation is repeated pixel by pixel, building an intensity matrix. By choosing a color scale, i.e. a function that assigns a color to the intensity values (for example, a 255-step gray scale) an elemental image may finally be obtained (see Fig. 4 for an example).

3 Tests and Results

Tests were performed to check the proper operation of the new DAQ system and the new image reconstruction software for IBA imaging, both developed specifically for MACHINA. In particular, it had to be verified that the scanning process correctly mapped physical distances onto a discrete data structure (without introducing artifacts or geometric inaccuracies), that the data acquisition program did not lose any PHA events due to artificial dead time introduced by the software, and that the acquisition could be performed using three detectors simultaneously. To achieve this, a number of scans were performed both on standard references and real samples of CH interest, with fine structures varying from a few mm down to about 500 μm.

For the tests presented, two different primary beams were used. The first, was an X-ray beam (to perform X-ray fluorescence measurements). The second, a proton beam. X-ray fluorescence test measurements were performed using a lightweight and small Moxtek MAGNUM Coolidge X-ray tube, with a 125 μm Be exit window, 40 kV of maximum voltage, 0.1 mA maximum anode current, and Rh anode. A brass housing (4 mm-thick case) surrounded the tube, so that only the forward-emitted X-rays passing through the collimator hole (diameters ranging from 1.5 mm down to 0.3 mm) could emerge out of the tube. This shielding fulfilled the radio-safety issues for both workers and public. These tests were designed to evaluate:

- single detector performances of the DAQ system,
- synchronization between the stage movement and the detector signal acquisition,
- software mapping capabilities of a real CH sample,
- multi-detector acquisition capabilities.

The first test was a scan of a metal grid using a single silicon detector and an X-ray source. The analysis demonstrated that the geometry of the grid could be correctly reconstructed from the information extracted from the elemental composition images.

The second test was a scan with a single silicon detector and an X-ray source of a painting with a complex structure, already analyzed with the acquisition program of the X-ray fluorescence scanner developed in the LABEC laboratory (see Fig. 4a). The composition and distribution of the pigments on the surface was reconstructed (see Figs. 4b, 4c, 4d). The agreement between the elemental composition maps obtained with the two acquisition software was verified.

The third test was a scan with an X-ray source of a standard for microscopy (see Fig. 5a). The DAQ was tested using three silicon detectors connected to two independent digitizers. It was verified that, as expected given the dimensions of the beam, the spatial characteristics of the standard are recognizable down to the dimensions of $\sim$ 500 μm and no artifacts or distortions were introduced by the imaging system (see Fig. 5).

The fourth and final test of the acquisition system was performed in a context emulating real situations with MACHINA by using a pulsed proton beam on the Defel line of LABEC Tandem accelerator [5,9]. The accelerator was set to produce a 2 MeV proton beam, which was extracted in air through a Si_3N_4 thin window by using the same exit nozzle used for MACHINA. In this fashion, it was possible to reproduce the experimental conditions in which MACHINA is going to be used. This final test consisted then in the scan of a modern replica of a fresco using a 2 MeV pulsed proton beam. The ability of the system to do IBA imaging on a sample of interest for CH was thus demonstrated.

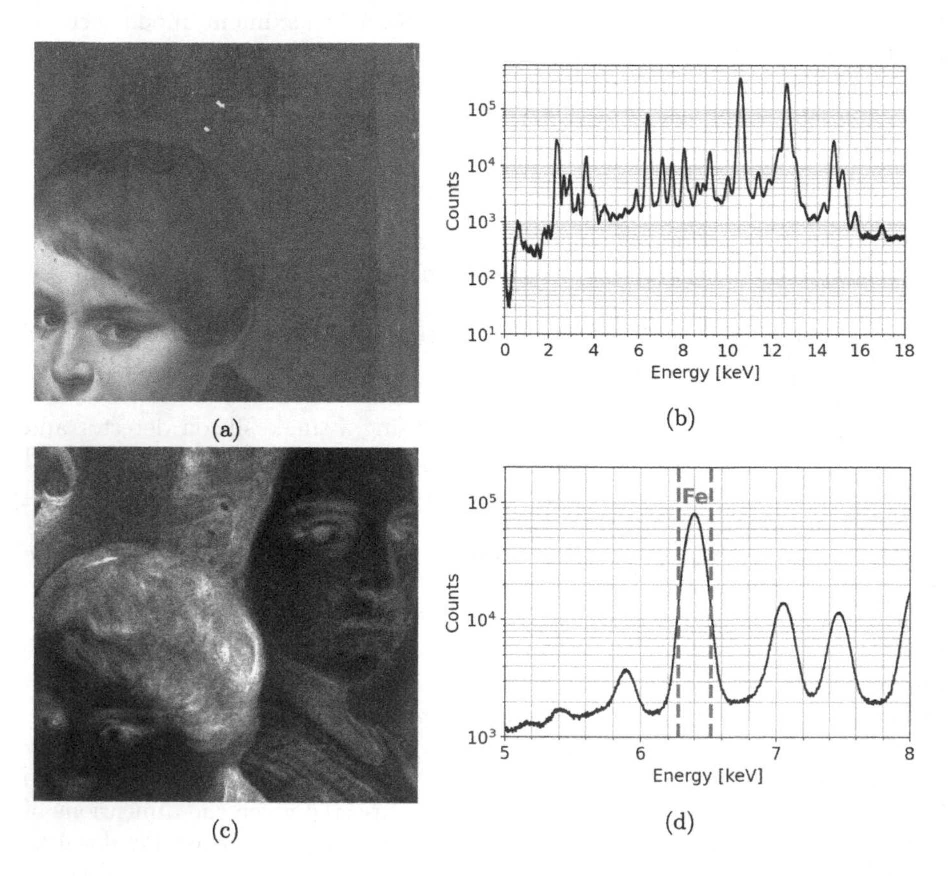

Fig. 4. (a) Detail of an oil on canvas used for the second test of the MACHINA imaging and scanning system. (b) Cumulative X-ray spectrum obtained by irradiating the region shown in (a) with an X-ray beam and by binning the trapezoid heights corresponding to each trigger event into a histogram. (c) Elemental map obtained for the scan area shown in (a) for a region of interest of the X-ray spectrum centered at the iron K_α peak. Brighter pixels indicate a higher number of iron K_α X-ray counts at the position corresponding to that pixel. (d) A blow-up of the X-ray spectrum shown in (b) in the energy interval about the iron K_α peak. The red lines show the bounds of the region of interest in the spectrum used to construct the elemental map shown in (c). (Color figure online)

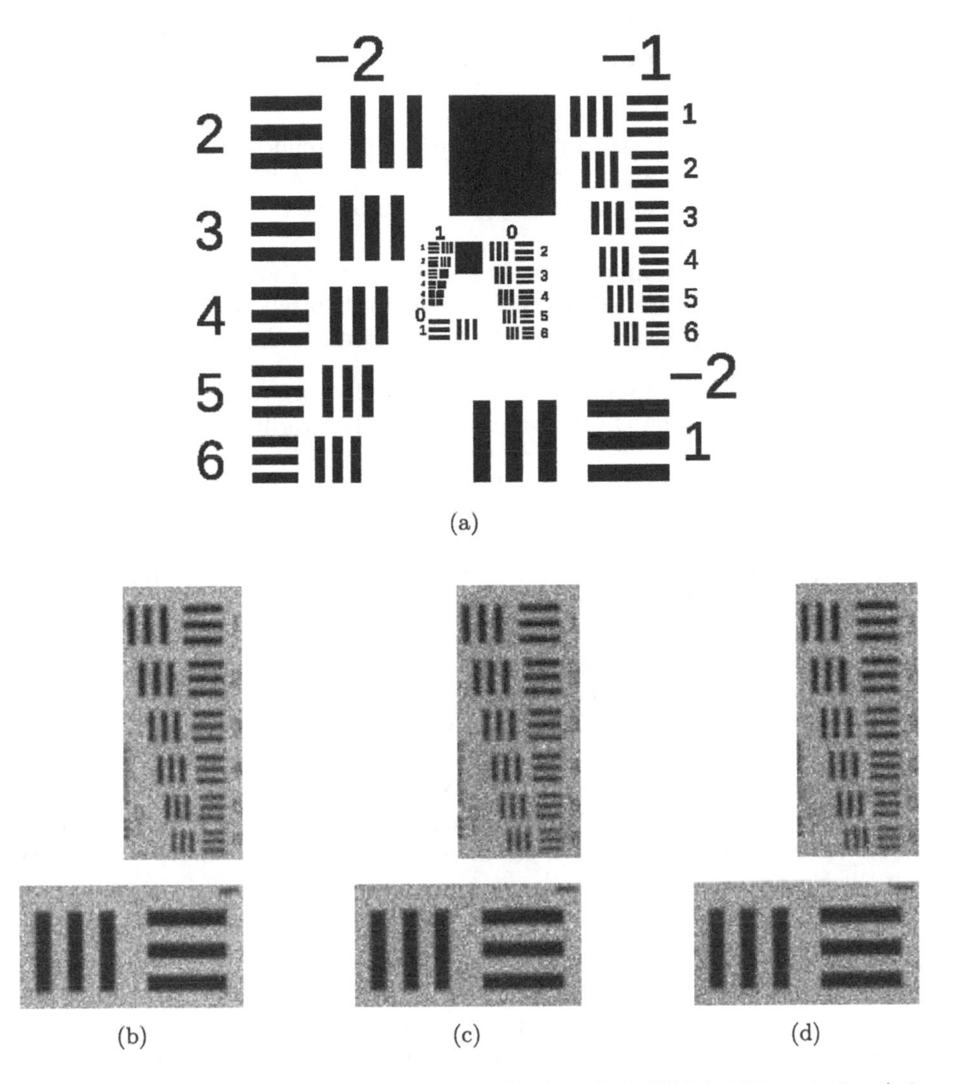

Fig. 5. (a) A diagram of the microscopy standard used (a USAF-1951 test chart) for tests of the multi-detector DAQ procedure. The test chart is a transparent glass slide, 2×3 inches, covered with a chromium mask. The standard consists of different groups, each consisting of 6 elements. Each element is a set of three horizontal bars and three vertical bars. For the tests, two scans were done: one of all elements in group -1 (with line widths ranging from 1000 µm down to 561 µm), and the other of element 1 in group -2 (line width of 2 mm). Data was acquired using three detectors simultaneously, and the corresponding chromium elemental distribution maps for each detector are shown in (b), (c), and (d).

4 Conclusions

The novel INFN-CERN MACHINA accelerator required a scanning and IBA imaging system for in-situ PIXE analysis of CH materials. This protocol would have to:

- handle the PHA data stream from at least three semiconductor detectors,
- command, program, and synchronize two independent digitizer boards,
- command a positioning and scanning system of the target in front of a fixed proton beam,
- synchronize all the components of the DAQ system (both hardware and software), to guarantee the correct spatio-temporal correlation of the PHA data read out with respect to the instantaneous position of the motorized stages.

A new DAQ system fulfilling the needs stated above was designed and constructed at LABEC. The DAQ system was tested to verify that:

- it reproduced the temporal distribution of PHA events recorded,
- it can be used to retrieve information of interest to the CH analysis field, and most importantly,
- the acquisition can be performed with multiple boards and detectors using an IBA source.

The results of the work will enable MACHINA to perform PIXE in-situ analysis in both point and imaging mode.

The final setup of MACHINA foresees the use of four semiconductor detectors in total: two for PIXE, one for Particle Induced Gamma-ray Emission (PIGE), and one for beam current normalization. As such, two natural next steps for this work are high-energy PIXE/PIGE imaging, and the implementation of offline current normalization algorithms. Needless to say, the system is yet to be tested with the MACHINA hardware.

References

1. Amat, A., Miliani, C., Brunetti, B.G.: Non-invasive multi-technique investigation of artworks: a new tool for on-the-spot data documentation and analysis. J. Cult. Herit. **14**(1), 23–30 (2013)
2. Artioli, G.: Scientific methods and cultural heritage: an introduction to the application of materials science to archaeometry and conservation science. OUP Oxford (2010)
3. Janssens, K., Van Grieken, R.: Non-destructive micro analysis of cultural heritage materials. Elsevier (2004)
4. Jeynes, C.: Chapter 10 - ion beam analysis for cultural heritage. In: Adriaens, M., Dowsett, M. (eds.) Spectroscopy, Diffraction and Tomography in Art and Heritage Science, pp. 335–364. Elsevier (2021)
5. Lagomarsino, S., et al.: The center for production of single-photon emitters at the electrostatic-deflector line of the tandem accelerator of labec (florence). Nucl. Instrum. Methods Phys. Res., Sect. B **422**, 31–40 (2018)

6. Mathot, S., et al.: The cern pixe-rfq, a transportable proton accelerator for the machina project. Nucl. Instrum. Methods Phys. Res., Sect. B **459**, 153–157 (2019)
7. Stokes, R.H., Wangler, T.P.: Radiofrequency quadrupole accelerators and their applications. Annu. Rev. Nucl. Part. Sci. **38**(1), 97–118 (1988)
8. Taccetti, F., et al.: Machina, the movable accelerator for cultural heritage in-situ non-destructive analysis: project overview. Rendiconti Lincei. Scienze Fisiche e Naturali **34**(2), 427–445 (06 2023)
9. Taccetti, N., et al.: The pulsed beam facility at the 3 mv van de graaff accelerator in florence: overview and examples of applications. Nucl. Instrum. Methods Phys. Res., Sect. B **188**(1–4), 255–260 (2002)
10. Van Grieken, R., Janssens, K.: Cultural heritage conservation and environmental impact assessment by non-destructive testing and micro-analysis. CRC Press (2004)

Abstracts Embeddings Evaluation: A Case Study of Artificial Intelligence and Medical Imaging for the COVID-19 Infection

Giovanni Zurlo[1] and Elisabetta Ronchieri[1,2](✉)

[1] Department of Statistical Sciences, University of Bologna, Bologna, Italy
giovanni.zurlo@studio.unibo.it
[2] INFN CNAF, Bologna, Italy
elisabetta.ronchieri@unibo.it

Abstract. During the COVID-19 pandemic, a huge amount of literature was produced covering different aspects of infection. The use of artificial intelligence (AI) in medical imaging has been shown to improve screening, diagnosis, treatment, and medication for the COVID-19 virus. Applying natural language processing (NLP) solutions to COVID-19 literature has contributed to infer significant COVID-19-related topics and correlated diseases. In this paper, we aim at evaluating biomedical transformer-based NLP techniques in COVID-19 research to understand if they are able to classify problems related to COVID-19. Particularly, once collected COVID-19 publications encompassing the terms AI and medical imaging, fifteen BERT-based models have been compared with respect to modality prediction and task prediction.

Keywords: Medical Imaging · COVID-19 · Infection · Artificial Intelligence · Embeddings

1 Introduction

The SARS-CoV-2 pandemic triggered unprecedented research efforts across various disciplines. Notably, the field of artificial intelligence (AI) applied to medical imaging has been prominently involved. Given the scarcity of resources in facing this devious disease, AI-based tools have emerged as potentially valuable assets to be harnessed. Natural Language Processing (NLP) offers a means to expedite the analysis of scientific articles on a larger scale, surpassing the constraints of manual analysis, and has long been recognized as a solution to mitigate information overload in biomedical research. Since the beginning of the pandemic, the NLP community has been consistently addressing the needs of domain experts by applying cutting-edge methods to enhance comprehension and knowledge discovery. Numerous NLP tasks are well studied, allowing existing systems to be readily adapted to address their respective tasks in the context of the COVID-19

G. L. Foresti et al. (Eds.): ICIAP 2023 Workshops, LNCS 14365, pp. 202–214, 2024.
https://doi.org/10.1007/978-3-031-51023-6_18

pandemic with relatively minimal effort. Many NLP methods rely on pre-existing resources, such as large corpora of related text or lists of concepts, names, and relationships, to help provide meaning and context. Given the novelty of both the virus and the disease, there was initially limited availability of relevant text, and the relationships between related concepts were largely unknown. The pandemic introduced completely new terminology that were provided by updates to ontologies, such as ICD-10 (*10th International Classification of Diseases*) [38] and MeSH (*Medical Subject Headings*) [25], requiring the existing NLP systems to incorporate these changes. Less overtly, the pandemic led to a significant domain shift in the frequency of existing terms (e.g., *Reverse Transcription Polymerase Chain Reaction (RT-PCR)* and *Quarantine*) in the scientific literature, necessitating updates to the systems to maintain performance integrity.

In a text-based visualization produced by González-Márquez, R. et al. [13], COVID-19 literature has been identified as "uniquely isolated" within the broader landscape of biomedical literature. The main COVID-19 cluster is surrounded by articles covering other epidemics, public health matters, and respiratory diseases. Moreover, within the clump itself, a rich internal structure has been observed, with multiple COVID-19-related topics distinctly segregated from one another.

The primary objective of this study is to assess the adequacy of commonly employed biomedical transformer-based models, trained on pre-pandemic corpora, in capturing the semantic features present in medical imaging literature. Concurrently, we aim to observe the potential advantage of citations supervision and the inclusion of COVID-19 literature in the pre-training phase. To accomplish this, we introduce a unique and independent test set specifically focused on the medical imaging domain. This novel dataset serves as a valuable resource for the extrinsic evaluation of contextual embeddings, comprising realistic text classification tasks based on 560 gold labels referred to two target variables: the clinical task and imaging modality.

2 Related Works

Embeddings refer to vector text representations that encode the semantic and syntactic similarities between textual segments. These can be computed at different levels of granularity, from word tokens or sentences to documents and beyond. Of course, there are dozens of different embedding methodologies, with contextualized vector representations recently emerging as an advancement in the field [7,29].

In the context of COVID-19 literature, both paper and concept embeddings have been used by several systems to support search [5] and retrieval [12] activities. Among these, SPECTER embeddings [9] have gained popularity in NLP and Information Retrieval (IR) studies [2,22] as precomputed vectors were made available along with the CORD-19 [36] dataset. SPECTER is a citation-informed transformer model specifically designed for scientific article processing.

While there has been significant progress in evaluating word embeddings [3, 30], evaluation of scientific document representations has remained limited.

Existing benchmarks either focus on document similarity [20,35] or exhibit a lack of diversity in tasks [9]. Further, it has been observed that models excelling in general-purpose text embedding benchmarks, such as the large Massive Text Embedding Benchmark (MTEB) [19], may not perform as effectively in scientific tasks [31]. Addressing these limitations, two recent ad-hoc benchmark suites, namely SciDoc, i.e., the Dataset Evaluation Suite for SPECTER [9], and SciRepEval, i.e., A Multi-Format Benchmark for Scientific Document Representations [31], have been proposed, with the latter including the former as a subset.

In accordance with the definition of extrinsic evaluation given in Bakarov, A. [3], any downstream task could be considered as an evaluation method, each reflecting practical applications of contextualized document representations. For instance, classifying papers into topical categories is a foundational task for document organization and discovery. Among its four task formats, SciRepEval propose several classification benchmark sets including a large-scale Field of Study (FoS) multi-label training set comprising over 500K papers with silver labels based on publication venues. This benchmark task bears resemblance to the one defined and employed for model comparison in the work done by González-Márquez, R. et al. [13].

3 Methodology

3.1 Dataset

We queried literature indexed in the National Institutes of Health (NIH)'s *iSearch* COVID-19 Portfolio [21], an interactive tool designed for portfolio analysis. This platform consolidates a comprehensive collection of COVID-19 publications and preprints from various sources, such as PubMed for articles and preprints from *arXiv*, *bioRxiv*, *ChemRxiv*, *medRxiv*, *Preprints.org*, *Qeios*, and *Research Square*. Unlike the well-known CORD-19 [36] dataset, which is no longer updated, the COVID-19 Portfolio is diligently maintained at the time of writing, with new literature selected daily by subject matter experts.

Our collection process is illustrated in Fig. 1 and shows two main streams of queries: a broad one [23] encompassing the terms `AI AND COVID-19 AND 'Medical Imaging'` and a modality-specific one [24] with `AI AND COVID-19 AND Lung AND (CT OR CXR OR US OR PET)`, where CT stands for Computerized Tomography, CXR denotes Chest X-Ray, US is Ultrasound, and PET corresponds to Positron Emission Tomography. These queries were extended with keywords synonyms to comprehensively capture relevant literature. Following the guidelines set in the Preferred Reporting Items for Systematic Reviews and Meta-Analyses (PRISMA) statement [26], we combined the results of both queries and filtered out 9 papers with missing abstracts and 13 records from excluded publication types (such as Published Erratum, News, Interview, Introductory Journal Article, Retraction Note and Retracted Article). Subsequently, a manual screening based on the title and abstract was performed. The inclusion criteria were followed and applied throughout this process. Ultimately, the investigation identified a total of 2476 papers focusing on AI in the context of lung imaging for COVID-19.

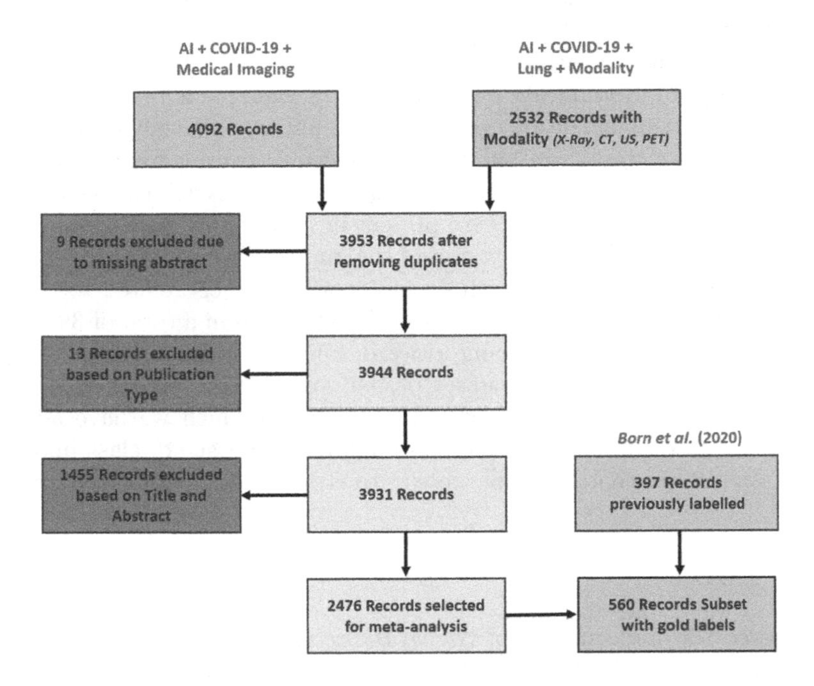

Fig. 1. PRISMA Flowchart Illustrating the Study Selection Used

This collection comprises papers spanning the period from 2020 to 2023, with the most recent entries extending up to our download date, on May 27th, 2023. The majority of the records (76%) are peer-reviewed articles from PubMed. Available metadata includes DOI and other identifiers from medical scientific archives, Publication Date and Type, Source and Journal Name, Title, Abstract, and Citations retrieved from PubMed. Data and code can be accessed at https://github.com/zurlog/abs-embeddings-eval.

3.2 Label Assignment

At this point in the study, we sought to assign appropriate labels to each research paper based on its domain-specific subject matter. In [13], the authors dealt with literature from the broad biomedical domain and extracted medical specialties and scientific branch labels from the journal names of each paper. However, considering the specific focus of our investigation and its strong multidisciplinary nature, we chose to adopt the 6-tasks and 4-modalities classification framework used in the review conducted by Born et al. [6]. Throughout this writing, we use 'modality' as a shorthand for imaging modality (i.e. *CT, CXR, Lung US* or *Multimodal*). The task categories include: Detection/Diagnosis, Monitoring/Severity

Assessment, Post-Hoc, Prognosis/Treatment, Review, Risk Identification, and Segmentation-only. The segmentation papers discuss localization of lung tissue or other disease features without direct applications to any clinically relevant downstream tasks.

In our research, we utilized a preexisting categorization that had been expertly curated by Born et al. [6]. Access to this valuable information was provided through a supplementary dataset of the paper, titled *Detailed results of systematic meta-analysis.* Note that, for publications covering multiple tasks, the primary task was the one considered. The merging process was carried out based solely on titles, as no identifiers were present in the aforementioned dataset. To facilitate this operation, we also explored the implementation of fuzzy-matching techniques. Ultimately, we successfully incorporated labels related to the "Primary Task," "Modality," and "Data Origin" features from a total of 397 articles published throughout 2020. Moving forward, for the remainder of the analysis, we hence focused on a curated subset of 560 gold-labeled records. This subset includes the original 397 articles reviewed in [6], to which we have added an additional 163 manually labeled papers to address the issue of class imbalance. This is a sample size that is comparable to the testing sets included in Singh, A. et al. [31].

3.3 Models

We compared the performance of twelve Bidirectional Encoder Representation (BERT) variants: the original BERT [11] in its *base* and *large* versions, SciBERT [4], BioBERT [17] (*base* and *large*), PubMedBERT [14] (*base* and *large*), CORD-19 BERT [10], COVID SciBERT [32], ClinicalCovidBERT [33], RadBERT [8] and BioCovidBERT [34] We compared three versions of SPECTER 2 [9] as well: the standard model, and two others with task-specific adapters [31]. SPECTER stands out from other candidates by incorporating citations as a supervision signal through a custom triplet-loss pre-training objective. This unique approach enhances its representation capability by leveraging inter-document context, since citing articles can indicate relatedness. We used the HuggingFace's `transformers` [37] library along with the publicly available checkpoints from the Model Hub (https://huggingface.co/models).

The models considered in this study share identical architectures, namely `bertbase` with 12 transformer layers, a hidden size of 768 and 110M parameters, or `bertlarge` with 24 transformer layers, a hidden size of 1024 and 340M parameters. However, their distinctions lie in three key aspects [40]: the pre-training dataset, the weight initialization, and the vocabulary (see Table 1).

Regarding the pre-training datasets, we may deal with in-domain vocabulary and structure, like in the case of biomedical scientific corpora that, to some extent, cover our topics of interest [4,14,17]; in-domain vocabulary only, exemplified by models trained solely on corpora of radiology reports [40]; and out-of-domain vocabulary, corresponding to benchmark BERT models. Secondly, we distinguish between two weight initialization approaches, either from-scratch pre-trainings that start with random weights or continuous pre-trainings, wherein the weights are initialized from another Pre-Trained Model (PTM) before resuming pre-training on a different corpus. From-scratch pre-training is definitely beneficial when a large amount of in-domain text is available [14], being this the case for biomedical literature. Furthermore, each PTM employs a predefined vocabulary, which can be freely selected to align with the pre-training dataset for all from-scratch pre-trainings, but it is dictated by the former pre-training vocabulary in the case of continuous pre-trainings.

Table 1. Summary of Models Used

Model	Training Corpus	Weights Initialization	Vocabulary	Details on Training Corpus
$BERT_{base}$	Wiki+Books	From-scratch	Derived from corpus	800M + 2.5B words, 1M steps
SciBERT	SemanticScholar Full-Texts	$BERT_{base}$	Derived from corpus	1.14M Full-Texts, 18% from computer science and 82% from broad biomedical domain
$BioBERT_{base}$	PubMed abstracts	$BERT_{Base}$	Same as $BERT_{Base}$	Updated 2019. 4.5B Words, 1M steps
$PubMedBERT_{base}$	PubMed abstracts + PMC Full-Texts	From-scratch	Derived from corpus	Updated Feb. 2020. 16.8B Words, 100K steps
CORD-19 BERT	CORD-19 dataset	$BERT_{Base}$	Same as $BERT_{Base}$	Updated Early 2020
CovidSciBERT	CORD-19 dataset	SciBERT	Extended from SciBERT	Updated Early 2020
ClinicalCovidBERT	CORD-19 dataset	Bio+Clinical BERT [1]	Same as $BERT_{Base}$	Full-Texts updated June 2020, 150K steps
RadBERT	Radiology Reports	$BioBERT_{base}$	Same as $BERT_{Base}$	4M reports from 600K unique patients treated at Stanford Health Care from 1992 to 2014
SPECTER 2	6M Triplets of Papers Citations	SciBERT	Same as SciBERT	Extended version of the `cite_prediction` dataset from [31]
$BERT_{large}$	Wiki+Books	From-scratch	Derived from corpus	800M + 2.5B words, 1M steps
$BioBERT_{large}$	PubMed abstracts	$BERT_{Large}$	Derived from corpus	Updated 2019. 4.5B Words, 1M steps
$PubMedBERT_{large}$	PubMed abstracts	From-scratch	Derived from corpus	Updated Feb. 2020. 3.2B Words, 100K steps
BioCovidBERT	CORD-19 dataset	$BioBERT_{large}$	Same as $BioBERT_{large}$	Full-Texts updated June 2020, 200K steps

Each Title + Abstract pair gets concatenated and split into a sequence of tokens, and any $BERT_{base}$ continual pre-training encodes it in a 768-dimensional latent space (or 1024-dimensional if based on $BERT_{large}$). The first token of a sequence is always a special classification token `[CLS]`, while the separator one `[SEP]` marks its end and also separates titles from abstracts. The context window size is fixed at 512 tokens for both architectures, resulting in the automatic truncation of longer inputs beyond this limit. This corresponds approximately to 300–400 words, and it has been reported that approximately 98% of all PubMed abstracts fall below this threshold [13]. While the large majority of abstracts in our dataset adhere to this constraint, few records required truncation.

Our goal is to obtain a singular vector representation for each Title + Abstract rather than one for each of its tokens. To achieve this, we compared three approaches commonly used in the literature: the first two involve extracting the final hidden state representation of the `[CLS]` token [11] and the trailing `[SEP]` token [13] respectively; the third one uses a so-called "mean-pooling" strategy on the second-to-last hidden states [11,39]. As for the mean-pooling strategy, we averaged the embeddings of all tokens returned by the second-to-last hidden layer, after excluding the vectors corresponding to the aforementioned special tokens.

3.4 Performance Metrics

In our study, we utilized the values of the features "Primary Task" and "Modality" which were gathered for a subset of AI for COVID-Imaging papers. These features served as the gold labels for conducting an embeddings comparison. Both the primary task and imaging modality information capture crucial and distinctive characteristics of each work within our data subset. Since our goal is to evaluate final document representations, rather than fine-tuning on individual downstream tasks, we follow [9,11] in directly applying the representations as features to the tasks.

In Singh et al. (2022) [31], embeddings are evaluated on classification by scoring their performance with features provided as input to linear support vector classifiers. Results for these tasks are evaluated by using F1 score (either binary- or macro-F1 depending on the dataset). We follow the approach of [13] instead, in which the accuracy (i.e. the fraction of correct predictions) of a k-Nearest Neighbors (kNN) classifier is used as a proxy for embedding quality as evaluated on a silver FoS labels classification task. All kNN-based metrics involved either $k = 6$ or $k = 13$ exact nearest neighbors and were obtained using the `KNeighborsClassifier` class from Scikit-Learn (version 1.2.2) with the arguments `algorithm='auto'`, `weights='distance'` and `distance= 'cosine'` [27]. All parameters were chosen after performing a cross-validated grid search.

To predict each test paper's label, our kNN classifier takes a weighted majority vote among the paper's nearest neighbors' labels in the training set. Neighbors are weighted by the inverse of their cosine distance, giving greater influence to those that are closer to the query point. The cosine distance is the standard choice when dealing with document vectors, as it provides a better way to capture semantic similarity. To measure accuracy, cross-validated values were averaged over the same 10-fold stratified split, in which the class distribution of the target variable is preserved across each fold. Additionally, a balanced version of accuracy was computed to control for the effect of class imbalance on the standard metric. In a multiclass classification setting, this score is equivalent to the average of Recall obtained in each class. The chance-level accuracies were obtained using the `DummyClassifier` from Scikit-Learn with `strategy='stratified'`. A DummyClassifier makes predictions that ignore the input features, so its accuracy values are exactly the same regardless of the embeddings provided. These scores serve as simple baselines to compare against other more complex classifiers.

4 Results and Discussions

The task of modality prediction is not significantly challenging, as citing the appropriate synonym within the textual input generally offers substantial evidence for most entries. Among the assessed models (as indicated in Table 2), the highest predictive capability was exhibited by SPECTER, particularly when utilizing its opening and trailing token embeddings. This observation can likely be attributed to the premise that sentence transformers were originally trained to optimize one specific representation. Notably, SPECTER employs the `[CLS]` token for this purpose. For consistency, the results for all three representations are presented in Table 2. Throughout this discussion, we will consistently refer to the standard SPECTER model. This choice is due to the observation that using the classification and proximity adapters from [31] yielded to sub-optimal performance outcomes within our experiments.

Table 2. Quality metrics for the embeddings (`Modality` Prediction) 10-fold kNN classification accuracy and balanced accuracy. Hyperparameters: $k = 13$, $weights = distance$, $distance = cosine$.

Model	Accuracy (%)			Balanced Accuracy (%)		
	`[CLS]`	`[SEP]`	`AVG`	`[CLS]`	`[SEP]`	`AVG`
BERT_{base}	54.3	57.7	61.3	38.6	43.3	49.9
SciBERT	58.2	56.4	63.6	43.9	41	48.2
BioBERT_{base}	53.2	65.2	61.1	36	52.6	46.2
PubMedBERT_{base}	57.7	74.5	64.3	42.9	58.6	50.1
CORD-19 BERT	56.4	*52.9*	60.2	43.0	*37.6*	44.9
CovidSciBERT	64.5	60.5	62.7	49.9	47.9	50.6
ClinicalCovidBERT	65.4	64.5	63.4	53.2	50.8	49.6
RadBERT	57.9	57.9	*58.2*	38	38	*39.7*
SPECTER 2	**82.5**	**83.8**	68.8	**75.3**	**76.7**	57.8
BERT_{large}	*50*	58.8	60.2	*34.7*	43.4	44.8
BioBERT_{large}	54.4	62.1	65.5	39.8	47.2	51.3
$\text{PubMedBERT}_{large}$	57	61.3	60.4	40.3	44.2	45.3
BioCovidBERT	69.5	64.8	**69.6**	53.8	49	**58.3**
Chance Level		35.5 ± 13			24.8 ± 9	

Nevertheless, SPECTER effectiveness diminished when employing the average pooling strategy, in which it is slightly outperformed by the BioCovidBERT. This last is a BioBERT_{large} variant that underwent continual pre-training on a corpus of articles related to COVID-19 from 2020. This additional optimization proved beneficial, as it slightly improved the performance compared to the weights-initialized model.

Predicting each paper's task presents higher difficulty due to potential class overlap or ambiguity; not by chance, the authors of [6] decided to record the primary task only. The balanced accuracy scores in Table 3 decreased steadly with respect to those observed in the modality label prediction. This decline can be attributed to the presence of stronger class imbalance and lower recall values for the under-represented classes, notably "post-hoc" and "risk identification".

Table 3. Quality metrics for the embeddings (`Task` prediction) 10-fold kNN classification accuracy and balanced accuracy. Hyperparameters: $k = 6$, $weights = distance$, $distance = cosine$.

Model	Accuracy (%)			Balanced Accuracy (%)		
	`[CLS]`	`[SEP]`	`AVG`	`[CLS]`	`[SEP]`	`AVG`
BERT_{base}	*59.6*	*58.2*	64.8	27	28.4	33.9
SciBERT	62.7	63	68.6	33.3	31.5	38.3
BioBERT_{base}	63.9	70.2	69.6	28.6	40.5	40.2
PubMedBERT_{base}	66.8	70.7	67.5	34.7	42.3	36.9
CORD-19 BERT	65.0	60.7	65.9	33.7	*25.9*	34.4
CovidSciBERT	70.2	70.2	71.8	42.3	42.4	45.1
ClinicalCovid BERT	70.9	71.3	70	43.6	46.7	40.9
RadBERT	60.9	60.9	*61.2*	26.5	26.5	*26.4*
SPECTER 2	**75.4**	**74.5**	**74.1**	**56.6**	**55.9**	**51.5**
BERT_{large}	60	64.1	66.6	*26.2*	34	38.5
BioBERT_{large}	62	68.6	67.3	28.7	37.7	36
$\text{PubMedBERT}_{large}$	63	67.7	68.9	30	36.1	38.4
BioCovidBERT	66.6	68.2	70.9	37	39.5	42.5
Chance Level		36.1 ± 6			14.8 ± 9	

Within this context, SPECTER consistently emerges as the best model, delivering superior accuracy and balanced accuracy in task prediction across diverse extraction strategies. Among the other models evaluated, CovidSciBERT and ClinicalCovidBERT exhibit the best performance, with no notable distinction between them. However, their competitive edge diminishes when predicting the rarer classes. Not surprisingly, the benchmark BERT models and RadBERT perform poorly. RadBERT constitutes a BioBERT_{base} variant continually pre-trained on a diverse corpus of radiological reports that include multiple imaging modalities across 36 body parts, indicating its broader applicability beyond pulmonary imaging. As a result, RadBERT is mainly optimized for processing such structured texts.

Overall, biomedical transformer models pre-trained on COVID-19 literature exhibit a pronounced advantage in both modality and task prediction. For example, the ClinicalCovidBERT model, which is initialized from a Biobert model and pre-trained on MIMIC clinical notes [15], can be compared against BioBERT in this regard. However, the CORD-19 BERT stands as an exception, showcasing the worst performance in three comparisons. This outcome could potentially be attributed to the lack of comprehensive biomedical content in its pre-training corpus, once again emphasizing the importance of careful pretraining and initialization choices.

Regarding model size, larger architectures do not exhibit a significant advantage over their smaller counterparts. This is in contrast with theoretical knowledge on pretrained language models [16] and experiments with fine-tuned models [11]. However, this may be due to our feature-based approach which already yielded mixed results when dealing with increasing model size [18,28]. Instead, with fine-tuning, the task-specific models can benefit from the larger, more expressive representations even when data is very small [11].

5 Conclusions

The ablation studies outlined in [31], which are based on the older SPECTER, provided mixed and mild support for the effectiveness of citation-supervised training. Similarly, in a pilot experiment detailed in [13], the embeddings derived from the same model yielded to suboptimal performance in a broad field of study classification task. We hypothesize that the semantic attributes of abstracts were already sufficient for categorizing into such broad subject areas.

Through our medical imaging test dataset, we were able to somewhat validate the findings presented in [9] concerning the competitive performance of SPECTER in a variety of document-level tasks. Given the inherent complexity of our domain-specific task, training with citation supervision proved to be considerably more beneficial. However, for rigorous investigations, along with updates of such analyses employing the enhanced SPECTER 2 model, we defer to future contributions.

We acknowledge that our results do not constitute a comprehensive evaluation of embeddings quality. This limitation arises because extrinsic evaluation fails if the embeddings are trained to serve in a wide range of different tasks, as performance scores across downstream tasks do not correlate between themselves [3]. Nevertheless, we hope that this new test set may serve as an addition to a comprehensive benchmark suite to improve cross-task and cross-domain generalization.

Looking ahead, we will leverage the insights gained from these experiments to facilitate the annotation process of our original dataset. Our approach will involve a combination of automated tools and manual assessment. We look forward to collecting more labelled records, potentially aiming at a valid training set with an adequate sample size.

References

1. Alsentzer, E., et al.: Publicly available clinical BERT embeddings. In: Proceedings of the 2nd Clinical Natural Language Processing Workshop, pp. 72–78. Association for Computational Linguistics, Minneapolis, Minnesota, USA, June 2019. https://doi.org/10.18653/v1/W19-1909
2. An, X., et al.: An active learning-based approach for screening scholarly articles about the origins of sars-cov-2. PLOS ONE **17**, e0273725 (2022). https://doi.org/10.1371/journal.pone.0273725
3. Bakarov, A.: A survey of word embeddings evaluation methods. CoRR abs/1801.09536 (2018)
4. Beltagy, I., et al.: SciBERT: a pretrained language model for scientific text. In: EMNLP. Association for Computational Linguistics (2019). https://www.aclweb.org/anthology/D19-1371
5. Bhatia, P., et al.: AWS CORD19-search: A scientific literature search engine for COVID-19. CoRR abs/2007.09186 (2020)
6. Born, J., et al.: On the role of artificial intelligence in medical imaging of covid-19. Patterns **2**(6), 100269 (2021). https://doi.org/10.1016/j.patter.2021.100269
7. Camacho-Collados, J., Pilehvar, M.T.: From word to sense embeddings: a survey on vector representations of meaning. J. Artif. Int. Res. **63**(1), 743–788 (2018). https://doi.org/10.1613/jair.1.11259
8. Chambon, P., et al.: Improved fine-tuning of in-domain transformer model for inferring covid-19 presence in multi-institutional radiology reports. J. Digit. Imaging **36**, 164–177 (2022)
9. Cohan, A., et al.: Specter: Document-level representation learning using citation-informed transformers (2020)
10. Deepset: covid_bert_base (2020). https://huggingface.co/deepset/covid_bert_base
11. Devlin, J., et al.: BERT: pre-training of deep bidirectional transformers for language understanding. CoRR abs/1810.04805 (2018)
12. Esteva, A., et al.: Co-search: COVID-19 information retrieval with semantic search, question answering, and abstractive summarization. CoRR abs/2006.09595 (2020)
13. González-Márquez, R., et al.: The landscape of biomedical research. bioRxiv (2023). https://doi.org/10.1101/2023.04.10.536208
14. Gu, Y., et al.: Domain-specific language model pretraining for biomedical natural language processing. CoRR abs/2007.15779 (2020)
15. Johnson, A.E., et al.: MIMIC-III, a freely accessible critical care database. Sci. Data **3**(1), May 2016. https://doi.org/10.1038/sdata.2016.35
16. Kaplan, J., et al.: Scaling laws for neural language models. CoRR abs/2001.08361 (2020)
17. Lee, J., et al.: BioBERT: a pre-trained biomedical language representation model for biomedical text mining. Bioinformatics **36**(4), 1234–1240 (2020)
18. Melamud, O., et al.: context2vec: learning generic context embedding with bidirectional LSTM. In: Proceedings of the 20th SIGNLL Conference on Computational Natural Language Learning. pp. 51–61. Association for Computational Linguistics, Berlin, Germany, August 2016. https://doi.org/10.18653/v1/K16-1006
19. Muennighoff, N., et al.: MTEB: massive text embedding benchmark. In: Proceedings of the 17th Conference of the European Chapter of the Association for Computational Linguistics, pp. 2014–2037. Association for Computational Linguistics, Dubrovnik, Croatia, May 2023. https://aclanthology.org/2023.eacl-main.148

20. Mysore, S., et al.: CSFCube - a test collection of computer science research articles for faceted query by example. In: Thirty-Fifth Conference on Neural Information Processing Systems Datasets and Benchmarks Track (Round 2) (2021). https://openreview.net/forum?id=8Y50dBbmGU
21. National Institutes of Health Office of Extramural Research: Open Mike: New NIH Resource to Analyze COVID-19 Literature: The COVID-19 Portfolio Tool. Retrieved April 2, 2021. https://nexus.od.nih.gov/all/2020/04/15/new-nih-resource-to-analyze-covid-19-literature-the-covid-19-portfolio-tool/ (2020)
22. Newton, A.J.H., et al.: A pipeline for the retrieval and extraction of domain-specific information with application to covid-19 immune signatures. BMC Bioinform. **24**(1), July 2023. https://doi.org/10.1186/s12859-023-05397-8
23. NIH OPA: iSearch COVID-19 Portfolio, Query#1 (2023). https://icite.od.nih.gov/covid19/search/#search:searchId=64b824d13089f55f525505be
24. NIH OPA: iSearch COVID-19 Portfolio, Query#2 (2023). https://icite.od.nih.gov/covid19/search/#search:searchId=647e4bf03089f55f5254e28b
25. NLM (U.S. Natl. Lib. Med.): COVID-19 and SARS-CoV-2 MeSH Terms - 2021. NLM Technical Bulletin, Dec. 04 (2020). https://www.nlm.nih.gov/pubs/techbull/nd20/nd20_mesh_covid_terms.html
26. Page, M.J., et al.: The prisma 2020 statement: an updated guideline for reporting systematic reviews. Systematic Rev. **10**(1), March 2021. https://doi.org/10.1186/s13643-021-01626-4
27. Pedregosa, F., et al.: Scikit-learn: machine learning in Python. J. Mach. Learn. Res. **12**, 2825–2830 (2011)
28. Peters, M.E., et al.: Dissecting contextual word embeddings: Architecture and representation. CoRR abs/1808.08949 (2018)
29. Reimers, N., et al.: Classification and clustering of arguments with contextualized word embeddings. In: Proceedings of the 57th Annual Meeting of the Association for Computational Linguistics, pp. 567–578. Association for Computational Linguistics, Florence, Italy, July 2019. https://doi.org/10.18653/v1/P19-1054
30. Schnabel, T., et al.: Evaluation methods for unsupervised word embeddings. In: Proceedings of the 2015 Conference on Empirical Methods in Natural Language Processing, pp. 298–307. Association for Computational Linguistics, Lisbon, Portugal, September 2015. https://doi.org/10.18653/v1/D15-1036
31. Singh, A., et al.: SciRepEval: a multi-format benchmark for scientific document representations. ArXiv abs/2211.13308 (2022)
32. Thakur, T.: Covid-scibert: a small language modelling expansion of scibert, a bert model trained on scientific text. https://github.com/lordtt13/word-embeddings/tree/master/COVID-19 (2020)
33. Tonneau, M.: clinicalcovid-bert-base-cased (2020). https://doi.org/10.57967/hf/0867
34. Tonneau, M.: biocovid-bert-large-cased (2023). https://doi.org/10.57967/hf/0869
35. Voorhees, E.M., et al.: TREC-COVID: constructing a pandemic information retrieval test collection. CoRR abs/2005.04474 (2020)
36. Wang, L.L., et al.: CORD-19: the COVID-19 open research dataset. In: Proceedings of the 1st Workshop on NLP for COVID-19 at ACL 2020. Association for Computational Linguistics, Online, July 2020. https://www.aclweb.org/anthology/2020.nlpcovid19-acl.1
37. Wolf, T., et al.: Huggingface's transformers: state-of-the-art natural language processing. CoRR abs/1910.03771 (2019)
38. World Health Organization: COVID-19 update for ICD-10. Publication (2020). https://www.who.int/publications/m/item/covid-19-update-for-icd-10

39. Xiao, H.: bert-as-service (2018). https://github.com/hanxiao/bert-as-service.git, read the documentation at: https://bert-as-service.readthedocs.io/en/latest/section/faq.html#frequently-asked-questions
40. Yan, A., et al.: RadBERT: adapting transformer-based language models to radiology. Radiol. Artif. Intell. **4**(4), July 2022. https://doi.org/10.1148/ryai.210258

Pigments and Brush Strokes: Investigating the Painting Techniques Using MA-XRF and Laser Profilometry

Valerio Graziani[1,2(✉)], Giulia Iorio[1,2], Stefano Ridolfi[3], Chiara Merucci[4], Paolo Branchini[1,2], and Luca Tortora[1,2,5]

[1] Roma Tre Department, National Institute for Nuclear Physics — INFN, Via della Vasca Navale 84, 00146 Rome, Italy
{valerio.graziani,giulia.iorio,paolo.branchini}@roma3.infn.it

[2] Roma Tre Surface Analysis Laboratory — LASR3, Via della Vasca Navale 84, 00146 Rome, Italy

[3] Ars Mensurae srl, Via Vincenzo Comparini 101, 00188 Rome, Italy
stefano@arsmensurae.it

[4] Gallerie Nazionali Barberini Corsini, Via delle Quattro Fontane 13, 00184 Rome, Italy
chiara.merucci@cultura.gov.it

[5] Department of Science, Roma Tre University, Via della Vasca Navale 84, 00146 Rome, Italy
luca.tortora@uniroma3.it

Abstract. To enhance the deep understanding of artists'mindset and technical practices the detailed characterization of paintings require the net of brushstrokes on the surface to be considered as informative. Laser profilometry is being coupled with a macro x-ray fluorescence (MA-XRF) device as a non-destructive, non-invasive, portable tool for such an investigation. A set of different brushstrokes was surveyed by testing the necessary sampling conditions and verifying the compatibility with the MA-XRF working conditions. Eight different surveyed sample brushstrokes proved to be clearly informative of a variety of features useful to characterize the gesturality of the artist's hand.

Keywords: Optical profilometry · MA-XRF imaging · painting analysis · surface digitalization · brushstrokes

1 Introduction

Studying the material part of ancient and modern paintings is part of the effort to understand the historical value of an artwork and of an artist's production. It is the basis for the correct preservation of whichever cultural asset, as well. To enhance the analytical capabilities in this field a large number of works are being developed in order to bring the traditional single-point based approach of several techniques to the "macro" level [1–4]. In the present work the coupling of two techniques (X-Ray Fluorescence and profilometry) on a common moving device for macro-analysis is presented.

G. L. Foresti et al. (Eds.): ICIAP 2023 Workshops, LNCS 14365, pp. 215–226, 2024.
https://doi.org/10.1007/978-3-031-51023-6_19

In the last decades, the use of handheld X-Ray Fluorescence (XRF) has significantly impacted the Cultural Heritage field, transforming the way researchers and conservators analyze and study artifacts and artworks [5]. In particular, macro-XRF imaging (MA-XRF) has now emerged as a powerful and innovative technique, revolutionizing the analysis of historical artifacts and artworks from the knowledge of the single-point composition to visualizing the distribution of the elements by full mapping of the painting surface. Such an approach to imaging is basically allowed by the non-destructivity, non-invasiveness (as does not require contact) and portability of the technique. Its advantages lie in the ability to quickly and accurately determine the elemental composition of a large number of materials without the need for sampling [1, 6].

The MA-XRF can be thus easily applied in-situ (e.g. museums, restoration laboratories or other preservation institutions and private collections), which is particularly important when handling artworks with large dimensions or non movable.

The absence of contact between the scanning device and the artwork is advantageous also in terms of risks as it minimizes any potential damage to delicate and irreplaceable objects [5, 7].

The elemental maps can offer insights into the artwork's layer structure, pigment palette and blendings, thickness of particular components, pentimenti, restoration attempts made over the time [8–12]. Armed with this knowledge, conservation scientists and restorers can identify potential risks such as chemical reactions and deteriorations, make informed decisions about appropriate preservation techniques and restoration materials, ensuring the longevity and integrity of cultural artifacts. The same information can enrich our understanding of the painting's evolution, being also informative about the artist's mind and cultural references if its devoted to the interpretation of the technical practices. Finally it can give clues for authentication, giving details which can be used as terms of comparison as well [13].

In this specific view, the (elemental) composition data can be relevant when studying paintings realized before the contemporary age because the artist's palette found in an artwork can be compared with that of other artists or with the palette of the same artist in a different period of his production. Conversely, in the case of contemporary art, the increasingly widespread use of synthetic organic pigments had made this kind of knowledge less informative. For this reason and for the fact that the forgers use the same materials available to artist being counterfeited, in contemporary art attribution studies and authentication studies those techniques providing process details are considered more informative.

An example of what is presented above is given by the MA-XRF imaging of the Fornarina by Raphael (ca. 1520), carried on in the frame of the MUltichannel Scanner for Artworks (MUSA) project during the 2019–2021 [5–7, 14–16].

The device developed in this project was basically obtained by coupling a XRF analytic head with a moving stage. The analytic head was composed by a x-ray tube and two detectors collimated to the same point by different angles. In addition, a camera and two position sensors were mounted as well. The X-ray tube was a silver target 4WMoxtek source, working at 38 kV and 38 mA. Dwell-time for each pixel was 350 ms. Two AMPTEK X-123 Peltier-cooled SDD were chosen for the detection of the excited signal [17].

The camera and the position sensors were added for precision pointing of the x-ray primary beam and for artwork safety respectively.

The overall acquisition system was fixed on a stage equipped with step-motors for plane parallel translation over the painted surface. The step was set to 1 mm and the resulting lateral resolution was about 1x1 mm^2/pixel.

A LabView built-in software ensured the synchronized translation of the supporting stage and collection of the fluorescence by the two detectors in parallel, while the continuous emission of the x-ray tube excited the materials. The data collected at each point were associated to the couple of position indexes, which could be used to organize the information in a hyperspectral datacube.

As a consequence of the physical dimension of the painting, the acquisition was divided into 29 sub portions, ensuring minimum overlapping between them for post-acquisition alignment (Fig. 1).

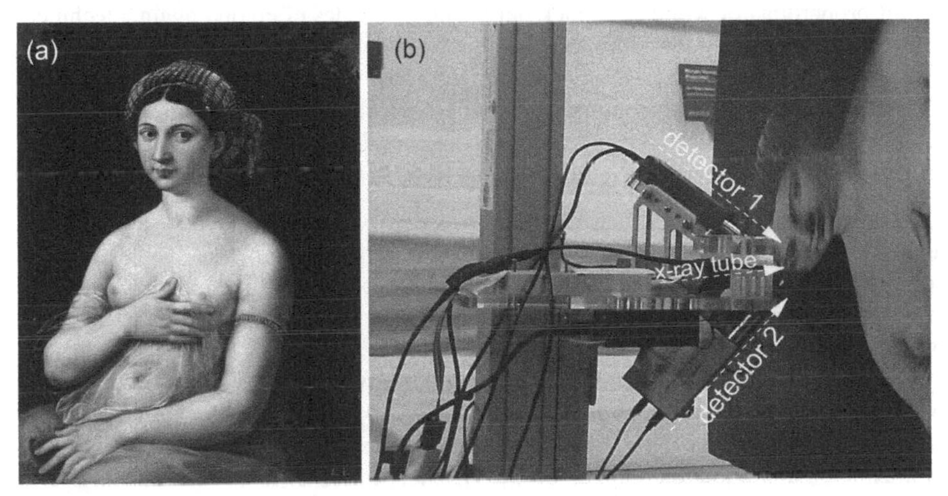

Fig. 1. (a) Fornarina, painting by Raphael, ca.1520; (b) XRF analytic head developed during the MUSA project.

Given the previous knowledge of the constituents of the painting [5], pigments palette was not the focus of the interpretation nor the presence of restoration interventions (even though not all were precisely mapped). What was more important in this case was obtaining clues about the composition of the scene and eventual changes depicting the artist's mindset. The detailed results of the analysis, obtained by comparisons between the visible image, the infrared riflectography and the several single element maps are reported elsewhere [5, 7] and could be here synthesized as follows:

- the image resulting from the various elemental maps revealed that the original concept of the artwork was more related to the portrait than to the representation of an ideal woman (Venus);
- the artwork had a complex creative process, due both to changes in the idea of the artist and to changes in the destination of the final work, and is to be considered a

"non finito": this can be seen in the cancellation of the background landscape, in the displacement of the volumes of Fornarina's left arm, in the rotation of the head from frontal to lateral, the ringlet hidden, the partially removed veil, and in the freedom with respect to the underdrawing;
- MA-XRF leads to an innovative approach to radiography in that elements emitting more than one detectable fluorescence line allow to select the imaging from different depths of the same painting, that is in approximation from different layers. This lead to appreciate details on pentimenti and modifications both in the composition and in the use of light and shades, which were not clearly distinguishable with the conventional approach due to the absorbing contribution of the wooden support;
- the distribution and association of the counts of specific elements suggested the presence of a coloured imprimatura, placing the artwork in a precise frame of technical practices acquired by the artist of Raphael's time; the same approach on the unglazed areas of body underpainting revealed that volumes were given by means of different proportions of white and black only, without shades of colour, again a technical solution common at that time;
- the overall perception of the artwork changed as the vegetal background likely underwent a colour change (darkening): signals of pigments enhancing volumetric variation in the bush were detected, suggesting a more complex relationship of the figure with its surroundings.

The overall information arising from the artwork gives the idea of a composite artwork, still not finished, which had a multi-step composition process. All the technical solutions found in the different part of the creation process (imprimatura, underdrawing, underpainting and volume handling, among the others) were coherent with the Raphael's production and could be detected by the MA-XRF [7].

In this deep understanding of the artwork evolution the missing element is the "hand" of artist, that is the way he worked on the panel. This would provide many other specific features which could be precious in understanding the artist work [18] but also in attribution and authentication studies [19]. Thus, in order to provide a novel useful tool in the field of studies presented above, coupling laser profilometry and XRF analysis is currently attempted. Unlike traditional contact-based methods, laser profilometry utilizes light to measure surface topography. This cutting-edge technology presents features similar to those of the MA-XRF described in terms of no-contact, non-destructivity, non-invasiveness, safety which make it suitable for implementation. Additionally, laser profilometry provides high-precision, 3D surface information with remarkable accuracy. The digital model of a painting produced by laser profilometry bears the intricate net of brushstrokes composing the painted layers but also the marks of texture alterations (panel warping, loose canvas, cracking, loss of the pictorial film, formation of new species on the surface), not only providing a detailed surface characterization but also shedding light on the artist's technique and the painting's history. Finally, the a digital model can be analyzed subsequently when the artwork under investigation is back in its conservation place.

In order to verify whether the technique can be applied efficiently to obtain a close reproduction of the painted surface thus allowing the analysis of the features described, some of the possible combinations of brushes and paints were tested under different scanning conditions.

2 Materials and Methods

An optical profilometer LJ-X800A by Keyence was chosen for the different acquisition tests. The emitting head is equipped with a 39 mm long array of LED, emitting a 405 nm laser at 10 mW. The working distance was in the range 72 ± 20 mm. The overall resolution in the resulting image depends on the direction considered. The detector resolution is 12.5 μm along the direction of the laser line (x-axis) whereas in the direction of the movement (y-axis) it depends on the relative velocity between the laser head and the sample. Along the z-axis the resolution is about 12.5 μm. Sampling frequency was set to 500 Hz and a batch number of profiles was fixed (10,000).

Two series of sample brushstrokes on cardboard were made to mimic the relief of a real paint surface using flat and rounded tip brushes, with different movements. The cardboard was prepared with a layer of homogeneous white paint to flatten as possible the surface. The coloured paint used for the brushstrokes samples was acrylic Permanent Alizarin Crimson by Maimeri.

The software employed for the acquisition of the profiles and the elaboration of the complete survey is LJ-X Navigator which could handle the overall process allowing for modifying batch number and sampling frequency as well. No filters and ground subtractions were applied.

3 Results and Discussion

As first step, obtaining the morphological profiles required the optimization of the translation velocity of the laser line on the sample surface to be surveyed (Fig. 2a) so that the resulting image showed the correct proportions with respect to object represented. This is achieved by making the x-direction resolution equal the distance between two consecutive profiles. Therefore, given the sampling frequency value at 500 Hz, the optimal scanning speed is equal to 6.250 mm/s. This speed range is not compatible with the one the stage developed during the MUSA project can reach nor with the acquisition times of the SDD (given a minimum acceptance value for the signal-to-noise ratio). The two instruments must be moved with very different velocities and as a consequence the two runs must be consecutive and not simultaneous. In any case, this is also required by the different indexing the stored profiles must have, as the resolution is higher in the case of the profilometry. However the improvement of the MA-XRF system developed during MUSA by coupling the profilometer is possible by the plug-and-play approach and the modularity of the acquisition head whereby the components required by x-ray spectroscopy can be quickly replaced by the profilometer [6, 7, 14]. Alternatively, it is necessary to have more than one independent acquisition system as represented in Fig. 2b.

(a)

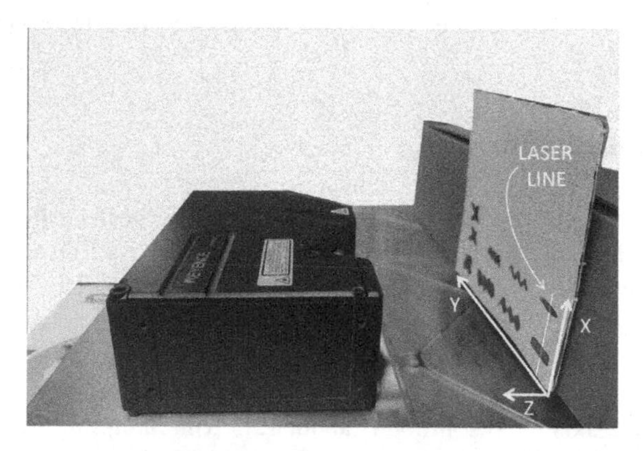

(b)

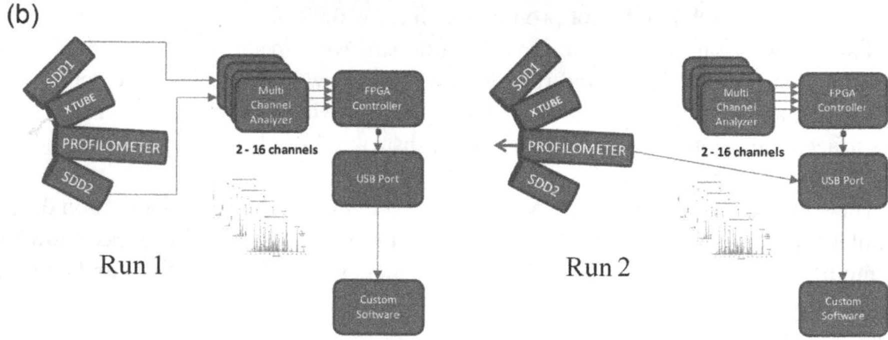

Fig. 2. (a) Profilometer setup during the brushstroke survey; (b) Schematic representation of the approach to keep for coupled analysis.

The eight profilometric surveys collected on the sample brushstrokes are shown in Fig. 3. The couples a-b, c-d, e-f, g-h correspond to the flat and rounded tips, respectively in single straight stroke, multiple continuous strokes, multiple discontinuous strokes, crossed strokes.

Comparing the two series Fig. 3a-c-e-g (flat tip) and Fig. 3b-d-f-h (rounded tip), it can be recognized how the brushstroke shapes are different and clearly distinguishable. Proportionally to the size, flat tip brush leaves less paint than the rounded one, likely as a consequence of the higher volume retained in the cylindrical shape in the second case.

The two brushes are characterized by a different distribution of the paint along the transverse direction of the stroke (from bottom to top) but similar along the longitudinal direction (direction of the hand movement):

- in the first direction, the flat tip releases less amount of paint and, on average, there is less difference between the elevation of the stroke centre with respect to the borders than that seen in the case of the rounded tip (a comparison for the two tips along the transverse direction approximately in the middle of the stroke is shown in Fig. 4a).
- in the second direction, the paint thickness in the rounded tip stroke ranges in the same values than in the flat tip one. This is likely because the paint is pressed in a similar way on the supporting surface by the two brushes, with exception of the

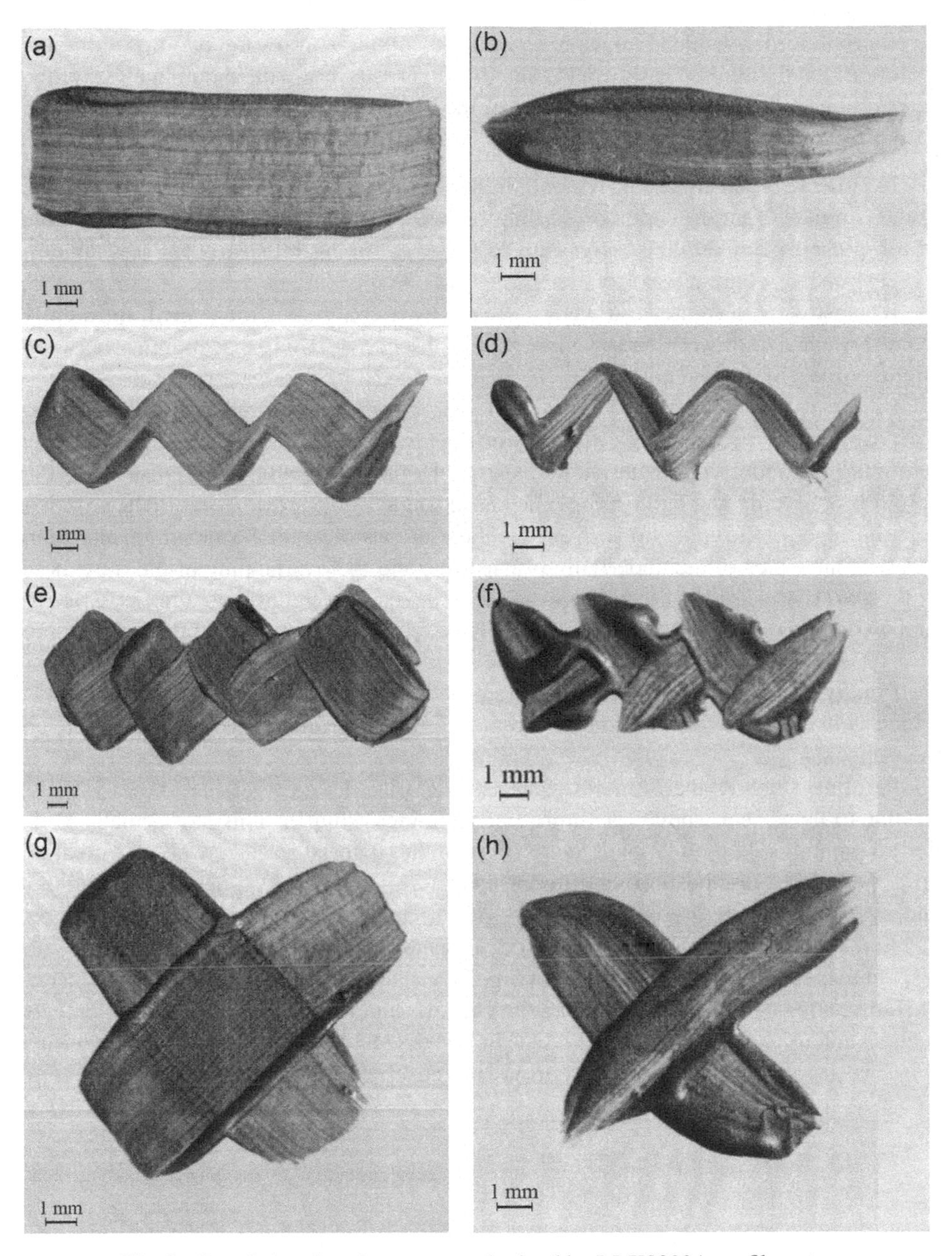

Fig. 3. Sample brushstrokes surveys obtained by LJ-X8000A profilometer

external borders which behave differently (a comparison for the two tips along the longitudinal direction is shown in Fig. 4b). Figure 3b shows the thinning effect of the paint layer due to the typical bending of the brush tip at the end of the movement on the right side.

In general, brushstrokes movement direction can be distinguished on the basis of the greater initial colour discharge decreasing towards the end of the movement (as visible in a-b-c-d-e-f-g) or on the stratigraphy of the overlapping brushstrokes (as visible in c-d-e-f): in these case it was from left to right.

Brushstrokes in figures c-d were obtained by a continuous movement, differently from those in e-f: this is clearly visible by the absence of strong discontinuity between single strokes in the first couple.

Figure g shows the effect of a wet-on-dry stroke whereas figure h shows the wet-on-wet case. The dry ground stroke in g allowed the upper one to cover it without completely removing the underlying material, due also to the fact that the flat brush distributes the colour more evenly and lightly along the entire length of the brushstroke: in the crossing area the edges of the upper brushstroke are therefore visible with the maximum elevation whereas the edges of the underlying brushstroke and the overlapping of the roughness of the two layers are in moderate elevation; in the case in figure h, it is not possible to find the elevation due to the underlying brushstroke within the trace of the overlapping one.

Finally, in Fig. 3g can be also distinguished the torsion effect given to the stroke, which is in this case from bottom to top giving strength (and releasing more paint) on the upper side.

Paintings such as the Fornarina can provide a large variety of features of interest similar to those seen above for future morphological studies. This is especially verified in paintings with remarkable materiality of the pictorial surface, when the creative process entailed composition and figure adaptation or when restoration interventions and overpaintings are present. In the case of Fornarina (Fig. 5) morphological features of interest concern key points of the work in which the subject has been adapted with respect to the original design (turbant in Fig. 5a, veil in Fig. 5b-c) or to a different concept of illumination of the modeling (elbow in Fig. 5b, shoulder in Fig. 5d). Surface defects such as colour changes (fruits and leaves in Fig. 5a and d) and restoration interventions (Fig. 5c) are also present in the Fornarina and must be efficiently monitored.

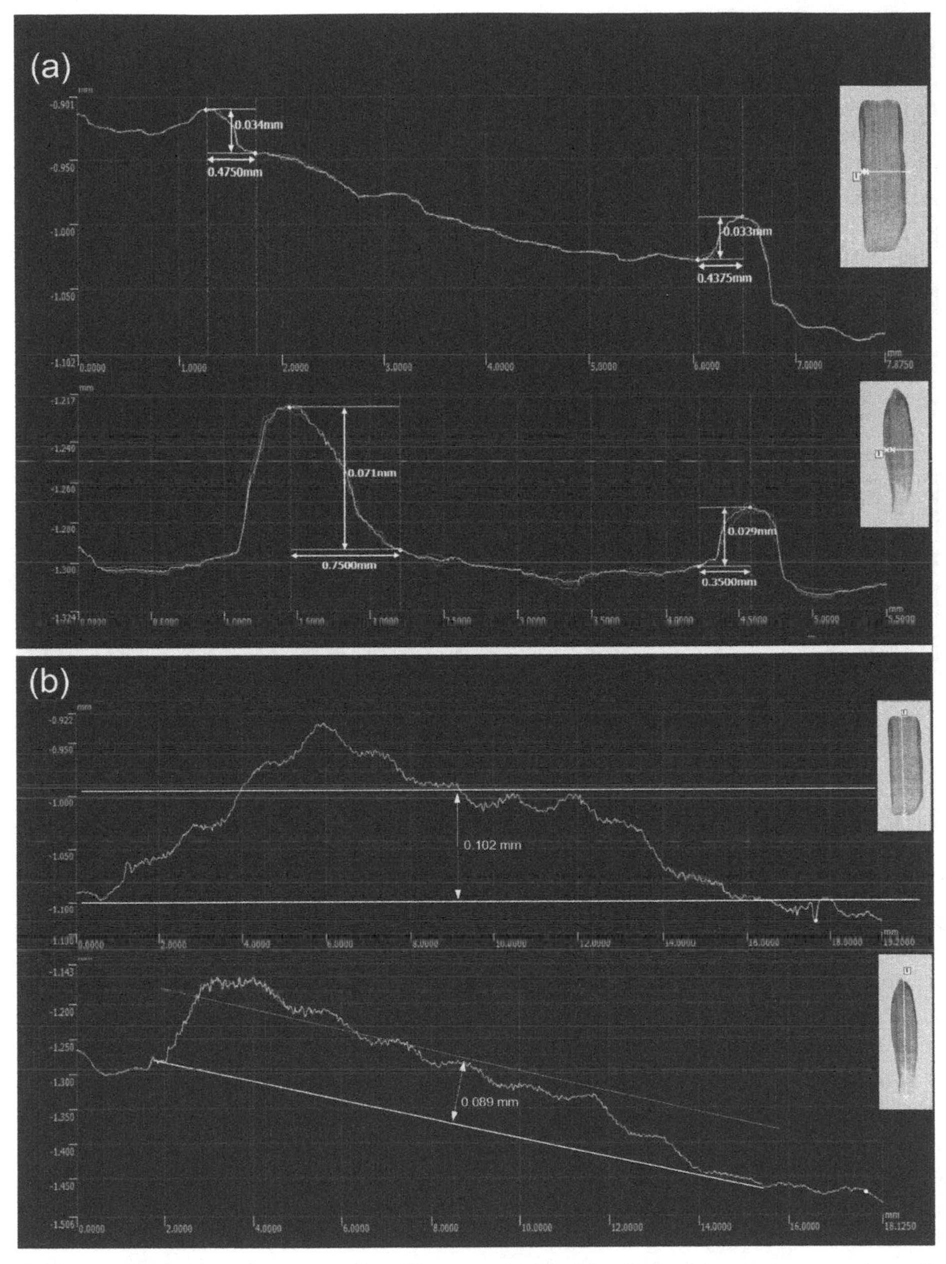

Fig. 4. Profiles related to the couple of brushstrokes in Fig. 3a-b: (a) transverse section; (b) longitudinal section

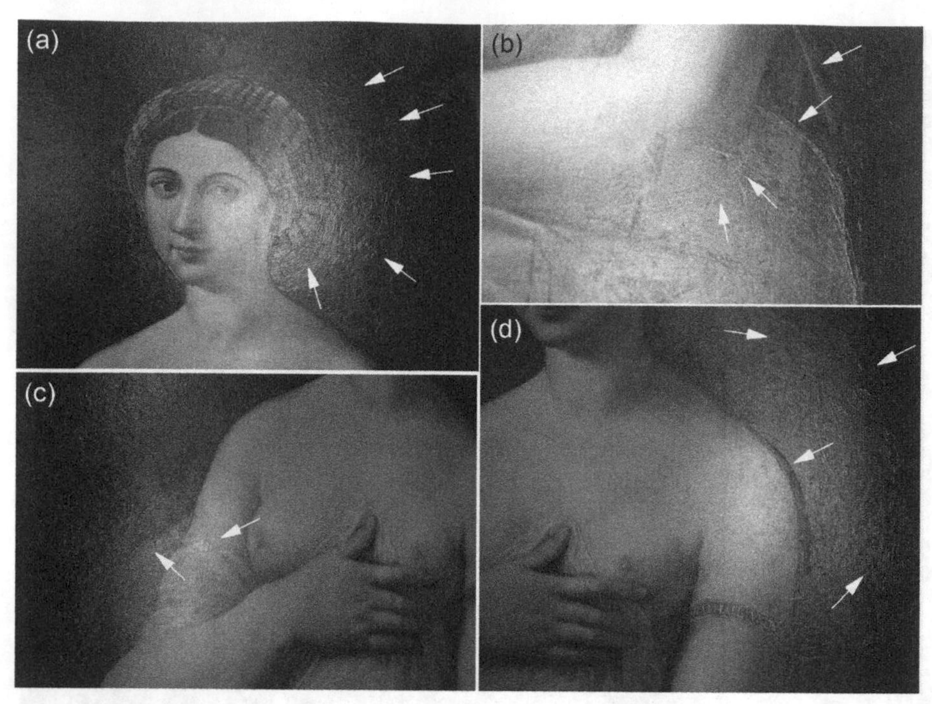

Fig. 5. Point of interest for morphological analysis on Fornarina surface under oblique light illumination during the project MUSA survey

4 Conclusions

In the field of Cultural Heritage, the prospects of innovative coupling of instruments are promising. Miniaturization and improved portability can enable in-situ (in museums or other preservation institutions) analyses, reducing the need for artwork transportation but also allowing the extended analysis on a larger (at least continuous) number of points.

MA-XRF proved to be informative about painting structure, pigment palette, pentimenti and technical practices as in the case study given by the Fornarina by Raphael in which it demonstrated to be a powerful tool for the deep understanding of the artist's mindset. Integrating non-destructive and non-invasive techniques to the already available MA-XRF devices could offer information complementary to the elemental composition. In this view, laser profilometry-MA-XRF iphenation represents a groundbreaking advancement in the study of paintings from whichever period in that profilometry can characterize accurately the gesturality of an artist by means of his brushstrokes (in terms of paint load, direction, inclination, elongation, stress, pressure given to the movement) and the relationship with the matter he was transforming. This potentiality was evident in the preliminary tests carried out on a set of acrylic sample strokes by means of two different brushes with a variety of hand movements and drying stages of the paint. Coupling the two techniques on the same scanning stage entails the difficulty to find appropriate accordance between the translation velocities required by the two as a consequence of the difference in the acquisition times and lateral resolutions. However, it is clear that in the next times the availability of digital models of the pictorial surfaces will become more and more frequent as the base point for deep analysis of the artwork both in the field of conservation and attribution studies. Moreover, the particular case study given by the Fornarina, which was exceptionally available outside its conservation climabox only for the MA-XRF during the project MUSA, remarks the importance of the availability of such digital models which can repeatedly explored just as in presence of free oblique light illumination without interfering with the conservation practices.

Acknowledgements. The work presented here was possible by the contribution of Regione Lazio and European Union, which funded the MUSA project and the PERSEPOLY project under the "POR FESR 2018 - progetti Gruppi di Ricerca 2020 n°B86C17000280002" and the "POR FESR Lazio 2014-2020 - progetti Gruppi di Ricerca 2020 CUP I85F21000930002" programmes respectively.

References

1. Alfeld, M.: MA-XRF for historical paintings: state of the art and perspective. Microsc. Microanal. **26**, 72–75 (2020)
2. Picollo, M., Cucci, C., Casini, A., Stefani, L.: Hyper spectral imaging technique in the cultural heritage field: new possible scenarios. Sensors **20**, 2843 (2020)
3. Botticelli, M., Risdonne, V., Visser, T., Young, C., Smith, M.J., Charsley., et al.: The Reflecting the past, imag(in)ing the past: macro-reflection imaging of painting materials by fast MIR hyperspectral analysis. Eur. Phys. J. Plus **138**, 432 (2023)
4. Simoen, J., De Meyer, S., Vanmeert, F., de Keyser, N., Avranovich, E., der Snickt, V., et al.: Combined Micro- and Macro scale X-ray powder diffraction mapping of degraded Orpiment paint in a 17th century still life painting by Martinus Nellius. Heritage Sci. **7**, 83 (2019)
5. Cesareo, R., Gigante, G.E., Castellano, A., Ridolfi, S., Barcellos Lins, S.A.: From Raphael (La Deposizione) to Raphael (La Fornarina): a circumnavigation accompanying the evolution of energy-dispersive XRF devices. Braz. J. Phys. **52**, 54 (2022)
6. Barcellos Lins, S.A., et al.: Modular MA-XRF scanner potentialities and further advances. In: Proceedings of the 2020 IMEKO TC-4 International Conference on Metrology for Archaeology and Cultural Heritage, October 22–24, pp. 496–500. Trento, Italy (2020)
7. Cosma, A., Merucci, C., Ridolfi, S.: Raffaello da vicino. Nuove indagini e nuove scoperte sulla Fornarina. Officina Libraria, Roma (2023)
8. Ferretti, M.: X-ray Fluorescence Applications for the Study and Conservation of Cultural Heritage. In: Radiation in Art and Archeometry, pp. 285–296, Elsevier Science (2000)
9. Barcellos Lins, S.A., Ridolfi, S., Gigante, G.E., Cesareo, R., Albini, M., Riccucci, C. et al.: Differential X-ray attenuation in MA-XRF analysis for a non-invasive determination of gilding thickness. Front. Chem. **8**, 175 (2020)
10. Szökefalvi-Nagy, Z., Demeter, I., Kocsonya, A., Kovács, I.: Non-destructive XRF analysis of paintings. Nucl. Instrum. Methods Phys. Res., Sect. B **226**, 53–59 (2004)
11. Mastrotheodoros, G.P., Asvestas, A., Gerodimos, T., Anagnostopoulos, D.F.: Revealing the materials, painting techniques, and state of preservation of a heavily altered early 19th century greek icon through MA-XRF. Heritage **6**, 1903–1920 (2023)
12. Mantler, M., Schreiner, M.: X-ray fluorescence spectrometry in art and archaeology. X-ray Spectrom. **29**, 3–17 (2000)
13. Saverwyns, S., Currie, C., Delgado, E.L.: Macro X-ray fluorescence scanning (MA-XRF) as tool in the authentication of paintings. Microchem. J. **137**, 139–147 (2018)

14. Barcellos Lins, S.A., Manso, M., Barcellos Lins, P.A., Brunetti, A., Sodo, A., Gigante, G.E.: Modular MA-XRF scanner development in the multi-analytical characterisation of a 17th century Azulejo from Portugal. Sensors **21**, 1913 (2021)
15. Giuntini, L., Castelli, L., Massi, M., Fedi, M., Czelusniak, C., Gelli, N.: Detectors and cultural heritage: the INFN-CHNet experience. Appl. Sci. **11**, 3462 (2021)
16. Cesareo, R., Gigante, G.E., Castellano, A., Ridolfi, S., Barcellos Lins, S.A.: Correction to: from Raphael (La Deposizione) to Raphael (La Fornarina): a circumnavigation accompanying the evolution of energy-dispersive XRF devices. Braz. J. Phys. **52**, 134 (2022)
17. Zuena, M., Baroni, L., Graziani, V., Iorio, M., Lins, S., Ricci, M.A., et al.: The techniques and materials of a 16th century drawing by Giorgio Vasari: a multi-analytical investigation. Microchem. J. **170**, 106757 (2021)
18. Sablatnig, R., Kammerer, P., Zolda, E.: Hierarchical classification of paintings using face- and brush stroke models. In: Proceedings of the Fourteenth International Conference on Pattern Recognition, 20–20 August. Brisbane, QLD, Australia (1998)
19. Li, J., Yao, L., Hendriks, E., Wang, J.Z.: Rhythmic brushstrokes distinguish van Gogh from his contemporaries: findings via automated brushstroke extraction. IEEE Trans. Pattern Anal. Mach. Intell. **34**, 1159–1176 (2012)

Automatic Affect Analysis and Synthesis (3AS)

Pain Classification and Intensity Estimation Through the Analysis of Facial Action Units

Federica Paolì[1(✉)], Andrea D'Eusanio[1], Federico Cozzi[1], Sabrina Patania[2], and Giuseppe Boccignone[2]

[1] Emotiva, Milan, Italy
{federica.paoli,andrea.deusanio,federico.cozzi}@emotiva.it
[2] PHuSe Laboratory-Dipartimento di Informatica, Università degli Studi di Milano, Via Celoria 18, 20133 Milan, Italy
{sabrina.patania,giuseppe.boccignone}@unimi.it
http://www.emotiva.it/

Abstract. This study focuses on using facial expressions to evaluate acute pain levels. We analyse videos by relying on an extended set of 17 Action Units (AUs) and head pose components. Multiple models are trained and compared to detect the presence of pain and classify its intensity on a 5-point scale, ranging from no pain to high pain. Validation studies were conducted on two publicly available datasets, evaluating both in within- and cross-dataset conditions. The experimental results show better pain classification performance when using both the extended AU set, instead of the restricted AU set related to pain expressions, and head pose information.

Keywords: Automatic Pain Assessment · Pain intensity estimation · Facial Action Units (AUs) · Cross-database evaluation · Healthcare

1 Introduction

The expression of pain has social implications as it elicits empathy and care. It is recognized as the 5*th* vital parameter by the American Pain Society [20], highlighting its significance in assessing health.

Under such circumstances the search for reliable and efficient methods for measuring pain intensity is cogent. While self-reported pain scales exist, they may not suit specific populations or disease contexts. Non-verbal measures, including facial expressions, are thus important in pain assessment [5,6] and have been utilized in automatic recognition of pain using image processing and computer vision techniques [26].

To this purpose, an overwhelming variety of visual features has been considered. A fine-grained analysis of facial behaviour is offered by the Facial Action Coding System (FACS, [9]), which provides so-called Action Units (AUs) as parametrical indicators of facial muscle activity. Such encoding allows for any

G. L. Foresti et al. (Eds.): ICIAP 2023 Workshops, LNCS 14365, pp. 229–241, 2024.
https://doi.org/10.1007/978-3-031-51023-6_20

visible facial expression to be decomposed into its constituent movements; it has been largely used in the pain and emotion fields, and an increasing number of researches have focused on facial behaviour analysis based on automatic AUs recognition [26,30] (cfr. Section 2).

Despite the considerable effort spent in this field, several research questions are important to address for designing effective automatic pain recognition systems that can perspectively afford to move from the lab to "in-the-wild" (e.g., clinical) assessment [26]. Among them, in this note, we specifically deal with the following:

1. Is considering a set of AUs larger than that only comprising pain-specific AUs beneficial for pain assessment?
2. Is the addition of other facial/head cues (e.g., head pose) and temporal dynamics relevant for recognition performance?
3. Can we simply and efficiently mitigate the inter-subject variability in pain response?

Further, beyond the above issues, it is overall mandatory to assess the external validity (generalization performance) of any lab study, a problem that is most often overlooked [26].

To make a step forward in contending with such questions, we conducted a study where we extracted and utilized AUs from video footage showing induced heat pain to train five different models: three linear models and two recurrent deep learning models (cfr. Sect. 3). These models were designed to predict not only the presence of pain but also five levels of pain intensity. Two of them explicitly rely on AUs dynamics in the temporal domain, by employing a sliding window approach. Also, we considered a wider set of AUs, than the set including those conventionally associated with pain; further, the head pose is taken into account. Eventually, we assessed performance on the primary dataset followed by cross-dataset tests to validate our findings. Results are reported in Sect. 3.3. Some preliminary conclusions are drawn in Sect. 4.

2 Background and Rationales

Background. Pain is a universal experience, which is challenging to define [6]. It can be generally stated as a "distressing experience associated with actual or potential tissue damage, involving sensory, emotional, cognitive, and social components" [28]. This complexity, reflected at the brain level, involves a network of cortical-subcortical-brainstem areas, with the cingulate and insula cortex serving as central hubs [4].

When an individual experiences pain, there is a disruption between the expected state of the body and the actual sensory information received from pain receptors. This mismatch triggers a cascade of neural processes that result in adaptive actions and regulation. One such result is the generation of facial expressions associated with acute pain [27].

Facial expressions, best described in terms of Action Units (AUs) based on Ekman and Friesen's Facial Action Coding System (FACS, [9]), have been identified to be specific to pain, although some problems have been noticed with such metrics (Prkachin and Solomon Pain Intensity -PSPI - metrics, e.g., [18]). Core AUs for adults in pain include brow lowering (AU4), cheek raise and lid tighten (AU6 and 7), nose wrinkle and upper lip raise (AU9 and 10), and eye closing (AU43) [7,18,19]. However, the PSPI metrics have limitations, as there are other facial expressions of pain such as raised eyebrows (AU1/2) or open mouth (AU25/26/27), which are not considered by PSPI. Additionally, facial expressions of emotions share AUs with PSPI [22].

Overall, the complexity of pain, even in controlled experiments, suggests nuanced and time-dependent facial expression behaviours that should be considered for automatic pain detection.

Related Work. In the last decade, automatic pain recognition changed from an idea to a research topic of considerable interest. It is out of the scope of the present work to discuss in detail published work in a fast-growing field that has witnessed significant progress. A recent review by Werner *et al.* [26] provides an extensive and in-depth discussion. For our purposes highlighting major trends and critical issues will suffice. The vast majority of pain recognition approaches to date have analyzed camera images containing facial expressions [26]; to such end, the classic computer vision pipeline, feature extraction → feature classification/regression, is applied. The choice of facial visual analysis is no surprise, given the above discussion, the non-intrusive character of this modality and its availability in currently released datasets. Beyond a variety of facial features that can be addressed [26], several approaches have considered, consistently with the psychological literature, the exploitation of AUs used as features for a subsequent pain recognition model since the early works of [1,12,14,15].

Yet, in whatever way facial analysis is conducted, some issues remain open in this field [26]. Among others:

- inter-individual differences often account for more variation than the signal of interest calling for person-specific calibration/normalization;
- using temporal information beyond frame-based analysis, to account for the dynamic nature of pain (facial muscles relax and contract at different intervals as the pain intensifies), whilst, related, exploiting motion-based cues (markedly, head-pose variation);
- investigating the nuanced role of a larger set of AUs, beyond traditionally pain-related AUs, due to the entanglement of pain and affective states or even to the "noise" induced by subjects' vocal actions [3] induced in pain conditions;
- measuring pain intensity is underexplored (in particular low-intensity pain, which only yields low amplitude responses);
- the generalization performance of methods is seldom addressed (namely, the estimate of how well we can expect the system to fulfil its recognition task on unseen data).

3 Materials and Methods

3.1 Datasets

BioVid Heat Pain Database [21] was created to improve pain measurement reliability and objectivity. It contains multimodal reactions from 87 subjects who underwent heat-induced pain (by a thermode). This study specifically examines a subset of the database, which includes videos labelled with pain/no pain or four different pain intensities. Our analysis focused on a subgroup of subjects who displayed visible reactions to the pain stimuli [25], excluding those who did not react visibly;

Pain E-motion Faces Database (PEMF) [10] consists of 272 micro-clips of 68 different identities. Each subject shows a neutral expression and three pain-related facial expressions: posed, induced via tonic spontaneous pain by algometer and phasic spontaneous pain by CO_2 laser.
The PEMF constitutes a novel, open and reliable set of dynamic pain expressions on which, to the best of our knowledge, no other research work has yet been conducted.

3.2 Methods

Models. To go beyond the limitation of considering only pain-related AUs, we adopted a broader set of AUs consisting of the most extensive and common set found annotated in the datasets: AU1, AU2, AU4, AU5, AU6, AU9, AU10, AU12, AU15, AU17, AU18, AU20, AU24, AU25, AU26, AU28 and AU43. The Emotiva[1] software was used to obtain AU activation levels. The detector assigns for each video frame confidence score in the range $[0, 1]$ to the ensemble of facial feature points representing different AUs. Eventually, at the frame level, this provides a 17-D feature vector $\mathbf{f}_{AU}$.

The software allows for head pose estimation via the measurement of three Euler rotation angles (yaw, pitch, and roll) thus tracking changes in head pose, which is deemed to provide valuable information for assessing pain. Thus, at the frame level this provides a 3-D head pose feature vector $\mathbf{f}_{HP}$.

Inter-individual variability was addressed via a simple individual-based normalization, by subtracting the baseline values from the AUs measured in each frame. The baseline values were determined by measuring the AUs in the first neutral-expression frame of each video. This approach is likely to have effectively mitigated the influence of individual differences in facial expressions.

Classification. The goal here is to infer the posterior probability $P(y \mid \mathbf{f}_{AU}, \mathbf{f}_{HP})$, y being a discrete random variable (RV) denoting pain level. To such end, we exploited the widely used Support Vector Machine (SVM), Random Forest (RF), and Multilayer Perceptron (MLP). Further, to explicitly account for

[1] Emotiva - Emotion AI Company, www.emotiva.it.

temporal information, we considered two variants of Recurrent Neural Networks (RNNs): Long Short-Term Memory (LSTM) and Gated Recurrent Unit (GRU).

The classification step considered a twofold outcome:

- **Binary classification**: to discriminate between pain and absence of pain; the case in which the painful stimulus is at its maximum intensity is considered pain, and the case of absent pain is when no painful stimulus is applied; thus, $y \in 0, 1$ a binary RV. Pain is detected when $P(y = 1 \mid \mathbf{f}_{AU}, \mathbf{f}_{HP}) > 0.5$.
- **Intensity level classification**: to discriminate between various levels of pain intensity; level of intensity was represented on a scale from 0 to 4, based on the intensity of the painful stimulus: (0) no pain, (1) low pain, (2) low-medium pain, (3) medium-high pain, and (4) high pain; thus, $y \in 1, \cdots, 5$. The estimated intensity level is obtained as $\hat{\ell} = \arg\max_{\ell}\{P(y = \ell \mid \mathbf{f}_{AU}, \mathbf{f}_{HP})\}$.

As to "static" classifiers (i.e., SVM, RF, MLP), classification was performed at the granularity of the frame. For what concerns time-dependent or "dynamic" classifiers (i.e., LSTM and GRU), we considered time windows of 5 and 10 frames (stride = 1).

As to static classifiers, the SVM with linear kernel was chosen and error penalty parameter C was found via grid-search along validation; RF used 100 trees, nodes are expanded until all leaves are pure, while the number of features to consider for each split was set to the recommended default; MLP is trained using Adam optimizer, ReLU activation function and a hidden layer with 100 neurons. The parameters chosen for the recurrent neural networks LSTM and GRU are summarized in Table 1.

The training phase utilized the BioVid Heat Pain Database [21], since including pain stimulus annotations. Thus, the stimulus, specifically temperature, served as the ground truth [10].

The splitting of the data into a train set, validation set, and test set was carried out as proposed in [11], so to retain almost the same percentage of the gender variation, the age and the "low-expressive" subjects [10].

Table 1. Parameters used for training LSTM and GRU recurrent neural networks.

	Number of classes to predict	Input size	Hidden size	Number of layers	Sequence length
Binary	1	17(20)	64	2	5 or 10
Intensity levels	5	17(20)	64	2	5 or 10

Data Analyses

1. *Within-dataset analysis.* The BioVid dataset [21] was considered. The complete dataset is divided into training, validation and test sets in such a way that 80% of the data is used for training, 10% for validation and the remaining 10% for testing [11]. In total, 10 sets of experiments were carried out, the performance of which, for each classifier, was quantified concerning the ground truth using the average values of accuracy and F1 score.

2. *Cross-dataset analysis.* The PEMF [10] was considered. Analyses were conducted both on the entire dataset comprising posed expressions and those resulting from two stimulus sources: laser and algometer. The dataset included annotations of pain intensity on a 9-point Likert-type scale for all stimuli. As the models taken into consideration for the experiments, these were trained on a pain scale from 0 to 4; the values of each intensity level on the scale taken as a reference in the BioVid-based training stage were also remapped for the cross-dataset analysis.

3.3 Results

Table 2. Results of binary pain detection experiments on the test set of the BioVid Heat Pain Database (the metrics that achieved the highest value are highlighted in bold).

	Pain-related AUs				17 AUs				
	Direct		Normalized		Direct		Normalized		
	Accuracy	F1-score	Accuracy	F1-score	Accuracy	F1-score	Accuracy	F1-score	Sliding window length
SVM	0.58	0.57	0.63	0.61	0.71	0.71	0.72	0.71	*None*
Random Forest	0.57	0.57	0.71	0.71	0.68	0.67	**0.87**	**0.87**	*None*
MLP	0.58	0.56	0.63	0.62	0.72	0.72	0.72	0.72	*None*
LSTM	0.72	0.64	0.73	0.68	0.72	0.67	0.82	0.81	5
LSTM	0.71	0.66	0.78	0.77	0.73	0.71	0.83	0.81	10
GRU	0.73	0.67	**0.80**	**0.78**	0.78	0.76	0.85	0.85	5
GRU	**0.74**	**0.69**	0.74	0.71	**0.78**	**0.77**	**0.87**	0.86	10

Table 3. Results of binary pain detection experiments on the BioVid Heat Pain Database test set with the head pose (the metrics that achieved the highest value are highlighted in bold).

	Pain-related AUs				17 AUs				
	Direct		Normalized		Direct		Normalized		
	Accuracy	F1-score	Accuracy	F1-score	Accuracy	F1-score	Accuracy	F1-score	Sliding window length
SVM	0.68	0.68	0.67	0.66	0.75	0.75	0.73	0.72	*None*
Random Forest	0.70	0.69	0.85	0.85	0.77	0.77	**0.90**	**0.90**	*None*
MLP	0.68	0.68	0.68	0.67	0.76	0.76	0.75	0.75	*None*
LSTM	0.73	0.69	0.82	0.81	0.74	0.71	0.85	0.84	5
LSTM	0.75	0.72	0.77	0.76	0.75	0.70	0.83	0.84	10
GRU	0.74	0.70	**0.86**	**0.86**	**0.82**	**0.82**	0.88	0.88	5
GRU	**0.77**	**0.73**	0.83	0.83	0.78	0.75	0.87	0.88	10

Within-Dataset. Tables 2 and 3 show the results obtained from the binary pain detection experiments directly, without applying normalization to the AU values, and with the latter. It can be appreciated that pain is recognized reasonably well in both linear and temporal deep-learning models. Table 2, in particular, shows how without normalization there is a minimum of 67% F1 score and 68%

accuracy up to 87% for both once normalization is applied to the AU values. Furthermore, Table 3 demonstrates that incorporating head pose features $\mathbf{f}_{HP}$ enhances performance up to 10%. Among the tested models, the RF model achieves the best performance on the test set, while LSTM and GRU models, with sliding windows of lengths 5 and 10, respectively, closely follow the RF model in performance.

Table 4. Results of pain intensity levels detection experiments on the test set of the BioVid Heat Pain Database (the metrics that achieved the highest value are highlighted in bold).

	Pain-related AUs				17 AUs				
	Direct		Normalized		Direct		Normalized		
	Accuracy	F1-score	Accuracy	F1-score	Accuracy	F1-score	Accuracy	F1-score	Sliding window length
SVM	0.25	0.24	0.28	0.27	0.37	0.38	0.43	0.43	*None*
Random Forest	0.23	0.22	0.42	0.42	0.38	0.38	**0.77**	**0.77**	*None*
MLP	0.24	0.24	0.27	0.26	0.33	0.33	0.33	0.33	*None*
LSTM	0.27	0.28	0.55	0.55	0.37	0.36	0.57	0.57	5
LSTM	0.28	0.28	0.58	0.58	0.37	0.37	0.60	0.60	10
GRU	**0.30**	**0.30**	0.63	0.63	0.36	0.36	0.74	0.74	5
GRU	0.29	0.28	**0.64**	**0.64**	**0.47**	**0.47**	0.69	0.70	10

Table 5. Results of pain intensity levels detection experiments on the test set of the BioVid Heat Pain Database with the head pose (the metrics that achieved the highest value are highlighted in bold).

	Pain-related AUs				17 AUs				
	Direct		Normalized		Direct		Normalized		
	Accuracy	F1-score	Accuracy	F1-score	Accuracy	F1-score	Accuracy	F1-score	Sliding window length
SVM	0.34	0.34	0.35	0.35	0.44	0.44	0.46	0.47	*None*
Random Forest	0.41	0.41	**0.69**	**0.69**	**0.54**	**0.53**	**0.83**	**0.83**	*None*
MLP	0.30	0.30	0.30	0.29	0.36	0.35	0.35	0.35	*None*
LSTM	0.44	0.45	0.53	0.53	0.36	0.35	0.60	0.60	5
LSTM	**0.45**	**0.45**	0.68	0.68	0.36	0.36	0.63	0.62	10
GRU	0.37	0.38	0.67	0.67	0.51	0.50	0.70	0.70	5
GRU	0.40	0.40	0.66	0.66	0.40	0.39	0.75	0.75	10

The pain intensity level detection experiments shown in Tables 4 and 5 describe a situation similar to the binary case. The only difference lies in the so-called "baseline" results, as without normalization or the addition of the head pose features, there is a very poor performance, however, reaching a 77% accuracy and F1 score with normalization alone and 83% with the head pose as well. It is still the linear Random Forest model, whose confusion matrix is shown in Fig. 1, that has the best performance, being immediately followed by LSTM and markedly by GRU.

Interestingly, when only pain-related AUs are considered, temporal models exhibit higher performance than static classifiers.

Comparison Against the State-of-the-Art. Eventually, to compare face-value results with those published in recent articles, the best scores achieved here and those of other state-of-the-art procedures reported in the literature are given in Table 6.

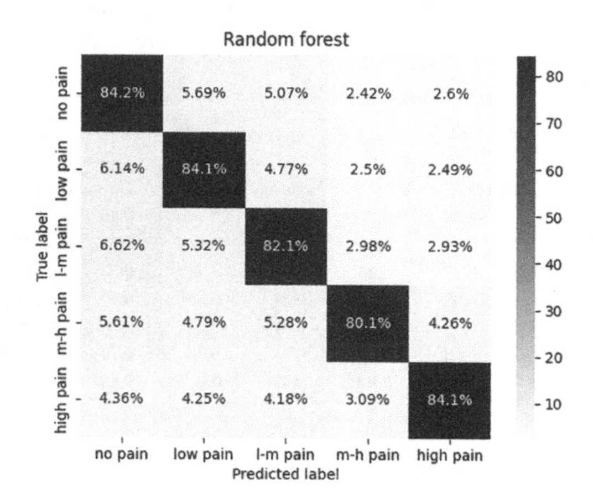

Fig. 1. Confusion matrix of the Random Forest model for pain intensity levels detection on the test set, with normalization of values and addition of head pose features

Table 6. Comparative analysis of accuracy and F1 score on different automatic pain detection techniques applied to the BioVid Heat Pain Database (the metrics not present in the cited works are marked with a dash). The one called "Our approach" refers to the Random Forest results on the BioVid Heat Pain Database test set using the head pose and the 17 AUs selected.

	Binary pain detection		Intensity levels pain detection	
	Accuracy	F1-score	Accuracy	F1-score
Werner et al. (2014) [24]	0.77	-	-	-
Werner et al. (2016) [23]	0.72	-	0.31	-
Yang et al. (2016) [29]	0.71	-	-	-
Othman et al. (2019) [16]	0.66	-	-	-
Das et al. (2020) [8]	0.68	-	-	-
Prajod et al. (2022) [17]	0.70	0.69	-	-
Our approach	**0.90**	**0.90**	**0.83**	**0.83**

A caveat is worthwhile: comparing results in pain detection research is challenging due to variations in experimental conditions and settings. Werner *et al.* [26] emphasize that different papers are difficult to compare, even with the same dataset, due to differences in prediction tasks, evaluation methods, performance measures, and data subsets. Addressing this overlooked problem is crucial for real progress in pain detection research.

Cross-Dataset. The cross-dataset condition, which is very demanding, showed, as expected, a loss of accuracy and F1 scores. For this evaluation, we considered, to match conditions with the BioVid dataset, actual stimulus-induced facial expressions, namely those elicited from phasic spontaneous pain by CO_2 laser and tonic spontaneous pain by algometer. Note that in general the key differences between tonic pain and phasic pain lie in their nature and duration. Phasic pain refers to episodic pain that comes and goes in distinct episodes whereas tonic pain is defined as a persistent, continuous pain that lasts for a prolonged period. For instance, algometers are often used in the study of conditions like fibromyalgia, chronic pain disorders, and neuropathic pain. Thus, even in a controlled lab setting, we expect differences in facial expression responses concerning the two stimuli (cfr., [10] for a discussion).

Results obtained in binary detection for stimulus types are reported in Tables 7 and 8; overall, the trained models generalized (above the chance level). By contrast, the performance of pain intensity level detection on the PEMF Database [10] is at chance level for all the adopted classifiers, and thus not reported here.

Here, we are not comparing with other work in the literature, since, to the best of our knowledge, we are the first to have results on this novel dataset [10].

Table 7. Results in binary pain detection obtained on heat stimulus (CO_2 laser) from the PEMF Database (the metrics that achieved the highest value are highlighted in bold).

	Laser								
	No head pose				With head pose				
	Direct		Normalized		Direct		Normalized		
	Accuracy	F1-score	Accuracy	F1-score	Accuracy	F1-score	Accuracy	F1-score	Sliding window length
SVM	0.34	0.33	**0.59**	**0.59**	0.64	**0.67**	0.53	0.53	*None*
Random forest	0.54	0.58	**0.59**	**0.59**	0.52	0.56	0.54	0.52	*None*
MLP	0.38	0.37	0.57	0.57	0.55	0.59	0.50	0.47	*None*
LSTM	**0.58**	**0.61**	0.51	0.50	**0.65**	0.64	**0.60**	**0.60**	5
LSTM	0.55	0.59	0.51	0.51	0.62	0.62	0.51	0.51	10
GRU	0.56	0.60	0.52	0.52	0.56	0.58	0.51	0.48	5
GRU	0.45	0.47	0.51	0.51	0.55	0.57	0.54	0.52	10

Table 8. Results in binary pain detection obtained on pressure stimulus (algometer) from the PEMF Database (the metrics that achieved the highest value are highlighted in bold).

	Algometer								
	No head pose				With head pose				
	Direct		Normalized		Direct		Normalized		
	Accuracy	F1-score	Accuracy	F1-score	Accuracy	F1-score	Accuracy	F1-score	Sliding window length
SVM	0.49	0.59	0.74	0.73	**0.72**	**0.78**	0.61	0.63	*None*
Random forest	**0.73**	**0.79**	**0.77**	**0.76**	0.58	0.68	**0.66**	**0.66**	*None*
MLP	0.47	0.57	0.70	0.70	0.62	0.71	0.42	0.41	*None*
LSTM	0.37	0.48	0.38	0.41	0.62	0.70	0.52	0.55	5
LSTM	0.37	0.48	0.38	0.39	0.58	0.68	0.49	0.52	10
GRU	0.35	0.47	0.40	0.43	0.63	0.71	0.52	0.53	5
GRU	0.25	0.30	0.35	0.37	0.60	0.69	0.54	0.56	10

In this case, first, one can appreciate a difference in performance concerning the stimulus considered. In the CO_2 laser condition, the LSTM classifier (5 sec. window) shows a slightly better average performance than SVM, RF, and MLP classifiers (Table 7); the opposite holds in the case of the algometer (Table 8). While the former result suggests that, *ceteris paribus*, explicitly accounting for dynamics can offer an advantage in a generalization task, the latter calls for further pondering. Statistical analyses conducted by the authors of the dataset [10], give evidence that pain-related faces elicited by algometer show higher pain intensity ($p = 0.012$) and higher arousal values ($p = 0.003$) than CO_2 laser pain-related faces. Thus, for the data handled here, algometer-induced pain intensity can be located between posed intensity and CO_2 laser-induced intensity. A further classification performance analysis (not fully reported here for space limitations) was thus performed on posed expression; the best results were achieved by the RF classifier (0.91 accuracy, 0.94 F1, with head pose and normalization). This result suggests that the RF classifier benefits from higher intensity expressions, which could explain the different classifier performance behaviour for the two types of stimuli.

A comment is deserved for the less prominent role played by normalization in the cross-dataset case. It is worth remarking that in the PEMF dataset, differently from BioVid where pain thresholds were calibrated for every participant and accordingly the four pain levels determined, the configuration parameters of the intensity of painful stimulation were selected according to data provided by previous studies.

4 Conclusion

In this note, we have addressed several problems that arise in acute pain automatic detection based on AUs.

The main results so far achieved concerning some crucial research questions in the field (cfr. Sect. 1) can be summarized as follows.

1. Considering a set of AUs larger than that only comprising pain-specific AUs is beneficial for pain detection performance both in the within- and cross-dataset conditions. These results are thus in line with current discussion that emerged in the psychological literature (e.g., [13])
2. Similarly, the addition of other facial/head cues, markedly head pose, improves recognition performance.
3. Addressing the problem of inter-subject variability in pain response significantly improves performance, even when a straightforward normalization procedure, such as the one employed here, is adopted; yet, this result might be affected by the pain-threshold calibration procedure adopted in the construction of the dataset.

As to the explicit use of time-dependent classifiers, as opposed to static ones we reported nuanced results. The performance obtained by the classic Random Forest classifier most often is slightly better than variations of recurrent neural network-based classifiers when the above improvements are considered; yet, the reverse holds when conditions are far from ideal. However, these preliminary results are not conclusive and do not entail a general performance trend, which, *prima facie*, might be conceptually counter-intuitive, given the dynamic nature of the facial expression of pain. For instance, the larger number of parameters to tune for the LSTM and GRU, given the size of the datasets, may have trivially contributed to their lower performance.

Overall, generalization performance confirms, at least in the limits of this study, to be a hard problem [26], though preliminary results achieved here are encouraging. In this respect, a cautionary note is due. Generalization assessment here refers to cross-dataset evaluation, but yet in the acute pain condition as elicited within the lab experimental setting. This step certainly is significant, however robust generalization to more complex forms of pain (e.g. chronic pain) and clinical settings can be considered still in its infancy and certainly it is likely to call for the involvement of additional modalities (e.g., physiological measurements, either with real or virtual sensors [2]) beyond facial expression analysis.

References

1. Bartlett, M.S., Littlewort, G.C., Frank, M.G., Lee, K.: Automatic decoding of facial movements reveals deceptive pain expressions. Curr. Biol. **24**(7), 738–743 (2014)
2. Boccignone, G., Conte, D., Cuculo, V., D'Amelio, A., Grossi, G., Lanzarotti, R.: An open framework for remote-PPG methods and their assessment. IEEE Access **8**, 216083–216103 (2020)
3. Bursic, S., Boccignone, G., Ferrara, A., D'Amelio, A., Lanzarotti, R.: Improving the accuracy of automatic facial expression recognition in speaking subjects with deep learning. Appl. Sci. **10**(11), 4002 (2020)
4. Chen, Z.S.: Hierarchical predictive coding in distributed pain circuits. Front. Neural Circ. **17**, 1073537 (2023)
5. Craig, K.D.: The facial expression of pain better than a thousand words? APS J. **1**(3), 153–162 (1992)

6. Craig, K.D., MacKenzie, N.E.: What is pain: are cognitive and social features core components? Paediatr. Neonatal Pain **3**(3), 106–118 (2021)
7. Craig, K.D., Prkachin, K.M., Grunau, R.V.: The Facial Expression of Pain. The Guilford Press (1992)
8. Das, P., Bhattacharyya, J., Sen, K., Pal, S.: Assessment of pain using optimized feature set from corrugator EMG. In: 2020 IEEE Applied Signal Processing Conference (ASPCON), pp. 349–353 (2020)
9. Ekman, P., Friesen, W.V.: Facial action coding system (1978)
10. Fernandes-Magalhaes, R., et al.: Pain E-motion Faces Database (PEMF): pain-related micro-clips for emotion research. Behav. Res. Methods, 1–14 (2022)
11. Gkikas, S.: Biovid holdouteval (2023). https://www.nit.ovgu.de/nit_media/Bilder/Dokumente/BIOVID_Dokumente/BioVid_HoldOutEval_Proposal.pdf
12. Hammal, Z., Cohn, J.F.: Automatic detection of pain intensity. In: Proceedings of the 14th ACM International Conference on Multimodal Interaction, pp. 47–52 (2012)
13. Kunz, M., Lautenbacher, S.: The faces of pain: a cluster analysis of individual differences in facial activity patterns of pain. Eur. J. Pain **18**(6), 813–823 (2014)
14. Lucey, P., et al.: Automatically detecting pain in video through facial action units. IEEE Transactions on Systems, Man, and Cybernetics, Part B (Cybernetics) **41**(3), 664–674 (2010)
15. Lucey, P., Howlett, J., Cohn, J., Lucey, S., Sridharan, S., Ambadar, Z.: Improving pain recognition through better utilisation of temporal information (2008)
16. Othman, E., Werner, P., Saxen, F., Al-Hamadi, A., Walter, S.: Cross-database evaluation of pain recognition from facial video. In: 2019 11th International Symposium on Image and Signal Processing and Analysis (ISPA), pp. 181–186 (2019)
17. Prajod, P., Huber, T., André, E.: Using explainable AI to identify differences between clinical and experimental pain detection models based on facial expressions. In: Þór Jónsson, B., et al. (eds.) MMM 2022. LNCS, vol. 13141, pp. 311–322. Springer, Cham (2022). https://doi.org/10.1007/978-3-030-98358-1_25
18. Prkachin, K., Solomon, P.: The structure, reliability and validity of pain expression: Evidence from patients with shoulder pain (2008)
19. Prkachin, K.M.: The consistency of facial expressions of pain: a comparison across modalities, pp. 297–306 (1992)
20. Nelson, R.: Decade of pain control and research gets into gear in USA (2003)
21. Walter, S., et al.: The biovid heat pain database data for the advancement and systematic validation of an automated pain recognition system. In: 2013 IEEE International Conference on Cybernetics (CYBCO), pp. 128–131 (2013)
22. Werner, P., Al-Hamadi, A., Limbrecht-Ecklundt, K., Walter, S., Gruss, S., Traue, H.C.: Automatic pain assessment with facial activity descriptors. IEEE Trans. Affect. Comput. **8**(3), 286–299 (2017)
23. Werner, P., Al-Hamadi, A., Limbrecht-Ecklundt, K., Walter, S., Gruss, S., Traue, H.C.: Automatic pain assessment with facial activity descriptors. IEEE Trans. Affect. Comput. **8**(3), 286–299 (2017)
24. Werner, P., Al-Hamadi, A., Niese, R., Walter, S., Gruss, S., Traue, H.C.: Automatic pain recognition from video and biomedical signals. In: 2014 22nd International Conference on Pattern Recognition, pp. 4582–4587 (2014)
25. Werner, P., Al-Hamadi, A., Walter, S.: Analysis of facial expressiveness during experimentally induced heat pain. In: 2017 Seventh International Conference on Affective Computing and Intelligent Interaction Workshops and Demos (ACIIW), pp. 176–180 (2017)

26. Werner, P., Lopez-Martinez, D., Walter, S., Al-Hamadi, A., Gruss, S., Picard, R.W.: Automatic recognition methods supporting pain assessment: a survey. IEEE Trans. Affect. Comput. **13**(01), 530–552 (2022)
27. Williams, A.C.C.: Facial expression of pain: an evolutionary account. Behav. Brain Sci. **25**(4), 439–455 (2002)
28. Williams, A.C.C., Craig, K.D.: Updating the definition of pain. Pain **157**(11), 2420–2423 (2016)
29. Yang, R., et al.: On pain assessment from facial videos using spatio-temporal local descriptors. In: 2016 Sixth International Conference on Image Processing Theory, Tools and Applications (IPTA), pp. 1–6 (2016)
30. Zhi, R., Liu, M., Zhang, D.: A comprehensive survey on automatic facial action unit analysis. Vis. Comput. **36**, 1067–1093 (2020)

Towards a Better Understanding of Human Emotions: Challenges of Dataset Labeling

Hajer Guerdelli[1,2](✉), Claudio Ferrari[3], Joao Baptista Cardia Neto[4], Stefano Berretti[1], Walid Barhoumi[2,5], and Alberto Del Bimbo[1]

[1] University of Florence, Florence, Italy
{hajer.guerdelli,stefano.berretti,alberto.delbimbo}@unifi.it
[2] University of Tunis El Manar, Tunis, Tunisia
[3] University of Parma, Parma, Italy
claudio.ferrari2@unipr.it
[4] FATEC Catanduva, Catanduva, Brazil
joao.cardia@fatec.sp.gov.br
[5] Université de Carthage, Ecole Nationale d'Ingénieurs de Carthage, Tunis, Tunisia
walid.barhoumi@enicarthage.rnu.tn

Abstract. A major challenge in automatic human emotion recognition is that of categorizing the very broad and complex spectrum of human emotions. In this regard, a critical bottleneck is represented by the difficulty in obtaining annotated data to build such models. Indeed, all the publicly available datasets collected to this aim are either annotated with (i) the six prototypical emotions, or (ii) continuous valence/arousal (VA) values. On the one hand, the six basic emotions represent a coarse approximation of the vast spectrum of human emotions, and are of limited utility to understand a person's emotional state. Oppositely, performing dimensional emotion recognition using VA can cover the full range of human emotions, yet it lacks a clear interpretation. Moreover, data annotation with VA is challenging as it requires expert annotators, and there is no guarantee that annotations are consistent with the six prototypical emotions. In this paper, we present an investigation aiming to bridge the gap between the two modalities. We propose to leverage VA values to obtain a fine-grained taxonomy of emotions, interpreting emotional states as probability distributions over the VA space. This has the potential for enabling automatic annotation of existing datasets with this new taxonomy, avoiding the need for expensive data collection and labeling. However, our preliminary results disclose two major problems: first, continuous VA values and the six standard emotion labels are often inconsistent, raising concerns about the validity of existing datasets; second, datasets claimed to be balanced in terms of emotion labels become instead severely unbalanced if provided with a fine-grained emotion annotation. We conclude that efforts are needed in terms of data collection to further push forward the research in this field.

Keywords: Emotion recognition · Valence-Arousal · Annotated dataset

G. L. Foresti et al. (Eds.): ICIAP 2023 Workshops, LNCS 14365, pp. 242–254, 2024.
https://doi.org/10.1007/978-3-031-51023-6_21

1 Introduction

Human-to-human interaction occurs between individuals and may include hearing, tact, and vision data that describe the interaction [25]. During an interaction, it is important that all the involved actors understand the meaning of what is happening, especially because part of the communication occurs in non-verbal ways. In this context, facial expressions play a very important role, because they bring a significant non-verbal meaning. So, facial expressions have a central role in manifesting the emotional state of people, and showing an expression can modify the mood of the interaction [7]. Given its importance in social interactions, understanding facial expressions and the related emotional states, using automatic systems based on computer vision is a task that has been studied since decades, with a vast literature [1,10,15,28,29,31]. Since it is present in most human interaction scenarios, facial expressions are relevant to understand non-verbal features. It is possible to apply a Facial Expression Recognition (FER) system in several scenarios, such as to measure tourist satisfaction [8], engagement evaluation [5], and human-robot affective interaction [24].

It is possible to divide FER systems into two main categories: *categorical* and *dimensional*. A system that operates in the first category, typically aims to classify a given facial expression into one of the six universal emotions plus the neutral one [6]. The main problem with this classification is that human emotions are more complex and subtle than those categories, and this rigid classification does not cover the real range of emotions. In dimensional emotion recognition [19,21,22], instead, a FER system usually extracts two values, *valence* and *arousal*, from a face image. Valence refers to a pleasure/displeasure degree, while arousal refers to the intensity of the displayed expression. This deals with the shortcoming of emotion recognition but the valence/arousal values are hard to interpret since they are not intuitive: while it is easy to understand what the label "happy" means, it is not so straightforward to understand what a value of 0.4 of valence and -0.2 of arousal does signify. Furthermore, a clear and unique mapping between the two representations has not been accurately defined.

Motivated by the above considerations, in this work we propose a way to transfer the continuous emotion description provided by VA values to a fine-grained human-readable taxonomy of expressions. The goal of this procedure is that of finding a way to avoid the burdensome process of collecting and labeling new data, while simultaneously analyzing the quality and reliability of existing annotation. The proposed procedure relies on two established studies in the emotion literature: the categorical/dimensional mapping of Russell *et al.* [22], and the tree-like emotion categorization proposed by Parrott [18].

Among the works of psychological constructionists, Russell *et al.* [22], identified 151 fine-grained emotional terms and also provided a map from each of them to a distribution in the VA plane. For example, the term "Excited" is associated with the values of 0.62 valence and 0.75 arousal, with a dispersion range of 0.25 and 0.2, respectively. Doing this mapping for all the terms resulted in a complete coverage of the valence/arousal plane, though the dispersion regions associated to each term largely overlap. Our idea here is to start from this fine mapping and define a way to decide the emotion labels at coarser levels, while keeping a quantitative mapping between textual labels and the valence/arousal values. For the aggregation of the terms, we followed the term classification proposed by Parrott [18], a Basic Emotion Theorist, where emotions were categorized into primary, secondary and tertiary layers. The first layer includes the six basic emotions, while the secondary and tertiary layers are derived from the first one, and include, respectively, 25 and 115 emotions (a total of 140 emotion labels were provided with this categorization). Overall, with this terms aggregation, we obtained an intermediate ground that can better describe the range of human emotions. In summary, the main contributions of this work are:

- We proposed an emotion classification based on merging Russell's 151 terms and the 140 terms of Parrott's classification; the intersection between the two resulted in a first-level 6 emotion terms (Love, Joy, Surprise, Anger, Sadness, Fear) taken from the primary emotions, and a second-level 32 terms from the secondary and tertiary emotions;
- We experimentally showed that classifying expressions according to the proposed fine terms still represents a difficult task for expression classification based on deep learning methods as originally developed for the classification task into the six prototypical emotions;
- From the experimental outcomes, we highlighted two problems: *(i)* existing datasets suffer from a severe class unbalance if provided with fine-grained annotations, and *(ii)* there exists an inconsistency between emotion labels and corresponding VA values.

2 Related Work

Works in the literature mainly focused on emotion recognition, spontaneous expression recognition, micro-expressions detection, action units detection, and valence/arousal estimation [12]. Given the focus of our work, we will discuss some related works, focusing mainly on two tasks: emotion recognition and valence/arousal estimation. We just mention that some works combined Action Units (AUs) to emotional states. An example is the work in [4], where authors simulated emotional facial expressions according to pleasure, arousal, and dominance (PAD) values, and animated virtual human's facial muscle AUs.

Emotion Recognition. In [13], a Multi-Scale Convolutional Neural Network was proposed that combined dilated convolutional kernels and automatic face correction aiming to improve the learned features from a CNN. Each expression was classified in one of the six universal expressions (*i.e.*, happiness, anger, sadness, disgust, surprise, fear) [6] or neutral. One shortcoming of this approach is the difficulty in correctly detecting the fear expression. To deal with pose and occlusion, the authors in [26] proposed a new approach called Region Attention Network (RAN), which captures the importance of facial regions in an adaptive manner. This aimed to increase robustness with respect to the aforementioned problems. To do so, RAN generated a compact representation aggregating and embedding a varied number of region features that come from a backbone CNN. A critical aspect of this approach is a region-biased loss, which increases the attention weights for essential regions. Spite being overall better than the baseline, using RAN the accuracy decreased on specific emotions. This decrease in performance was mainly observed for *disgust* on the Occlusion-AffectNet data, for *disgust* and *neutral* on Pose-AffectNet, and for *fear* on Pose-FERPlus. In [30], two methods were proposed for facial expression recognition: the Double-channel Weighted Mixture Deep Convolution Neural Network, and the Deep Convolution Neural Network Long Short-Term Memory network of double-channel weighted mixture. The former focused on static images being able to quickly recognize expressions and provide static image features. The latter utilized static features to extract temporal features and precisely recognize facial expressions. The work in [14] proposed an approach with an end-to-end network utilizing attention mechanisms. It comprised four modules designed, respectively, for feature extraction, attention, reconstruction, and classification. The method combined Local Binary Pattern images with the attention modules to enhance the attention model, focusing on useful features for facial expressions recognition.

The main problem with emotion recognition is that it summarizes human emotion into seven classes, not being able to represent the diversity in which humans portray their emotions. In this sense, a more fine-grained representation of human emotion would be required.

Dimensional Emotion Recognition. In [12], a CNN-RNN based approach was proposed to dimensional emotion recognition, defined as the prediction of facial affection. In this case, VA varied in a continuous space [12]. A set of RNN subnets exploited the low-, mid- and high-level features from the trained CNN. The work in [2] utilized a Convolutional Autoencoder (CAE) that learned a representation from facial images, while keeping a low dimensional size for the features. A CNN is initially trained on a facial expression recognition dataset and the weights of the pre-trained network were used to initialize the convolutional layers in the proposed CAE. CAE was then used as an encoder-decoder to learn the latent representation of the data, which in turn was used to train a support vector regressor to infer the VA values. In [23], the EmoFAN network was proposed that combined networks performing facial alignment with expression recognition. In this way, as emotion classification, it was possible to obtain VA as well as fiducial points on the face. Instead of utilizing an attention mechanism, the keypoints of the face were used to focus on relevant regions of the surface.

Different from emotion recognition, dimensional emotion recognition is less straightforward to interpret. Annotating data in the dimensional space is also harder and more prone to error given the subjective nature of the human emotional state. One way to deal with such a problem is to provide a middle ground that motivated us to exploit the terms taxonomy defined by Russell [22], and the hierarchical label categories proposed by Parrott [18]. This is relevant because it generates a more realistic way of representing human emotions: it is simpler to interpret the results than dimensional emotion recognition, while increasing the diversity of representations, differing from the basic 6 classes used in emotion recognition. Though we did not delve into this, the proposed fine grained terms can be further extended following our approach, by finding additional correspondences between Russell's and Parrott's terms.

3 Proposed Emotion Taxonomy

In this section, we discuss the selection processes and the taxonomy we chose for our work.

Russell Vs. Parrott. Describing an emotional state of a person with the six prototypical emotions, *i.e.*, joy, sadness, disgust, fear, anger and surprise, provides us with her/his expression but does not give us the person's emotions. In the work of Russell [22], a set of 151 terms were provided with the corresponding distributions of *valence*, *arousal* and *dominance* values (this is given as a mean value plus a standard deviation term). The 151 terms not only describe the emotions in a precise way but also include terms that represent emotional interaction between two individuals. Russell defined the emotional state by three independent and bipolar dimensions, pleasure-displeasure, degree of arousal, and dominance-submissiveness. He defined it as a result of a study on 200 subjects and 42 verbal-report emotion scales. In our work, we use the *dimensional* representation of emotions based on VA, without using the dominance dimension. This is motivated by a subsequent work of Russell [20] that introduced the *circumplex* model of emotion, where the dimensions of excitation and valence were distributed in a two-dimensional circular space. Most of facial expression datasets that are labeled with the dimensional representation use these two dimensions.

In the same context, Parrott [18] defined a taxonomy for emotion-related terms in a tree-like structure starting with six primary emotions (*i.e.*, love, joy, surprise, anger, pain and fear), then secondary emotions and tertiary emotions, for a total of 140 terms. Although the Parrott's hierarchy accurately describes the emotion of a person, it does not provide the measure of VA. Hence, one founding idea of our work is that of taking the common terms between Russell's work and Parrott's classification to derive the VA of each emotion.

Terms Selection and Mapping. To define a larger set of emotional terms with respect to the six basic emotions, we first merged the terms of Russell and Parrott; then, we selected the subset of most similar terms to get a mapping from Parrott's structure to Russell's values for VA. For example, the terms *joy* and *joyful* appear in both taxonomies, and the related (*valence,arousal*) pair given by Russell for joyful is $(0.76, 0.48)$. So, we can both exploit the values mapping and the Parrott's hierarchical structure. This process is depicted in Fig. 1.

This process, resulted into 32 terms that we organized as reported in Table 1. In addition, we also included the *neutral* emotion to the selected set of terms. In Parrott's classification of emotions, 3 positive and 3 negative terms were proposed, respectively, for primary emotion, 11 positive and 14 negative terms

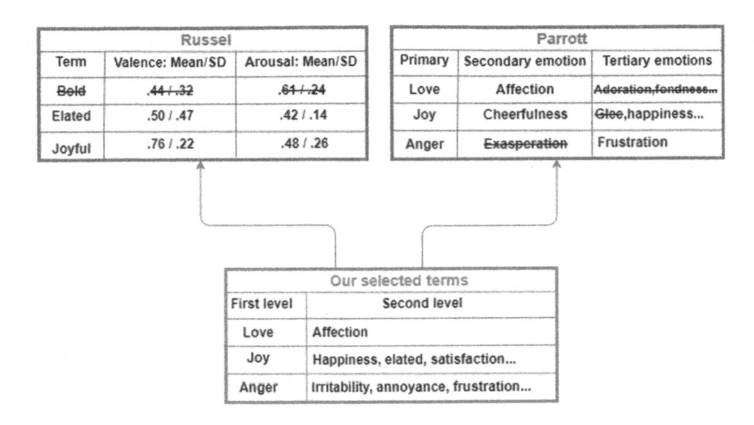

Fig. 1. Mapping between emotion definitions.

Table 1. Our selected emotional terms (6 and 32, respectively. at the first and second level of the categorization) plus neutral.

First level emotions	Second level emotions
Love	Affection
Joy	Happiness, Elation, Satisfaction, Excitement, Triumph
Surprise	Astonishment
Anger	Irritability, Annoyance, Frustration, Rage, Hostility, Hatred, Scorn, Disgust, Contempt
Sadness	Depression, Despair, Displeasure, Shame, Guilt, Regret, Refeatism, Embarrassment,
	Humiliation, Insecurity, Isolation, Loneliness, Rejection
Fear	Terror, Anxietty, Distress

for secondary emotion, and 42 positive and 73 negative terms for tertiary emotion, respectively. With this classification of emotions, we can see an imbalance between positive and negative terms. Since our classification is derived from Parrott's, we can observe the dominance of negative emotions over positive ones in the proposed 32 terms: 3 positive and 3 negative terms, and 7 positive and 25 negative terms for the first and second level emotions, respectively.

Terms Taxonomy. After merging Russell's and Parrott's terms, the intersection between the two classifications yields 38 terms (see Table 1) from the primary, secondary, and tertiary layers of the Parrott term representation. Taking Parrott's classification as a guideline for our classification, we divided the 38 terms into two levels: the first level includes 6 terms (*love*, *joy*, *surprise*, *anger*, *sadness*, *fear*); the second level comprises 32 terms plus the neutral state (*neutral* is positioned in the origin of the VA plane). The choice of having only two levels in the hierarchy, unlike Parrott's classification, is due to the fact that most of the selected terms come from Parrott's tertiary layer. In the experiments, we used only the terms in the second level to re-label the dataset because the terms in the first level are independent emotions, intense, long-lasting and irreversible: even in case of reversing them, they need longer period. Contrarily, emotions in the second class are less intense and dependent upon primary emotions.

4 Experiments

The objective of our evaluation is to demonstrate the applicability of the proposed relabeling to a benchmark dataset. In the following, we considered the AffectNet dataset [16]. Due to the limited number of datasets annotated with both emotions and VA [9], the AffectNet dataset was chosen.

4.1 Relabeling the AffectNet Dataset

AffectNet [16] is a large facial expression dataset acquired in the wild, with around 0.4 million images manually annotated for the presence of six categorical facial emotions (*i.e.*, *happy*, *sad*, *surprise*, *fear*, *disgust*, *angry*), plus contempt and neutral. Values of VA are also provided. In addition, the three non-emotional labels of *none*, *uncertain*, and *no-face* are used. In particular, the *none* ("none of the eight emotions") category is an expression/emotion (such as sleepy, bored, tired, seducing, confuse, shame, focused, *etc.*), that could not be assigned by annotators to any of the six basic emotions, contempt or neutral. However, VA could be assigned to these images. The relabeling process was performed according to the VA annotations in AffectNet. These values were used to link each image with one of the proposed 32 emotional terms plus neutral. The relabeling is as follows:

Validation Set - 19 terms were used in the relabeling (Insolent, Anxious, Disgusted, Insecure, Self-satisfied, Frustrated, Astonished, Depressed, lonely, Shamed, Excited, Affectionate, Elated, Hate, Defeated, Hostile, Irritated, Enraged, Happy). For a total of 248 images (6,20% of the relabeled images) no matching terms were found;

Training Set - 21 terms were used in the relabeling (Insolent, Affectionate, Self-satisfied, Anxious, Disgusted, Insecure, Shamed, Elated, Depressed, Hate, Excited, Astonished, Irritated, lonely, Happy, Frustrated, Defeated, Enraged, Regretful, Hostile, Despairing). In this case, 44,871 images (15,60% of the relabeled images) were not relabeled.

Figure 2 shows the multi-term relabeling process for an example image from AffectNet. With the proposed terms, emotions can be described with more than the basic terms: for example "sad" has a variety of terms, such as depression, despair, displeasure, shame, guilt, regret, embarrassment that describe the emotion more accurately. More in detail, the image on the left was originally annotated in AffectNet with VA values of, respectively, −0.793651 and −0.373016, and "Sadness" as expression label. With the proposed labeling, we can associate the point in the VA plane with first structure emotion "Sadness" and second structure emotion "Shame". Applying the proposed terms hierarchy, we can derive a coarser expression characterization with 6 first class emotions (love, joy, surprise, anger, sadness, fear) and 32 s class emotions, that are more descriptive of the main emotion.

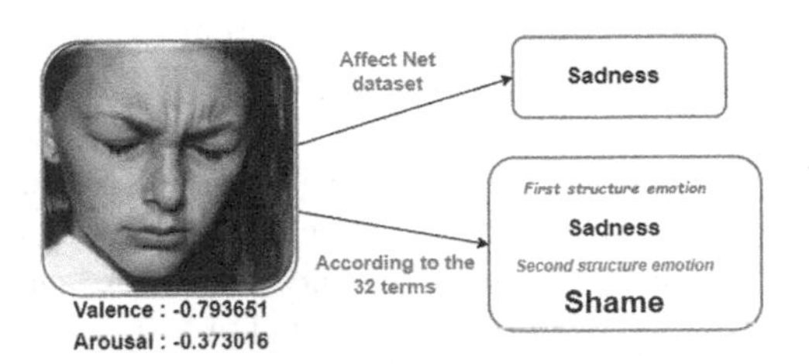

Fig. 2. Example of relabeling an image from the AffectNet dataset according to the proposed approach.

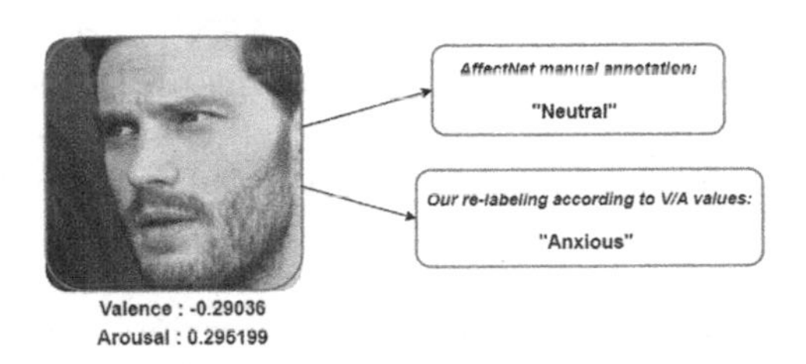

Fig. 3. Difference between AffectNet and our re-labeling.

It turns out that AffectNet is manually annotated and labeled with a balanced number of classifications for the emotions (500 images for each emotion). But when we re-labeled the AffectNet images based on VA values, we found a large imbalance in the re-labeling results. This is due to the large difference between annotating a dataset according to the VA values and according to manual labeling. An example showing the efficacy of our proposed terms is illustrated in Fig. 3. In this case, an image with valence -0.29036 and arousal 0.295199 was originally labeled in AffectNet as "Neutral", though the "Neutral" label is the non-emotional state associated with the origin of the VA plane (0.0 value for both dimensions). The relabeling process applied to this image resulted in a relabeling with the "Anxious" term. According to Russell's work, the distributions for VA associated to this term are as follow: mean valence of 0.01 ± 0.45, and arousal of 0.59 ± 0.31.

Our observation here, is that the VA values provide a viable way to expand the categorical emotion representation with 7 (6 basic plus neutral) or 8 (with contempt) terms to a finer grained set of emotional states. Proposing the use of Russell's mapping and the relation between Russell's and Parrott's terms, we contribute a motivated and flexible way to include more terms in the classification of expression datasets fostering the design of more effective and realistic classification solutions.

4.2 Results

In this section, we aim at evaluating the proposed classification emotion annotations. We are interested in estimating the consistency of our mapping with the manual annotations for the basic 8 emotion classes. With this aim, we used the Distract your Attention Network (DAN) [27]. The DAN model was proposed for facial expression recognition, while we used it as classifier using our proposed annotations as classes to perform emotion recognition. We used 287,651 images for the training set and 3,999 images for the validation set, according to the annotation of VA values provided by AffectNet. DAN includes three key components: a Feature Clustering Network (FCN) extracts robust features by adopting a large-margin learning objective to maximize class separability; a Multi-head cross Attention Network (MAN) instantiates a number of attention heads to simultaneously attend to multiple facial areas and build attention maps on these regions; finally, an Attention Fusion Network (AFN) distracts these attentions to multiple locations before fusing the attention maps to a comprehensive one.

In AffectNet, a total of 291,650 images manually annotated with eight labels were released for public research. We used the validation set including 3,999 images (the test set is not released). Results of our experimentation (Table 2) shows an accuracy of **27.94%** compared to the original accuracy of **61%** obtained by running DAN as classifier on the original 8 expression labels of AffectNet. Observing the distribution of the 32 emotions over the whole dataset, and the lack of homogeneity between the number of images for each term, an increased difficulty of the classification task is revealed. We believe this result can open the way to new challenges in the task of emotion recognition. Especially in the availability of datasets that are annotated not only with more than the 6 basic emotions but also with the values of VA.

Table 2. Accuracy on the AffectNet dataset obtained by using DAN with our 32 terms in comparison with the 8 original terms.

Dataset	Original terms	32 terms
AffectNet [16]	61%	27.94%

Unbalanced Data – In this paragraph, we address the problem of class unbalance in facial recognition datasets, and how this problem affects the final performance. Without handling unbalance during data collection results in classifications that are biased toward the majority class with poor accuracy for the minority classes [11]. The class unbalance problem represents one of eight different unbalance problems [17]; it happens when there is a noticeable disparity in the number of examples that belong to the various classes. The unbalance in datasets is not only evident in terms of emotions, but also in the distribution of positive and negative emotions. We refer further to the unbalance in Parrott's classification regarding positive and negative emotions.

AffectNet's validation set, which includes 500 images for each emotion manually annotated, is unbalanced when considering the VA value of each image. However, AffectNet's training set is out of balance in terms of VA value. In Fig. 4, we show the impact of class unbalance of the AffectNet training set on our relabeling: the relabeling results shift towards the majority of classes already distributed in AffectNet: see the over-represented class "happy", with 13,4915 images, versus the under-represented class "contempt" with 4,250 images. This induces not only unbalance in the data but also inconsistency between the manual labeling and the assigned VA values. This leads us to one outcome of our work: data unbalance results into poor accuracy, even when data annotation is improved with the proposed 32 classes.

Inconsistency Between Emotion and VA Labels – As presented in Sect. 4.1 and in Fig. 3, VA values are attributed regardless of the actual value of the labeled emotion, and this is due to the manual annotation of the dataset (the manual annotation in AffectNet was done by one annotator for each image). For "neutral" the attributed VA values are not only 0.0 but also assume other values such as (V,A)=(−0.176846,-0.176846), (V,A)=(−0.135501,0.00483933). This presents a challenge in relabeling an existing benchmark dataset according to its valence and arousal values, while the emotions are labelled manually.

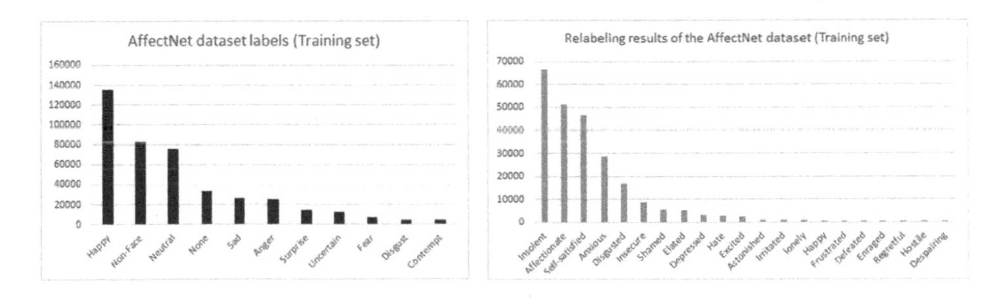

Fig. 4. Impact of class unbalance on: AffectNet training set (left); our relabeling (right).

5 Conclusions

We believe that emotions are more extensive than the six basic categorical facial expressions, and more complicated than describing a state of a person by only "sad", "happy", "fear", *etc.* In our work, we proposed a classification of emotions according to the dimensional representation of emotions based on VA proposed by Russell [22], and Parrott's classification of emotions [18]. Taking inspiration from both the models, and combining a subset of the Parrott's hierarchical categorization of emotions with the Russell's mapping between the categorical and the dimensional domain resulted in 38 terms, which describe more precisely the state of a person. We divided the terms into two classes: a primary class of emotions with 6 terms, and a secondary class of emotions that depend on the first class with 32 terms. We re-labeled the AffectNet dataset according to the 32 proposed terms.

Despite the fact that the results show an accuracy of 27.94% compared to the 61% obtained for the original dataset annotation with 8 emotions, they revealed the gap between manual annotation and annotating according to the VA values. The validation set of AffectNet is balanced in terms of annotating through judging if the image is happy, sad, surprise, fear, disgust, angry, contempt, presenting 500 images for each emotion, but looking at the VA value of each image, we found an unbalance in attributing the emotions in the dataset, and that is the reason behind the unbalance in the relabeling of the dataset.

Labeling a dataset manually or based on VA values are two different ways that are still difficult to compare. As future work, we will annotate other datasets that include both VA and categorical emotion labels, like the OMGEmotion Challenge dataset [3] with the overall set of terms in our proposed taxonomy.

References

1. Abdullah, S.M.S., Abdulazeez, A.M.: Facial expression recognition based on deep learning convolution neural network: a review. J. Soft Comput. Data Min. **2**(1), 53–65 (2021)
2. Allognon, S.O.C., Britto, A.S., Koerich, A.L.: Continuous emotion recognition via deep convolutional autoencoder and support vector regressor. In: Intetnational Joint Conference on Neural Networks (IJCNN), pp. 1–8. IEEE (2020)
3. Barros, P., Churamani, N., Lakomkin, E., Siqueira, H., Sutherland, A., Wermter, S.: The omg-emotion behavior dataset. In: IEEE Int. Joint Conference on Neural Networks (IJCNN), pp. 1–7 (2018)
4. Boukricha, H., Wachsmuth, I., Hofstätter, A., Grammer, K.: Pleasure-arousal-dominance driven facial expression simulation. In: IEEE International Conference on Affective Computing and Intelligent Interaction, pp. 1–7 (2009)
5. Carlotta Olivetti, E., Violante, M.G., Vezzetti, E., Marcolin, F., Eynard, B.: Engagement evaluation in a virtual learning environment via facial expression recognition and self-reports: a preliminary approach. Appl. Sci. **10**(1) (2020)
6. Ekman, P., Friesen, W.V.: Constants across cultures in the face and emotion. J. Pers. Soc. Psychol. **17**(2), 124 (1971)
7. Feng, W., Kannan, A., Gkioxari, G., Zitnick, C.L.: Learn2Smile: learning non-verbal interaction through observation. In: IEEE/RSJ International Conference on Intelligent Robots and Systems (IROS), pp. 4131–4138 (2017)
8. González-Rodríguez, M.R., Díaz-Fernández, M.C., Gómez, C.P.: Facial-expression recognition: an emergent approach to the measurement of tourist satisfaction through emotions. Telematics Inform. **51**, 101404 (2020)
9. Guerdelli, H., Ferrari, C., Barhoumi, W., Ghazouani, H., Berretti, S.: Macro-and micro-expressions facial datasets: a survey. Sensors **22**(4), 1524 (2022)

10. Hu, Y., Zeng, Z., Yin, L., Wei, X., Zhou, X., Huang, T.S.: Multi-view facial expression recognition. In: IEEE International Conference on Automatic Face & Gesture Recognition, pp. 1–6 (2008)
11. Huang, C., Li, Y., Loy, C.C., Tang, X.: Deep imbalanced learning for face recognition and attribute prediction. IEEE Trans. Pattern Anal. Mach. Intell. **42**(11), 2781–2794 (2019)
12. Kollias, D., Zafeiriou, S.P.: Exploiting multi-CNN features in CNN-RNN based dimensional emotion recognition on the omg in-the-wild dataset. IEEE Trans. Affect. Comput. (2020)
13. Lai, Z., Chen, R., Jia, J., Qian, Y.: Real-time micro-expression recognition based on ResNet and atrous convolutions. J. Ambient Intell. Hum. Comput., 1–12 (2020)
14. Li, J., Jin, K., Zhou, D., Kubota, N., Ju, Z.: Attention mechanism-based CNN for facial expression recognition. Neurocomputing **411**, 340–350 (2020)
15. Li, Y., Wang, S., Zhao, Y., Ji, Q.: Simultaneous facial feature tracking and facial expression recognition. IEEE Trans. Image Process. **22**(7), 2559–2573 (2013)
16. Mollahosseini, A., Hasani, B., Mahoor, M.H.: AffectNet: a database for facial expression, valence, and arousal computing in the wild. IEEE Trans. Affect. Comput. **10**(1), 18–31 (2017)
17. Oksuz, K., Cam, B.C., Kalkan, S., Akbas, E.: Imbalance problems in object detection: a review. IEEE Trans. Pattern Anal. Mach. Intell. **43**(10), 3388–3415 (2020)
18. Parrott, W.G.: Emotions in social psychology: essential readings. Psychology Press (2001)
19. Plutchik, R.: Emotion, a psychoevolutionary synthesis (1980)
20. Russell, J.A.: A circumplex model of affect. J. Pers. Soc. Psychol. **39**(6), 1161 (1980)
21. Russell, J.A.: Core affect and the psychological construction of emotion. Psychol. Rev. **110**(1), 145 (2003)
22. Russell, J.A., Mehrabian, A.: Evidence for a three-factor theory of emotions. J. Res. Pers. **11**(3), 273–294 (1977)
23. Toisoul, A., Kossaifi, J., Bulat, A., Tzimiropoulos, G., Pantic, M.: Estimation of continuous valence and arousal levels from faces in naturalistic conditions. Nat. Mach. Intell. **3**(1), 42–50 (2021)
24. Val-Calvo, M., Álvarez-Sánchez, J.R., Ferrández-Vicente, J.M., Fernández, E.: Affective robot story-telling human-robot interaction: exploratory real-time emotion estimation analysis using facial expressions and physiological signals. IEEE Access **8**, 134051–134066 (2020)
25. Valtakari, N.V., Hooge, I.T., Viktorsson, C., Nyström, P., Falck-Ytter, T., Hessels, R.S.: Eye tracking in human interaction: possibilities and limitations. Behav. Res. Methods, 1–17 (2021)
26. Wang, K., Peng, X., Yang, J., Meng, D., Qiao, Y.: Region attention networks for pose and occlusion robust facial expression recognition. IEEE Trans. Image Process. **29**, 4057–4069 (2020)
27. Wen, Z., Lin, W., Wang, T., Xu, G.: Distract your attention: multi-head cross attention network for facial expression recognition. arXiv:2109.07270 (2021)
28. Xiang, J., Zhu, G.: Joint face detection and facial expression recognition with MTCNN. In: International Conference on Information Science and Control Engineering (ICISCE), pp. 424–427 (2017)
29. Yang, H., Ciftci, U., Yin, L.: Facial expression recognition by de-expression residue learning. In: IEEE Conference on Computer Vision and Pattern Recognition (CVPR) (2018)

30. Zhang, H., Huang, B., Tian, G.: Facial expression recognition based on deep convolution long short-term memory networks of double-channel weighted mixture. Pattern Recogn. Lett. **131**, 128–134 (2020)
31. Zhang, L., Tjondronegoro, D.: Facial expression recognition using facial movement features. IEEE Trans. Affect. Comput. **2**(4), 219–229 (2011)

Video-Based Emotion Estimation Using Deep Neural Networks: A Comparative Study

Leonardo Alchieri[1(✉)], Luigi Celona[2], and Simone Bianco[2]

[1] USI Università della Svizzera Italiana, Via la Santa 1, Lugano, Switzerland
leonardo.alchieri@usi.ch
[2] University of Milano-Bicocca, Viale Sarca 336, Milano, Italy
{luigi.celona,simone.bianco}@unimib.it

Abstract. In this study we investigate the effectiveness of deep neural networks in predicting valence and arousal solely from visual information of video sequences. Several recent Convolutional Neural Network (CNN) and Transformer architectures are used as backbone of the proposed model. We also assess the impact of pretraining on model performance by comparing the results of trained from scratch versus pre-trained models. Experimental results on the One-Minute Gradual Emotion Recognition Challenge dataset suggest that pre-training on emotion recognition datasets is beneficial for most models. Comparison with the state-of-the-art reveals similar performance on valence Concordance Correlation Coefficient (CCC) and lower performance on arousal CCC. However, the predictions in our experiments are not statistically different in most cases. The study concludes by emphasizing the complexity of video emotion recognition and the need for further research to enhance the robustness and accuracy of emotion recognition models. The source code used for the experiments is made publicly available.

Keywords: Video emotion recognition · Convolutional neural networks · Transformers · Valence · Arousal

1 Introduction

The use of emotions is a fundamental aspect of interpersonal interactions among humans. Affective computing has emerged as a field of computer science that aims to understand, model, and simulate human emotions [28]. The goal of affective computing is to create intelligent systems that can interact with humans in a more natural and intuitive way, by recognizing and responding to our emotional states. Emotion recognition from images and videos is a central topic in this field, as it has the potential to provide valuable insights into human behavior and improve human-computer interaction.

Emotions are complex and multidimensional phenomena that involve subjective experiences, physiological responses, and behavioral expressions [16]. Different approaches have been proposed to classify emotions, including discrete

G. L. Foresti et al. (Eds.): ICIAP 2023 Workshops, LNCS 14365, pp. 255–269, 2024.
https://doi.org/10.1007/978-3-031-51023-6_22

emotional theories [10], which posits the existence of six basic emotions, and dimensional emotion theories, such as the Circumplex Model of Affect [30], which represents emotions along two dimensions, namely valence and arousal. Researchers have developed various methods for emotion recognition, including facial expression analysis [29], speech analysis [15], and physiological signal analysis [4].

Video emotion recognition is an unobtrusive way to capture emotional responses and it is similar to how humans perceive emotions. Effectiveness of deep learning models for image recognition makes them suitable also for video emotion recognition. Several datasets have been developed for emotion recognition from images and videos aimed at facilitating research and benchmarking, e.g., EmotiW [7–9], Aff-wild [35], AffectNet [26]. One of the most widely used datasets is the One-Minute Gradual-Emotion Recognition Challenge [1] (shortly OMG-Emotion), which contains a large collection of videos of people expressing various emotions, and annotated with valence and arousal. This dataset has been used by many researchers to develop and evaluate different emotion recognition models [19,20,32].

In this study, we investigate the effectiveness of deep learning methods in predicting valence and arousal solely from visual information of video sequences. Specifically, we start by replicating the results of the top-ranked methods at the OMG-Emotion Challenge, namely Peng *et al.* [27] and Zheng *et al.* [37]. We then explore the use of recent Deep Neural Networks (DNNs) as the backbone for the Peng *et al.* method. In particular we evaluate the use of both CNNs, namely ConvNext [25], Distract your Attention Network (DAN) [33], and ResNet-50 [14], and Transformers, i.e. Former-DFER [36] and SwinT [23]. Finally, we assess the impact of pretraining on model performance by comparing the results of trained from scratch versus pre-trained models. This work aims to provide insights into the most effective models presented at the OMG-Emotion Challenge for emotion recognition from videos and to facilitate future research in this field. We provide the code of our experiments and the replication of previous models in an open source repository.[1]

2 Related Work

Most of the models developed and evaluated on the OMG-Emotion dataset were proposed during the challenge in 2018. Peng *et al.* [27] presented a method involving two separate CNN models for encoding video and audio signals, respectively. The pre-processed face sequence is encoded using a SphereFace20 [21] and a Bidirectional Long Short-Term Memory (Bi-LSTM) [31] for temporal aggregation. The spectrogram of the audio signal is instead encoded with a VGG16 plus average pooling for temporal aggregation. The feature vectors for the two signals are then concatenated and passed to a fully connected layer for valence and

[1] https://github.com/LeonardoAlchieri/OMGEmotionOverview (last access: 04/07/2023).

Fig. 1. Preprocessed sample frames extracted from the OMG-Emotion dataset.

arousal estimation. Triantafyllopoulos *et al.* [32] used OpenSmile [12] for estimating arousal and valence from the audio signal and a VGGFace with Bi-LSTM for estimating valence from the video signal. The predictions of the two models for valence are averaged to obtain the final score. Zheng *et al.* [37] exploited a VGG16 for encoding a 64-frame sequence of pre-processed faces. OpenSmile is used for obtaining the acoustic representation of the video clip. Multi-modal feature fusion is achieved by concatenating the feature vectors for the two modalities, then a Support Vector Regressor with Radial Basis Function kernel predicts the valence and arousal values.

Barros *et al.* [2] proposed P-AffMem, an autoencoder capable of synthesizing faces whose expression is conditioned by arousal and valence values given in input together with the face image. For the purpose of estimating the valence and arousal values, the P-AffMem encoder estimates a feature vector which is processed by a discriminator for the estimation of the two values. Kollias *et al.* [20] presented an ensemble methodology to combine features extracted from two VGGFace-based models. Temporal aggregation is achieved by Recurrent Neural Networks (RNNs). Recently, Foteinopoulou and Patras [13] presented a model analyzing only the visual signal for estimating valence and arousal. They proposed a two-stage attention architecture that uses features from the neighbourhood of the clips to introduce context information in the feature extraction. Furthermore, they proposed a new loss function using the distance between label vectors to learn intra-batch latent representation similarities.

3 The OMG-Emotion Dataset

The OMG-Emotion dataset is introduced herein, as it serves as the basis for the experiments we conduct on valence-arousal estimation from video visual information. This dataset is a widely-used benchmark for video-based emotion recognition in the literature. The OMG-Emotion dataset [1] includes 567 videos with different duration and frame rates, which were collected from YouTube using a keyword-based search that focused on the word “monologue”. The average length of each video is approximately one minute, and a total of 5291 utterances were identified in the videos, each displaying a single emotion. Figure 1 shows frames randomly extracted and preprocessed from utterances.

Each utterance was manually annotated by an average of 5.4 people for valence and arousal values using the Circumplex Model of Affect [30]. The

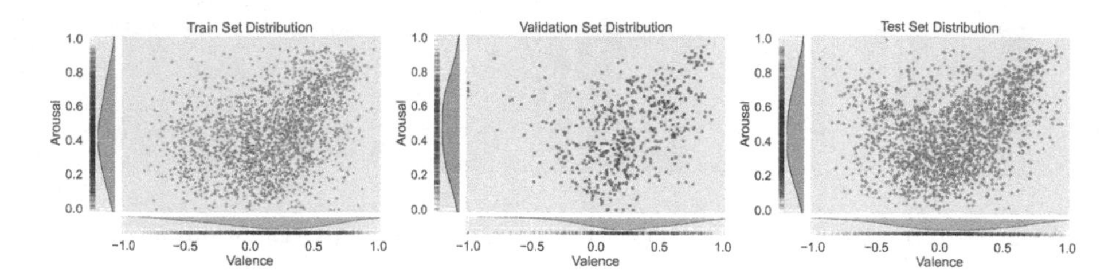

Fig. 2. Distribution in the 2-D Valence-Arousal space of utterance-level annotations of the training (left plot, circles), validation (center plot, squares) and test (right plot, diamonds) OMG-Emotion sets.

valence scale is from -1 to 1, while arousal is from 0 to 1. Besides, six discrete emotion labels based on Ekman's theory [11] and a neutral label were annotated.

The dataset is divided into three sets, train, validation and test, with 2443, 618 and 2230 samples, respectively. We show in Fig. 2 the distribution of labels for the valence-arousal dimensions in the OMG-Emotion dataset, for the three splits (train, validation, and test). We notice that it is similar for the train and test splits, with the majority of the labels clustered around the center of the valence-arousal space. However, there are some differences in the distribution in the validation set compared to the other two splits, especially on the valence dimension. Also, some utterances in the validation set have a valence score of -1, which is not present in the train and test splits. Additionally, there are fewer utterances with values in the upper left quadrant, which represents emotions with valence less than 0 and arousal greater than 0.5. These differences in the distribution of labels between the validation set and the other two splits could be an important factor to consider when evaluating models.

4 The Video Emotion Estimation Pipeline

Figure 3 shows a general pipeline for video emotion estimation, whose implementation of building blocks will be evaluated. Given a video sequence as input, the model estimates valence and arousal values by first pre-processing the frames, then feeding them to a model consisting of a backbone, a temporal aggregator, and a Multi-Layer Perceptron (MLP). The backbone encodes the spatial information of the frames. The temporal aggregator combines the frame-level features before passing them to the MLP, which maps them into valence and arousal.

4.1 Pre-processing

We implement pre-processing steps to extract facial information from the videos, similarly to what performed in [1,21]. Frames are first extracted from each utterance and then cropped around the detected face and aligned using the RetinaFace model [6]. A total of 471,983 frames is obtained for the train set, 120,647 for the validation set, and 440,025 for the test set.

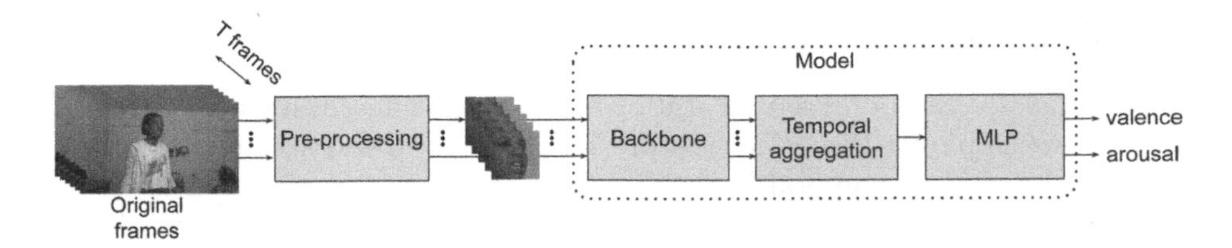

Fig. 3. Pipeline for video-based facial expression recognition.

Table 1. Datasets exploited for pre-training model backbones. For all models except ResNet-50, the pre-trained weights were downloaded from open-source repositories.

Model name	Dataset	Weights source
Former-DFER	DFEW [17]	https://github.com/zengqunzhao/Former-DFER
Convnext small	Imagenet [5]	https://github.com/facebookresearch/ConvNeXt
SphereFace20	Casia-WebFace [34]	https://github.com/yangyucheng000/sphereface
DAN	AffectNet [26]	https://github.com/yaoing/DAN
Swin3D Tiny	Imagenet [5]	https://github.com/SwinTransformer/Video-Swin-Transformer
ResNet-50	AffectNet [26]	Local Training

4.2 Backbone

We identify six modern DNNs that serve as the backbone, with four being based on CNNs and the remaining two utilizing Transformer architecture. The CNN-based architectures are ResNet-50 [14], ConvNeXt small [25], SphereFace20 [21], and Distract your Attention Network (DAN) [33], all of which are widely recognized and used in the field of image recognition. The Transformer-based architectures are Former-DFER [36] and Swin3D Tiny [24]. ConvNeXt small [25] and Swin3D Tiny [24] are recent architectures that have shown state-of-the-art performance on several image recognition tasks. Both are designed to capture multi-scale features and can effectively handle variations in object size and position. SphereFace20 [21] is the model used as the backbone in the second-ranked method of the OMG-Emotion Challenge [27]. It is a DNN architecture that aims to improve the discriminative power of the embedding features by enhancing angular separability between different classes. DAN [33] has been specifically designed for facial expression recognition. The model exploits self-attention to identify informative facial regions for emotion recognition and performs well on multiple expression datasets. Finally, we choose Former-DFER [36] for its effectiveness in capturing long-term dependencies in sequential data, as shown by [36] on the DFEW dataset [17]. All of these models provide pre-trained weights that can be leveraged for transfer learning. Table 1 outlines the dataset on which each model was pre-trained.

4.3 Temporal Aggregation

To aggregate the temporal representation of multiple frames into spatio-temporal features, we employ a Bi-LSTM network for the CNN-based architectures, similarly to [27], and a temporal Transformer for the Transformer-based architecture. The Bi-LSTM network is effective at modeling temporal dependencies in sequential data [3,22]. The temporal Transformer can efficiently capture temporal information in a non-sequential manner by processing all frames of the video simultaneously. We choose these architectures as they are among the most commonly used methods for temporal modeling in video-based emotion recognition [18].

4.4 Multi-layer Perceptron

Spatio-temporal representations, obtained from the temporal aggregation module, are then fed into an MLP, whose depth is empirically set to 1. Therefore, it consists of a linear layer that maps the N-dimensional representation of the utterance into 2 values, namely valence and arousal. The linear layer is followed by a Sigmoid to limit the predicted value for arousal in the range [0, 1] and the activation function TanH to limit the value for the valence in the range [−1, 1].

4.5 Loss Functions

The optimization of the model is carried out by using one of the following two loss functions, namely Mean-Squared Error (MSE) and Concordance Correlation Coefficient (CCC). Let x_i and y_i, with $i \in \{v, a\}$, be the labels and predicted values for valence and arousal, respectively. Given the MSE for each dimension defined as $e_i = (x_i - y_i)^2$, the total loss based on MSE is defined as follows:

$$\mathcal{L}_{MSE} = \frac{e_a + e_v}{2}. \tag{1}$$

The CCC takes values in $[-1, 1]$ and is defined as follows:

$$\rho_i = \frac{2s_{i,xy}}{s_{i,x}^2 + s_{i,y}^2 + (\bar{x}_i - \bar{y}_i)^2}, \tag{2}$$

where $s_{i,x}$ and $s_{i,y}$ are the variances of the valence/arousal labels and predicted values, $\bar{x}_i$ and $\bar{y}_i$ are the corresponding mean values, and $s_{i,xy}$ is the covariance value. Therefore, the resulting CCC loss for both valence and arousal is the following:

$$\mathcal{L}_{CCC} = 1 - \frac{\rho_a + \rho_v}{2}, \tag{3}$$

where ρ_a and ρ_v are the CCC for the arousal and valence.

5 Experiments and Results

5.1 Ablation Analysis

We conduct an ablation analysis to identify significant hyperparameters and configurations in the deep learning models trained on the OMG-Emotion dataset. Our analysis focused on: (i) the use of activation functions for clipping the range of score predictions; (ii) the number of frames to sample from each utterance; (iii) the use of Bi-GRU instead of Bi-LSTM for temporal aggregation. Given the impracticality of performing this analysis on all presented models, we decide to perform it on the model proposed by Peng *et al.* [27]. The SphereFace backbone is initialized with the weights trained on the Casia-WebFace dataset and a CCC loss function is exploited for all the ablation experiments. We sequentially conduct each ablation analysis step, with the best configuration from the previous step being carried over to the subsequent step. In instances where parameters were yet to be tested, we choose the following default values: 16 frames and Bi-LSTM for temporal aggregation. We use the SGD optimizer and train each model for a maximum of 30 epochs on the train set. We report the results on the validation set by initializing the model with the epoch weights that obtained the highest mean CCC on the validation set.

Predicted Score Clipping. The use of activation functions (i.e., TanH and Sigmoid) to limit the predicted score ranges shows no increase in valence (0.4344 compared to 0.4396 obtained without activation functions) but a performance gain of +13% for arousal (i.e., 0.2920 without activations and 0.3284 with activations).

Number of Frames. We evaluate the sampling of utterances with a different number of frames. For utterances of 8 frames, 0.2914 of arousal and 0.4406 of valence are obtained; for 32 frames, we achieve 0.2741 and 0.4674; for 16 frames, we achieve the performance of 0.3284 and 0.4344. Previous results demonstrate that the use of sequences of 16 frames obtains a slightly better result (+0.01 on valence-arousal average) but at a lower computational cost.

Temporal Aggregator. For the time aggregator it is evaluated the use of a Bi-LSTM and a Bi-GRU with the same configuration. When training with a Bi-LSTM, valence and arousal CCCs of 0.3284 and 0.4344 are respectively obtained, compared to 0.2385 and 0.3896 when using a Bi-GRU. Therefore, it is possible to state that Bi-LSTM is more effective than Bi-GRU.

5.2 Model Comparison

In this section, we present the results achieved by using the considered backbone models and time aggregators. Specifically, we train ConvNext Small, SphereFace20, ResNet-50 and DAN with Bi-LSTM as temporal aggregator, while

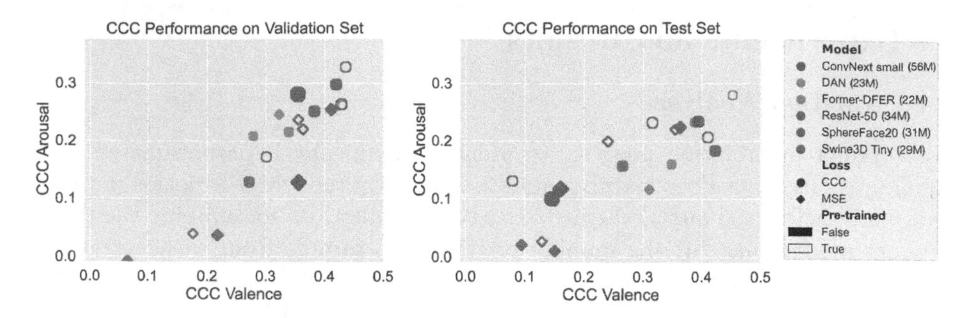

Fig. 4. Results for valence and arousal on the validation (left) and test (right) sets. The marker size is proportional to the number of parameters in the model (shown in the legend in parentheses).

Former-DFER and Swin3D-Tiny are trained with temporal Transformers. The models are trained on the train set for a maximum of 30 epochs with a batch size of 32 utterances, except for ConvNext Small, which is trained with a batch size of 8 due to memory constraints. For each utterance, a sequence of 16 frames is sampled. We evaluate each model using four different combinations. They are: (i) pre-trained weights with CCC loss; (ii) pre-trained weights with MSE loss; (iii) random initialization with CCC loss; (iv) random initialization with MSE loss.

In Fig. 4, we present the results of evaluating each model on both the validation and test sets, for all possible combinations. The results shown are picked, for each model, from the top-performing epochs, based on the highest average CCC on the validation set. The models starting from pre-trained weights achieve the highest performance on both datasets, particularly when using CCC as loss function. The three best-performing models, namely SphereFace20, ResNet-50, and Former-DFER, are all trained with pre-trained weights and CCC loss. It is noteworthy that the performance on the test set of the models based on SphereFace20 and ResNet-50 is inferior to that on the validation set. This disparity in performance could be attributed to potential overfitting on the validation set, which is the basis for the reported results. Some models do not converge when trained with the CCC loss function, and as such are not displayed. A detailed report of the results for all four training setups is provided in Appendix A.

Our results suggest that pre-trained weights can improve the performance of almost all models. Furthermore, using CCC as the loss function can lead to higher evaluation results (for those models that converge using it). The architectures that consistently performed well across different conditions are SphereFace20, ResNet-50, and Former-DFER. This can be due to a combination of architecture and datasets on which they were pre-trained.

5.3 State-of-the-Art Comparison

In this section, we compare our best CNN- and Transformers-based models with state-of-the-art models. In our comparison we include state-of-the-art models that exploit the analysis of only the visual signal of a video. These methods are: Zheng *et al.* [37], Foteinopoulou and Patras [13], Peng *et al.* [27], Kollias *et al.* [20], and P-AffMem [2]. Previous models were trained using the train and validation sets together, rather than using the train set exclusively, as was done in the previous section. For a fair comparison with our methods, we train our methods ResNet-50 with Bi-LSTM and FormerDFER with Transformer in the same setup. We also train state-of-the-art methods that have publicly available code. We manage to successfully retrain the model proposed by Peng *et al.* [27][2]. The source code provided by Zheng *et al.* [37][3] is instead incomplete and does not allow its reproducibility. For this reason, important changes had to be made to train the model in the same way as the authors.

We show the comparison results in Table 2. Several considerations can be made. First, P-AffMem and Kollias *et al.* achieve the best average performance of about 0.08 higher than the other methods. However, we highlight that Kollias *et al.* uses an ensemble of models and P-AffMem is an encoder trained to reconstruct emotion-conditioned faces, therefore both models differ substantially from the others. Second, the high performance of Zheng *et al.* and Peng *et al.* reported in the OMG-Emotion Challenge is not replicated in our training (the results reported in the rows with asterisks). Although Zheng *et al.* method required a lot of reimplementation, for Peng *et al.* we use the original code. The drop in performance equal to 0.06 for Peng *et al.* is therefore attributable to some difference in the dataset. Finally, our proposals based on ResNet-50 and Former-DFER achieve comparable results on arousal (about 0.29), while for valence Former-DFER achieves a CCC of 0.07 higher than that of ResNet-50. Although our results are significantly lower than those of Kollias and P-AffMem (i.e., 0.12 lower on average), they are in line with the most recent study by [13]. They are also very close to those obtained by replicating Peng *et al.* training.

The statistical t-test conducted by comparing the predictions of three models, namely our trained Peng *et al.* model, our ResNet-50 based model and our FormerDFER based model, reveals that only the Peng *et al.* model has significantly different results for arousal, while there are no significant differences in valence between the three models. Although Peng *et al.* model achieves higher CCC scores on valence, the results suggest that these differences may be due to stochastic effects.

---

[2] https://github.com/pengsongyou/OMG-ADSC (last access: 04/07/2023).

[3] https://github.com/ewrfcas/OMGEmotionChallengeCode (last access: 04/07/2023).

Table 2. Comparison of state-of-the-art methods against proposed methods in terms of CCC on the test set of the OMG-Emotion dataset.

Method	Backbone	T-Agg	Arousal	Valence	Mean
Zheng *et al.* (2018) [37]*	VGGFace16	CNN	-0.0066	0.0114	0.0537
Zheng *et al.* (2018) [37]	VGGFace16	CNN	0.2440	0.4378	0.3409
Foteinopoulou and Patras (2022) [13]	ResNet-50	Attention-based	0.2600	0.4800	0.3700
Peng *et al.* (2018) [27]*	SphereFace20	Bi-LSTM	0.2405	0.5070	0.3738
Peng *et al.* (2018) [27]	SphereFace20	Bi-LSTM	0.3612	0.4988	0.4300
Kollias *et al.* (2020) [20]	CNN	RNN	0.3650	**0.5350**	0.4500
P-AffMem (2019) [2]	Encoder-Decoder	None	**0.4300**	0.5300	**0.4800**
Our (this paper)	ResNet-50	Bi-LSTM	0.2994	0.3707	0.3350
Our (this paper)	Former-DFER	Transformer	0.2823	0.4427	0.3625

* Methods retrained by us by exploiting the public code released by the authors.

Table 3. P-values for the t-test result comparing predictions, for arousal and valence, from our implementations of Peng *et al.* (2018) [27]. "Model 1" and "Model 2" refer to the backbones used in the implementations (order denotes display, not importance).

Model 1	Model 2	Arousal p-value	Valence p-value
ResNet-50	SphereFace20	0.000*	0.084
Former-DFER	SphereFace20	0.000*	0.898
Former-DFER	ResNet-50	0.781	0.051

6 Conclusions

In this study, we examine the task of video emotion recognition using only visual information from the OMG-Emotion dataset [1]. Our primary objective is to investigate the performance of various models on this task, and to determine the relevance of pre-training for emotion recognition. Additionally, we seek to replicate state-of-the-art models in this domain.

To this end, we replicate two of the best-performing models from the OMG-Emotion Challenge, namely a SphereFace20-based model by [27] and a VGG-16-based one by [37]. Our analysis also encompasses modifying the model by [27], specifically changing its backbone and temporal aggregation. For backbone, we try ResNet-50 [14], Former-DFER [36], DAN [33], ConvNext Small [25], and Swin3D Tiny [24] architectures. For temporal aggregation, we use either Bi-LSTM for convolutional based models and temporal Transformers for Transformer-based ones. They are evaluated in terms of performance when trained using CCC or MSE loss, and when using pre-trained weights or random initialization.

Our findings suggest that pre-training is beneficial for most models, and particularly on those based on SphereFace20, ResNet-50, and Former-DFER. Indeed, we achieve increase in CCC of maximum 25% on the validation set and 40% on the test set, achieved when using ResNet-50 as backbone. However, it remains unclear whether this performance boost is due to the emotional context in the datasets or the specific architectures of the neural networks.

We also manage to replicate the results by [27], while we do not manage with the model by [37]. When comparing with the state-of-the-art, our models or our retrains achieve similar performance on valence CCC (our best 0.5070 vs. state-of-the-art 0.5300) and lower performance on arousal CCC (our best 0.2994 vs. state-of-the-art 0.4300). However, we find that, when training the model by [27] with SphereFace20, ResNet-50, and Former-DFER as backbones, the performance is statistically similar ($p > 0.05$) on valence predictions and different only for the SphereFace20-based model on arousal.

Overall, our study underscores the complexity of video emotion recognition and the necessity for continued progress in this field. While our findings are encouraging, additional research is warranted to further elucidate the impact of different architectures on emotion recognition, and to identify more robust and accurate approaches to this challenging task (Tables 4, 5, 6, 7).

Acknowledgment. Financial support from ICSC - Centro Nazionale di Ricerca in High Performance Computing, Big Data and Quantum Computing, funded by European Union - NextGenerationEU.

A Appendix

We report the results for all training of the 6 models presented in Sect. 4.2, with all 4 possible combinations implemented: either CCC or MSE as loss functions; and with or without pre-trained weights. Each model is trained for a maximum of 30 epochs on the train set, and the results of the best epochs are reported. Values of `NaN` indicate that the model could not converge with the given configuration.

Table 4. Results, in term of CCC, for experiments with pre-trained models and using **CCC** as a **loss** function. Highlighted in bold are the best results, for each dimension, on both the Validation Set and the Test Set. A value of NaN indicates that the model could not converge.

Model	Validation Set		Test Set	
	CCC		CCC	
	Arousal	Valence	Arousal	Valence
SphereFace20	**0.3284**	**0.4344**	0.2070	0.4097
ResNet-50	0.2636	0.4275	0.2323	0.3150
DAN	NaN	NaN	NaN	NaN
ConvNext small	NaN	NaN	NaN	NaN
Former-DFER	0.2621	0.429	**0.2803**	**0.4520**
Swin3d Tiny	0.1727	0.3004	0.1308	0.0782

Table 5. Results, in term of CCC, for experiments with pre-trained models and using **MSE** as a **loss** function. Highlighted in bold are the best results, for each dimension, on both the Validation Set and the Test Set. All models converged using this loss function.

Model	Validation Set		Test Set	
	CCC		CCC	
	Arousal	Valence	Arousal	Valence
SphereFace20	0.2514	0.382	0.1836	**0.4222**
ResNet-50	**0.2982**	**0.4184**	**0.2342**	0.3932
DAN	0.2157	0.3388	0.1603	0.3480
ConvNext small	0.2810	0.3537	0.1003	0.1452
Former-DFER	0.2089	0.2784	0.2329	0.3872
Swin3d Tiny	0.1292	0.2714	0.1572	0.2646

Table 6. Results, in term of CCC, for experiments with models trained from skratch (no pre-train) and using **CCC** as a **loss** function. Highlighted in bold are the best results, for each dimension, on both the Validation Set and the Test Set. A value of NaN indicates that the model could not converge.

Model	Validation Set		Test Set	
	CCC		CCC	
	Arousal	Valence	Arousal	Valence
SphereFace20	**0.2370**	**0.3549**	0.2202	**0.3545**
ResNet-50	0.2203	0.3628	0.1993	0.2397
DAN	NaN	NaN	NaN	NaN
ConvNext small	NaN	NaN	NaN	NaN
Former-DFER	NaN	NaN	NaN	NaN
Swin3d Tiny	0.1727	0.3003	0.1308	0.0782

Table 7. Results, in term of CCC, for experiments with pre-trained models and using **MSE** as a **loss** function. Highlighted in bold are the best results, for each dimension, on both the Validation Set and the Test Set. A value of NaN indicates that the model could not converge.

Model	Validation Set		Test Set	
	CCC		CCC	
	Arousal	Valence	Arousal	Valence
SphereFace20	0.0371	0.2176	0.0104	0.1508
ResNet-50	**0.2546**	**0.4110**	**0.2244**	**0.3634**
DAN	0.2462	0.3231	0.1166	0.3106
ConvNext small	0.1284	0.3563	0.1173	0.1604
Former-DFER	NaN	NaN	NaN	NaN
Swin3d Tiny	-0.0085	0.0652	0.02052	0.0944

References

1. Barros, P., Churamani, N., Lakomkin, E., Siqueira, H., Sutherland, A., Wermter, S.: The omg-emotion behavior dataset. In: International Joint Conference on Neural Networks (IJCNN), pp. 1–7. IEEE (2018)
2. Barros, P., Parisi, G., Wermter, S.: A personalized affective memory model for improving emotion recognition. In: ICML, pp. 485–494. PMLR (2019)
3. Bin, Y., Yang, Y., Shen, F., Xie, N., Shen, H.T., Li, X.: Describing video with attention-based bidirectional LSTM. IEEE Trans. Cybern. **49**(7), 2631–2641 (2018)
4. Bota, P.J., Wang, C., Fred, A.L., Da Silva, H.P.: A review, current challenges, and future possibilities on emotion recognition using machine learning and physiological signals. IEEE Access **7**, 140990–141020 (2019)
5. Deng, J., Dong, W., Socher, R., Li, L.J., Li, K., Fei-Fei, L.: Imagenet: a large-scale hierarchical image database. In: CVPR, pp. 248–255. IEEE (2009)
6. Deng, J., Guo, J., Ververas, E., Kotsia, I., Zafeiriou, S.: Retinaface: Single-shot multi-level face localisation in the wild. In: CVPR, pp. 5203–5212. IEEE/CVF (2020)
7. Dhall, A.: Emotiw 2019: automatic emotion, engagement and cohesion prediction tasks. In: International Conference on Multimodal Interaction, pp. 546–550 (2019)
8. Dhall, A., Kaur, A., Goecke, R., Gedeon, T.: Emotiw 2018: audio-video, student engagement and group-level affect prediction. In: International Conference on Multimodal Interaction, pp. 653–656. ACM (2018)
9. Dhall, A., Sharma, G., Goecke, R., Gedeon, T.: Emotiw 2020: driver gaze, group emotion, student engagement and physiological signal based challenges. In: International Conference on Multimodal Interaction, pp. 784–789 (2020)
10. Ekman, P., Friesen, W.V.: Constants across cultures in the face and emotion. J. Pers. Soc. Psychol. **17**(2), 124 (1971)
11. Ekman, P., Oster, H.: Facial expressions of emotion. Annu. Rev. Psychol. **30**(1), 527–554 (1979)

12. Eyben, F., Wöllmer, M., Schuller, B.: Opensmile: the munich versatile and fast open-source audio feature extractor. In: International Conference on Multimedia, pp. 1459–1462. ACM (2010)
13. Foteinopoulou, N.M., Patras, I.: Learning from label relationships in human affect. In: International Conference on Multimedia, pp. 80–89. ACM (2022)
14. He, K., Zhang, X., Ren, S., Sun, J.: Deep residual learning for image recognition. In: CVPR, pp. 770–778. IEEE (2016)
15. Jain, M., Narayan, S., Balaji, P., Bhowmick, A., Muthu, R.K., et al.: Speech emotion recognition using support vector machine. arXiv preprint arXiv:2002.07590 (2020)
16. James, W.: What is an Emotion? Simon and Schuster (2013)
17. Jiang, X., et al.: Dfew: a large-scale database for recognizing dynamic facial expressions in the wild. In: International Conference on Multimedia, pp. 2881–2889. ACM (2020)
18. Khan, S., Naseer, M., Hayat, M., Zamir, S.W., Khan, F.S., Shah, M.: Transformers in vision: a survey. ACM Comput. Surv. (CSUR) **54**(10s), 1–41 (2022)
19. Kollias, D., Zafeiriou, S.: A multi-component cnn-rnn approach for dimensional emotion recognition in-the-wild. arXiv preprint arXiv:1805.01452 (2018)
20. Kollias, D., Zafeiriou, S.: Exploiting multi-CNN features in CNN-RNN based dimensional emotion recognition on the omg in-the-wild dataset. IEEE Trans. Affect. Comput. **12**(3), 595–606 (2020)
21. Liu, W., Wen, Y., Yu, Z., Li, M., Raj, B., Song, L.: Sphereface: deep hypersphere embedding for face recognition. In: CVPR, pp. 212–220. IEEE (2017)
22. Liu, X., Li, Y., Wang, Q.: Multi-view hierarchical bidirectional recurrent neural network for depth video sequence based action recognition. Int. J. Pattern Recogn. Artif. Intell. **32**(10), 1850033 (2018)
23. Liu, Z., et al.: Swin transformer: hierarchical vision transformer using shifted windows. In: ICCV, pp. 10012–10022. IEEE/CVF (2021)
24. Liu, Z., et al.: Video swin transformer. In: CVPR, pp. 3202–3211. IEEE/CVF (2022)
25. Liu, Z., Mao, H., Wu, C.Y., Feichtenhofer, C., Darrell, T., Xie, S.: A convnet for the 2020s. In: CVPR, pp. 11976–11986. IEEE/CVF (2022)
26. Mollahosseini, A., Hasani, B., Mahoor, M.H.: Affectnet: a database for facial expression, valence, and arousal computing in the wild. IEEE Trans. Affect. Comput. **10**(1), 18–31 (2017)
27. Peng, S., Zhang, L., Ban, Y., Fang, M., Winkler, S.: A deep network for arousal-valence emotion prediction with acoustic-visual cues. arXiv preprint arXiv:1805.00638 (2018)
28. Picard, R.W.: Affective computing. MIT press (2000)
29. Rázuri, J.G., Sundgren, D., Rahmani, R., Cardenas, A.M.: Automatic emotion recognition through facial expression analysis in merged images based on an artificial neural network. In: Mexican International Conference on Artificial Intelligence, pp. 85–96. IEEE (2013)
30. Russell, J.A.: A circumplex model of affect. J. Pers. Soc. Psychol. **39**(6), 1161 (1980)
31. Schuster, M., Paliwal, K.K.: Bidirectional recurrent neural networks. IEEE Trans. Signal Process. **45**(11), 2673–2681 (1997)
32. Triantafyllopoulos, A., Sagha, H., Eyben, F., Schuller, B.: audeering's approach to the one-minute-gradual emotion challenge. arXiv preprint arXiv:1805.01222 (2018)

33. Wen, Z., Lin, W., Wang, T., Xu, G.: Distract your attention: multi-head cross attention network for facial expression recognition. arXiv preprint arXiv:2109.07270 (2021)
34. Yi, D., Lei, Z., Liao, S., Li, S.Z.: Learning face representation from scratch. arXiv preprint arXiv:1411.7923 (2014)
35. Zafeiriou, S., Kollias, D., Nicolaou, M.A., Papaioannou, A., Zhao, G., Kotsia, I.: Aff-wild: valence and arousal'in-the-wild'challenge. In: CVPR, pp. 34–41. IEEE (2017)
36. Zhao, Z., Liu, Q.: Former-dfer: dynamic facial expression recognition transformer. In: International Conference on Multimedia, pp. 1553–1561. ACM (2021)
37. Zheng, Z., Cao, C., Chen, X., Xu, G.: Multimodal emotion recognition for one-minute-gradual emotion challenge. arXiv preprint arXiv:1805.01060 (2018)

International Contest on Fire Detection (ONFIRE)

ONFIRE Contest 2023: Real-Time Fire Detection on the Edge

Diego Gragnaniello[1], Antonio Greco[1], Carlo Sansone[2], and Bruno Vento[2](✉)

[1] Department of Information and Electrical Engineering and Applied Mathematics (DIEM), University of Salerno, Via Giovanni Paolo II, 132 84084 Fisciano, SA, Italy
{digragnaniello,agreco}@unisa.it

[2] Department of Electrical Engineering and Information Technology (DIETI), University of Naples, Via Claudio, 21, 80125 Naples, NA, Italy
{carlo.sansone,bruno.vento}@unina.it

Abstract. ONFIRE Contest 2023 is a competition, organized within ICIAP 2023 conference, among methods based on deep learning, aimed at the recognition of fire from videos in real-time on edge devices. This topic is inspiring various research groups for the underlying security reasons and for the growing necessity to realize a system that allows to safeguard the territory from the enormous damage that fires can cause. The participants are required to design fire detection methods, starting from a training set that consists of videos in which fire (flames and/or smoke) is present (positive samples), and others (negative samples) that do not contain a fire. The videos have been collected from existing datasets by selecting as positive videos only those that really frame a fire and not flames and smoke in controlled conditions, and as negative videos the ones that contain moving objects that can be confused with flames or smoke. Since the videos are collected in different conditions, the dataset is very heterogeneous in terms of image resolution, illumination, pixel size of flame or smoke, background activity, scenario (urban or wildfire). The submitted methods are evaluated over a private test set, whose videos are different from the ones available in the training set; this choice allows to test the approaches in realistic conditions, namely in unknown operative scenarios. The proposed experimental protocol allows to measure not only the accuracy but also the computational resources required by the methods, so that the top-rank approaches will be both effective and suited for real-time processing on the edge.

Keywords: Contest · ONFIRE · Fire detection · Deep Learning · Edge

1 Introduction

The recent development of smart cameras, that are today able to run computer vision processes thanks to small graphics accelerators, has allowed to exploit

G. L. Foresti et al. (Eds.): ICIAP 2023 Workshops, LNCS 14365, pp. 273–281, 2024.
https://doi.org/10.1007/978-3-031-51023-6_23

them for various video-surveillance tasks [12]. Among them, fire detection is an application that is rising a great interest, because fires can cause enormous damage to the population and to the environment [14]. However, it is a challenging computer vision problem, because flames and smoke can be confused with fog or fire colored objects present in the environment [2]. Given the great interest shown by the scientific community, demonstrated by the high number of methods developed in recent years [15], and considering that it is not trivial to design an approach for fire detection that is both effective and fast, we decided to organize a contest on this topic.

ONFIRE 2023 is an international competition among methods, executable on board of smart cameras or embedded systems, for real-time fire detection from videos acquired by fixed CCTV cameras. To evaluate their real-time processing capability on board of devices with limited resources, the performance of the competing methods will be evaluated in terms of fire detection accuracy and processing resources. As for the former, we consider both the detection errors (using Precision and Recall) and the notification speed (i.e. the delay between the manually labelled fire start, either its ignition or appearance on scene, and the fire notification). Regarding the latter, the processing frame rate and the memory usage are taken into account. In this way, we evaluate not only the ability to detect fires and avoid false alarms of the proposed approaches, but also their promptness in notification and the computational resources needed for real-time processing.

To allow the participants to train their methods, we provide a dataset including 322 videos collected from publicly available fire detection datasets [8,13,19]; all the positive video clips are annotated with the instant in which the fire begins. The accuracy of the competing methods is evaluated on a private test set composed by unpublished videos that are different from the ones available in the training set.

Therefore, by providing a standard dataset and a well-defined experimental protocol that measures not only the accuracy but also the computational resources required by the methods, we propose precise guidelines for the evaluation of fire detection approaches. In addition, being the test set not known, we evaluate the generalization capability of the competing approaches in a realistic way, as if these were installed in a real application in an unknown scenario.

2 Related Works

Several fire detection methods have been proposed in recent years, demonstrating a great interest for this application in the scientific community [14]. The processing pipeline is typically based on two steps: the localization of the fire and its recognition [15]. The former is not a mandatory step but allows to simplify the subsequent recognition phase [31]; however, some approaches address the problem through image classification [22]. The latter may be applied on a single frame [5] or on a temporal sequence of images [10].

Obviously as for the localization step, the resolution plays a fundamental role; on one hand, a higher resolution allows to detect a more distant fire; on the other hand, it requires more computational capabilities. For this reason, it is often necessary to find a good trade-off and to reduce the resolution. The input size of existing methods varies from 48×48 [16,33] until over 600×600 pixels [17,31]. Clearly, using a lower resolution implies to have less pixels available for localization and recognition. This limitation can substantially penalize the system, especially when it has to identify fire or smoke at a considerable distance. To preserve the loss of useful details, various researchers proposed methods that identify candidate fires at high resolution and subsequently classify the candidate patch [4,23]. Some approaches for the localization of the fire are based on background subtraction, which analyzes a sequence of frames to detect moving objects in the scene such as flames or smoke [4,11,13]. However, various works in the literature avoid this localization step by carrying out only pre-processing operations [25,27,28] or using attention mechanisms [5,20,22,26].

For the recognition step, there is a wide range of available techniques. Various methods for smoke detection exploit the appearance and movement [24], while others focus on features such as color and texture [11]. More recently, several CNN models are adopted in the approaches. Clearly, such models differ in the computational resources they require [5,22,23]. To overcome the problem of resource consumption, lightweight CNNs have been proposed [1,20,35]. Among them, there are algorithms that simultaneously locate and classify the object of interest [6,21,34]. In the literature, we can also find methods based on temporal analysis of a sequence of frames, that are able to analyze the temporal evolution of flame and/or smoke [7,11,32]. These approaches are based on modern 3D-CNNs [3,18,31] or on recurrent neural networks (RNNs) [9,29,30].

3 Contest Dataset and Task

The interest of the scientific community has not only been focused on the design of innovative methodologies, but also on the creation of datasets suitable for these purposes. In particular, we are interested in video clips acquired with static CCTV cameras, available in the following datasets. MIVIA Fire [13] contains 31 videos with resolutions ranging from 320×240 to 800×600. Most of the videos are recorded outdoor and flames are present in 15 videos, while smoke is present in 28 of these. MIVIA Smoke [13] includes 76 smoke videos out of 149 total videos. The videos are all recorded outdoor with a low activity environment and a long distance from the camera. D-Fire [8] contains 100 videos with perfect balance between positive and negative samples. The videos have the same features with minimal variability and are all recorded outdoor and the positive videos contain smoke. KMU dataset [19] includes 38 low resolution videos, of which 10 are recorded indoor and 28 outdoor. 14 videos resume fire at a short distance, while 24 at a long distance.

For the training set of the contest, we collected video clips from all the above mentioned datasets. It consists of 322 videos, but a fire (flames and/or smoke) is present in 218 videos (159 contain smoke only, 33 fire only and 26 both), while the others (negative) do not contain a fire. The video clips last between a few seconds and 3 min. More details are reported in Fig. 1.

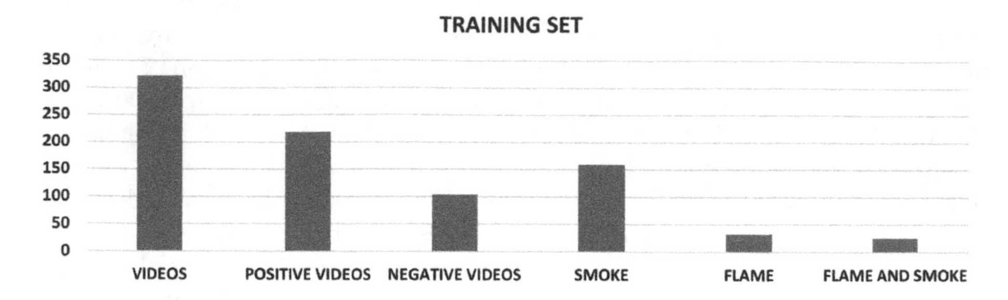

Fig. 1. Statistics of the training set given to ONFIRE 2023 participants.

We carefully selected as positive videos only those that really frame a fire and not flames and smoke in controlled conditions, and as negative videos the ones that contain moving objects that can be confused with flames or smoke. The contest participants receive a training set, which contains all the videos and the annotations regarding the ignition start time. The participants can extract their own validation set from the training set. Examples of images extracted from the training videos are depicted in Fig. 2.

Since the videos are collected in different conditions, the dataset is very heterogeneous in terms of image resolution, illumination, pixel size of flame or smoke, background activity, scenario (urban or wildfire). It will likely be, to the best of our knowledge, the largest and most miscellaneous fire detection video dataset ever made available; in addition, the annotation of the fire ignition time is something that is typically not available in online public datasets.

The test set is private and is not provided to the participants, so as to perform a realistic evaluation of the generalization capability of the methods. The submitted methods will be evaluated on all the test videos and the ranking of the contest will be defined according to the criteria described in the next section.

4 Evaluation Metrics

The fire detection accuracy of the competing methods will be evaluated in terms of Precision and Recall. To formalize these metrics, it is necessary to define the sets of true positives (TP), false positives (FP) and false negatives (FN). Our test set contains both positive and negative samples, i.e. video depicting fire events or not, respectively. Each positive video shows one fire event only. Each prediction is evaluated by comparing the fire start instant g with the detection one p. We indicate as the fire start instant the timestamp of the first frame in which it is visible. It is manually labelled by the human operator after visioning the whole video, thus it is a ground truth. Notwithstanding this, we consider that automatic methods could anticipate human detection by a few seconds.

Fig. 2. Examples of images captured from the training videos available for the contest. The dataset includes samples containing flames and smoke framed at various distances from the camera and with variable background human activity.

Therefore, we define:

- TP: all the detections occurring in positive videos at $p \geq \max(0, g - \Delta t)$;
- FP: all the detections occurring at any time in negative videos or in positive videos at $p < \max(0, g - \Delta t)$;
- FN: the set of positive videos for which no fire detection occurs.

with Δt equal to 5 s.

Defined these sets for TP, FP and FN, we can compute the Precision (P) and Recall (R) with respect to the number of true positives ($|TP|$), false positives ($|FP|$) and false negatives ($|FN|$):

$$P = \frac{|TP|}{|TP| + |FP|} \tag{1}$$

$$R = \frac{|TP|}{|TP| + |FN|} \tag{2}$$

Precision assumes values in the range [0,1] and measures the capability of the methods to reject false positives; the higher is P, the higher the specificity of the method. Recall assumes values in the range [0,1] and evaluates the sensitivity of the method to detect fire; the higher is R, the higher the sensitivity of the method.

The fire notification time of a method for the i-th true positive is p_i, while the fire start time is g_i. Therefore, the notification delay d_i on the i-th sample is:

$$d_i = |p_i - g_i| \tag{3}$$

The average notification delay D is defined as the ratio between the sum of all the notification delays d_i and the total number of true positives $|TP|$:

$$D = \frac{\sum_i^{|TP|} d_i}{|TP|} \tag{4}$$

The smaller is D, the faster the notification of a fire. To obtain a value in the range [0,1], we can compute the normalized average notification delay D_n as follows:

$$D_n = \frac{max(0; 60 - D)}{60} \tag{5}$$

Then, we compute the processing frame rate (PFR), namely the average number of frames processed by the method in one second (on a target GPU that we will adopt for our experiments). In particular, being N the total number of frames processed in the test set and t_i the processing time in seconds for a single frame, the PFR is computed as follows:

$$PFR = \frac{1}{\frac{\sum_i^N t_i}{N}} \tag{6}$$

The higher is PFR, higher is the processing speed of the method. To normalize this value with respect to the minimum processing frame rate needed to achieve real-time performance, namely PFR_{target}, we compute the PFR_{delta} score as follows:

$$PFR_{delta} = max(0; \frac{PFR_{target}}{PFR} - 1) \tag{7}$$

Finally, we measure the memory usage (MEM), namely the memory in GB occupied by the method on the target GPU. The lower is MEM, lower is the necessary memory on the processing device. To normalize this value with respect to the maximum GPU memory available on the target processing device, namely MEM_{target}, we compute the MEM_{delta} score as follows:

$$MEM_{delta} = max(0; \frac{MEM}{MEM_{target}} - 1) \tag{8}$$

We define the ranking of the contest according to Fire Detection Score (FDS), computed as follows:

$$FDS = \frac{P \cdot R \cdot D_n}{(1 + PFR_{delta}) \cdot (1 + MEM_{delta})} \tag{9}$$

The method which achieves the highest FDS will be the winner of the ONFIRE Contest 2023, since it will demonstrate the best trade-off between fire detection accuracy, notification promptness and required processing resources.

5 Conclusion

ONFIRE Contest 2023 represents an opportunity to train and test methods for fire detection designed to process video sequences in real-time on edge devices. The dataset made available for the contest consists of hundreds of videos containing frames in which flames or smoke are visible. These videos have been collected indoor or outdoor, wildfire or urban scenarios, with variations in terms background activity. Fires are framed at a short and long distance at different resolutions, sometimes with occlusions. The choice of videos that are collected in heterogeneous scenarios allows the participants to train and validate methods that may be robust in different operative conditions. The submitted methods are evaluated not only in terms of accuracy and notification promptness, but also for their processing speed and required computational resources. The experimental protocol proposed for the contest is a novel contribution, since it is the first time, to the best of our knowledge, that the evaluation of fire detection methods takes also into account the efficiency of the approaches. We are confident that this contest can foster the development of effective and efficient fire detection approaches able to quickly and reliably notify the presence of fires.

References

1. Ayala, A., Fernandes, B., Cruz, F., Macêdo, D., Oliveira, A.L., Zanchettin, C.: KutralNet: a portable deep learning model for fire recognition. In: 2020 International Joint Conference on Neural Networks (IJCNN), pp. 1–8. IEEE (2020)
2. Bu, F., Gharajeh, M.S.: Intelligent and vision-based fire detection systems: a survey. Image Vis. Comput. **91**, 103803 (2019)
3. Cao, Y., Tang, Q., Lu, X.: STCNet: spatiotemporal cross network for industrial smoke detection. Multimedia Tools Appl. **81**(7), 10261–10277 (2022)
4. Cao, Y., Tang, Q., Wu, X., Lu, X.: EFFNet: enhanced feature foreground network for video smoke source prediction and detection. IEEE Trans. Circ. Syst. Video Technol. **32**, 1820–1833 (2021)
5. Cao, Y., Tang, Q., Xu, S., Li, F., Lu, X.: QuasiVSD: efficient dual-frame smoke detection. Neural Comput. Appl. **34**(11), 8539–8550 (2022)
6. Chaoxia, C., Shang, W., Zhang, F.: Information-guided flame detection based on faster R-CNN. IEEE Access **8**, 58923–58932 (2020)
7. Chen, X., An, Q., Yu, K., Ban, Y.: A novel fire identification algorithm based on improved color segmentation and enhanced feature data. IEEE Trans. Instrum. Meas. **70**, 1–15 (2021)

8. De Venâncio, P.V.A., Rezende, T.M., Lisboa, A.C., Barbosa, A.V.: Fire detection based on a two-dimensional convolutional neural network and temporal analysis. In: 2021 IEEE Latin American Conference on Computational Intelligence (LA-CCI), pp. 1–6. IEEE (2021)
9. Dewangan, A., et al.: FigLib & SmokeyNet: dataset and deep learning model for real-time wildland fire smoke detection. Remote Sensing **14**(4), 1007 (2022)
10. Di Lascio, R., Greco, A., Saggese, A., Vento, M.: Improving fire detection reliability by a combination of videoanalytics. In: Campilho, A., Kamel, M. (eds.) ICIAR 2014. LNCS, vol. 8814, pp. 477–484. Springer, Cham (2014). https://doi.org/10.1007/978-3-319-11758-4_52
11. Dimitropoulos, K., Barmpoutis, P., Grammalidis, N.: Spatio-temporal flame modeling and dynamic texture analysis for automatic video-based fire detection. IEEE Trans. Circuits Syst. Video Technol. **25**(2), 339–351 (2014)
12. Foggia, P., Greco, A., Roberto, A., Saggese, A., Vento, M.: A social robot architecture for personalized real-time human-robot interaction. IEEE Internet Things J. **10**, 1 (2023)
13. Foggia, P., Saggese, A., Vento, M.: Real-time fire detection for video-surveillance applications using a combination of experts based on color, shape, and motion. IEEE Trans. Circuits Syst. Video Technol. **25**(9), 1545–1556 (2015)
14. Gaur, A., Singh, A., Kumar, A., Kumar, A., Kapoor, K.: Video flame and smoke based fire detection algorithms: a literature review. Fire Technol. **56**(5), 1943–1980 (2020)
15. Geetha, S., Abhishek, C., Akshayanat, C.: Machine vision based fire detection techniques: a survey. Fire Technol. **57**(2), 591–623 (2021)
16. Gu, K., Xia, Z., Qiao, J., Lin, W.: Deep dual-channel neural network for image-based smoke detection. IEEE Trans. Multimedia **22**(2), 311–323 (2019)
17. Huang, J., He, Z., Guan, Y., Zhang, H.: Real-time forest fire detection by ensemble lightweight YOLOX-L and defogging method. Sensors **23**(4), 1894 (2023)
18. Huo, Y., Zhang, Q., Zhang, Y., Zhu, J., Wang, J.: 3DVSD: an end-to-end 3D convolutional object detection network for video smoke detection. Fire Saf. J. **134**, 103690 (2022)
19. Ko, B.C., Ham, S.J., Nam, J.Y.: Modeling and formalization of fuzzy finite automata for detection of irregular fire flames. IEEE Trans. Circ. Syst. Video Technol. **21**(12), 1903–1912 (2011). https://cvpr.kmu.ac.kr/
20. Li, S., Yan, Q., Liu, P.: An efficient fire detection method based on multiscale feature extraction, implicit deep supervision and channel attention mechanism. IEEE Trans. Image Process. **29**, 8467–8475 (2020)
21. Li, Z., Mihaylova, L., Yang, L.: A deep learning framework for autonomous flame detection. Neurocomputing **448**, 205–216 (2021)
22. Majid, S., Alenezi, F., Masood, S., Ahmad, M., Gündüz, E.S., Polat, K.: Attention based CNN model for fire detection and localization in real-world images. Expert Syst. Appl. **189**, 116114 (2022)
23. Nguyen, M.D., Vu, H.N., Pham, D.C., Choi, B., Ro, S.: Multistage real-time fire detection using convolutional neural networks and long short-term memory networks. IEEE Access **9**, 146667–146679 (2021)
24. Prema, C.E., Suresh, S., Krishnan, M.N., Leema, N.: A novel efficient video smoke detection algorithm using co-occurrence of local binary pattern variants. Fire Technol. **58**(5), 3139–3165 (2022)
25. Pundir, A.S., Raman, B.: Dual deep learning model for image based smoke detection. Fire Technol. **55**(6), 2419–2442 (2019)

26. Shahid, M., Hua, K.l.: Fire detection using transformer network. In: Proceedings of the 2021 International Conference on Multimedia Retrieval, pp. 627–630 (2021)
27. Sheng, D., Deng, J., Xiang, J.: Automatic smoke detection based on SLIC-DBSCAN enhanced convolutional neural network. IEEE Access **9**, 63933–63942 (2021)
28. Shi, J., Wang, W., Gao, Y., Yu, N.: Optimal placement and intelligent smoke detection algorithm for wildfire-monitoring cameras. IEEE Access **8**, 72326–72339 (2020)
29. Tao, H., Lu, M., Hu, Z., Xin, Z., Wang, J.: Attention-aggregated attribute-aware network with redundancy reduction convolution for video-based industrial smoke emission recognition. IEEE Trans. Industr. Inf. **18**(11), 7653–7664 (2022)
30. Tao, H., Xie, C., Wang, J., Xin, Z.: CENet: a channel-enhanced spatiotemporal network with sufficient supervision information for recognizing industrial smoke emissions. IEEE Internet Things J. **9**(19), 18749–18759 (2022)
31. de Venâncio, P.V.A., Campos, R.J., Rezende, T.M., Lisboa, A.C., Barbosa, A.V.: A hybrid method for fire detection based on spatial and temporal patterns. Neural Comput. Appl. **35**(13), 9349–9361 (2023)
32. Xie, Y., Zhu, J., Guo, Y., You, J., Feng, D., Cao, Y.: Early indoor occluded fire detection based on firelight reflection characteristics. Fire Saf. J. **128**, 103542 (2022)
33. Yuan, F., Zhang, L., Wan, B., Xia, X., Shi, J.: Convolutional neural networks based on multi-scale additive merging layers for visual smoke recognition. Mach. Vis. Appl. **30**, 345–358 (2019)
34. Zeng, J., Lin, Z., Qi, C., Zhao, X., Wang, F.: An improved object detection method based on deep convolution neural network for smoke detection. In: 2018 International Conference on Machine Learning and Cybernetics (ICMLC), vol. 1, pp. 184–189. IEEE (2018)
35. Zhang, J., Zhu, H., Wang, P., Ling, X.: ATT squeeze U-Net: a lightweight network for forest fire detection and recognition. IEEE Access **9**, 10858–10870 (2021)

FIRESTART: Fire Ignition Recognition with Enhanced Smoothing Techniques and Real-Time Tracking

Luca Zedda, Andrea Loddo(✉), and Cecilia Di Ruberto

Department of Mathematics and Computer Science,
University of Cagliari, Cagliari, Italy
{luca.zedda,andrea.loddo,cecilia.dir}@unica.it

Abstract. Fires can potentially cause significant harm to both people and the environment. Recently, there has been a growing interest in real-time fire and smoke detection to provide practical assistance. Detecting fires in outdoor areas is crucial to safeguard human lives and the environment. This is especially important in situations where more than traditional smoke detectors may be required. In this work, we propose FIRESTART, which aims to achieve accurate and robust ignition detection for prompt identification and response to fire incidents. The proposed framework utilizes a lightweight deep learning architecture and post-processing techniques for fire-starting interval detection. Its evaluation was conducted on the ONFIRE dataset, comparing it with several state-of-the-art methods. The results are encouraging, particularly from computational and real-time use perspectives.

Keywords: Deep Learning · Computer Vision · Image Processing · Vision Transformers · Fire Detection

1 Introduction

Fires can be incredibly destructive, posing a risk to people and the environment. Detecting fires in outdoor areas is crucial for ensuring safety. In recent years, there have been many large-scale wildfires and forest fires across the globe, such as 2017, 2018, and 2020 California fire seasons, the Australian bushfires that began in 2019 and continued until March 2020, and the 2021 Planargia-Montiferru mega-fire. These fires have killed billions of animals and numerous people [5].

In recent years, fire detection systems have received increased attention due to their vital role in protecting people and property from fire hazards. These sensor detection systems can detect different fire aspects, including light, heat, and smoke [13,22].

Multiple fire detection systems have been created to reduce the harm caused by fires, each utilizing distinct technologies. Conventional fire detection techniques involve inexpensive and user-friendly sensors that can sense smoke, fire size, initial flame location, and atmospheric temperature [4,13,19,20,22].

G. L. Foresti et al. (Eds.): ICIAP 2023 Workshops, LNCS 14365, pp. 282–293, 2024.
https://doi.org/10.1007/978-3-031-51023-6_24

Although smoke detectors are commonly used for fire safety, they may have limitations such as delayed triggering of alarms, limited coverage area, and signal transmission issues. These detectors can also produce false alarms, which can be especially problematic in small areas. As a result, traditional smoke detectors may not be adequate in large open spaces like stadiums, public places, and aircraft hangers. In addition, many sensors require proximity to the source of the fire or smoke and are typically installed on the ceiling, which can lead to delays in detecting smoke. This can defeat the purpose of early warning [16].

Using cameras in fire and smoke detection systems is more effective than relying on point sensors. Cameras can quickly scan large areas and connect to local processing devices that use automatic detection algorithms. As a result, they are a valuable tool for detecting fire and smoke [19], gaining the Computer Vision (CV) community's attention. In this context, Convolutional Neural Networks (CNNs) have been demonstrated to be outstanding in image classification and have been applied to detect fires and smoke in images and videos [7].

In general, fire detection systems can be divided into four categories based on how they obtain and use features in their decision-making process. These categories include (i) systems that use handcrafted (HC) features [17], (ii) systems based on deep features, such as those extracted from CNNs or long short-term memory networks (LSTM) [9], (iii) hybrid systems that use both deep and HC features [19,21] and (iv) end-to-end deep learning-based systems [1,4,13,20,22].

Researchers previously relied on image descriptors to identify and analyze fire attributes such as color, shape, and motion. However, these methods have proven to be unreliable in complex scenarios. For example, color-based descriptors are affected by light reflections and shadows. In contrast, motion-based descriptors may not be able to accurately differentiate between fire areas and other moving objects that resemble fire [4,17].

One way to improve results is by utilizing deep features. This can be done using off-the-shelf pre-trained models, a common practice adopted to avoid training complex networks from scratch. An alternative is the employment of ad hoc networks. For instance, Liu et al.developed a multi-sensor fire detection method that employs LSTM to minimize false alarms and enhance accuracy [9].

With the improvement of processing capacity and the generation of large amounts of data in the big data era, the computer vision community has focused on improving the reliability of previous fire detection methods. In particular, they have utilized CNNs to achieve this goal [1,4,13,19–22].

Several studies have introduced new CNNs that are either lightweight and suitable for mobile and embedded applications with favorable performance for real-time fire detection application [1]. Other kinds of CNNs were also proposed. For example, Yin et al.used standard CNN for smoke detection [22], while Majid et al.utilized attention-based CNNs for fire detection [13]. A deep learning-based object detection system was proposed in [20], while Daoud et al.designed ad hoc CNN to detect fire using real-world images [4].

In this context, hybrid methods were proposed as well. For example, in [21], a novel CNN was proposed with the integration of HC features, while a hybrid method composed of a CNN followed by a temporal analysis technique was proposed to reduce the false positive rates of single-stage CNNs in [19].

Although significant progress has been made in fire detection methods, there are still challenges to be addressed. One major obstacle in developing deep learning methods for fire, flame, and smoke detection using convolutional neural networks is the need for more curated datasets for training and testing. Most research has relied on publicly available images and videos, which require substantial resources for annotation. Moreover, detecting multiple instances of smoke and fire flames simultaneously is complex due to the presence of multiple spots with significant variability in the image scene. The lack of standardized datasets also makes it difficult to determine which detection method is superior or should be used. Customized datasets may result in more accurate fire detection for one method but less accurate smoke detection, and vice versa. Despite these challenges, efforts are ongoing to reduce false positives and improve accuracy in fire detection systems [7].

Real-time detection of fires is a significant challenge in video surveillance applications, especially from image sequences. This feature is highly desirable as it can prevent environmental disasters, continuously monitor urban environments, and protect forests. Intelligent cameras equipped with onboard video analytics algorithms capable of detecting fires (flames and smoke) in real time are being deployed across remote areas. These cameras are self-sufficient from a computational perspective and require a small embedded system to process the image sequence using the fire detection algorithm. It is important to find a balance between accurate fire detection, timely notifications, and efficient use of processing resources, as methods requiring extensive processing resources are impractical for this application.

In general, the detection of fires in outdoor environments is a significant concern for ensuring the safety of human life. Fires pose a severe hazard to industries, crowded events, and highly populated areas worldwide as fire incidents have the potential to cause harm to property, the environment, and living beings. The impact of such incidents can be severe and wide-ranging, affecting human and animal life, financial infrastructure, and the environment. However, immediate action can help minimize the damage caused. To that end, a vision-based automated system can be beneficial in detecting fires.

This work proposes a deep learning-based fire detection framework named FIRESTART, based on a lightweight deep learning architecture followed by a post-processing technique to improve accuracy in fire-starting interval detection. Our primary goal is to achieve accurate and robust ignition detection, which can help promptly identify and respond to such incidents.

The main contributions of our work are listed as follows:

1. we have designed a new deep learning (DL)-based framework for real-time fire detection from images and videos;
2. we have thoroughly evaluated our approach using the ONFIRE dataset, released for the related competition;

3. we have provided a comprehensive comparison between several state of the art deep learning-based architectures on the same dataset.

The rest of the manuscript is organized as follows. Section 2 illustrates the dataset, the main concepts of object detection and attention mechanism techniques, along with our approach and the evaluation metrics. In Sect. 3, the experimental results are given and discussed. Finally, in Sect. 4, we draw the findings and directions for future works.

2 Materials and Methods

In this section, in Sect. 2.1, we provide information about the public dataset used in this study. Then, we summarize some basic concepts about CNNs and ViTs in Sect. 2.2 and 2.3 respectively. Our proposed approach is presented in Sect. 2.4, and finally, we describe the chosen evaluation metrics Sect. 2.5.

2.1 Dataset

We utilized the official dataset from the ONFIRE 2023 contest [8] for the scope of this work. This dataset contains 329 videos of varying sizes, both spatially and temporally. Of these videos, 104 depict real-life events with no fire incidents, while 219 showcase a diverse range of fire events. Each video has a label indicating whether it depicts a non-fire event labeled as *Normal* or a fire event labeled as *Fire*, *Smoke*, or *Both*. Examples of these different events can be seen in Fig. 1. Also, the contest organizers provided the starting second for the ignition event.

To train our classification models, we extracted video frames at intervals of 10 frames, labeling them as "Normal" if the video contains no fire event or if the current frame's timestamp is earlier than the fire event's starting timestamp. Otherwise, we label the frames according to the video's specified label.

(a) Normal example video frame

(b) Fire example video frame

(c) Smoke example video frame

(d) Smoke and Fire example video frame

Fig. 1. Samples from the challenge dataset for each provided class

2.2 Convolutional Neural Networks

The field of computer vision has been transformed by CNNs, which have shown remarkable performance in tasks like object detection and image classification. CNNs are particularly adept at learning spatial features and handling large datasets of high-dimensional inputs.

CNNs usually consist of multiple layers, including convolutional, pooling, and fully connected layers. In the convolutional layers, the network utilizes filters (known as kernels) to process the input image and create feature maps that emphasize different aspects of the image. Pooling layers then decrease the spatial dimensions of the feature maps, increasing efficiency and decreasing the risk of overfitting. Finally, the fully connected layers classify the input image based on the features learned in the previous layers.

One of the most significant benefits of CNNs is their ability to learn hierarchical representations of images. The lower layers concentrate on simple features like edges and corners, while the higher layers focus on more complex features like object parts and textures.

CNNs are used in various computer vision tasks, including image classification, object detection, and semantic segmentation.

2.3 Vision Transformers

Vision Transformers (ViTs) have recently been getting a lot of attention from the CV community. They are a type of neural network that uses the transformer model, initially developed for natural language processing (NLP). ViTs have shown promising results in tasks involving image recognition [6].

One of the critical features of ViT is its use of self-attention mechanisms, which help to capture interactions between different regions of an image. Attention mechanisms have become popular in NLP and CV, thanks partly to the transformer architecture and multi-head self-attention (MHSA) [18]. These mechanisms allow models to focus on important information while ignoring irrelevant data. In detail, the MHSA is defined in Equation (1):

$$\text{Attention}(Q, K, V) = \text{softmax}\left(\frac{QK^T}{\sqrt{d_k}}\right)V \tag{1}$$

where Q, K, and V are the query, key, and value vectors, respectively, and d_k is the dimension of the key vectors. The softmax function is applied along the rows of the matrix $\frac{QK^T}{\sqrt{d_k}}$ to compute the attention weights.

To achieve multi-head attention, the input vectors are divided into h groups, and the attention function is applied independently to each group (see Equation (4)):

$$\text{MultiHead}(Q, K, V) = \text{Concat}(head_1, head_2, \ldots, head_h)W^O \tag{2}$$

$$\text{where} \quad head_i = \text{Attention}(QW_i^Q, KW_i^K, VW_i^V) \tag{3}$$

$$\text{and} \quad W^O = \text{learned projection matrix} \tag{4}$$

Table 1. Architectures employed in this work, with their sizes and number of parameters in millions.

Architecture name	Size	Params (M)
ConvNextv2 [12]	Tiny	39
Dino [3]	Small	21
MobileViTv2 [14]	Small	17
Swinv2 [10]	Tiny	27
ViT [6]	Base	86

In this formulation, $head_i$ represents the attention output for the i-th head and W^O is a learned projection matrix that maps the concatenated outputs of the different heads to the output dimension.

The self-attention mechanism enables the network to weigh the importance of different information in a sequence based on its relevance to other parts. In the context of ViT, self-attention is utilized to model the relationships between image patches, treating them as a sequence.

Compared to traditional convolutional neural networks, which rely on spatial convolutions for feature extraction, ViT proves more effective in capturing long-range dependencies and representing global image features. Moreover, ViT has been extended to other computer vision tasks, such as object detection and semantic segmentation [15], showcasing promising outcomes.

One limitation of ViT is its computational cost when processing large images with numerous patches. To mitigate this issue, several techniques have been proposed to reduce the number of patches required, including overlapping patches or hierarchical patch representations [11].

2.4 The Proposed Approach

Our method, named FIRESTART, consists of two phases. First, we use a deep learning architecture to assign a classification label and identify a possible time frame for a fire to start. The second phase involves a post-processing technique that smooths out the detected fire-starting interval, thus improving accuracy. Our primary goal is to achieve accurate and robust ignition detection.

To achieve our objective and implement the first phase of our approach, we tested and compared various off-the-shelf transformers and convolutional-based architectures as trainable models. The complete list is provided in Table 1. When selecting the models, we considered their scope and the number of trainable parameters to ensure their suitability for real-time applications.

In particular, the models were trained to classify four different fire event scenarios: *Smoke*, *Fire*, *Smoke and Fire*, and *No Fire*. Our method centers on predicting class scores for each frame while monitoring previous predictions. We utilize a moving average to smooth out the predicted class scores and distinguish

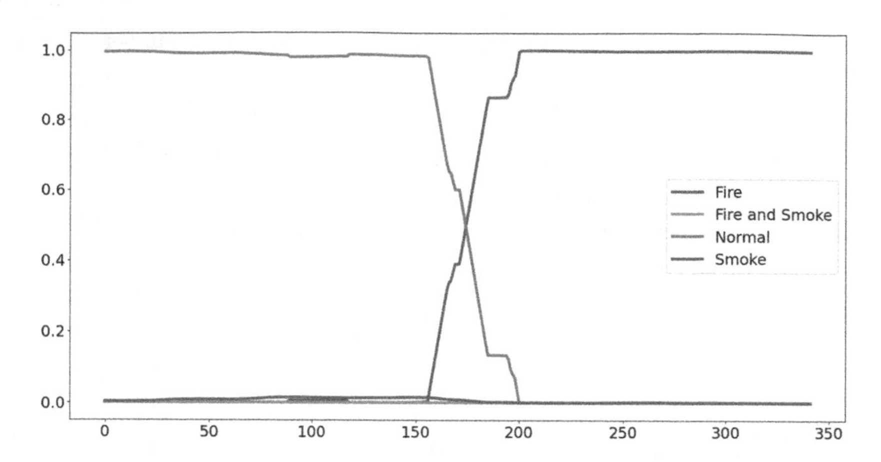

Fig. 2. Visualization of the produced model output after probabilities smoothing

the accurate fire event starting frame from outliers. The size of the moving average filter is determined by the equation stated in Equation (5), which specifies the range of the moving average. Here, **FPS** represents the input video frames per second, **seconds** represents the total number of seconds of the input video, and **k** is a scalar value. In the final predictions, **k** is set to *0.08*.

$$\text{range size} = \text{FPS} \times \text{seconds} \times \text{k} \tag{5}$$

To determine when a fire starts, we choose the second at which the likelihood of a fire occurring becomes greater than that of a non-fire event. If the video shows a fire at the beginning, we consider the starting second to be zero. You can find an example of this process in Fig. 2.

2.5 Metrics

To evaluate the results obtained by the proposed method, we employed a subset of the evaluation metrics of the ONFIRE 2023 contest. The performance of the competing methods is evaluated in terms of Precision (P), Recall (R), and Average notification delay (D), and the sets of true positives (TP), false positives (FP) and false negatives (FN) are defined based on the comparison between the fire start instant g and the detection instant p. TP is defined as all the detections occurring in positive videos at $p \geq \max(0, g - \Delta t)$, FP as all the detections occurring at any time in negative videos or in positive videos at $p < \max(0, g - \Delta t)$, and FN as the set of positive videos for which no fire detection occurs, with Δt equal to 5 s.

- **Precision** refers to the model's ability to accurately identify fire event starting seconds, defined in Equation (6):

$$P = \frac{TP}{TP + FP} \tag{6}$$

- **Recall** measures the model's ability to identify all relevant cases, among all ground-truth fire event starting seconds, as defined in Equation (7):

$$R = \frac{TP}{TP + FN} \tag{7}$$

- **Average notification delay (D)** is defined as the ratio of the sum of all notification delays d_i over the total number of true positive detections, as defined in Equation (9) where p_i is the predicted starting second and g_i is the ground truth

$$d_i = |p_i - g_i| \tag{8}$$

$$D = \frac{\sum_{i=1}^{n} d_i}{n} \tag{9}$$

- **Fire Detection Score (FDS)** is the contest ranking metric, representing the best trade-off between fire detection accuracy, notification promptness and required processing resources (see Equation (10)).

$$FDS = \frac{P \cdot R \cdot D_n}{(1 + PFR_{delta}) \cdot (1 + MEM_{delta})} \tag{10}$$

where:

$$D_n = \frac{max(0; 60 - D)}{60} \tag{11}$$

$$PFR_{delta} = max(0; \frac{PFR_{target}}{PFR} - 1) \tag{12}$$

$$MEM_{delta} = max(0; \frac{MEM}{MEM_{target}} - 1) \tag{13}$$

$$PFR = \frac{1}{\frac{\sum_{i=1}^{N} t_i}{N}} \tag{14}$$

MEM is the memory usage in GB of the method on the target GPU.

3 Experimental Results and Discussion

3.1 Experimental Setup

The experiments were conducted on a workstation equipped with the following hardware specifications: an *Intel(R) Core(TM) i9-8950HK* CPU operating at *2.90 GHz*, *32 GB* RAM, and an *NVIDIA GTX1050 Ti GPU* with *4 GB* memory. During the contest's final evaluation procedure on the private test set, the organizers set MEM_{target} to 4 and PFR_{target} to 10.

Training Details: We utilized pre-trained models from the ImageNet dataset and employed the *AdamW* optimizer with a learning rate of 0.001, momentum of 0.9, and weight decay of 0.0001. Each model underwent 30 epochs, with a batch size of 32, for every sampled frame class. The official training set was divided into three parts: training (70%), validation (10%), and testing (20%).

Table 2. Image augmentation parameters adopted for models training.

Augmentation	Parameters	Probability
HorizontalFlip	-	0.5
VerticalFlip	-	0.5
RandomRotate90	[90, 180, 270] degrees	0.5
RandomResizedCrop	[0.5, 1.0] of original size	0.5

Data Augmentation: we employed a preliminary step of data augmentation to overcome the dataset size limitations, address data imbalance, strengthen models against object rotations, enhance diversity, and enable targeted generalization capabilities. We utilized the *Albumentation* library [2], which provides a wide range of transformations. We implemented an oversampling strategy in two steps to address the class imbalance in the training set. Firstly, we repeated the underrepresented samples until an equal number of samples were obtained for each class. Secondly, we augmented the repeated samples online using various geometric transformations while preserving important visual features to enhance the diversity of the images' representations (see Table 2).

3.2 Experimental Results

Quantitative Results. Table 3 summarizes the results of our experiments. Specifically, we present the performance obtained by the selected architectures over the test set. Based on the experimental results, the Average notification delay has a considerably low distribution across various architectures. MobileViTv2 and ViT, both transformer-based architectures, have a distribution of 0. MobileViTv2 provides the best Precision and FPS metrics, with a score of 93.2% and 25, respectively. Despite having a larger input image size of 256, MobileViTv2 has an impressively high FPS compared to other architectures with an input size of 224. This is due to its specific formulation for real-time applications, particularly on mobile devices.

The ConvNextv2 architecture has the best Recall results and the second-best Precision metric. However, it has a more significant Average notification delay and almost half of the FPS compared to MobileViTv2.

The Swinv2 and ViT architectures have similar results, with the only notable difference being their FPS. This difference is linked to the number of trainable parameters, with ViT having the most significant number among the selected architectures, resulting in the lowest value among all models.

Based on the obtained results, our proposed approach uses MobileViTv2 as the founding model. The results on the private test set are reported in Table 4 and are encouraging for the adoption of transformer-based architectures on the task at hand because D_n reached an outstanding value of 0.96, showing that it can detect a fire almost immediately. Moreover, it must be noted that the memory usage of the proposed approach was $< 4GB$, and the elaboration frame rate was

Table 3. Experimental results obtained on the testing set. The reported performance metrics, calculated using the macro average, include Precision, Recall, Average notification delay, and number of processing FPS. Additionally, we provided information on the backbone used and the input image size for each model tested. The best results are highlighted in bold.

Architecture	Size	Image size	$P\uparrow$	$R\uparrow$	$D\downarrow$	$D_n\uparrow$	$FPS\uparrow$
ConvNextv2	Tiny	224	0.927	**0.932**	0.078	0.998	18
Dino	Small	224	0.909	0.902	0.1	0.998	19
MobileViTv2	Small	256	**0.932**	0.921	**0**	**1**	**25**
Swinv2	Tiny	256	0.891	0.881	0.082	0.998	20
ViT	Base	224	0.891	0.881	**0**	**1**	13

Table 4. Experimental results obtained on the private testing set. The reported performance metrics are Precision, Recall, D, and FDS.

Approach	Architecture	$P\uparrow$	$R\uparrow$	$D_n\uparrow$	$FDS\uparrow$
FIRESTART	MobileViTv2	0.315	0.531	0.960	0.161

$>10FPS$, enabling FIRESTART to be lightweight and suitable for real-time usage. However, the proposed approach seems to be quite sensitive as the low precision value suggests the identification of many false positives. Additionally, the recall is not so high because some positive videos trigger the fire notification more than 5 s before the actual fire ignition, which could be due to false positives. According to contest regulations, these errors are not considered true positives.

Qualitative Results. The visual results of our study are presented in Fig. 3. We show the performances of the trained MobileViTv2 architecture to recognize a fire event. Figure 3 depicts three different frames of a positive video where the actual fire event begins. We also show the predictions of the model frame by frame, demonstrating its strong capabilities even in low-resolution frames.

4 Conclusions

Our main goal was to find a solution to the challenge of simultaneously detecting the nature of the fire event and determining the exact second where the event begins in a video. We implemented several off-the-shelf deep learning architectures, which have proved to be highly effective, providing an excellent trade-off between processing speed and performance. Our approach helps to address the issue related to spontaneous and artificial fire events, which could result in the loss of resources and, most importantly, human lives. Despite the results on the private test set, FIRESTART had low memory usage and acceptable frame-rate elaboration, enabling it to be lightweight and suitable for real-time use.

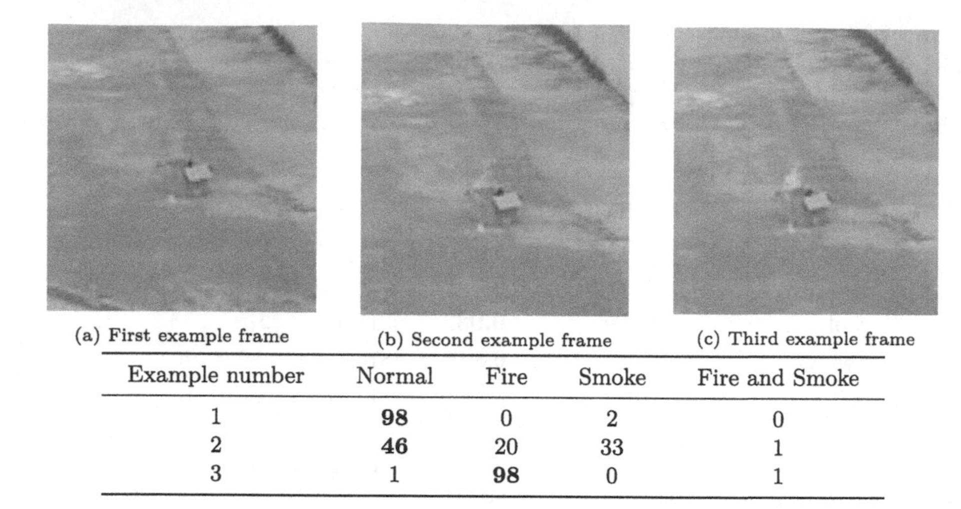

(a) First example frame (b) Second example frame (c) Third example frame

Example number	Normal	Fire	Smoke	Fire and Smoke
1	**98**	0	2	0
2	**46**	20	33	1
3	1	**98**	0	1

Fig. 3. Class probabilities prediction extracted from the trained MobileVitv2 model for each example frame

Our aim is to enhance our methodology for detecting fire events and pinpointing their starting time more accurately through ongoing research. While we have achieved promising results with the contest dataset, we are committed to refining it to operate effectively on real-world images and cross-dataset scenarios. This will enable it to handle environmental variations between different datasets. Ultimately, we aim to extend our approach to combine expert solutions and ensemble techniques to maximize performance and real-time capabilities.

References

1. Almeida, J.S., Huang, C., Nogueira, F.G., Bhatia, S., de Albuquerque, V.H.C.: Edgefiresmoke: a novel lightweight CNN model for real-time video fire-smoke detection. IEEE Trans. Industr. Inf. **18**(11), 7889–7898 (2022)
2. Buslaev, A., Iglovikov, V.I., Khvedchenya, E., Parinov, A., Druzhinin, M., Kalinin, A.A.: Albumentations: fast and flexible image augmentations. Inf. **11**(2), 125 (2020)
3. Caron, M., et al.: Emerging properties in self-supervised vision transformers (2021)
4. Daoud, Z., Ben Hamida, A., Ben Amar, C.: Fireclassnet: a deep convolutional neural network approach for PJF fire images classification. Neural Computing and Applications, pp. 1–17 (2023)
5. Dong, C.: The season for large fires in southern california is projected to lengthen in a changing climate. Commun. Earth Environ. **3**(1), 22 (2022)
6. Dosovitskiy, A., et al.: An image is worth 16x16 words: transformers for image recognition at scale. CoRR abs/2010.11929 (2020)
7. Geetha, S., Abhishek, C., Akshayanat, C.: Machine vision based fire detection techniques: a survey. Fire Technol. **57**, 591–623 (2021)

8. Gragnaniello, D., Greco, A., Sansone, C., Vento, B.: Onfire contest 2023: Real-time fire detection on the edge. ONFIRE Contest, International Conference on Image Analysis and Processing (ICIAP) (2023)
9. Liu, P., Xiang, P., Lu, D.: A new multi-sensor fire detection method based on LSTM networks with environmental information fusion. In: Neural Computing and Applications, pp. 1–15 (2023)
10. Liu, Z., et al.: Swin transformer V2: scaling up capacity and resolution. In: IEEE/CVF Conference on Computer Vision and Pattern Recognition, CVPR 2022, New Orleans, LA, USA, June 18–24, 2022, pp. 11999–12009. IEEE (2022)
11. Liu, Z., et al.: Swin transformer: Hierarchical vision transformer using shifted windows. In: 2021 IEEE/CVF International Conference on Computer Vision, ICCV 2021, Montreal, QC, Canada, October 10–17, 2021, pp. 9992–10002. IEEE (2021)
12. Liu, Z., Mao, H., Wu, C., Feichtenhofer, C., Darrell, T., Xie, S.: A convnet for the 2020s. In: IEEE/CVF Conference on Computer Vision and Pattern Recognition, CVPR 2022, New Orleans, LA, USA, June 18–24, 2022, pp. 11966–11976. IEEE (2022)
13. Majid, S., Alenezi, F., Masood, S., Ahmad, M., Gündüz, E.S., Polat, K.: Attention based CNN model for fire detection and localization in real-world images. Expert Syst. Appl. **189**, 116114 (2022)
14. Mehta, S., Rastegari, M.: Separable self-attention for mobile vision transformers (2022)
15. Thisanke, H., Deshan, C., Chamith, K., Seneviratne, S., Vidanaarachchi, R., Herath, D.: Semantic segmentation using vision transformers: a survey (2023)
16. Töreyin, B.U.: Smoke detection in compressed video. In: Applications of digital image processing XLI. vol. 10752, pp. 896–900. SPIE (2018)
17. Toulouse, T., Rossi, L., Celik, T., Akhloufi, M.: Automatic fire pixel detection using image processing: a comparative analysis of rule-based and machine learning-based methods. SIViP **10**, 647–654 (2016)
18. Vaswani, A., et al.: Attention is all you need. In: Guyon, I., von Luxburg, U., Bengio, S., Wallach, H.M., Fergus, R., Vishwanathan, S.V.N., Garnett, R. (eds.) Advances in Neural Information Processing Systems 30: Annual Conference on Neural Information Processing Systems 2017, December 4–9, 2017, Long Beach, CA, USA, pp. 5998–6008 (2017)
19. de Venâncio, P.V.A., Campos, R.J., Rezende, T.M., Lisboa, A.C., Barbosa, A.V.: A hybrid method for fire detection based on spatial and temporal patterns. Neural Comput. Appl. **35**(13), 9349–9361 (2023)
20. de Venancio, P.V.A., Lisboa, A.C., Barbosa, A.V.: An automatic fire detection system based on deep convolutional neural networks for low-power, resource-constrained devices. Neural Comput. Appl. **34**(18), 15349–15368 (2022)
21. Xie, Y., et al.: Efficient video fire detection exploiting motion-flicker-based dynamic features and deep static features. IEEE Access **8**, 81904–81917 (2020)
22. Yin, Z., Wan, B., Yuan, F., Xia, X., Shi, J.: A deep normalization and convolutional neural network for image smoke detection. IEEE Access **5**, 18429–18438 (2017)

Rapid Fire Detection with Early Exiting

Grace Vincent[(✉)], Laura Desantis, Ethan Patten, and Sambit Bhattacharya

Fayetteville State University, Fayetteville, NC 28301, USA
{gvincent,sbhattac}@uncfsu.edu, {ldesantis,epatten}@broncos.uncfsu.edu

Abstract. Efficient and effective fire detection has proven critical and if not achieved it can pose significant ecological and economic challenges. By introducing early exits into fire video processing using MSDNet, our approach enables quick identification of fires and smoke, ensuring a prompt response to potential fire incidents. Emphasizing efficiency, our method is tailored for resource-constrained edge devices, providing a practical solution for fire-prone regions and enhancing overall fire detection and prevention efforts. Investigating different model sizes yielded accuracy ranging from 86% to 94%, with smaller models outperforming larger models. The adoption of MSDNet allowed for the achievement of an F1-Score of 0.2. This preliminary work has shown the value of small models in the robust detection of fires and introducing early exits can further performance.

Keywords: Fire Detection · Video Processing · Early Exits

1 Introduction

As global temperatures continue to rise and soil moisture decreases, the detection of fires has become an increasingly critical mission. The occurrence of fires in remote and unmanaged areas of forests, where vast amounts of fuel are present (such as trees and leaves), poses significant ecological and economic challenges [3]. Rapid fire detection is of paramount importance in enabling a swift response and effective mitigation efforts. However, current solutions relying on sensor-based techniques suffer from limitations in scope, range, and alert time. These sensors require close proximity to the fire and rapid transportation of carbon particles for detection [6]. The ability to use visual-based detection methods allows cameras to act as volume sensors for detecting smoke or flame in an image/frame. As this fire information is transmitted at high speeds it removes the delay of appearance in these "sensors". Yet, these surveillance cameras are often deployed on edge devices with limited computing capabilities and often have to downlink information for processing.

Thus, harnessing the power of efficient and accurate surveillance video processing opens up new possibilities for rapid fire detection, offering the potential for enhanced responsiveness and early intervention in fire-prone areas [5,7]. With

G. L. Foresti et al. (Eds.): ICIAP 2023 Workshops, LNCS 14365, pp. 294–301, 2024.
https://doi.org/10.1007/978-3-031-51023-6_25

edge computing, the need to downlink is removed and data can be processed and analyzed directly onboard. Introducing early exits into fire video processing allows a model to quickly detect whether there is a fire or not in a frame without fully processing the video. This is done by creating classification heads, or exit points, throughout the model so that if a frame is easily and confidently identifiable it would not need to use as many resources. This work leverages the power of early exits, implementing MSDNet can quickly identify fires and smoke, ensuring a prompt response to potential fire incidents. Additionally, we emphasize the efficiency of our approach, making it suitable for resource-constrained edge devices, providing a practical solution for fire-prone regions, and enhancing the overall effectiveness of fire detection and prevention efforts.

In summary, the contributions of this work are

1. Analyzing the role of the model architecture and the impact of ImageNet pretraining on fire identification.
2. Investigated the implementation of MSDNet for rapid fire detection in an effective and efficient manner.

The rest of the paper is organized as follows. Section 2 highlights current progress in fire detection and describes early-exit techniques. Section 3 details the information about the videos and their annotation process (Sect. 3.1) and summarizes the technical approach (Sect. 3.2). Section 4 presents in detail the preliminary results of the contributions listed above.

2 Related Works

2.1 Fire Detection

Fire detection and management are essential to preventing fire-driven catastrophes, as fires not only consume any vegetation but man-made constructions as well. In a rapidly developing world, the need for efficient and accurate fire detection models to minimize the effects of such fires has never been higher. In previous decades, fire detection has been conducted on networks of satellite imagery, allowing for a view of large portions of the Earth [14]. However, a fire must be substantial enough to be seen from such an altitude. By the time a fire reaches that point in an urban setting, the damage done to buildings and human lives can be catastrophic.

2.2 Advantages of Deep Learning

Deep learning has the advantage of being able to detect patterns in complex data by learning from examples automatically. In contrast, traditional remote sensing algorithms for detecting wildfires often require the establishment of complex functional relationships between numerous data parameters. Additionally, wildfires are complex physical processes that involve different interrelated factors, making them difficult to model using traditional methods [12]. Therefore, deep learning algorithms that can automatically uncover complex, interconnected patterns in the data with less need for human interference are a good choice for the task of wildfire detection.

2.3 Early Exiting

Integrating early exit strategies with models can significantly enhance the efficiency of fire detection models. It functions by identifying and filtering out samples at the earlier layers of the model if a certain confidence threshold is met, saving computational resources and time [4]. As a result, only challenging samples that may require more complex analysis will continue through the deeper layers of the model. This approach allows fire detection models to achieve faster response times, making them suitable for real-time applications. One such application can be found within video surveillance systems, where timely and accurate fire detection is of utmost importance to prevent disasters and mitigate potential damages.

3 Methodology

3.1 Dataset

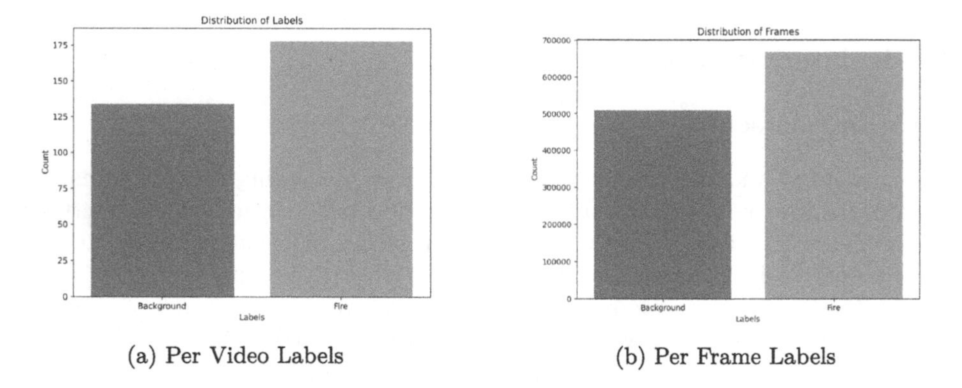

(a) Per Video Labels (b) Per Frame Labels

Fig. 1. The distribution of annotations of fire or background at the video and frame level. There are 219 videos with fire and 113 videos with no instance of fire or smoke. Each video has a varying length.

The dataset used throughout this work is a curated Fire detection video dataset consisting of 322 surveillance videos [8]. The videos have been collected from existing datasets (i.e., MIVIA [11], RISE [1], KMU [2], Wildfire, D-Fire [15]) and annotations correspond to if a fire is in the video and what time the fire ignites or appears in the video. Through examination and calculation, the corresponding frame for each ignition time was identified for processing of videos frame-by-frame. The distributions of labels on the video level and on the frame level are depicted in Fig. 1.

Fig. 2. Sample frames from videos that depict both flame and smoke. These sample frames highlight the diversity of videos compiled in this dataset.

These surveillance videos capture fires under different conditions. The videos not only vary in proximity to the fire, but also in the environment, lighting, density of fire/smoke, and fire and video length. This wide range of videos produces a robust fire dataset with large complexities. Sample frames from different videos showcase the variety in the dataset (note each frame has either smoke or flame present) seen in Fig. 2.

3.2 Model Architectures

In this work, we train a variety of classification networks for the binary task of detecting fire or background frames within the videos. The models and their number of parameters are summarized in Table 1. The investigated architectures varied in size (number of parameters) to evaluate the impact on performance and evaluation speed. These models were both trained, in a supervised manner, with randomly initialized weights and supervised ImageNet weights [13]. The choice of ImageNet pretrained weights used to initialize the backbones has been shown to play an important role in downstream task performance, i.e., classification. The classifiers learned with a batch size of 64, a learning rate of 10^{-4}, and a Step Learning Rate scheduler.

One of the other main architectures we used was MSDNet. MSDNet stands for Multi-Scale Dense Network. MSDNet is a dense convolutional neural network, and the purpose of MSDNet is to allocate computational resources effectively when

Table 1. Model characteristics. For each architecture, the number of parameters and the required memory is provided.

Model Name	Params (M)	Memory (MB)
MobileNet-V2	3.4	13
EfficientNet-B0	4.0	16
ResNet-50	23.5	91

classifying images and allow for early exit. To do this, MSDNet has multiple classifiers interspersed throughout its architecture, which lead to different branches and exits, and this is the mechanism that allows MSDNet to exit early [9].

Through training, a confidence threshold is set up for MSDNet. When going through the process of classifying an image, the classifiers in MSDNet's architecture check to see if the confidence level for that image is above the threshold or not. If the confidence is higher than the threshold, then MSDNet will take the exit at that classifier (Fig. 3).

Fig. 3. The image on the left depicts an "easy" image of a horse with distinguishable features, on the right is a more "difficult" image of a horse. This example is adapted from Huang et al. [9].

This means that for images that are easy for MSDNet to classify (meaning that the confidence threshold is reached early in the model architecture), like the image on the right, MSDNet will exit and use minimal computational resources to identify that image. However, with a more complex or difficult image to classify, like the one on the right, MSDNet will continue deeper into the network until the confidence threshold is reached, or it fails to classify the image. This way, it uses less computational resources on images that are "easy" and more on images that are "difficult" [9].

4 Preliminary Results

To evaluate the importance of model size, highlighting the number of parameters, we implement the four models in Table 1. Training these models with both randomly initialized weights and ImageNet pretrained weights, investigates ImageNet's ability to generalize to tasks outside of its domain. In general, the ImageNet pretrained weights yield the highest accuracy across the different model

Table 2. The Accuracy, F1-Score, Recall, and Precision performance on the hold-out test set are presented for each of the previously discussed models. Note that the first three models have random weight initialization and the last three are ImageNet pretrained.

Model	Weight Init	Accuracy	F1-Score	Recall	Precision
EfficientNet-B0	Random	0.866	1.0	1.0	1.0
MobileNet-V2	Random	0.890	0.8	1.0	0.667
ResNet-50	Random	0.870	0.8	1.0	0.667
EfficientNet-B0	ImageNet Pretrain	0.943	1.0	1.0	1.0
MobileNet-V2	ImageNet Pretrain	0.924	0.889	1.0	0.8
ResNet-50	ImageNet Pretrain	0.863	0.8	0.8	0.8

sizes. Further, it is seen that smaller models (i.e., MobileNet V2 and EfficientNet-B0) outperform larger models (i.e., ResNet-50) for this frame-by-frame classification task. Yet, the ImageNet pretrained EfficientNet-B0 produces the highest accuracy of 94% with only 4 million parameters as seen in Table 2. We note that EfficientNet-B0 also has the highest performance across F1-Score, Recall, and Precision.

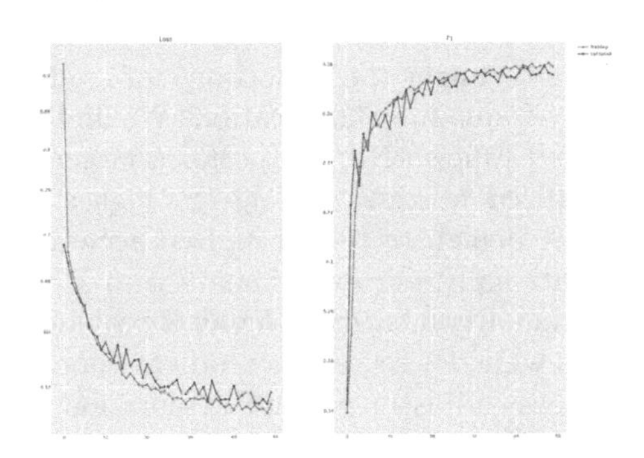

Fig. 4. The loss and F1 score for initial MSDNet experiment. Y axis is Accuracy, X axis is Epochs

The two metrics that we used to measure our results for MSDNet are the loss and the F1 score, which measures the precision and recall of the model [10]. Our initial experiment with MSDNet was on a subset of about 10,000 images that were randomly selected, and this resulted in a rather poor performance compared to some of the other architectures we experimented with. The initial experiment showed a loss of about 0.5 and an F1 score of about 0.28 for both the train and validation sets, as shown in Fig. 4.

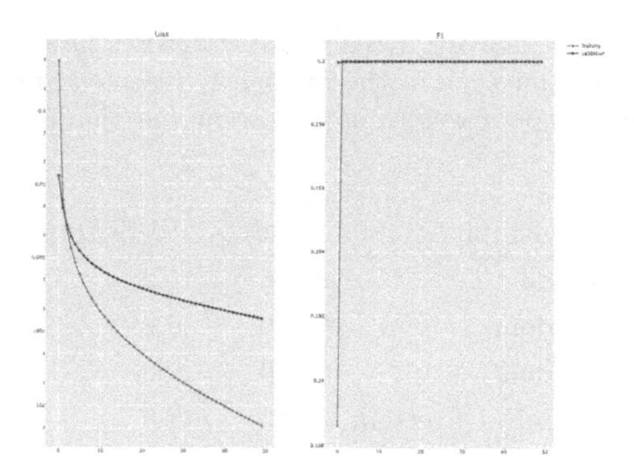

Fig. 5. The loss and F1 score for second MSDNet experiment. Y axis is Accuracy, X axis is Epochs.

With a more curated dataset, the loss went from 0.5 to less than 0.0013, however, the F1 stayed at 0.2 for all of the epochs as shown in Fig. 5. While the loss improved dramatically, the F1 score may indicate that the model is overfitting to the training set, and thus the dataset should be further adjusted.

5 Conclusions

These preliminary results highlight the importance of model size and initialization for efficient frame-by-frame fire classification. We find that smaller models, such as MobileNet V2 and EfficientNet, outperform larger models like ResNet-50, with randomly initialized weights yielding the highest accuracy. ImageNet pretrained EfficientNet-B0 demonstrates the highest accuracy of 94% with just 4 million parameters. Additionally, our experiments with MSDNet reveal the significance of dataset curation in combating imbalance problems, showing dramatic improvements in the loss while F1 scores may indicate potential overfitting. For future work, further dataset adjustments and exploration of advanced training techniques could enhance the robustness and generalization capabilities of the models. Additionally, investigating the impact of colorspace or change detection could offer valuable insights for improved capabilities.

Acknowledgements. This paper is based upon work which was partially supported by the National Aeronautics and Space Administration (NASA) under Grant No. 80NSSC21M0312 and 80NSSC23M0054.

References

1. https://www.ri.se/en/what-we-do/services/rise-fire-database
2. https://cvpr.kmu.ac.kr/Dataset/Dataset.htm, April 2012
3. Alkhatib, A.A.: A review on forest fire detection techniques. Int. J. Distrib. Sens. Netw. **10**(3), 597368 (2014)

4. Banijamali, E., Kharazmi, P., Eghbali, S., Wang, J., Chung, C., Choudhary, S.: Pyramid dynamic inference: encouraging faster inference via early exit boosting. In: ICASSP 2023–2023 IEEE International Conference on Acoustics, Speech and Signal Processing (ICASSP), pp. 1–5. IEEE (2023)
5. Çetin, A.E., et al.: Video fire detection-review. Digit. Signal Process. **23**(6), 1827–1843 (2013)
6. Chen, T.H., Wu, P.H., Chiou, Y.C.: An early fire-detection method based on image processing. In: 2004 International Conference on Image Processing, 2004. ICIP 2004. vol. 3, pp. 1707–1710. IEEE (2004)
7. Gaur, A., Singh, A., Kumar, A., Kumar, A., Kapoor, K.: Video flame and smoke based fire detection algorithms: a literature review. Fire Technol. **56**, 1943–1980 (2020)
8. Gragnaniello, D., Greco, A., Sansone, C., Vento, B.: Onfire contest 2023: real-time fire detection on the edge. In: ONFIRE Contest, International Conference on Image Analysis and Processing (ICIAP) 2023 (2023)
9. Huang, G., Chen, D., Li, T., Wu, F., Van Der Maaten, L., Weinberger, K.Q.: Multi-scale dense networks for resource efficient image classification. arXiv preprint arXiv:1703.09844 (2017)
10. Korstanje, J.: The f1 score, August 2021. https://towardsdatascience.com/the-f1-score-bec2bbc38aa6
11. Mivia: Fire detection dataset. https://mivia.unisa.it/datasets/video-analysis-datasets/fire-detection-dataset/
12. Rashkovetsky, D., Mauracher, F., Langer, M., Schmitt, M.: Wildfire detection from multisensor satellite imagery using deep semantic segmentation. IEEE J. Sel. Top. Appl. Earth Observ. Remote Sens. **14**, 7001–7016 (2021)
13. Russakovsky, O., et al.: ImageNet large scale visual recognition challenge. Int. J. Comput. Vision **115**, 211–252 (2015)
14. Seydi, S.T., Saeidi, V., Kalantar, B., Ueda, N., Halin, A.A.: Fire-net: a deep learning framework for active forest fire detection. J. Sens. **2022**, 1–14 (2022)
15. de Venancio, P.V., Campos, R., Rezende, T., Lisboa, A., Barbosa, A.: A hybrid method for fire detection based on spatial and temporal patterns (2023). https://doi.org/10.1007/s00521-023-08260-2

Recent Advances in Digital Security: Biometrics and Forensics (BIOFORM)

Morphing-Attacks Against Binary Fingervein Templates

Tobias Mitterreiter, Jutta Hämmerle-Uhl, and Andreas Uhl(✉)

Visual Computing and Security Lab (VISEL), Department of Artificial Intelligence and Human Interfaces, University of Salzburg, Salzburg, Austria
uhl@cs.sbg.ac.at

Abstract. For the first time, the feasibility of creating morphed templates for attacking vascular biometrics is investigated, in particular finger vein recognition schemes generating binary vascular patterns are addressed. A conducted vulnerability analysis reveals that (i) the extent of vulnerability, (ii) the type of most vulnerable recognition scheme, and (iii) the preferred way to construct the morphed template for a given target template depends on the employed sensor. It turns out that targeted template doppelgaenger selection is important for an attack success. The identified threat level in terms of IAPMR is often found to be > 0.8 for several sensor/template generation scheme/morphing technique combinations. Thus, the risk as imposed by such attacks can be said to be considerable.

1 Introduction

Since the introduction of the "magic passport" [1] concept, the threat of using morphed facial portrait images in ID documents has been discussed in depth. As this threat has been considered a serious one since, we have observed an explosion of work dedicated to face morphing (detection) consequently [2,3]. Apart from the face modality, the threat originating from morphed samples or templates of other modalities is less obvious, as there is no connection with ID documents and no corresponding inclusion of morphed sample image data. As a consequence, we have seen only a single proposal for fingerprint morphing using traditional model-based techniques [4] and its potential detection [5,6], and a second proposal for fingerprint morphing using learning-based schemes (i.e. GANs [7]). For iris recognition, a first work deals with the construction of morphed iris codes [8], later also image-level iris morphing has been demonstrated [9]. Also, a suggestion for systematic analysis of biometric system vulnerability with respect to morphing attacks [10] addressed face and iris morphing attacks. Recently, sample-oriented generation of morphed fingervein sample data has been explored together with a demonstration of the feasibility of using these data as presentation attack artefacts [11].

In this work, we investigate the feasibility of creating morphed vascular binary templates, in particular we deal with finger vein recognition systems. Based on the morphed finger vein templates we conduct a vulnerability analysis of five different recognition systems. The actual threat of such data is illustrated in Fig. 1 - the most efficient attacks inject such morphed templates into the database, replacing the templates of a

G. L. Foresti et al. (Eds.): ICIAP 2023 Workshops, LNCS 14365, pp. 305–317, 2024.
https://doi.org/10.1007/978-3-031-51023-6_26

legitimate user, thus allowing the legitimate user as well as the attacker to authenticate to the biometric system based on the morphed templates.

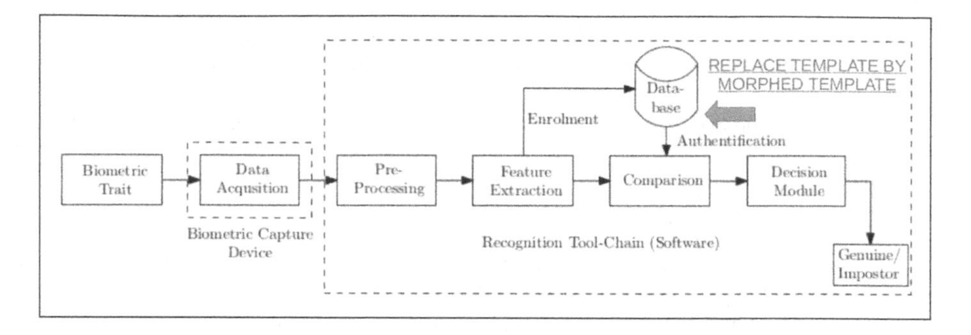

Fig. 1. Main point of morphed template attack against a biometric system.

The principle of the attack is visualised in Fig. 2: The attacker's template is f_1 and there is no corresponding template stored in the database. Therefore the attacker can not authenticate him/herself with the system via f_1 ("failed"). Template f_2 is computed from a sample of a legitimate user. The authentication is successful because the user's template t_1 is stored in the database and the similarity score derived in the comparison is higher than the threshold. As the attacker has full access to all database templates, he/she aims to morph his/her template f_1 with a template derived from a legitimate users' sample, in this case, t_1.

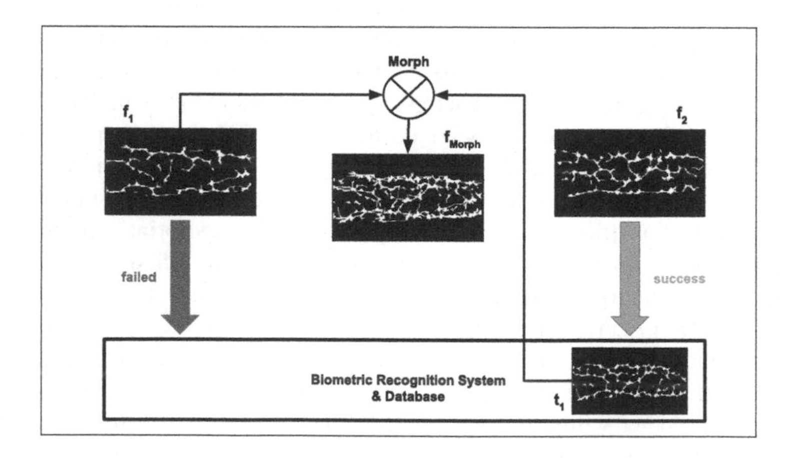

Fig. 2. Working principle of attacking the template database.

The attacker then replaces template t_1 with the resulting morphed template f_{Morph} in the database. This approach allows both, the attacker and the legitimate user, to authenticate successfully via their finger veins in the future, as both extracted templates f_1 and f_2 should be close enough to the morphed template f_{Morph} in the database.

Another attack option is to replace a template retrieved from the database for the comparison process by a morphed template achieving the same effect, but this attack is much less static and requires dynamic injection into the data transfer from the database or into an ongoing template comparison process, respectively. Also, for this attack we do not really require a morphed template, injecting the attackers template directly is sufficient to make the attack work.

The remainder of the paper is organised as follows. In Sect. 2, we will demonstrate how a digitally morphed template can be created from two binary finger vein templates. Section 3 explains the experimental setup to conduct the vulnerability analysis, including the definition of the used recognition software and finger vein datasets, respectively, and defining the way how to actually assess the vulnerability. This section also contains an explanation how we aim to reveal the existence of morphed templates in a database. Experimental results are presented and discussed in Sect. 4, while we conclude the paper in Sect. 5.

2 Morphing of Binary Finger Vein Templates

Morphing is defined originally as the transformation of one image into another and involves two parts: cross dissolving and warping. Cross dissolving is linear interpolation to fade from one image to another in terms of grayscale or colour value. Considering two samples $Sample1$ and $Sample2$, we interpolate a value from 0 to 1 and use $Sample1 * \alpha + Sample2 * (1 - \alpha)$ as the value of the new pixel in the morphed sample. α is called "blending factor" and defines the respective contribution of $Sample1$ and $Sample2$ to the morphed sample (this has been applied to finger vein samples in [11]).

However, we aim at binary templates but not at grayscale samples, and the original concept of morphing needs to be adapted correspondingly as we cannot rely on particular landmarks and interpolation of binary values is hardly possible in meaningful manner. In particular, it makes sense to consider the way how the similarity of two binary templates is determined during template comparison. Contrasting to e.g. iris code comparisons, where binary codes are compared under left-right shifting them against each other and taking the minimum resulting Hamming distance as their similarity value, in comparing vascular binary features typically the "Miura Matcher" [12] is being employed. In this comparison algorithm, two binary templates are correlated against each other computing the maximum among two-dimensional shifts of rotated template versions. The correlation is computed on the center region of the templates, the so-called "kernel" (see Fig. 4).

We define the following template morphing approaches (in fact, these are more template fusion schemes):

- **Template OR (XOR)**: This approach applies a logical OR to the two vascular samples f_1 and f_2: $f_{morph} = f_1 \vee f_2$ without any alignment between the two templates. Note that $\vee$ is the fundamental operation for all template morphing variants, the sole difference is the type of alignment that is applied between the two. Figure 3 illustrates the simple XOR process and outcome.

- **(Template) Rotation**: The Miura Matcher uses a rotation compensation in the template comparison process, which we use to determine under which rotation parameter the similarity between the two templates to be morphed is maximal. Thus we rotate one of the two templates by that parameter, and use nearest neighbour interpolation and cropping to result in an identical template size. Finally, the rotated template and the second one are fused by logical OR.

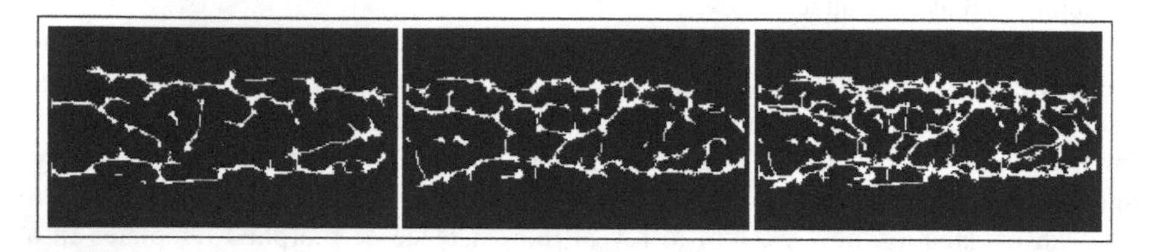

Fig. 3. Two MC binary templates from the MMCBNU dataset and their XOR morphed template.

- **(Kernel) Alignment**: The approach is to determine how to align the kernel (i.e. central area according to the Miura Matcher) of a template to a second template, so that the similarity is highest. To do this, we employ the Miura Matcher to compute the convolution matrix and filter the result for the highest value, corresponding to the alignment with the highest similarity, and to return the optimal shift parameters (in two directions). This process is repeated for different rotations to obtain the optimal kernel alignment and shift parameters for the subsequent morphing process. Figure 4 illustrates the kernel of a template and its aligned OR fusion with another template.

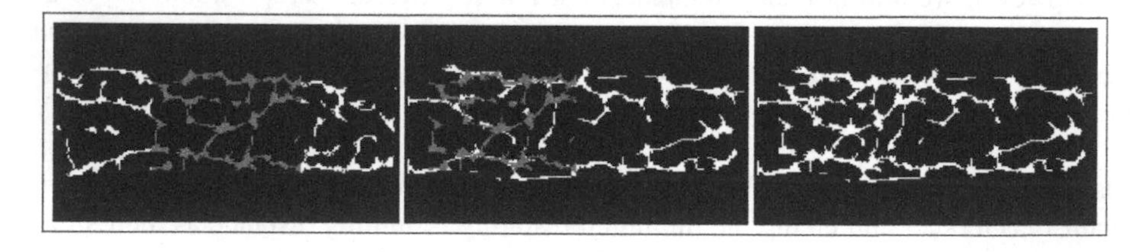

Fig. 4. The kernel of a MC template from the MMCBNU dataset (in blue), and its morphing with a second template (in blue overlay and binary). (Color figure online)

- **F(ull) Alignment**: This approach is an extended version of the previous (Kernel) Alignment approach. The difference is that we no longer consider the kernel alone but rotate and shift the entire template to optimise similarity, before template OR is being applied. Figure 5 illustrates the process and highlights the difference to (Kernel) Alignment.

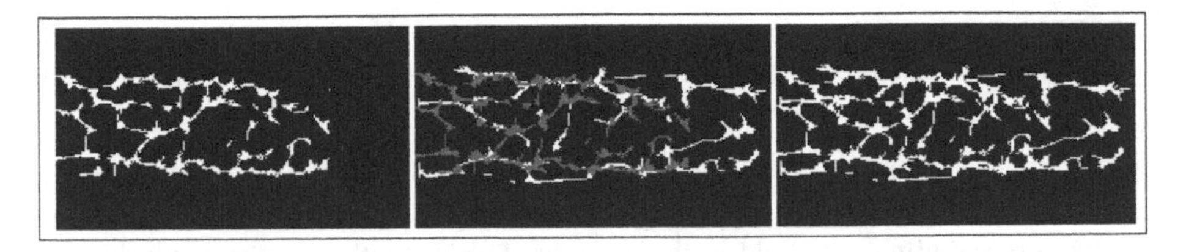

Fig. 5. A rotated and shifted MC template from the MMCBNU dataset, and its morphing with a second template (in blue overlay and binary). (Color figure online)

For the envisioned attack, typically an attacker does not just morph his/her template with an arbitrary template in the database. We have an attacker template, say f_1, and need to select a suited template t_1 in the database from a legitimate subject to result in the best possible recognition result for both subjects. There is work on this topic for facial portrait data called "how to find the suited doppelgaenger" [13], but in the finger vein setting, we only need to consider a smaller set of requirements for a suited "doppelgaenger" finger vein template. In order to investigate the role this selection plays, we have chosen two approaches: First, in "similar" mode, we select t_1 as the closest template of a different subject contained in the dataset determined in terms of Miura Matcher template comparison score using a particular recognition system. Second, in "unsimilar" mode, we select t_1 as the most distinct sample to f_1 in the same sense.

3 Experimental Settings

3.1 Assessment Criteria

The vulnerability of a biometric recognition system to attacks is determined by the Impostor Attack Presentation Match Rate (IAPMR) introduced in ISO/IEC 30107–3 [14]. IAPMR is defined as the proportion of attack presentations using the same type of presentation attack instruments in which the target reference matches. This general measure has been adapted to the specific morphing scenario [15] resulting in the Mated Morph Presentation Match Rate (MMPMR), which covers the fact that not one target subject (contained in the morphed reference) is compared to others - but for a successful morph attack, both data subjects that previously contributed to the morphed image are expected to match. However, as we have found both involved subjects to be symmetrically represented (which is to be expected due to the symmetric XOR construction of the morphs), we resort to the simpler IAPMR for result reporting.

To investigate the importance of the actual template used to create the morph, we discriminate three IAPMR variants:

- IAPMR_1 determines IAPMR by considering template comparison scores for which the morphed template is compared to template t_1 only, i.e. the template of the legitimate user that has actually been used to create the morph.
- IAPMR_n determines IAPMR by considering template comparison scores for which the morphed template is compared to all templates of the subject from which t_1 has been derived.

- $IAPMR_{n-1}$ determines IAPMR by considering template comparison scores for which the morphed template is compared to all templates of the subject from which t_1 has been acquired *except* for template t_1.

For defining a "successful" template comparison in the context of IAPMR, we first compute the EER of the corresponding dataset/recognition scheme combination and use the corresponding threshold in the decision. Subsequently, we start with the first template of the first subject, determine its most similar and dissimilar template in the databasc (from different subjects) and generate the corresponding morphs. Then we compute the template comparisons between all templates of the first subject and the generated morphs and check if the result adheres to the threshold (i.e. a successful attack has been conducted). The results increase the correspondings counters and we proceed to the next template of the first subject. This procedure is conducted for all subjects.

3.2 Data and Recognition Software

For the experiments, four publicly available finger vein databases were used. The data sets under investigation are:

- *The Finger Vein Universiti Sains Malaysia Database* (**FV-USM** [16]): Contains 5904 palmar finger vein images, exhibiting a resolution of 640×480 pixels, acquired from 123 subjects. All of them participated in 2 acquisition sessions where each time 4 fingers per subject and 6 images per finger were captured by a custom built capturing device.
- *The Multimedia Chonbuk National University Database* (**MMCBNU_6000** [17]): The 6000 palmar light transmission finger vein images, exhibiting a resolution of 640×480 pixels, contained in this dataset were acquired from 100 subjects. From all of them 6 fingers per subject and 10 images per finger were captured in a single session utilizing a capturing system based on a modified webcam.
- The *University of Twente Finger Vascular Pattern Database* (**UTFVP** [18]) contains six fingers (ring, middle and index finger from both hands) from 60 volunteers in two sessions. At each session, two palmar samples per finger were captured (resulting in 4 samples per finger). The samples have an original resolution of 672×380 pixels, while their region of interest (RoI) is 672×285 pixels.
- The *PLUSVein-FV3 Palmar LED Finger Vein Data Set* (**PLUS** [19]) contains palmar images from the ring, middle and index finger of the left and right hand (5 samples per finger) and have been acquired using an open access capturing device [20]. Here, only LED illuminated images are used, the resolution of the single finger RoI cropped from the 3-finger capture is 736×192 pixels.

The finger detection, finger alignment and RoI extraction for UTFVP and PLUS is done as described in [21]. After pre-processing and feature extraction, the resulting binary templates are used to perform the experiments. We conducted these experiments by applying the PLUS OpenVein Finger- and Hand-Vein Toolkit (http://www.wavelab.at/sources/OpenVein-Toolkit/ [22]). We selected five techniques based on the binary vessel structure. The extraction schemes used are *Wide Line Detector (WLD)* [23],

Isotropic Undecimated Wavelet Transform (IUWT) [24], *Gabor Filter (Gabor)* [25], *Maximum Curvature (MC)* [12], and *Principal Curvature (PC)* [26]. These binary feature templates are subsequently compared using a correlation-based approach proposed in [12], the so called Miura Matcher.

4 Experimental Results

The experimental section is split into two parts - first, we conduct a threat analysis, i.e., we experiment if the generated morphed templates are a real threat to the biometric system in question, and here we discriminate different data sets and template generation schemes. Second, we investigate if a database maintainer can check the database for eventual morphs, i.e. if we can reliably discriminate morphs from real legitimate templates.

4.1 Threat Evaluation

We explain the results for each IAPMR variant looking at Table 1 (left). The results are based on the FV-USM dataset (123 subjects and 12 samples per finger). For feature extraction we use Maximum Curvature, and for the morphing procedure, we use the XOR Approach. For IAPMR_1, we note that the attack is successful in all cases, no matter if the most similar or dissimilar template has been used (the accordance of the template in the morph and the attacking one is sufficient to guarantee the attack is working).

Table 1. IAPMR results of three datasets using XOR Morphing.

FV-USM (MC Rec.)			MMCBMU (PC Rec.)			UTFVP (PC Rec.)		
	Similar	Unsimilar		Similar	Unsimilar		Similar	Unsimilar
IAPMR_1	1	1	IAPMR_1	1	1	IAPMR_1	1	1
IAPMR_{n-1}	0.449	0.263	IAPMR_{n-1}	0.180	0.139	IAPMR_{n-1}	0.965	0.918
IAPMR_n	0.461	0.279	IAPMR_n	0.193	0.154	IAPMR_n	0.967	0.922

The situation is different for IAPMR_{n-1} and IAPMR_n. While both values are rather similar (slightly higher for IAPMR_n as the identical template as used in the morph is also considered in the comparison process, among the other ones), there is a clear difference between similar and unsimilar template selection in the morph construction, and IAPMR differs by a factor a bit lower than 2. Therefore, in the following, we will present results for IAPMR_n only but discriminate between the similar and unsimilar template selection process, respectively.

In Fig. 6, we present the overall results for the FV-USM dataset. For the similar doppelgaenger template selection, results follow a clear trend: MC, IUWT, and WLD template generation techniques exhibit a lower IAPMR value (still around 0.40 - 0.45) while PC and GF are most subsceptible to the morphing attack (top IAPMR values are between 0.6 and 0.7). There is also a clear ranking with respect to successful morphing techniques: XOR and Rotation work best, while the Kernel Alignment approach is worst.

For the unsimilar doppelgaenger selection scheme, results are different. While the most vulnerable template generation schemes are still PC and GF, their highest IAPMR values are around 0.4. Contrasting to before, in unsimilar mode the best morphing techniques are Alignment and Falignment. Thus, results clearly confirm that the doppelgaenger selection strategy is of high importance for a successful attack, and that different template generation schemes are fairly different in how far they are vulnerable to template morphing attacks. The former fact also implies, that the attack can be made much more effective if an entire dataset is compromised (as we can select the most similar doppelgaenger), as opposed to the case if only a single template is compromised.

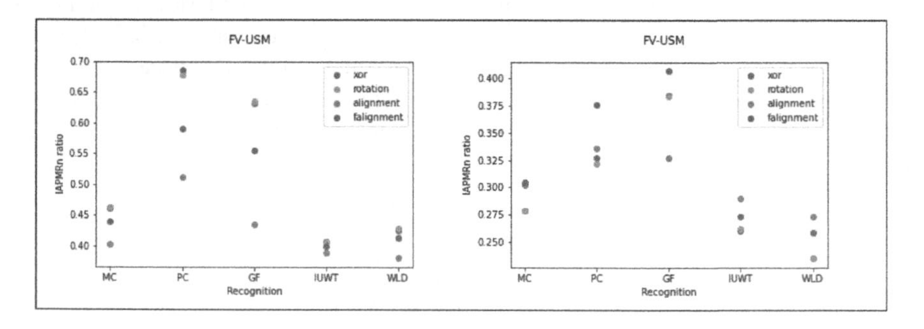

Fig. 6. FV-USM dataset: IAPMR_n results of all template types and morphing approaches in similar (left) and unsimilar (right) mode, respectively.

Now let us look into the question if the results are stable across different datasets (i.e. finger vein sensors). Table 1 (middle, for MMCBMU data) reveals a different behaviour as compared to Table 1 (left). We notice rather low IAPMR_{n-1} and IAPMR_n values and the difference between similar and unsimilar template selection for the morphing process is rather negligible.

Figure 7 shows the overview results of the MMCBMU dataset. We clearly observe, that the situation is different as compared to the FV-USM dataset. Here, it is only GF template generation which is highly vulnerable by the morphing attack (with IAPMR_n being almost 1.0), and PC is much less vulnerable (actually, here PC is the least vulnerable template generation scheme). For the similar doppelgaenger template selection scheme, there is no significant winner in terms of best morphing approach.

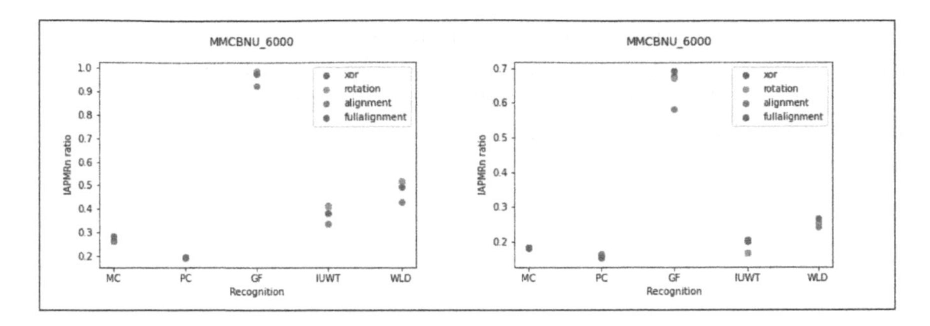

Fig. 7. MMCBMU dataset: IAPMR_n results of all template types and morphing approaches in similar (left) and unsimilar (right) mode, respectively.

When comparing the similar to the unsimilar template selection results, the general trend is almost identical. GF is most vulnerable (with IAPMR_n at 0.7), while the other template generation schemes are between 0.1 and 0.3 in terms of IAPMR_n. It is also interesting to note that for the MMCBMU dataset, the doppelgaenger template selection variant chosen is by far less important for the resulting threat (except for GF), as compared to the FV-USM dataset.

Table 1 (right) shows exemplary results for the UTFVP dataset, shown in similar way as Table 1 for the FV-USM and MMCBMU datasets. respectively. Contrasting to the results for FV-USM (but in accordance to those for MMCBMU), here we do not observe a large difference between IAPMR for the similar and unsimilar template selection approach, respectively, while all displayed IAPMR variants are on a very high level (which on the other hand does not correspond to results for MMCBMU data).

Again, a summary of the results for the UTFVP dataset is displayed in Fig. 8, we again observe different behaviour as compared to the previous dataset. PC is most subsceptible (with IAPMR_n close to 1.0), while IUWT and GF still reach IAPMR_n of 0.45 - 0.6. For the similar template selection mode there is no clear winner in terms of morphing generation (Rotation is often pretty well performing).

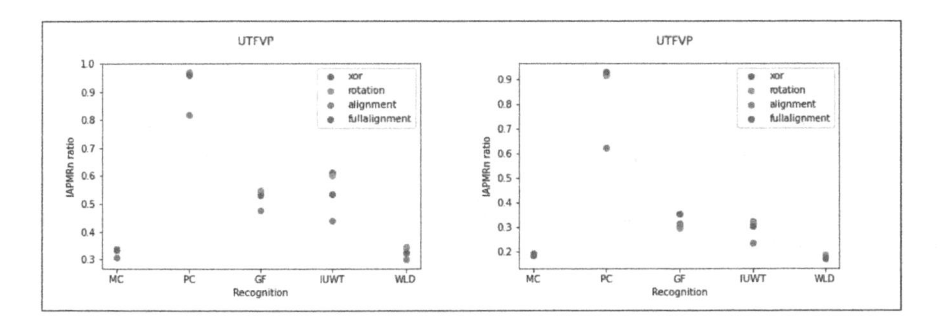

Fig. 8. UTFVP dataset: IAPMR_n results of all template types and morphing approaches in similar (left) and unsimilar (right) mode, respectively.

The unsimilar template selection mode results in lower $IAPMR_n$ values, but not as clear as e.g. for FV-USM data. In this setting, Falignment is the best performing morphing approach. PC can still reach $IAPMR_n$ of > 0.9, so for this setting an arbitrary template can be selected for the morph.

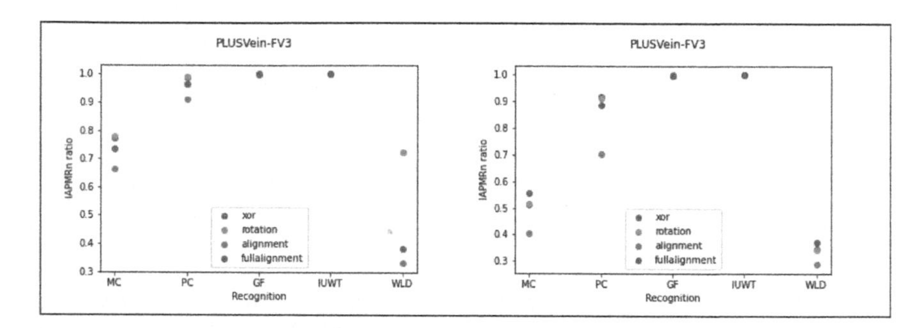

Fig. 9. PLUS dataset: $IAPMR_n$ results of all template types and morphing approaches in similar (left) and unsimilar (right) mode, respectively.

The last dataset we consider is the PLUS dataset as shown in Fig. 9. We notice that these data is most vulnerable to the attacks considered. In similar as well as in unsimilar template selection mode, $IAPMR_n$ of 0.9–1.0 is achieved for PC, GF, and IUWT template generation schemes. MC is still between 0.4 and 0.8 $IAPMR_n$ for all variants and only WLD exhibits some resistance against morphing attacks, still with an $IAPMR_n$ between 0.3 and 0.4 for all settings investigated.

Overall, it is difficult to identify overall trends. MC and WLD template generation techniques seem to be least subsceptible to the morphing attacks under investigation, but there is a clear interference between dataset properties and most vulnerable template generation scheme. While a best morphing approach is difficult to figure out, the Alignment approach is often seen as the worst performing one. In general, targeted template selection to identify the most similar doppelgänger template pays off in most cases.

4.2 Detecting Morphed Templates

This subsection deals with the vulnerability assessment of the database itself. As the logical OR operation used in all morphing approaches increases the number of white pixels, this number could serve as a simple criterion to identify morphed templates. We consider to following criterion for the number of white pixels (wPixels) with x serving as variable threshold.

$$\#wPixels \leq x \cdot StdDev(\#wPixels) + Mean(\#wPixels)$$

When applying this criterion with $x = 2$, the number of false positive morph detections is low for all template generation schemes (averaged across all datasets, compare Table 2: 4% - 5%).

Table 2. The average probabilities across all datasets that a feature is falsely detected as morph at a threshold of $x = 2$.

	MC	PC	GF	IUWT	WLD
False positives	0.04	0.04	0.05	0.05	0.05

On the other hand, the correct morph identification rate is very high for all template generation schemes as displayed in Table 3. For the XOR, Rotation and Falignment approaches we detect more than 0.96 of all morphs. This shows that we can easily identify morphs by looking at the number of white pixels. In contrast, morphs that we generate using the Alignment approach can "only" be identified with a probability of 0.66. This is due to the fact that in this technique, we only fuse the kernel with the input template which reduces the number of white pixels in the generated morph.

Table 3. The probabilities across all datasets that a morph is correctly detected at a threshold of $x = 2$.

Approach	MC	PC	GF	IUWT	WLD
XOR	1.00	0.97	0.98	0.97	0.99
Rotation	1.00	0.97	0.98	0.98	0.99
Alignment	0.84	0.55	0.62	0.63	0.68
Full Alignment	1.00	0.93	0.93	0.94	0.98

Based on these results, the database maintainer is able to run regular checks across the database to identify morphed templates of the type discussed. Therefore, the construction of morphed templates needs to be refined in order to mitigate the problems caused by the highly increased number of white pixels.

5 Conclusion and Future Work

We have investigated the feasibility of creating morphed templates for attacking finger vein recognition schemes by replacing templates in the database by morphed ones. A conducted vulnerability analysis reveals that (i) the extent of vulnerability and (ii) the type of most vulnerable template generation scheme depends on the employed sensor. We have also found that the similarity of the two templates involved in the morph is crucial, so a random selection should be avoided. The optimal method how to generate the morph for a given target template is also found to be sensor dependent. Thus, there is no general rule for an attacker how to conduct an attack of the described type, but for most sensor/template generation scheme we were able to identify a morphing scheme with a significant threat level.

Future work includes the refinement of the morphing techniques to avoid the considerable increase of white pixels (which can be exploited to identify morphed templates of the type discussed). Also the consideration of other template generation schemes, including deep-learning based ones, is of importance as the current investigation is restricted to binary template types.

Acknowledgements. This work has been partially funded by the Austrian Science Fund projects P32201 and I4232, respectively.

References

1. Ferrara, M., Franco, A., Maltoni, D.: The magic passport. In: IEEE International Joint Conference on Biometrics, pp. 1–7 (Sept 2014)
2. Scherhag, U., Rathgeb, C., Merkle, J., Breithaupt, R., Busch, C.: Face recognition systems under morphing attacks: a survey. IEEE Access **7**, 23012–23026 (2019)
3. Venkatesh, S., Ramachandra, R., Raja, K., Busch, C.: Face morphing attack generation & detection: a comprehensive survey. IEEE Trans. Technol. Society **2**(3), 128–145 (2021)
4. Ferrara, M., Cappelli, R., Maltoni, D.: On the feasibility of creating double-identity fingerprints. IEEE Trans. Inf. Forensics Secur. **12**(4), 892–900 (2017)
5. Goel, I., Puhan, N.B., Mandal, B.: Deep convolutional neural network for double-identity fingerprint detection. IEEE Sensors Letters **4**(5), 1–4 (2020)
6. Satapathy, G., Bhattacharya, G., Puhan, N.B., Ho, A.T.S.: Generalized benford's law for fake fingerprint detection. In: 2020 IEEE Applied Signal Processing Conference (ASPCON), pp. 242–246 (2020)
7. Makrushin, A., Trebeljahr, M., Seidlitz, S., Dittmann, J.: On feasibility of gan-based fingerprint morphing. In: 2021 IEEE 23rd International Workshop on Multimedia Signal Processing (MMSP), 1–6. (2021)
8. Rathgeb, C., Bush, C.: On the feasibility of creating morphed iris-codes. In: 2017 IEEE International Joint Conference on Biometrics (IJCB) (2017)
9. Sharma, R., Ross, A.: Image-level iris morph attack. In: 2021 IEEE International Conference on Image Processing (ICIP), 3013–3017 (2021)
10. Gomez-Barrero, M., Rathgeb, C., Scherhag, U., Busch, C.: Predicting the vulnerability of biometric systems to attacks based on morphed biometric information. IET Biometrics **7**(4), 333–341 (2018)
11. Aydemir, A.K., Hämmerle-Uhl, J., Uhl, A.: Feasibility of morphing-attacks in vascular biometrics. In: 2021 IEEE/IAPR International Joint Conference on Biometrics (IJCB'21), pp. 1–7 (2021)
12. Miura, N., Nagasaka, A., Miyatake, T.: Extraction of finger-vein patterns using maximum curvature points in image profiles. IEICE Trans. Inf. Syst. **90**(8), 1185–1194 (2007)
13. Roettcher, A., Scherhag, U., Busch, C.: Finding the suitable doppelgaenger for a face morphing attack. In: 2020 IEEE International Joint Conference on Biometrics (IJCB) (2020)
14. ISO/IEC JTC1 SC37 Biometrics: Information technology - biometric presentation attack detection - part 3: Testing and reporting. ISO ISO/IEC IS 30107–3:2017, International Organization for Standardization, Geneva, Switzerland (2017)
15. Scherhag, U., et al.: Biometric systems under morphing attacks: assessment of morphing techniques and vulnerability reporting. In: 2017 International Conference of the Biometrics Special Interest Group (BIOSIG), pp. 149–159 (2017)
16. Asaari, M.S.M., S. A. Suandi, B.A.R.: Fusion of band limited phase only correlation and width centroid contour distance for finger based biometrics. Expert Syst. Appl. **41**(7) (2014) 3367–3382 (2014)

17. Lu, Y., Xie, S.J., Yoon, S., Wang, Z., Park, D.S.: An available database for the research of finger vein recognition. In: Image and Signal Processing (CISP), 2013 6th International Congress on. Volume 1, pp. 410–415 IEEE (2013)
18. Ton, B., Veldhuis, R.: A high quality finger vascular pattern dataset collected using a custom designed capturing device. In: International Conference on Biometrics, ICB 2013, IEEE (2013)
19. Kauba, C., Prommegger, B., Uhl, A.: Focussing the beam - a new laser illumination based data set providing insights to finger-vein recognition. In: 2018 IEEE 9th International Conference on Biometrics Theory, Applications and Systems (BTAS), Los Angeles, California, USA , 1–9 (2018)
20. Kauba, C., Prommegger, B., Uhl, A.: Openvein - an open-source modular multipurpose finger vein scanner design. In: Uhl, A., Busch, C., Marcel, S., Veldhuis, R. (eds.) Handbook of Vascular Biometrics, pp. 77–111. Springer Nature Switzerland AG, Cham, Switzerland (2019)
21. Lu, Y., Xie, S., Yoon, S., Yang, J., Park, D.: Robust finger vein roi localization based on flexible segmentation. Sensors **13**(11), 14339–14366 (2013)
22. Kauba, C., Uhl, A.: An available open-source vein recognition framework. In: Uhl, A., Busch, C., Marcel, S., Veldhuis, R. (eds.) Handbook of Vascular Biometrics, pp. 113–142. Springer Nature Switzerland AG, Cham, Switzerland (2019)
23. Huang, B., Dai, Y., Li, R., Tang, D., Li, W.: Finger-vein authentication based on wide line detector and pattern normalization. In: 2010 20th International Conference on Pattern Recognition (ICPR), pp. 1269–1273, IEEE (2010)
24. Starck, J., Fadili, J., Murtagh, F.: The undecimated wavelet decomposition and its reconstruction. IEEE Trans. Image Process. **16**(2), 297–309 (2007)
25. Kumar, A., Zhou, Y.: Human identification using finger images. IEEE Trans. Image Process. **21**(4), 2228–2244 (2012)
26. Choi, J.H., Song, W., Kim, T., Lee, S.R., Kim, H.C.: Finger vein extraction using gradient normalization and principal curvature. In: Image Processing: Machine Vision Applications II. Volume 7251 of Proc.SPIE, pp. 359–367 (2009)

A Robust Approach for Crop Misalignment Estimation in Single and Double JPEG Compressed Images

Giovanni Puglisi[1(✉)] and Sebastiano Battiato[2]

[1] University of Cagliari, Cagliari, Italy
puglisi@unica.it
[2] University of Catania, Catania, Italy
battiato@dmi.unict.it

Abstract. In forensics investigation, information about the crop performed after a JPEG compression can be useful exploited to reconstruct the manipulation history of the analyzed images and localize forgeries. Statistics computed from AC histograms, obtained performing an additional JPEG compression with a set of constant quantization matrices, are exploited to design a robust and effective algorithm to retrieve information about the employed crop. Finally, the effectiveness of the proposed solution has been demonstrated through a series of tests conducted considering different patch sizes, quantization matrices, and comparisons with a state-of-the-art solution in single and double JPEG compression scenario.

Keywords: Multimedia Forensics · Non-aligned Double JPEG Compression · Misalignment Estimation

1 Introduction

Today lots of images are acquired by a digital camera, often edited, and uploaded to Instant Messaging platforms and Social Networks. Single and multiple JPEG compressions [15,19] are then applied to these images and part of the information contained in the original picture is lost. In forensics investigation, could be useful to reconstruct the history of the image under analysis [1,2,11,12], recovering as example information about the camera source device or the camera model [20,25] that took the picture. Double compression traces could be detected [14], information related to camera model through First Quantization Estimation (FQE) [3–5,13,16] retrieved and the source exploiting Photo Response Non Uniformity (PRNU) analysis [20,24] identified. Image life-cycle typically involves multiple JPEG compressions. In literature two scenarios have been studied, aligned and non-aligned double JPEG compression based on the presence of misalignment between consecutive quantizations due to image cropping. The estimation of horizontal and vertical shifts can be then employed to perform tampering localization [7,27] and first quantization matrix estimation in NA-DJPEG (non-aligned double JPEG) scenarios [6,10,28].

G. L. Foresti et al. (Eds.): ICIAP 2023 Workshops, LNCS 14365, pp. 318–329, 2024.
https://doi.org/10.1007/978-3-031-51023-6_27

Misalignment estimation in a NA-DJPEG scenario has been usually studied considering pixel domain. Blocking artifact characteristics matrix (BACM) has been introduced in [18] and the analysis of its symmetry exploited to detect shifts between consecutive compressions. High-pass filters in pixel domain have been also employed in several approaches [8,9] and improved considering only homogeneous regions [28] to avoid being deceived by image content. Several frequency domain based approaches have been also designed to perform misalignment estimation. These approaches, usually, crop the input image considering all the possible shifts, perform a further compression and exploit DCT (Discrete Cosine Transform) coefficient analysis to obtain the estimation [6,21,22]. Although the aforementioned frequency domain approaches usually provide better results than the spatial domain ones, they usually require information about the last compression (e.g., the last quantization matrix).

To overcome this issue, a robust and effective solution able to estimate the crop misalignment parameters in scenarios where information about the last compression is not available (e.g., final saving in lossless format) has been designed. Specifically, the effect of additional compressions performed with a set of constant quantization matrices has been analyzed to implement the proposed solution.

The remainder of this paper is structured as follows: Sect. 2 describes the proposed approach, Sect. 3 reports experimental results and comparisons with a state-of-the-art method in different scenarios. Finally, Sect. 4 concludes the paper.

2 Proposed Approach

Starting from an input image I and a 8×8 quantization matrix, JPEG compression [26] can be defined as a function $f_Q(I)$ that provides as output a JPEG compressed image I'. Specifically, RGB input image I is converted to a different color space (YCbCr) and split into non-overlapping blocks 8×8. Moreover, DCT is applied to these blocks, the result divided by Q values, rounded and finally encoded by entropy based engine. In this paper only the luminance channel Y is considered. We also denote QF as the quality factor associated to standard quantization matrices [26] and QF_i as the quality factor used in the i-th JPEG compression.

Double JPEG compressed images can be obtained applying the aforementioned function twice $I'' = f_{Q_2}(f_{Q_1}(I))$, with Q_1 and Q_2 quantization matrices employed for the first and the second compression respectively. Depending on the presence of cropping between consecutive compressions, two cases are usually studied in literature: A-DJPEG and NA-DJPEG, i.e., aligned and non-aligned scenarios [20].

In order to detect misalignment parameters, a third compression with $Q_3 = Q_2$ tacking into account all the possible shift pairs (r, c) has been employed in frequency domain state-of-the-art solutions [21,22]. These approaches, exploit some

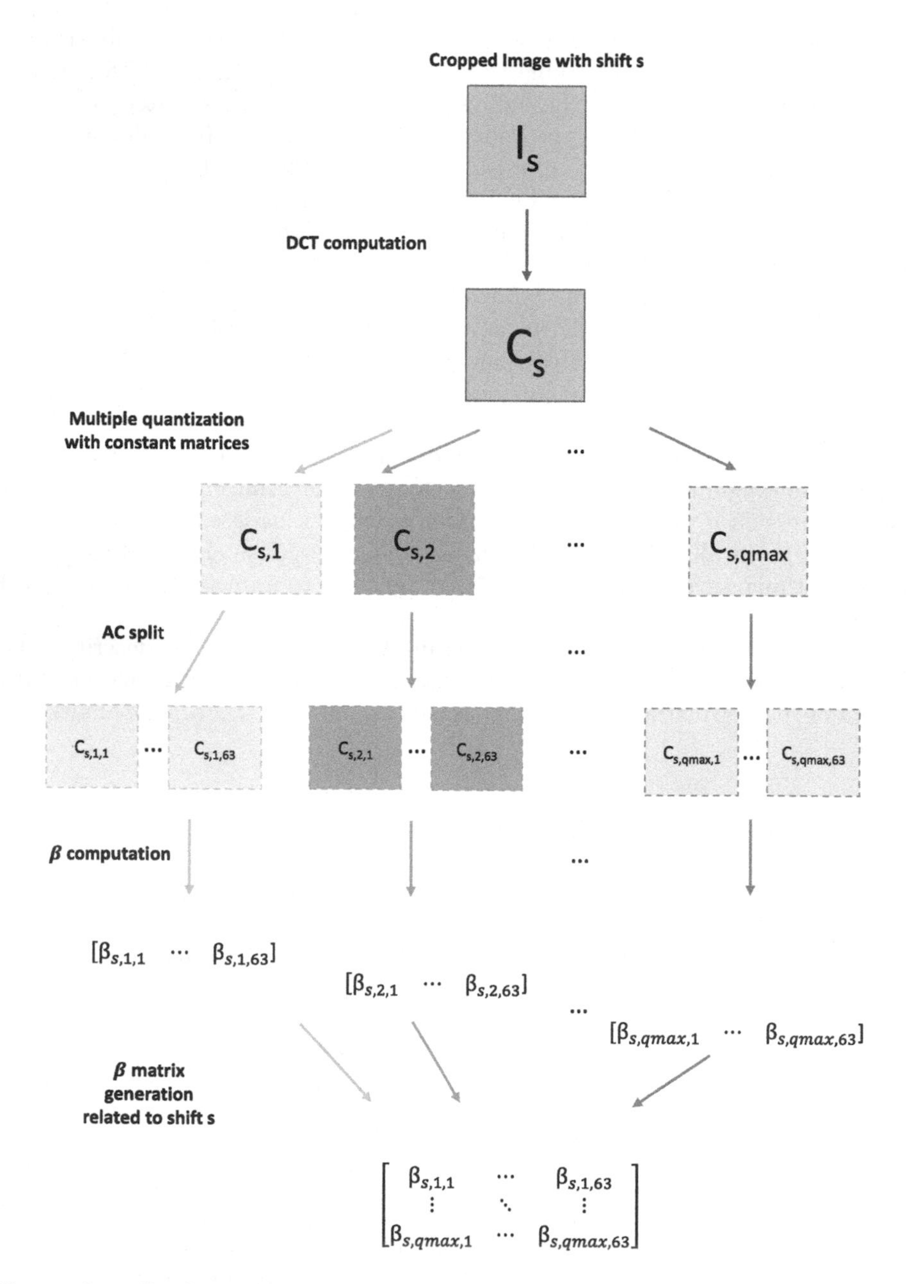

Fig. 1. Overall scheme of the feature matrix computation from shifted image I_s. At first, DCT is applied to I_s and C_s is generated. This transformed image is then compressed with a set of constant matrices with quantization factors $q \in \{1, 2, \ldots, qmax\}$. Data contained in these transformed and compressed images $C_{s,q}$ are split with respect to AC frequencies. Starting from $C_{s,q,cf}$, β values are computed and the $qmax \times 63$ β matrix related to a specific shift s is obtained.

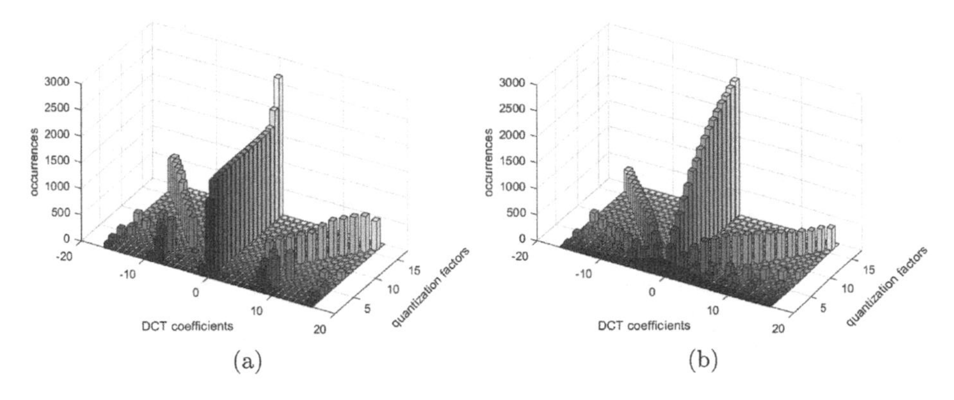

Fig. 2. Central portion of AC histograms ($cf = 3$) computed from two shifted images I_{19} (a) and I_{27} (b) compressed with a set of constant matrices. Specifically, these cropped images have been obtained from a single compressed 512×512 Lena image ($QF = 70$) with ground truth shift $s = 19$ (i.e., $r = 2$, $c = 3$).

differences in the statistics computed from DCT histograms obtained considering first and third block grids aligned with respect to the other ones computed with grid misalignment.

In this paper a scenario where the last compression matrix is not available (i.e., after JPEG compression, the input image is saved in a lossless format) is studied and an effective algorithm exploiting a set of constant quantization matrices has been designed. At first, 64 cropped images I_s are extracted from the JPEG compressed input with $s = r \times 8 + c$ ($r, c \in \{0, 1, \dots 7\}$). For each shift s, DCT is applied to I_s and C_s is generated. This transformed image is then compressed with a set of constant matrices with quantization factors $q \in \{1, 2, \dots, qmax\}$. Data contained in these transformed and compressed images $C_{s,q}$ are split with respect to AC frequencies ($cf \in \{1, 2, \dots, 63\}$). Zero-centred Laplace distribution [17] is then considered to model AC coefficients:

$$f(x) = \frac{1}{2\beta} \exp\left(-\frac{|x|}{\beta}\right) \tag{1}$$

with β scale parameter computed by maximum likelihood estimation close form solution as follows:

$$\beta_{s,q,cf} = \frac{1}{N_r \times N_c} \sum_{ki=1}^{N_r} \sum_{kj=1}^{N_c} |C_{s,q,cf}(ki, kj)| \tag{2}$$

and N_r and N_c number of DCT blocks extracted from I_s in the row and column directions.

Considering then the model (1), each $C_{s,q,cf}$ has been described with the related $\beta_{s,q,cf}$ value (see Fig. 1). Finally, tacking into account all the 64 cropped images extracted from the input compressed image, a 3D feature volume is obtained.

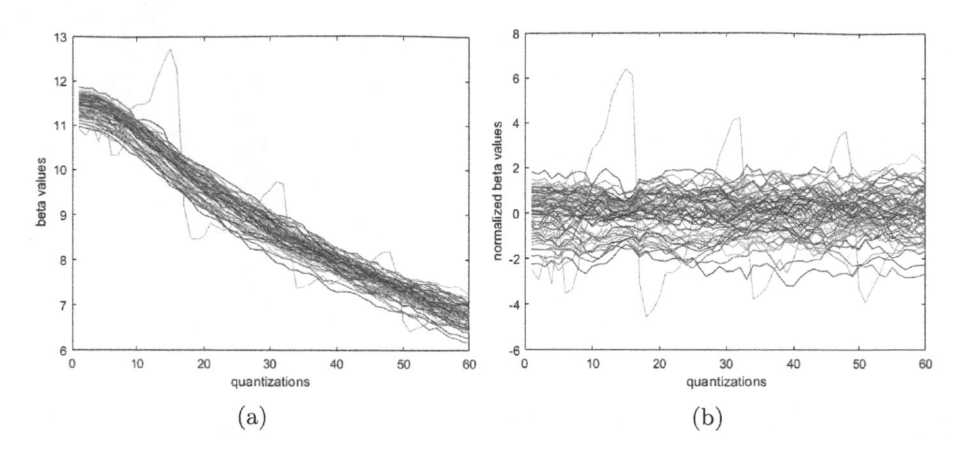

(a) (b)

Fig. 3. β values variation at increasing of the employed quantization factor q for all the 64 shifts (a). Note that the curve obtained considering the correct shift (in blue), has a periodic component. A normalized (z-score normalization) version able to remove the main decreasing trend, underlines this component (b). (Color figure online)

To properly study the behavior of the computed features (i.e., the 3D volume $\beta_{s,q,cf}$) a simple test has been conducted considering as input the 512×512 Lena image compressed with $QF = 70$ and cropped with $s = 19$ ($s = r \times 8 + c$, with $r = 2$ and $c = 3$). Specifically, the histograms generated with $cf = 3$ (zig-zag order) related to I_{19} (i.e., the correct shift) and I_{27} are studied at varying of the quantization factor used to generate the constant compression matrices. As can be seen from Fig. 2, where a window around the center is visualized (i.e., AC values $c \in [-17, 17]$), histogram values in Fig. 2(b) smoothly collapse into the zero bin at increasing of the quantization factor. On the contrary, histogram values related to the correct shift show a different behavior (see Fig. 2(a)). To properly understand these differences, the effect of the rounding function applied in the additional compression must be studied:

$$c_q = \left[\frac{c}{q}\right] q \tag{3}$$

where $q \in [1, 2, \ldots, 2q_1 + 1]$ in the example depicted in Fig. 2 with q_1 quantization factor employed in the first compression. In the considered example Lena image has been compressed with $QF = 70$, and the quantization factor related to the third AC in zig-zag order is $q_1 = 8$. In the histogram generated considering the correct shift and $q = 1$ (i.e., no further quantization) AC coefficients are mainly clustered into specific bins multiple of q_1. Analyzing the behavior of the elements in the bin equals to q_1, they shift around the initial position, reach the maximum value with $q = 2q1$ (i.e., their $c_q = 16$) and are set to zero with $q = 2q1 + 1$. This behavior, due to (3), can be useful exploited employing

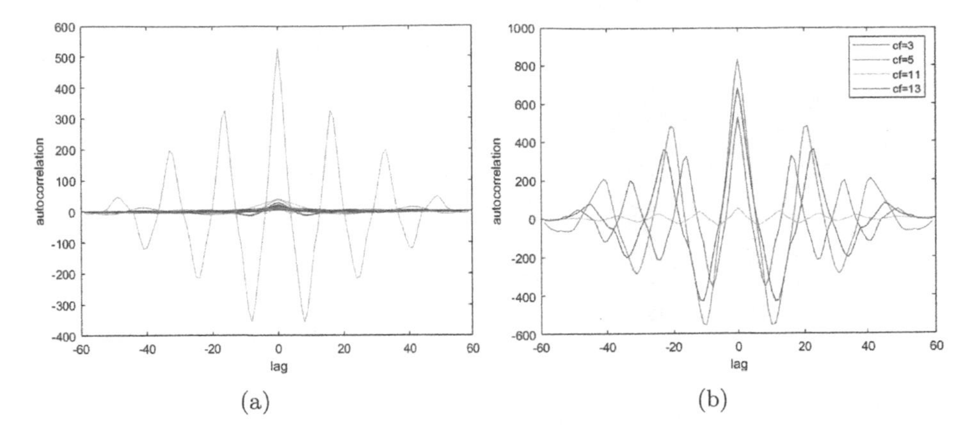

Fig. 4. Autocorrelation function computed from the normalized data considering all the 64 shifts (a). It is worth noting that the periodicity of the signal is strictly related to the quantization factor employed in the first compression ($q_1 = 8$ in this example). This behavior is visible also considering other AC frequencies related to the correct shift (b).

a simple feature such as the β parameter of the related distribution (1) where a simple average of the absolute input values is actually computed (2).

Fixed a specific AC frequency $cf = 3$, β values relate to all the possible shifts $s \in \{0, 1, \ldots, 63\}$ are then computed at varying of the quantization factor $q \in \{1, 2, \ldots, qmax\}$. As reported in Fig. 3(a) the curve related to the correct shift (in blue) has a periodic component not visible in the curves related to the other shifts. To further underline this component removing the main decreasing trend due to the increase of the intensity of the applied quantization, data have been normalized with z-score normalization (see Fig 3(b)). Finally, autocorrelation function is computed from the normalized data to better underline this periodic component. It is worth noting that the periodicity of the signal is strictly related to the quantization factor employed in the first compression (see Fig. 4(a)). This behavior is also visible considering other AC frequencies as depicted in Fig. 4(b).

Fixed a specific AC frequency cf, a symmetric matrix M^{cf}_{dist} containing all the possible distances $d_{si,sj}$ between each pair of normalized β curves can be computed considering a selected metric (e.g., Euclidean distance). As depicted in Fig. 5(a), the row (or column) of M^{cf}_{dist} containing the highest distance values identifies the correct shift $\hat{s}$. A projection along columns (or rows) of M^{cf}_{dist} can be then employed to simplify the estimation task as follows:

$$Pm^{cf}_{dist}(sj) = \frac{1}{64} \sum_{si=0}^{63} M^{cf}_{dist}(si, sj) \quad sj \in \{0, 1, \ldots 63\} \tag{4}$$

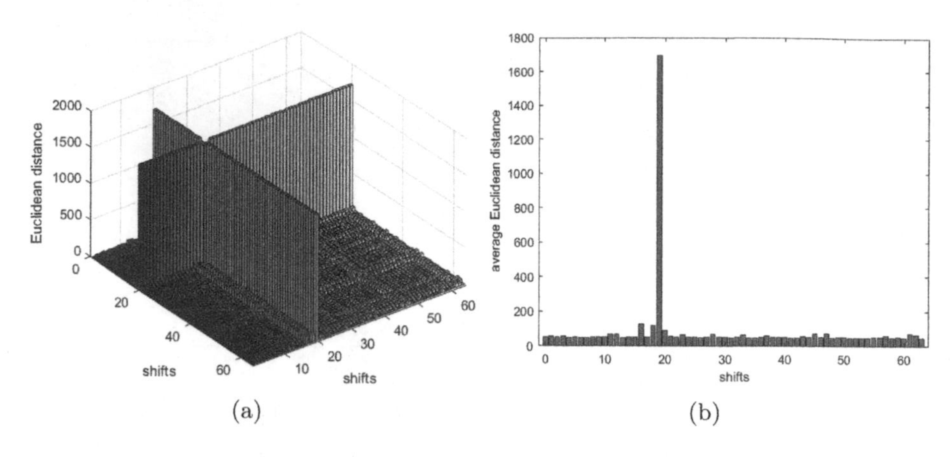

Fig. 5. Euclidean distances among normalized curve of β values obtained from different shifts (a) and its projections with respect to columns (b). Third AC frequency (i.e., $cf = 3$) is considered.

Once the vector Pm_{dist}^{cf} is computed (see Fig. 5(b)) the shift related to the maximum value of Pm_{dist}^{cf} can be easily found as:

$$\hat{s}^{cf} = \arg\max_{s} Pm_{dist}^{cf}(s) \quad s \in \{0, 1, \dots 63\} \tag{5}$$

To increase the robustness of the proposed solution also in challenging conditions (e.g., small patches), all the information contained in the input image must be properly exploited. The final estimation $\hat{s}$ is then obtained through a simple voting strategy based on a histogram built considering all the shifts $\hat{s}_{cf}$.

So far, only single compressed and cropped images have been considered in our analysis. In order to take into account a more complex scenario, several studies have been conducted considering non-aligned double JPEG compressed images. Specifically, the pipeline described before has been applied to a 512×512 Lena image compressed twice ($QF_1 = 80$, $QF_2 = 90$) with misalignment $s = 19$. Both autocorrelation function related to $cf = 4$ and the related distance matrix M_{dist}^{4} is reported in Fig. 6. In this double compression scenario, two β curves related to $s = 0$ (no shift) and $s = 19$ (correct shift) show a periodic pattern. This behavior depends on the further compression (i.e., the second one) applied to the cropped image and can mislead the estimation of the correct shift (see Fig. 6(b)).

To cope with this additional peak at $s = 0$ in the final histogram employed to perform the shift estimation $\hat{s}$ built with the contribution of all frequencies cf, an effective strategy is designed. Specifically, if $\hat{s} = 0$ the ratio r_{12} between the occurrences related to the first and the second most voted shifts is computed. As can be seen in Fig. 7 this ratio can be then used as a simple feature to discriminate aligned (a) versus non-aligned case (b) and allow us to select the proper values of the shift (i.e., $s = 0$ or s related to the second most voted misalignment).

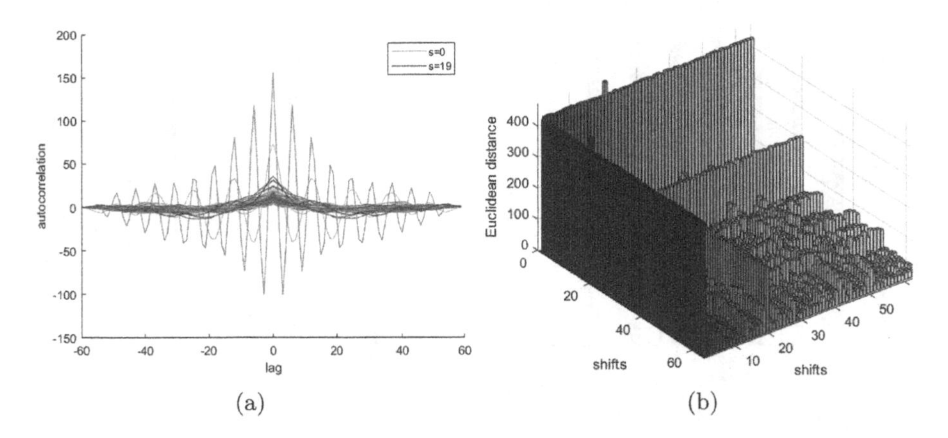

Fig. 6. Autocorrelation function computed from the normalized data considering all the 64 shifts and the 512 × 512 double compressed Lena image as input (a). Due to the multiple compressions, two curves related to $s = 0$ (i.e., no shift) and $s = 19$ (the correct one) contains a periodic component. This behaviour is also visible in the related Euclidean distance map M^4_{dist}(b).

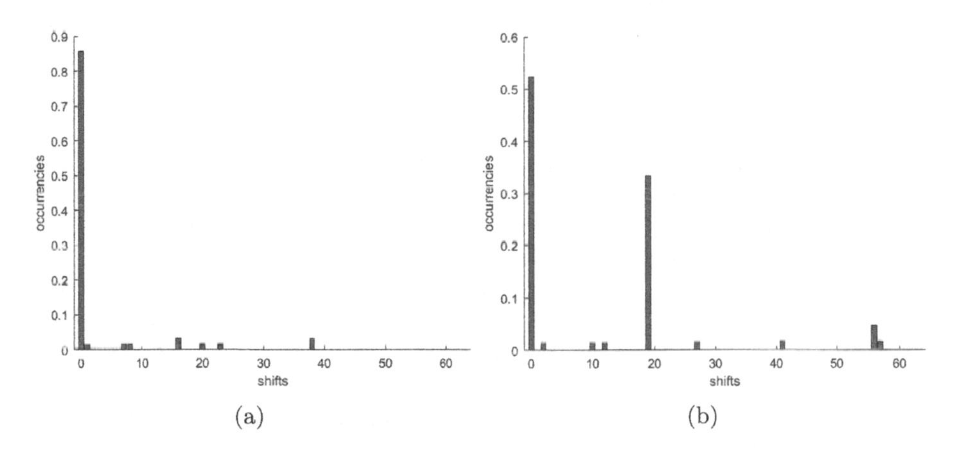

Fig. 7. Final histograms employed to compute the grid misalignment considering a double compressed ($QF_1 = 80$, $QF_2 = 90$) 512 × 512 Lena image as input with grid misalignment equals to 0 (a) and 19 (b)). Although in both cases the most voted bin is $s = 0$, the ratio r_{12} between the occurrences related to the first and second most voted misalignment is 27 and 1.28 respectively.

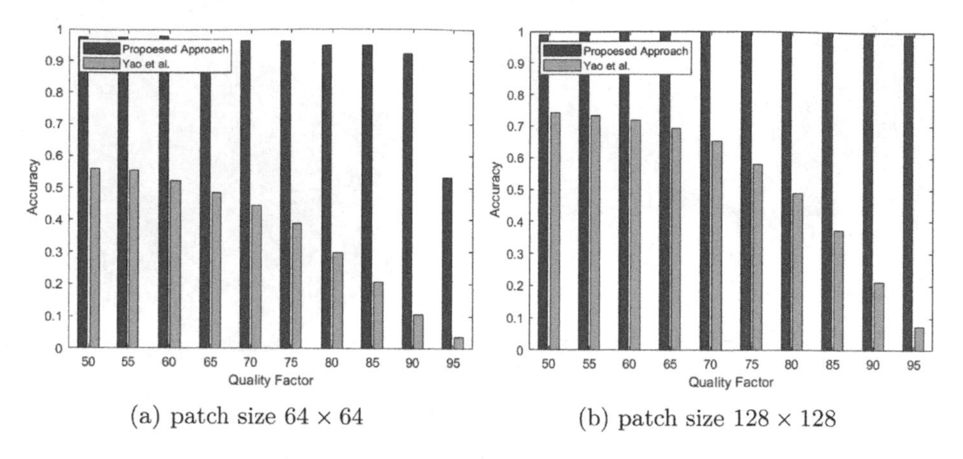

(a) patch size 64 × 64

(b) patch size 128 × 128

Fig. 8. Comparison between the proposed solution and Yao et al. [28] at varying of $QF \in \{50, 55, 60, 65, 70, 75, 80, 85, 90, 95\}$, and patch size in single compression scenario.

3 Experimental Results

To assess the performance of the proposed approach, two scenarios have been considered: single JPEG compression followed by a crop operation and double JPEG compression with misalignment. In these scenarios, the images have been finally saved in a lossless format (i.e., png) and information about the employed quantization matrices is actually lost. Several test datasets, D'_{test} and D''_{test} related to single and double JPEG compression scenario respectively, have been built from UCID [23] collection. Patches $d \times d$ are cropped from 100 UCID images with $d \in \{64, 128\}$, single and double JPEG compressed with QF_1 from 50 to 95 at step of 5 and $QF_2 = 90$. Considering all the 64 possible shifts each test dataset consists of 128000 $(2 \times 100 \times 10 \times 64)$ patches. State-of-the-art solution [28] has been considered in our tests reimplementing the algorithm described in the related paper. It is worth noting that, methods developed in [21,22] cannot be considered in these comparisons due to the lack of information about the last compression matrix. The proposed solution outperforms state-of-the-art method [28], in terms of accuracy in the misalignment value estimation task, by a large margin in all the considered scenarios involving both single (Fig. 8) and double JPEG compression (Fig. 9). Note that a larger margin is achieved in the challenging scenario reported in Fig. 9(a) (smaller patches, double JPEG compression with QF_1 close to QF_2).

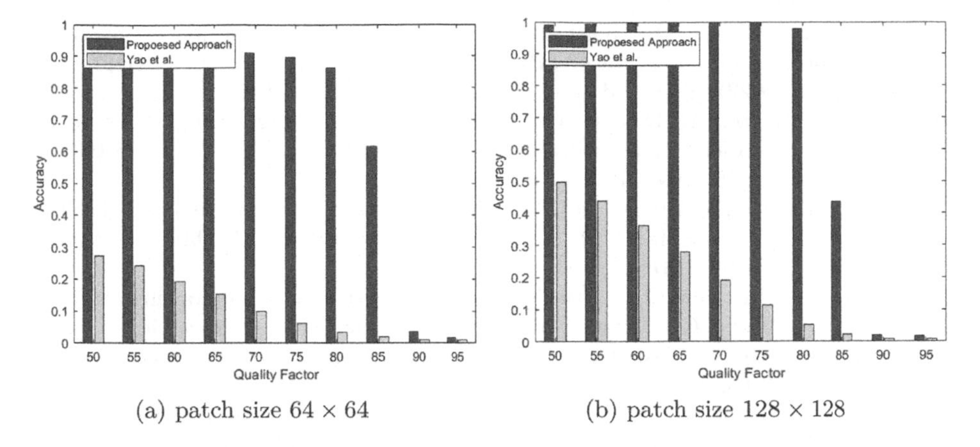

(a) patch size 64×64 (b) patch size 128×128

Fig. 9. Comparison between the proposed solution and Yao et al. [28] at varying of $QF_1 \in \{50, 55, 60, 65, 70, 75, 80, 85, 90, 95\}$ and patch size in double compression scenario ($QF_2 = 90$)

4 Conclusions

Information about the crop performed after a JPEG compression can be useful exploited in forensics investigation to reconstruct the manipulation history of the analyzed images. Statistics computed from AC histograms obtained performing an additional JPEG compression employing a set of constant quantization matrices are exploited to design an effective misalignment estimation solution. To assess the performance of the proposed approach, a series of tests and comparisons with a state-of-the-art solution in single and double compressed scenarios have been performed. Finally, future works will be devoted to improve the accuracy of the proposed method also in $QF_1 \geq QF_2$ scenario.

Acknowledgements. Acknowledge financial support from: PNRR MUR project PE0000013-FAIR.

References

1. Battiato, S., Giudice, O., Paratore, A.: Multimedia forensics: discovering the history of multimedia contents. In: Proceedings of the 17th International Conference on Computer Systems and Technologies 2016, pp. 5–16. ACM (2016)
2. Battiato, S., Messina, G.: Digital forgery estimation into DCT domain: a critical analysis. In: Proceedings of the First ACM Workshop on Multimedia in Forensics, pp. 389–399. 25, 6 (2010)
3. Battiato, S., Giudice, O., Guarnera, F., Puglisi, G.: Estimating previous quantization factors on multiple JPEG compressed images. EURASIP J. Inf. Secur. **2021**(8), 1–11 (2021). https://doi.org/10.1186/s13635-021-00120-7
4. Battiato, S., Giudice, O., Guarnera, F., Puglisi, G.: First quantization estimation by a robust data exploitation strategy of DCT coefficients. IEEE Access **9**, 73110–73120 (2021). https://doi.org/10.1109/ACCESS.2021.3080576

5. Battiato, S., Giudice, O., Guarnera, F., Puglisi, G.: CNN-based first quantization estimation of double compressed JPEG images. J. Vis. Commun. Image Represent. **89**, 103635 (2022). https://doi.org/10.1016/j.jvcir.2022.103635
6. Bianchi, T., Piva, A.: Detection of nonaligned double JPEG compression based on integer periodicity maps. IEEE Trans. Inf. Forens. Secur. **7**(2), 842–848 (2012). https://doi.org/10.1109/TIFS.2011.2170836
7. Bianchi, T., Piva, A.: Image forgery localization via block-grained analysis of JPEG artifacts. IEEE Trans. Inf. Forens. Secur. **7**(3), 1003–1017 (2012). https://doi.org/10.1109/TIFS.2012.2187516
8. Bruna, A.R., Messina, G., Battiato, S.: Crop detection through blocking artefacts analysis. In: Maino, G., Foresti, G.L. (eds.) Image Analysis and Processing—ICIAP 2011. LNCS, vol. 6978, pp. 650–659. Springer, Heidelberg (2011). https://doi.org/10.1007/978-3-642-24085-0_66
9. Dalmia, N., Okade, M.: A novel technique for misalignment parameter estimation in double compressed JPEG images. In: 2016 Visual Communications and Image Processing (VCIP), pp. 1–4 (2016). https://doi.org/10.1109/VCIP.2016.7805446
10. Dalmia, N., Okade, M.: Robust first quantization matrix estimation based on filtering of recompression artifacts for non-aligned double compressed JPEG images. Signal Process. Image Commun. **61**, 9–20 (2018)
11. Fan, Z., De Queiroz, R.: Identification of bitmap compression history: JPEG detection and quantizer estimation. IEEE Trans. Image Process. **12**(2), 230–235 (2003)
12. Farid, H.: Digital image ballistics from JPEG quantization: a followup study. Department of Computer Science, Dartmouth College, Tech. Rep. TR2008-638 (2008)
13. Galvan, F., Puglisi, G., Bruna, A.R., Battiato, S.: First quantization matrix estimation from double compressed JPEG images. IEEE Trans. Inf. Forens. Secur. **9**(8), 1299–1310 (2014)
14. Giudice, O., Guarnera, F., Paratore, A., Battiato, S.: 1-D DCT domain analysis for JPEG double compression detection. In: Proceedings of International Conference on Image Analysis and Processing. LNCS, vol. 11752, pp. 716–726. Springer, Cham (2019). https://doi.org/10.1007/978-3-030-30645-8_65
15. Giudice, O., Paratore, A., Moltisanti, M., Battiato, S.: A classification engine for image ballistics of social data. In: Battiato, S., Gallo, G., Schettini, R., Stanco, F. (eds.) Image Analysis and Processing— ICIAP 2017. LNCS, vol. 10485, pp. 625–636. Springer, Cham (2017). https://doi.org/10.1007/978-3-319-68548-9_57
16. Kee, E., Johnson, M.K., Farid, H.: Digital image authentication from JPEG headers. IEEE Trans. Inf. Forens. Secur. **6**(3), 1066–1075 (2011)
17. Lam, E.Y., Goodman, J.W.: A mathematical analysis of the DCT coefficient distributions for images. IEEE Trans. Image Process. **9**(10), 1661–1666 (2000)
18. Luo, W., Qu, Z., Huang, J., Qiu, G.: A novel method for detecting cropped and recompressed image block. In: 2007 IEEE International Conference on Acoustics, Speech and Signal Processing (ICASSP 2007), vol. 2, pp. II-217–II-220 (2007). https://doi.org/10.1109/ICASSP.2007.366211
19. Moltisanti, M., Paratore, A., Battiato, S., Saravo, L.: Image manipulation on Facebook for forensics evidence. In: Murino, V., Puppo, E. (eds.) Image Analysis and Processing—ICIAP 2015. LNCS, vol. 9280, pp. 506–517. Springer, Cham (2015). https://doi.org/10.1007/978-3-319-23234-8_47
20. Piva, A.: An overview on image forensics. ISRN Signal Process. **2013**, 22 (2013). https://doi.org/10.1155/2013/496701

21. Puglisi, G., Battiato, S.: A robust misalignment estimation approach in non-aligned double JPEG compression scenario. In: 2022 IEEE International Conference on Image Processing (ICIP), pp. 3096–3100 (2022). https://doi.org/10.1109/ICIP46576.2022.9897626
22. Puglisi, G., Battiato, S.: Misalignment estimation in non-aligned double JPEG scenario based on AC histogram analysis. In: Rousseau, J.J., Kapralos, B. (eds.) Pattern Recognition, Computer Vision, and Image Processing. ICPR 2022 International Workshops and Challenges. LNCS, vol. 13644, pp. 338–346. Springer, Cham (2023). https://doi.org/10.1007/978-3-031-37742-6_26
23. Schaefer, G., Stich, M.: UCID: an uncompressed color image database. In: Storage and Retrieval Methods and Applications for Multimedia 2004, vol. 5307, pp. 472–480. International Society for Optics and Photonics (2003)
24. Stamm, M.C., Wu, M., Liu, K.J.R.: Information forensics: an overview of the first decade. IEEE Access **1**, 167–200 (2013)
25. Verdoliva, L.: Media forensics and deepfakes: an overview. IEEE J. Select. Topics Signal Process. **14**(5), 910–932 (2020)
26. Wallace, G.K.: The JPEG still picture compression standard. Commun. ACM **34**(4), 30–44 (1991)
27. Wu, L., Kong, X., Wang, B., Shang, S.: Image tampering localization via estimating the non-aligned double JPEG compression. In: Alattar, A.M., Memon, N.D., Heitzenrater, C.D. (eds.) Media Watermarking, Security, and Forensics 2013, vol. 8665, pp. 260–266. International Society for Optics and Photonics, SPIE (2013). https://doi.org/10.1117/12.2003695
28. Yao, H., Wei, H., Qin, C., Zhang, X.: An improved first quantization matrix estimation for nonaligned double compressed JPEG images. Signal Process. **170**, 107430 (2020)

Human-in-the-Loop Person Re-Identification as a Defence Against Adversarial Attacks

Rita Delussu[1](✉), Lorenzo Putzu[1], Emanuele Ledda[1,2], and Giorgio Fumera[1]

[1] Department of Electrical and Electronic Engineering, University of Cagliari, Cagliari, Italy
{rita.delussu,lorenzo.putzu,fumera}@unica.it, emanuele.ledda@uniroma1.it
[2] Department of Informatics, Bioengineering, Robotics, and Systems Engineering, University of Genoa, Genoa, Italy

Abstract. Person re-identification (Re-Id) is a computer vision task useful to security-related applications of video surveillance systems. Recently it has been shown that Re-Id systems, currently based on deep neural networks, are vulnerable to adversarial attacks, some of which are based on manipulating the query image to prevent other images of the same individual from being retrieved. Whereas some ad hoc defence strategies have been proposed so far against different implementations of this kind of attack, we argue that the human-in-the-loop (HITL) approach, originally proposed for retrieval systems (including Re-Id) to improve retrieval accuracy under normal operational conditions, can also act as an effective and general defence strategy, with the notable advantage that it does not degrade accuracy in the absence of attacks, contrary to ad hoc defences. We provide empirical evidence of this fact on several benchmark data sets and state-of-the-art Re-Id models, using a simple HITL implementation based on relevance feedback algorithms.

Keywords: Person Re-Identification · Adversarial Attack · Defence Strategy · Human-in-the-loop

1 Introduction

Person re-identification (Re-Id) consists of recognising images of a person of interest across different and non-overlapping cameras. This is a challenging task due to, e.g., pose variations, low camera resolution, occlusions, and different camera views. Re-Id systems devised for offline investigations (e.g., for law enforcement agencies) are typically made up of the following functional modules: a pedestrian detector processes recorded videos to extract bounding boxes (BBs) of all pedestrians; the extracted BBs are stored in a database, called *gallery* set; during operation, the user/operator can select the image (BB) of a person of interest (*query*) from a video, and the task of the Re-Id system is to

G. L. Foresti et al. (Eds.): ICIAP 2023 Workshops, LNCS 14365, pp. 330–342, 2024.
https://doi.org/10.1007/978-3-031-51023-6_28

retrieve all the images of the query identity from the gallery; to this aim, a feature a extractor (e.g., a Deep Neural Network, DNN) is run on both the gallery images and the query, a matching process is carried out, and a ranked list of gallery images is returned, ordered by the decreasing similarity to the query. This can considerably reduce the search effort required to the user. State-of-the-art Re-Id methods based on DNN models achieved remarkable performance on benchmark data sets, especially in a fully-supervised scenario. Recently, several authors started investigating whether existing Re-Id methods are vulnerable to adversarial attacks (e.g., [22]), that have already been shown to be a serious threat for deep learning-based systems in different computer vision tasks, as well as in other application domains (e.g., natural language processing) [3,5].

In terms of the adversary's *model* [3], in the context of Re-Id systems, the *goal* of an attacker is to prevent the retrieval of the gallery images of a specific query identity (*targeted* attack) or of any query identity (*indiscriminate* or *untargeted* attack), by pushing them to the bottom ranks; in the latter case, the Re-Id system becomes ineffective, and thus unusable. Specific attacks devised so far in the literature to achieve these goals can be categorised as physical [23,24] and digital [1,4,6,9,11,14,18,19,22,26,27]. The former are generally implemented assuming the adversary's *capability* of placing in the scenes acquired by the cameras objects exhibiting a precise texture or template, e.g., wearing a t-shirt with a specific "patch", which may allow to evade a pedestrian detector, or to prevent a person wearing it from appearing in the top ranks, when chosen as a query [24]. Digital attacks assume instead that the adversary has access to the Re-Id system, e.g., to manipulate the query image selected by the user, by keeping the changes imperceptible to human eyes; to this aim, several image manipulation *strategies* have been proposed, such as universal perturbation (e.g., [9]) or changing the image colours (e.g., [27]). Both physical and digital attacks require the adversary some *knowledge* of the underlying Re-Id system, for instance, the model (e.g., architecture) of the pedestrian detector or the DNN used for feature extraction. Defence strategies proposed so far were developed ad hoc for specific attacks [1,4,11,14,27]. Moreover, analogously to defences developed in the adversarial machine learning literature for other tasks, they worsen the system's performance in the absence of attacks. Therefore, a trade-off between the security of a Re-Id system and its performance in the absence of attacks is required.

In this work, we focus on the human-in-the-loop (HITL) approach, which had been originally adopted in several applications of machine learning, such as robotics, automatic engineering, as well as Re-Id [13,20,21], for improving the effectiveness of a system by exploiting human (user) feedback, especially in the case of large-scale video surveillance system [20]. In particular, in Re-Id systems HITL methods exploit the inherent interaction with a skilled law enforcement operator, who is already in the loop, since it is involved in the retrieval process, by asking for feedback on the top-ranked gallery images retrieved in response to a given query; appropriate algorithms then use the feedback to update the ranked list, by pushing gallery images of the query individual (if any) toward the top ranks. In particular, in retrieval tasks (of which Re-Id is a particular case), HITL aims at reducing the *semantic gap* between the concept expressed by the user with the query image and the low-level image representation computed by the machine [7,10,12,25].

We argue that the HITL approach can also act as a *generic* (not ad hoc) defence strategy against adversarial attacks to Re-Id systems based on manipulating the query image. The rationale is that, under the perspective of semantic gap reduction, HITL could also allow reducing the *additional* gap introduced by an adversarial attack (see Sect. 3). Under the viewpoint of an adversarial defence, HITL also exhibits two very interesting features: (i) some HITL implementations do not require any re-training or fine-tuning step of the underlying Re-Id model, and can therefore be easily embedded into any Re-Id system; (ii) contrary to ad hoc defence strategies developed in the adversarial machine learning field (e.g. adversarial re-training), including the Re-Id task, it does not worsen the system's performance under normal conditions (i.e., in absence of attacks), as its original goal is to *improve* performances under such conditions. We also point out that a similar HITL approach to adversarial defence as the one we focus on in this work can be applied to any image retrieval system beside Re-Id.

The remainder of this work is as follows. We review related work in Sect. 2, and then we describe the proposed HITL-based defence strategy in Sect. 3. In Sect. 4, we describe the experimental setup, and we report and discuss the results. Finally, in Sect. 5 we draw conclusions and discuss some directions for future works.

2 Related Work

In this section, we provide an overview of Re-Id, and on existing adversarial attacks to Re-Id systems.

Person Re-Identification. Many different Re-Id methods have been proposed so far in the literature. Early ones were based on manually defined features and image similarity measures, and on metric learning [2], whereas state-of-the-art ones are mostly based on ad hoc DNN models. Different approaches have also been proposed to address the main issues of the Re-Id task (such as pose variations), as well as different application scenarios, e.g., supervised and unsupervised domain adaptation [28]. As an example, we summarise here the state-of-the-art Re-Id methods that we used in the experiments reported in Sect. 4 [8,15,31–33]. Most of them use Generative Adversarial Networks (GANs) [8,31–33], intending to address two main issues, i.e., the domain gap between source and target data [8,32] and the lack of data [31,33]. For instance, in [8], an image-image translation method is used to generate images with target domain style while preserving identity information; a local max pooling operation is also used to improve the feature representation and reduce the influence of noisy translated images. In [31], a GAN is used for augmenting training data and a label smoothing regularisation technique is developed to label generated (unlabelled) data. The embedded GAN improves the model's generalisation capability.

Adversarial Person Re-Identification. As mentioned in Sect. 1, attacks against Re-Id systems developed so far in the literature can be categorised into physical and digital attacks. In this work we focus on the latter [1,4,6,9,11,14, 18,19,22,26,27], because they are simpler to reproduce. We leave the evaluation of HITL against physical attacks to future work.

Some authors proposed both a specific attack and a corresponding defence strategy [1,4,11,14,27], while other proposed only an attack [6,9,18,19,22,26]. The most common attack approaches aim at generating a universal adversarial perturbation (UAP) [6,9,11,19,26,27], which consists of suitably modifying the image pixels (usually the query), such that the manipulation is imperceptible to human eyes. Some authors evaluated the attack transferability by using a different data set for testing [9,26], whereas others focused on a specific scene [1, 4,6,19]. Recently, some methods based on GANs have been proposed [14,18,22], with the aim of generating realistic adversarial images.

Among the proposed defence strategies [1,4,11,14,27], the most common is adversarial training, which consists of including in training data either simulated attack samples [1,4,14] or other information, e.g., greyscale or sketch images [11].

3 The HITL Approach as an Adversarial Defence

We focus on adversarial attacks to Re-Id systems based on manipulating the query image to push gallery images of the same identity toward the bottom ranks, whose goal is to prevent the user from retrieving images of the person of interest. To this aim, the manipulation should reduce the similarity values that are computed during the matching phase between the query and gallery images of the same identity. Different attack strategies have been proposed to achieve such a goal, as well as some ad hoc defence strategies (see Sect. 2).

Motivations. In this context, we argue that the HITL strategy, originally proposed for a different purpose, can also be effective as a defence strategy against this kind of attack. In image retrieval systems, the HITL approach is usually implemented by asking the user for feedback on the *relevance* to the query of some of the top-ranked retrieved images; the ranked list is then updated based on the acquired feedback, with the aim of pushing relevant images toward the top ranks. This process can be repeated for several iterations (feedback rounds) [17]. In Re-Id systems, the feedback typically consists of selecting one or more true (if any) or wrong matches, i.e., gallery images that do or do not correspond to the query identity [7,13,20,21]. In a retrieval task, the rationale of the HITL approach based on user feedback is to reduce the semantic gap between the concept expressed by the user through the query image (in a Re-Id system, the identity of the corresponding individual) and the low-level image representation computed by the machine [7,10,12,25]. Under this perspective, the kind of adversarial attack on Re-Id systems described above can be seen as purposely introducing an *additional* semantic gap aimed at reducing the low-level image similarity computed by the system between the query and the corresponding gallery images. Accordingly, the *same* HITL approach used in Re-Id systems under normal operational conditions may also be beneficial to counteract the effect of the considered adversarial attack, at least by mitigating the corresponding additional semantic gap.

We point out that, in contrast to ad hoc defences proposed in previous works against specific attacks based on manipulating the query image (see Sect. 2), HITL would be a *generic* defence strategy, as it is not devised against a specific attack (i.e., a specific manipulation strategy of the query image). Moreover, the HITL approach would exhibit a notable advantage with respect to the well-known drawback of defence strategies specifically developed for adversarial settings: whereas the latter typically worsen the performance in the absence of attacks, HITL is originally aimed at *improving* the retrieval accuracy under normal operational conditions [7,13,20,21]. Therefore, no trade-off between accuracy under normal conditions and under attack would be required.

Implementation. To experimentally evaluate the effectiveness of the HITL approach as a defence against adversarial attacks to Re-Id systems based on manipulating the query image, we choose a simple implementation based on relevance feedback (RF) algorithms originally developed in the context of image retrieval systems. We have shown in previous work that RF algorithms are also effective for HITL Re-Id (not in adversarial settings) [7]. Regardless of the specific RF algorithm, in our HITL Re-Id implementation, the feedback consists in asking the user to select *all* the true matches (also called *positive* images) among the top-K gallery images, for a given K value; the remaining images among the top-K ones are automatically labelled as negative. RF algorithms then use positive (P) and negative (N) examples (images) to update the ranked list; each one performs this process differently. The RF algorithms we considered in our previous work are the well-known Query Shift (QS) [12], Relevance Score (RS) [10] and Passive Aggressive (PA) [16], as well as a modified version of RS, called Mean RS (M-RS) [7], that we devised specifically for Re-Id systems. **QS** is based on moving the query in feature space, by computing a new query vector towards the Euclidean centre of P and farther from N, such that in the next round, a larger number of positive images can be found in the top ranks:

$$\overline{x}_{\mathrm{q}} = x_{\mathrm{q}} + \frac{1}{|P|} \sum_{\mathrm{p} \in P} x_{\mathrm{p}} - \frac{1}{|N|} \sum_{\mathrm{n} \in N} x_{\mathrm{n}} \ . \tag{1}$$

RS computes a *relevance score* for each retrieved image, based on its distance to the nearest positive and nearest negative neighbouring images in P and N, $x_{\mathrm{p}}^{\mathrm{NN}}$ and $x_{\mathrm{n}}^{\mathrm{NN}}$:

$$s_{\mathrm{NN}}(x_i) = \frac{\|x_i - x_{\mathrm{n}}^{\mathrm{NN}}\|}{\|x_i - x_p^{\mathrm{NN}}\| + \|x_i - x_{\mathrm{n}}^{\mathrm{NN}}\|} \ , \tag{2}$$

where $||\cdot||$ is the distance metric, typically the Euclidean distance. Obviously, such a score is higher when the distance from the nearest positive image decreases compared to the distance from the nearest negative one. Also, **M-RS** computes a *relevance score*, but it assumes that images of the same identity and of different identities form two distinct clusters. Accordingly, the score is computed with respect to the centroids μ_P and μ_N of the positive and negative clusters:

$$s_\mu(x_i) = \frac{||x - \mu_N||}{||x_i - \mu_P|| + ||x_i - \mu_P||} . \tag{3}$$

Similarly to RS and M-RS, **PA** computes a score for each retrieved image but using a pattern classification approach, that assumes that P and N are linearly separable in feature space:

$$s(x_i) = w \cdot x_i , \tag{4}$$

where w is a query-specific weight vector that should provide higher scores for positive images than for negative ones. Accordingly, it is obtained as the solution to the following optimisation problem:

$$w = \arg\max_v \sum_{\forall p \in P} \sum_{\forall n \in N} v(x_p - x_n) . \tag{5}$$

We finally point out that the above implementation of HITL for Re-Id systems is model-agnostic, i.e., it does not make any assumption on the underlying Re-Id model, and thus it can be embedded on top of any Re-Id system.

4 Experiments

In the following, we first describe the data sets and the experimental set-up, and then present and discuss the results.

Data Sets. We considered two widely used data sets: Market-1501 [30] and DukeMTMC-Re-Id [29] (for short Market and Duke). Market consists of 1,501 identities (IDs) and 32,668 images acquired from six cameras in front of a supermarket. This data set is divided into 751 IDs (12,936 images) for training and 750 IDs (the remaining images) for testing. The gallery contains 19,732 images. Duke consists of 1,404 IDs and 36,411 images acquired from eight cameras in a campus. It is split into 702 IDs (16,522 images) for training and 702 IDs (the remaining images) for testing. The gallery contains 17,661 images.

Experimental Setting. To implement the considered adversarial attack, we used an existing manipulation strategy of the query image whose source code has been made available by the authors [22]; to our knowledge, no defence strategy has been proposed so far against it. For our experiments, we selected the following Re-Id models, against which the above attack has already been successfully tested: AlignedReId [15], LSRO [31], CamStyle [33], HHL [32], and

Table 1. Re-identification accuracy of the five models with no attack, under attack with no defence, and using the HITL-based defence to counter the attack, after three feedback rounds of each of the considered RF algorithms. Best results attained by HITL-based defence algorithm (for each model) are highlighted in bold.

Model	Data set		Market-1501					DukeMTMC-reid				
			mAP	Rank-1	Rank-5	Rank-10	Rank-20	mAP	Rank-1	Rank-5	Rank-10	Rank-20
Alig. ReId [15]	No attack		79.1	91.8	97.0	98.1	99.0	69.7	82.1	91.7	94.4	95.8
	Under attack		2.4	1.7	4.0	5.8	8.9	16.7	19.9	31.6	38.2	45.7
	HITL-D	RS	35.42	37.32	37.86	38.42	39.31	74.56	78.41	78.82	78.95	79.35
		MM	35.49	37.17	37.35	37.41	37.83	74.22	78.82	79.22	79.44	79.8
		QS	26.66	29.78	30.64	30.73	30.73	64.23	74.01	76.93	77.11	77.15
		PA	**38.77**	**39.79**	**39.82**	**39.93**	**39.96**	**76.46**	**79.58**	**79.85**	**79.89**	**79.94**
LSRO [31]	No attack		77.2	89.9	96.1	97.4	98.5	55.2	72.0	85.7	89.5	92.1
	Under attack		9.3	10.7	19.9	25.1	31.6	10.1	13.1	23.6	28.8	35.4
	HITL-D	RS	53.74	56.03	56.56	57.13	58.08	**60.11**	**67.15**	**67.86**	**68.63**	**69.57**
		MM	**57.07**	**59.98**	**60.75**	**61.13**	**61.73**	57.69	65.57	67.15	67.46	68.13
		QS	38.8	43.82	44.89	44.98	45.01	33.64	43.0	46.86	47.53	47.85
		PA	55.09	59.26	59.74	59.92	60.24	55.53	63.51	64.81	65.26	65.98
CamStyle [33]	No attack		70.8	86.6	95.0	96.6	97.9	58.1	76.5	86.8	90.0	92.7
	Under attack		7.6	8.3	14.9	19.1	23.8	12.5	14.9	25.7	30.5	37.3
	HITL-D	RS	53.67	57.21	57.81	58.34	59.12	69.35	**77.29**	**78.28**	**78.59**	**79.35**
		MM	**55.59**	**59.62**	**60.21**	**60.42**	**60.78**	66.59	74.15	75.31	75.54	75.72
		QS	38.99	46.44	48.78	48.9	48.93	44.81	57.14	61.09	61.49	61.8
		PA	52.97	55.61	55.97	56.03	56.15	**70.31**	77.11	77.92	78.1	78.32
HHL [32]	No attack		64.3	82.3	92.6	95.4	96.8	51.8	71.4	83.5	87.7	90.9
	Under attack		9.3	9.5	17.7	23.0	28.5	9.4	11.8	19.0	22.8	28.5
	HITL-D	RS	52.84	57.1	57.81	58.52	59.09	**54.02**	**61.54**	**62.57**	**63.29**	**64.63**
		MM	**55.27**	**60.9**	**61.88**	**62.29**	**62.8**	51.57	60.19	61.45	61.76	62.21
		QS	35.15	43.2	46.5	46.97	47.12	31.45	41.52	45.2	45.83	46.41
		PA	54.58	60.24	61.07	61.22	61.52	51.33	59.38	60.19	60.64	60.95
SPGAN [8]	No attack		66.6	84.3	94.1	96.4	97.8	54.6	73.5	85.1	88.9	91.7
	Under attack		4.2	3.8	8.8	12.0	15.7	12.0	15.0	23.8	29.8	36.0
	HITL-D	RS	**47.32**	**51.31**	**51.96**	**52.67**	**53.53**	**67.95**	**74.87**	**75.72**	**76.35**	**76.66**
		MM	45.69	49.5	50.24	50.53	50.68	65.16	72.8	74.01	74.37	74.87
		QS	29.28	36.49	39.19	39.49	39.61	41.81	53.9	57.05	57.85	58.26
		PA	45.79	48.84	49.32	49.38	49.47	66.51	72.76	73.25	73.47	73.7

SPGAN [8]. We trained them using the hyper-parameters values recommended in the respective works.

We then evaluated the Re-Id performance attained by the above models in the absence of attacks (i.e., on the original testing sets) and under attack (i.e., by manipulating all the query images of the testing sets). In both cases, we evaluated the performance of both original Re-Id models and the same models empowered with the HITL approach for each of the RF algorithms described in Sect. 3. As in our previous work [7], we carried out three feedback rounds for each RF algorithm, asking the user feedback on the top $K = 50$ retrieved images. Note that since visual inspection of the top-ranked images is part of the task of the law enforcement operator and since, in real application scenarios, the gallery can contain a thousand images or more, asking the user/operator to select the top-50 images corresponding to the query identity does not add any significant effort. Given the considerable size of query sets, we simulated the user/operator feedback using the ground truth identity labels as in [7]. We use two common metrics to evaluate the results: mean Average Precision (mAP) and Cumulative Matching Curve (CMC) at ranks $k = 1, 5, 10, 20$.

Experimental Results. Table 1 shows the results obtained by each Re-Id model in the absence of attacks ("No attack"), under attack without any defence strategy ("Under attack"), and under attack when the proposed HITL defence ("HITL-D", for each RF algorithm) is used, after the third feedback round. The accuracy of all Re-Id models drastically worsens under attack when no defence is used. On the other hand, the proposed HITL-D strategy mitigates the drop in performance on Market and, to a much larger extent, on Duke; in particular, for some models and RF algorithms, it is even capable of outperforming the mAP and rank-1 CMC values attained *in absence* of attacks (see, e.g., SPGAN+MM or CamStyle+PA). Table 2 shows the results attained after each feedback round for more fine-grained analysis. As can be expected, the performance always improves as the number of rounds of the HITL-D strategy increases, even if with different trends for the different RF algorithms, and in particular, the performance drop under attack is mitigated since the first round.

For completeness, we report in Table 3 the results obtained by the HITL approach in the absence of attacks. For brevity, we only consider the model that attained the *worse* performance in terms of mAP in the previous experiments. It can be seen that the HITL approach always improves the performance of the underlying model also under normal operational conditions, in agreement with its original goal, and also in this case, this happens, with only a few exceptions, since the *first* feedback round. The above results provide evidence that the HITL approach can be an effective defence strategy against the considered kind of attack on Re-Id systems and that, contrary to ad hoc defences [1], it also *improves* the performance under normal operating conditions.

Ablation Study. The considered RF algorithms make use of the query image, as a positive example, for re-ranking gallery images after *each* feedback round. Taking into account that the considered attack manipulates the query image to prevent gallery images of the same identity from being retrieved, one may think that the effectiveness of the HITL-D approach under attack can be further improved by *not* using the query image during feedback rounds. On the other hand, this might decrease its performance in the absence of attacks, as happens to ad hoc defences. To verify the above assumption, we repeated the above experiments using, for the sake of brevity, only one representative Re-Id model, LSRO [31], and only the RF algorithm that achieved the best performance in most of the previous experiments, i.e., RS, which we modified in such a way that the query image is not used during feedback rounds.

Table 4 compares the results achieved in the absence of attack without HITL, by the original RS algorithm ("query-dependent", QD-RS), and by its modified version ("non-query-dependent", NQD-RS). Contrary to what one may expect, not using the query image during feedback rounds did not significantly improve

Table 2. Re-identification accuracy of the five models under attack with no defence, and using the HITL-based defence to counter the attack, after each feedback round of the considered RF algorithms.

			Data set	Market-1501					DukeMTMC-reid				
Model			Round	mAP	Rank-1	Rank-5	Rank-10	Rank-20	mAP	Rank-1	Rank-5	Rank-10	Rank-20
Aligned ReId [15]	Under attack			2.4	1.7	4.0	5.8	8.9	16.7	19.9	31.6	38.2	45.7
	HITL-D	RS	1	21.08	24.79	25.83	26.54	27.52	61.7	72.26	73.61	74.06	74.73
			2	29.28	31.56	32.45	33.02	34.0	70.66	75.9	76.35	76.57	77.02
			3	35.42	37.32	37.86	38.42	39.31	74.56	78.41	78.82	78.95	79.35
		MM	1	21.89	24.76	26.16	26.84	27.97	61.87	71.45	73.92	74.24	74.69
			2	30.26	31.86	32.24	32.66	33.58	69.96	75.31	75.76	76.03	76.57
			3	35.49	37.17	37.35	37.41	37.83	74.22	78.82	79.22	79.44	79.8
		QS	1	19.41	23.69	24.26	24.5	25.12	56.31	68.36	72.67	73.52	74.15
			2	25.49	29.16	29.96	30.08	30.2	62.93	73.03	76.39	76.57	76.66
			3	26.66	29.78	30.64	30.73	30.73	64.23	74.01	76.93	77.11	77.15
		PA	1	21.28	24.41	26.01	27.17	29.54	60.24	71.63	73.61	73.97	74.82
			2	34.62	36.67	36.91	37.05	37.29	72.32	77.15	77.51	77.6	78.01
			3	**38.77**	**39.79**	**39.82**	**39.93**	**39.96**	**76.46**	**79.58**	**79.85**	**79.89**	**79.94**
LSRO [31]	Under attack			9.3	10.7	19.9	25.1	31.6	10.1	13.1	23.6	28.8	35.4
	HITL-D	RS	1	37.32	42.04	43.44	44.39	45.75	33.72	42.46	45.69	47.44	49.6
			2	46.73	49.44	50.21	50.8	51.66	49.66	58.35	59.74	60.91	62.16
			3	53.74	56.03	56.56	57.13	58.08	**60.11**	**67.15**	**67.86**	**68.63**	**69.57**
		MM	1	38.02	42.43	43.91	45.07	46.35	34.16	42.37	46.05	48.16	51.71
			2	48.13	51.16	52.67	53.44	55.49	49.85	59.16	61.62	62.52	63.6
			3	**57.07**	**59.98**	**60.75**	**61.13**	**61.73**	57.69	65.57	67.15	67.46	68.13
		QS	1	33.32	40.74	42.07	42.46	42.84	27.4	36.98	41.11	42.19	43.27
			2	38.15	43.56	44.6	44.71	44.74	32.72	42.37	46.1	47.13	47.53
			3	38.8	43.82	44.89	44.98	45.01	33.64	43.0	46.86	47.53	47.85
		PA	1	24.03	28.41	38.15	41.27	43.94	23.48	30.12	39.72	43.04	46.95
			2	44.12	50.36	53.12	53.92	55.2	43.89	54.67	58.03	59.29	60.73
			3	55.09	59.26	59.74	59.92	60.24	55.53	63.51	64.81	65.26	65.98
CamStyle [33]	Under attack			7.6	8.3	14.9	19.1	23.8	12.5	14.9	25.7	30.5	37.3
	HITL-D	RS	1	35.39	41.42	42.99	43.94	45.55	44.78	55.07	57.68	59.65	61.27
			2	46.03	50.24	50.98	51.87	53.09	59.74	68.49	70.24	71.1	72.44
			3	53.67	57.21	57.81	58.34	59.12	69.35	**77.29**	**78.28**	**78.59**	**79.35**
		MM	1	36.08	40.91	43.08	44.24	45.84	45.04	54.31	57.68	58.93	60.82
			2	47.4	51.75	53.03	53.8	54.99	59.05	67.41	69.84	70.6	71.41
			3	**55.59**	**59.62**	**60.21**	**60.42**	**60.78**	66.59	74.15	75.31	75.54	75.72
		QS	1	30.62	38.45	40.53	41.24	42.49	38.1	50.45	54.22	55.57	57.0
			2	37.83	46.14	48.1	48.49	48.63	43.64	56.33	59.92	60.55	60.82
			3	38.99	46.44	48.78	48.9	48.93	44.81	57.14	61.09	61.49	61.8
		PA	1	27.92	33.55	40.02	42.07	44.3	39.0	48.34	55.57	58.21	60.55
			2	45.46	50.42	51.72	52.2	52.82	60.1	69.75	72.76	73.74	74.42
			3	52.97	55.61	55.97	56.03	56.15	**70.31**	77.11	77.92	78.1	78.32
HHL [32]	Under attack			9.3	9.5	17.7	23.0	28.5	9.4	11.8	19.0	22.8	28.5
	HITL-D	RS	1	35.14	43.05	44.42	45.22	46.94	32.21	40.98	43.9	45.47	47.98
			2	45.9	50.77	51.84	52.46	53.71	45.46	53.86	55.57	56.82	57.81
			3	52.84	57.1	57.81	58.52	59.09	**54.02**	**61.54**	**62.57**	**63.29**	**64.63**
		MM	1	36.02	42.93	44.86	46.02	47.98	32.38	40.39	44.03	45.65	47.44
			2	47.33	53.59	54.87	55.85	57.16	45.13	54.13	56.28	57.0	58.03
			3	**55.27**	**60.9**	**61.88**	**62.29**	**62.8**	51.57	60.19	61.45	61.76	62.21
		QS	1	29.28	39.46	41.81	42.76	43.29	26.04	36.09	39.36	40.66	41.43
			2	34.29	42.99	45.96	46.5	46.73	30.45	40.75	44.25	45.02	45.51
			3	35.15	43.2	46.5	46.97	47.12	31.45	41.52	45.2	45.83	46.41
		PA	1	21.26	27.2	36.49	39.55	43.05	22.55	28.86	37.66	40.71	44.61
			2	43.6	51.81	55.26	56.47	57.45	41.67	51.48	55.34	56.06	56.96
			3	54.58	60.24	61.07	61.22	61.52	51.33	59.38	60.19	60.64	60.95
SPGAN [8]	Under attack			4.2	3.8	8.8	12.0	15.7	12.0	15.0	23.8	29.8	36.0
	HITL-D	RS	1	25.27	30.31	32.19	33.88	36.31	41.13	51.48	54.17	56.01	58.84
			2	38.65	43.47	44.54	45.16	46.32	57.58	65.71	68.22	69.3	70.87
			3	**47.32**	**51.31**	**51.96**	**52.67**	**53.53**	**67.95**	**74.87**	**75.72**	**76.35**	**76.66**
		MM	1	25.73	29.9	31.47	32.78	35.01	41.4	51.26	54.04	56.55	59.16
			2	38.03	42.52	43.68	44.27	45.52	57.44	66.88	68.76	69.57	70.42
			3	45.69	49.5	50.24	50.53	50.68	65.16	72.8	74.01	74.37	74.87
		QS	1	20.93	27.4	29.48	30.46	31.89	34.51	46.59	50.45	51.84	53.37
			2	27.93	35.84	38.72	39.1	39.22	40.67	52.78	56.15	56.78	57.32
			3	29.28	36.49	39.19	39.49	39.61	41.81	53.9	57.05	57.85	58.26
		PA	1	19.11	23.22	29.33	31.32	33.85	36.19	46.1	53.1	55.3	58.08
			2	37.39	42.76	44.54	44.98	45.67	56.76	65.71	68.27	68.99	70.33
			3	45.79	48.84	49.32	49.38	49.47	66.51	72.76	73.25	73.47	73.7

Table 3. Re-identification accuracy of the HHL model with no attack, without using HITL (first row), and using each of the considered RF algorithms, after each feedback round (next rows). Best results for each column are highlighted in bold.

Model			Data set	Market-1501					DukeMTMC-reid				
			Round	mAP	Rank-1	Rank-5	Rank-10	Rank-20	mAP	Rank-1	Rank-5	Rank-10	Rank-20
HHL [32]	No HITL			64.3	82.3	92.6	95.4	96.8	51.8	71.4	83.5	87.7	90.9
	HITL	RS	1	90.04	98.16	98.46	98.55	98.66	80.32	94.88	95.83	96.14	96.36
			2	94.0	98.81	98.96	99.02	99.08	87.58	96.95	97.13	97.22	97.44
			3	95.67	**99.23**	99.29	99.29	99.32	**90.58**	**97.71**	**97.85**	**97.94**	**97.94**
		MM	1	83.94	97.68	98.43	98.63	98.72	74.61	93.27	95.02	95.6	95.83
			2	85.84	97.51	98.75	98.81	98.93	79.97	94.79	96.54	96.86	97.17
			3	89.17	97.74	98.93	98.99	99.08	83.97	95.83	97.22	97.31	97.49
		QS	1	80.03	93.94	97.62	98.13	98.52	68.18	86.27	93.0	94.43	95.11
			2	81.47	94.36	98.16	98.57	98.75	70.46	87.61	94.3	95.33	95.92
			3	81.68	94.42	98.13	98.63	98.78	70.83	87.84	94.48	95.69	96.01
		PA	1	53.13	61.19	77.32	82.45	86.79	55.07	67.86	83.17	87.25	89.95
			2	91.88	97.57	98.81	98.96	99.02	84.58	95.11	96.41	96.77	96.99
			3	**95.9**	**99.23**	**99.35**	**99.35**	**99.35**	89.88	97.31	97.71	97.76	97.76

the performance of the HITL-D strategy: indeed, the performance gap between QD-RS and NQD-RS is almost negligible for each feedback round.

Table 4. Performance of the LSRO model in absence of attacks, without using HITL (first row), and using HITL with the RS algorithm, either including the query image in each feedback round as a positive example (QD-RS), or not including it (NQD-RS).

Model		Data set	Market-1501					DukeMTMC-reid				
		Round	mAP	Rank-1	Rank-5	Rank-10	Rank-20	mAP	Rank-1	Rank-5	Rank-10	Rank-20
LSRO [31]	No HITL		77.2	89.9	96.1	97.4	98.5	55.2	72.0	85.7	89.5	92.1
	QD-RS	1	95.42	99.44	99.47	99.5	99.52	83.25	95.33	96.1	96.45	96.81
		2	97.35	99.58	99.58	99.61	99.64	89.88	97.22	97.49	97.62	97.76
		3	98.11	99.67	99.7	99.7	99.7	92.3	98.03	98.07	98.16	98.25
	NQD-RS	1	95.43	99.44	99.5	99.52	99.55	83.25	95.38	96.1	96.45	96.81
		2	97.37	99.58	99.58	99.61	99.64	89.91	97.22	97.53	97.67	97.8
		3	98.18	99.73	99.73	99.73	99.73	92.34	98.07	98.11	98.2	98.29

5 Conclusions

We have shown that the HITL approach to Re-Id, originally proposed to improve the performance of a given model by exploiting user feedback on the identity of the top-ranked retrieved images, can also be an effective defence strategy against adversarial attacks based on manipulating the query image to push images of the same individual toward bottom ranks. Notably, we achieved this result under a simple implementation of the HITL approach that is independent on the underlying Re-Id model and does not require its retraining or fine-tuning. With respect to ad hoc defences, a HITL-based defence also improves performance in the absence of attacks instead of worsening it. Taking into account that in our experiments, the considered HITL defence was sometimes capable only of mitigating the drastic performance drop due to adversarial attacks, an interesting direction for future work is to investigate its *combination* with ad hoc defence strategies, which may allow to improve robustness further and to avoid at the same time a performance reduction in the absence of attacks.

Acknowledgments. This work was supported by the projects: "Law Enforcement agencies human factor methods and Toolkit for the Security and protection of CROWDs in mass gatherings" (LETSCROWD), EU Horizon 2020 programme, grant agreement No. 740466; "IMaging MAnagement Guidelines and Informatics Network for law enforcement Agencies" (IMMAGINA), European Space Agency, ARTES Integrated Applications Promotion Programme, contract No. 4000133110/20/NL/AF.

References

1. Bai, S., Li, Y., Zhou, Y., Li, Q., Torr, P.H.S.: Adversarial metric attack and defense for person re-identification. IEEE Trans. Pattern Anal. Mach. Intell. **43**(6), 2119–2126 (2021). https://doi.org/10.1109/TPAMI.2020.3031625
2. Bedagkar-Gala, A., Shah, S.K.: A survey of approaches and trends in person re-identification. Image Vis. Comput. **32**(4), 270–286 (2014). https://doi.org/10.1016/j.imavis.2014.02.001
3. Biggio, B., Roli, F.: Wild patterns: Ten years after the rise of adversarial machine learning. Pattern Recognit. **84**, 317–331 (2018). https://doi.org/10.1016/j.patcog.2018.07.023
4. Bouniot, Q., Audigier, R., Loesch, A.: Vulnerability of person re-identification models to metric adversarial attacks. In: CVPR W, pp. 794–795 (2020)
5. Chakraborty, A., Alam, M., Dey, V., Chattopadhyay, A., Mukhopadhyay, D.: A survey on adversarial attacks and defences. CAAI Trans. Intell. Technol. **6**(1), 25–45 (2021). https://doi.org/10.1049/cit2.12028
6. Chang, H., Li, Y., Si, N., Zhang, H.: A targeted adversarial attack method against person re-identification model. In: ICFTIC, pp. 313–316 (2021)
7. Delussu, R., Putzu, L., Fumera, G.: Human-in-the-loop cross-domain person re-identification. Expert Syst. Appl. **226**, 120216 (2023). https://doi.org/10.1016/j.eswa.2023.120216
8. Deng, W., Zheng, L., Ye, Q., Kang, G., Yang, Y., Jiao, J.: Image-image domain adaptation with preserved self-similarity and domain-dissimilarity for person re-identification. In: CVPR, pp. 994–1003 (2018). https://doi.org/10.1109/CVPR.2018.00110
9. Ding, W., Wei, X., Ji, R., Hong, X., Tian, Q., Gong, Y.: Beyond universal person re-identification attack. IEEE Trans. Inf. Forensics Secur. **16**, 3442–3455 (2021). https://doi.org/10.1109/TIFS.2021.3081247
10. Giacinto, G.: A nearest-neighbor approach to relevance feedback in content based image retrieval. In: CIVR, pp. 456–463 (2007). https://doi.org/10.1145/1282280.1282347
11. Gong, Y., Huang, L., Chen, L.: Person re-identification method based on color attack and joint defence. In: CVPR W, pp. 4312–4321 (2022). https://doi.org/10.1109/CVPRW56347.2022.00477

12. Lin, W., Chen, Z., Ke, S., Tsai, C., Lin, W.: The effect of low-level image features on pseudo relevance feedback. Neurocomputing **166**, 26–37 (2015). https://doi.org/10.1016/j.neucom.2015.04.037
13. Liu, C., Loy, C.C., Gong, S., Wang, G.: POP: person re-identification post-rank optimisation. In: ICCV, pp. 441–448 (2013). https://doi.org/10.1109/ICCV.2013.62
14. Liu, D., et al.: Generative metric learning for adversarially robust open-world person re-identification. ACM Trans. Multim. Comput. Commun. Appl. **19**(1), 20:1–20:19 (2023). https://doi.org/10.1145/3522714
15. Luo, H., Jiang, W., Zhang, X., Fan, X., Qian, J., Zhang, C.: Alignedreid++: Dynamically matching local information for person re-identification. Pattern Recognit. **94**, 53–61 (2019). https://doi.org/10.1016/j.patcog.2019.05.028
16. Piras, L., Giacinto, G., Paredes, R.: Passive-aggressive online learning for relevance feedback in content based image retrieval. In: ICPRAM, pp. 182–187 (2013)
17. Putzu, L., Piras, L., Giacinto, G.: Ten years of relevance score for content based image retrieval. In: Machine Learning and Data Mining in Pattern Recognition, pp. 117–131 (2018). https://doi.org/10.1007/978-3-319-96133-0_9
18. Subramanyam, A.: Meta generative attack on person reidentification. IEEE Trans Circuits Syst, Video Technol (2023)
19. Verma, A., Subramanyam, A.V., Shah, R.R.: Wasserstein metric attack on person re-identification. In: MIPR, pp. 234–239 (2022). https://doi.org/10.1109/MIPR54900.2022.00049
20. Wang, H., Gong, S., Zhu, X., Xiang, T.: Human-in-the-loop person re-identification. In: ECCV, pp. 405–422 (2016). https://doi.org/10.1007/978-3-319-46493-0_25
21. Wang, H., Zhu, X., Gong, S., Xiang, T.: Person re-identification in identity regression space. Int. J. Comput. Vis. **126**(12), 1288–1310 (2018). https://doi.org/10.1007/s11263-018-1105-3
22. Wang, H., Wang, G., Li, Y., Zhang, D., Lin, L.: Transferable, controllable, and inconspicuous adversarial attacks on person re-identification with deep mis-ranking. In: CVPR, pp. 339–348 (2020). https://doi.org/10.1109/CVPR42600.2020.00042
23. Wang, X., Zheng, X., Du, P., Liu, L., Ma, H.: Occlusion resilient adversarial attack for person re-identification. In: MASS, pp. 527–535 (2021). https://doi.org/10.1109/MASS52906.2021.00071
24. Wang, Z., Zheng, S., Song, M., Wang, Q., Rahimpour, A., Qi, H.: advpattern: Physical-world attacks on deep person re-identification via adversarially transformable patterns. In: ICCV, pp. 8340–8349 (2019). https://doi.org/10.1109/ICCV.2019.00843
25. Wu, A., et al.: Unsupervised person re-identification by camera-aware similarity consistency learning. In: ICCV, pp. 6921–6930 (2019). https://doi.org/10.1109/ICCV.2019.00702
26. Yang, F., et al.: Learning to attack real-world models for person re-identification via virtual-guided meta-learning. In: AAAI, pp. 3128–3135 (2021)
27. Yang, F., et al.: Towards robust person re-identification by defending against universal attackers. IEEE Trans. Pattern Anal. Mach. Intell. **45**(4), 5218–5235 (2023). https://doi.org/10.1109/TPAMI.2022.3199013
28. Ye, M., Shen, J., Lin, G., Xiang, T., Shao, L., Hoi, S.C.H.: Deep learning for person re-identification: a survey and outlook. IEEE Trans. Pattern Anal. Mach. Intell. **44**(6), 2872–2893 (2022). https://doi.org/10.1109/TPAMI.2021.3054775

29. Zhang, Z., Wu, J., Zhang, X., Zhang, C.: Multi-target, multi-camera tracking by hierarchical clustering: Recent progress on dukemtmc project. CoRR abs/1712.09531 (2017)
30. Zheng, L., Shen, L., Tian, L., Wang, S., Wang, J., Tian, Q.: Scalable person re-identification: A benchmark. In: ICCV, pp. 1116–1124 (2015). https://doi.org/10.1109/ICCV.2015.133
31. Zheng, Z., Zheng, L., Yang, Y.: Unlabeled samples generated by GAN improve the person re-identification baseline in vitro. In: ICCV, pp. 3774–3782 (2017). https://doi.org/10.1109/ICCV.2017.405
32. Zhong, Z., Zheng, L., Li, S., Yang, Y.: Generalizing a person retrieval model hetero- and homogeneously. In: ECCV. Lecture Notes in Computer Science, vol. 11217, pp. 176–192 (2018). https://doi.org/10.1007/978-3-030-01261-8_11
33. Zhong, Z., Zheng, L., Zheng, Z., Li, S., Yang, Y.: Camera style adaptation for person re-identification. In: CVPR, pp. 5157–5166 (2018). https://doi.org/10.1109/CVPR.2018.00541

Generalized Deepfake Detection Algorithm Based on Inconsistency Between Inner and Outer Faces

Jie Gao[1,2](✉), Sara Concas[2], Giulia Orrù[2], Xiaoyi Feng[1], Gian Luca Marcialis[2], and Fabio Roli[3]

[1] Northwestern Polytechnical University, Xi'an, China
jie_gao@mail.nwpu.edu.cn, fengxiao@nwpu.edu.cn
[2] University of Cagliari, Cagliari, Italy
{sara.concas90c,giulia.orru,marcialis}@unica.it
[3] University of Genova, Genova, Italy
fabio.roli@unige.it

Abstract. Deepfake refers to using artificial intelligence (AI) and machine learning techniques to create compelling and realistic media content, such as videos, images, or recordings, that appear real but are fake. The most common form of deepfake involves using deep neural networks to replace or superimpose faces in existing videos or images on top of other people's faces. While this technology can be used for various benign purposes, such as filmmaking or online education, it can also be used maliciously to spread misinformation by creating fake videos or images. Based on the classic deepfake generation process, this paper explores the Inconsistency between inner and outer faces in fake content to find synthetic defects and proposes a general deepfake detection algorithm. Experimental results show that our proposed method has certain advantages, especially regarding cross-method detection performance.

Keywords: deepfake detection · generalization · manipulations

1 Introduction

Deepfake refers to manipulating facial identities through automated algorithms to forge videos/images. The earliest deepfake was [10] created in 2017 by a Reddit user named "deepfakes". Since then, this technology has become more advanced and accessible, increasing the creation and dissemination of deepfakes while raising concerns about their potential harm to individuals and society.

Although there are already multiple deepfake detection methods [16], the generalization problem is still open and crucial. To the best of our knowledge, state-of-the-art (SOTA) methods have achieved high levels of in-domain performance. However, these methods are helpless in the face of cross-domain, which refers to detecting unknown manipulations, different from those present in the training data [7,11]. Given the shortcomings of existing methods, this work aims

G. L. Foresti et al. (Eds.): ICIAP 2023 Workshops, LNCS 14365, pp. 343–355, 2024.
https://doi.org/10.1007/978-3-031-51023-6_29

to study the inconsistency between the foreground and the background of faces in images to detect deepfakes since, in manipulated images, the information of the faces of a source and a target subject coexist. Regardless of the manipulation technique used, the coexistence of this dual information exists in a deepfake. For this reason, designing a system capable of detecting it corresponds to designing a detector capable of generalizing even on unknown manipulations.

The contributions of this work are as follows:

(1) Considering that the forged mask is difficult to obtain in the deepfake content, we directly use facial landmarks to extract the mask and segment the inner face and outer face during the feature extraction process, thereby reducing the dependence on prior knowledge.
(2) We design two different encoders to extract features of inner and outer faces separately and build a consistency loss to distinguish real from fake. Furthermore, we build a parameter-sharing decoder to force it to recover images from low-rank feature vectors.
(3) We further constrain the encoder to extract rich features with reconstruction loss, resulting in improved generalization. The experimental results reveal that our proposed technique is competitive in cross-method experiments.

The remaining chapters of the paper are arranged as follows. Section 2 introduces the existing deepfake detection methods. Section 3 explains the architecture and details of our proposed method. In Sect. 4, we show the experimental setup and results. Finally, Sect. 5 discusses the proposed approach and draws some conclusions.

2 Related Works on Deepfake Detection

As one of the significant threats to face recognition technology, deepfake technology has been widely spread in social media. Recently, many works [15] have been dedicated to resisting the harm brought by deepfake technology spread [12].

Although detection technologies are increasingly precise and various techniques are exploited [4–6], many methods are trained for specific deepfake manipulations, resulting in a substantially decreased cross-method detection performance.

To solve this problem, some deepfake detection methods focusing on improving cross-method generalization [1] have been proposed. For example, Face X-ray [8] proposed a new way to judge whether a face image has been replaced. They designed a detector based on the observation that there must be a blending boundary in the forgery step to distinguish authenticity by judging whether a face has a blending border. Kaede *et al.* [14] present novel synthetic training data called self-blended images (SBIs) to detect deepfakes. Kim *et al.* [7] proposed a new domain adaptation framework. Specifically, they constructed a student and teacher models based on knowledge distillation and representation learning. The student model can quickly adapt to novel deepfakes by extracting knowledge from the pre-trained teacher model and applying transfer learning without using source domain data during domain adaptation.

Although good generalization performance was reported in the above papers, they redesigned fake synthetic data or complex models for training, increasing the computational complexity and difficulty of training. In order to further solve the above challenges, this paper proposes a simple deepfake detection method, which can reduce the dependence on the training set while improving the generalization ability.

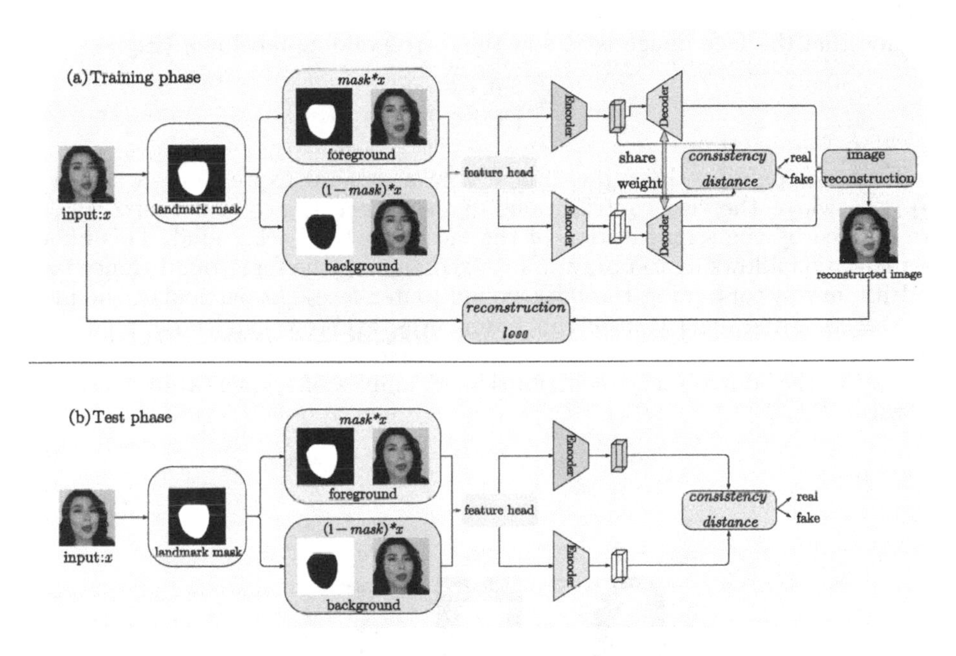

Fig. 1. Proposed architecture. The training step is depicted in (a): the input image is divided into two parts based on face landmarks: foreground and background. The foreground (inner face) is derived from the target image, while the background (outer face) is derived from the source image. The image reconstruction loss function is used to force different encoders to extract detailed information. The consistency loss function is used to differentiate between the consistency of the inner and outer faces. The test phase pipeline is depicted in (b). In the test stage, we only use the feature vector distance obtained by the two encoders for inference.

3 The Proposed Method

To address the generalization problem, we investigate the nature of the face as a biometric trait. In fact, source and target faces differ from individual identities' perspectives in manipulated content. Based on this phenomenon, this work exploits inner and outer faces inconsistencies in forged faces to detect manipulated content.

Considering the feasibility of the above analysis, we need to assume that any input image consists of two parts (inner face and outer face). Specifically, we consider an input image composed of two components, one related to the source image and the other to the target image. If the image is real, the two components are consistent because the image is associated with a single individual and has not been manipulated. In deepfakes, on the other hand, these two components are very different because they belong to two different identities.

In order to describe an input face image more concretely and intuitively, we assume that the face image is x, and the corresponding label is y (Eq. (1)).

$$y = \begin{cases} 0, x \in real \\ 1, x \in fake \end{cases} \tag{1}$$

First, we express the input image x in a general expression as shown in Eq. (2), where the two contributions of the target x_{target} and source x_{source} are divided by the segmentation of the face from the background. This binary segmentation allows us to obtain a mask containing the foreground (inner face) and its inverse containing the background (outer face). In particular, the inner face refers to the inner contour, including eyes, eyebrows, and nose, relative to the target image contribution; the outer face refers to the rest of the area and is relative to the source image contribution. Examples of segmentation masks are shown in Fig. 2.

$$x = mask \odot x_{target} + (1 - mask) \odot x_{source} \tag{2}$$

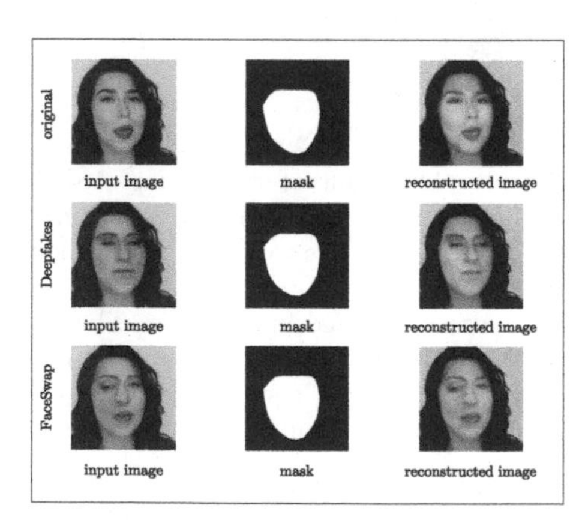

Fig. 2. Examples of the input image processing of the proposed architecture: the input image is segmented and reconstructed by the autoencoder. The reconstructed image is then compared with the original.

According to Eq. (2), in order to represent all input images as the relationship between x_{source} and x_{target}, we consider that a real input image satisfies:

$x_{source} = x_{target}$, and a deepfake image satisfies: $x_{source} \neq x_{target}$. Considering that x_{source} and x_{target} are difficult to describe intuitively, we try to map them to distance space to explore their similarity. Therefore, the intuitive idea is distinguishing real images from deepfakes based on consistency loss.

Figure 1 shows the schematic diagram of our proposed deepfake detection method. As shown in Fig. 1, we can assume that the input image consists of a source image x_{source} and a target image x_{target}. First, we use a facial landmark detection algorithm to obtain the foreground mask, then perform inner and outer face segmentation based on it. Second, we construct a feature head to initialize the masked inner and outer face regions and further construct two different encoders to extract the features of the inner and outer faces, respectively. Our aim is to map these features into the distance space and construct an inconsistency loss function to perform real-fake discrimination. Finally, inspired by the deepfake synthesis process (different encoders, same decoder) and in order to reduce computational burden and enable faster reconstruction of learned distinct facial features for both inner and outer faces, we designed a parameter-shared decoder. We use a direct adding strategy to reconstruct the image since we obtain corresponding inner and outer face features based on landmark mask segmentation. Moreover, we further set the reconstruction loss function as an additional constraint to enable the encoder to extract detailed information. The encoder and decoder structure of the proposed method is shown in Table 1 and Table 2.

Table 1. The encoder of the proposed method consists of four convolutional and pooling layers, followed by a fully connected (FC) layer and a batch normalization (BN) step.

Layer Name	Kernel	Input	Output
Input	–	224 × 224 × 3	224 × 224 × 3
Feature Head	3 × 3, 2 filters	224 × 224 × 3	224 × 224 × 64
Conv1	3 × 3, 2 filters	224 × 224 × 64	224 × 224 × 64
MaxPool1	2 × 2, stride 2	224 × 224 × 64	112 × 112 × 64
Conv2	3 × 3, 2 filters	112 × 112 × 64	112 × 112 × 128
MaxPool2	2 × 2, stride 2	112 × 112 × 128	56 × 56 × 128
Conv3	3 × 3, 2 filters	56 × 56 × 128	56 × 56 × 256
MaxPool3	2 × 2, stride 2	56 × 56 × 256	28 × 28 × 256
Conv4	3 × 3, 2 filters	28 × 28 × 256	28 × 28 × 512
MaxPool4	2 × 2, stride 2	28 × 28 × 512	14 × 14 × 512
FC	–	14 × 14 × 512	512
BN	–	512	512

For the distance between x_{source} and x_{target}, we define it as $D(x_{target}, x_{source})$, and set the threshold τ as the basis for the binary classi-

Table 2. The decoder of the proposed method consists of four layers of convolution and unpooling, followed by a fifth level of convolution for the fusion of the inner and outer face information for image reconstruction.

Layer Name	Kernel	Input	Output
Input	–	$14 \times 14 \times 512$	$14 \times 14 \times 512$
Conv1	3×3, 5 filters	$14 \times 14 \times 512$	$14 \times 14 \times 512$
UnPool1	2×2, stride 2	$14 \times 14 \times 512$	$28 \times 28 \times 512$
Conv2	3×3, 3 filters	$28 \times 28 \times 512$	$28 \times 28 \times 256$
UnPool2	2×2, stride 2	$28 \times 28 \times 256$	$56 \times 56 \times 256$
Conv3	3×3, 3 filters	$56 \times 56 \times 256$	$56 \times 56 \times 128$
UnPool3	2×2, stride 2	$56 \times 56 \times 128$	$112 \times 112 \times 128$
Conv4	3×3, 2 filters	$112 \times 112 \times 128$	$112 \times 112 \times 64$
UnPool4	2×2, stride 2	$112 \times 112 \times 64$	$224 \times 224 \times 64$
Conv5	3×3, 3 filters	$224 \times 224 \times 64$	$224 \times 224 \times 3$

fications, which can be described as Eq. (3).

$$\begin{cases} D\left(x_{target}, x_{source}\right) < \tau, y = 0, x \in real \\ D\left(x_{target}, x_{source}\right) > \tau, y = 1, x \in fake \end{cases} \tag{3}$$

Inspired by cosine similarity, we utilize cosine distance between x_{source} and x_{target} as the metric function and normalize it to [0,1], as shown in Eq. (4).

$$D\left(x_{target}, x_{source}\right) = 0.5 * (1 - cos(x_{target}, x_{source})) \tag{4}$$

In addition, we adopt the traditional cross-entropy loss function as the loss of the classifier, which is expressed by Eq. (5). Among them, $\hat{y}$ is represented by the distance in Eq. (4), and y refers to the ground truth label.

$$L_{cls}(y, \hat{y}) = -y \log \hat{y} + (1 - y) \log(1 - \hat{y}) \tag{5}$$

Moreover, to ensure that the encoder can extract enough detailed information, we design a decoder for image reconstruction and construct a reconstruction loss function as a constraint.

Pixel loss is commonly used in the field of image reconstruction, where the traditional approach is to use mean squared error (MSE) loss as a constraint term. However, MSE loss averages over all pixels, assuming that each pixel in the image has equal learning capability. In contrast, perceptual loss is a commonly used loss function in deep learning-based image style transfer methods. Compared to traditional MSE loss, perceptual loss focuses more on the perceptual quality of the image, aligning with the human visual perception of image quality and better capturing the learning of pixels that impact visual perception. Therefore, to ensure that the reconstructed image does not alter the visual perception of the input image by the human eye, we employ both pixel loss and perceptual loss as reconstruction loss.

Among them, the reconstruction loss $L_{rec}(x, \hat{x})$ consists of two parts, namely perceptual loss $L_{per}(x, \hat{x})$ and pixel loss $L_{pix}(x, \hat{x})$, as shown in Eq. (8). Here, we use the pre-trained VGG16 as a measure of perceptual loss, as shown in Eq. (7), where i represents the i_{th} pixel in the image x, and i belongs to [1, n], n is the total number of pixels in x. And $\hat{x}$ represents the reconstructed image. We adopt the mean square error loss function for the pixel loss, as shown in Eq. (6).

$$L_{pix}(x, \hat{x}) = \|x - \hat{x}\|_2^2 \tag{6}$$

$$L_{per}(x, \hat{x}) = \frac{1}{n}\sum_{i=1}^{n}\left(vgg16\left(x_i\right) - vgg16\left(\hat{x}_i\right)\right)^2 \tag{7}$$

$$L_{rec}(x, \hat{x}) = L_{pix}(x, \hat{x}) + L_{per}(x, \hat{x}) \tag{8}$$

Finally, the total loss L consists of three parts, as shown in Eq. (9).

$$L = L_{cls}(y, D\left(x_{target}, x_{source}\right)) + L_{pix}(x, \hat{x}) + L_{per}(x, \hat{x}) \tag{9}$$

4 Experimental Analysis and Results

4.1 Datasets

In this paper, we aim to study the cross-method generalization of the deepfake detection model, that is, whether the model trained based on a specific forgery method can obtain good results on the fake content synthesized by other forgery methods. Based on the above purpose, in this paper, we choose the **FaceForensics++** [13] dataset, which contains multiple forgery methods, to confirm the efficiency of our proposed method. This large-scale dataset aims to advance the development of deepfake detection and defence technologies. It contains 5,000 video clips covering a variety of deepfake techniques, including facial synthesis and expression exchanges. The dataset includes five forgery methods: Deepfakes, Face2Face, FaceSwap, FaceShifter, and NeuralTextures. Each synthetic forgery method contains 1000 fake videos. In particular:

(1) **Deepfakes** is formed by replacing the faces in the source video or images with the faces in the target sequence. The logic is based on two autoencoders trained to reconstruct the source and target faces separately.
(2) **Face2Face** is a facial reenactment system that transfers the expression of a source video to a target video while maintaining the identity of the target.
(3) **FaceSwap** is a graphic-based approach which transfers the face region from a source video to a target video. This approach extracts face regions and fits 3D templates based on sparsely detected facial landmarks.

(4) **FaceShifter** is a face swap algorithm based on a two-stage method with high realism and occlusion awareness. The first stage generates high-realistic replacement faces, and the second stage deals with face occlusion, the self-supervised improvement of unnatural face regions through the constraints of multiple loss functions.
(5) **NeuralTextures** uses raw video data to learn the neural texture of the target person to achieve the purpose of facial reenactment, and only the facial expressions corresponding to the mouth area are modified in this dataset.

Table 3. Accuracy (ACC) results based on different τ of FF++ dataset.

Train Set (FF++)	Test Set (ACC-%)						Average
	Deepfakes	Face2Face	FaceSwap	FaceShifter	NeuralTextures	FF++	
$\tau = 0.1$	50.56	50.56	50.53	50.56	50.44	50.54	50.53
$\tau = 0.2$	54.39	54.08	53.39	54.48	52.98	53.92	53.87
$\tau = 0.3$	60.43	58.49	56.46	60.19	55.96	58.57	58.35
$\tau = 0.4$	65.19	61.28	57.64	64.40	57.00	61.58	61.18
$\tau = 0.5$	67.91	63.05	57.98	67.03	57.36	63.06	62.73
$\tau = 0.6$	67.99	61.84	56.97	67.29	56.17	62.19	62.08
$\tau = 0.7$	65.77	60.56	54.34	64.99	55.49	60.88	60.34
$\tau = 0.8$	60.08	56.76	51.87	58.03	53.06	56.86	56.11
$\tau = 0.9$	52.57	51.14	50.08	50.78	50.23	51.48	51.04

4.2 Experimental Setup

In this paper, we divide the training set and test set according to 75% and 25%. The specific settings are as follows:

- For the intra-method experiments, we use the single forge method as the training set and synthesized data based on the same method as the test set. For example, we choose 750 Deepfakes videos as the training set and the remaining 250 Deepfakes videos as the testing set. We arranged other fake datasets in the same way.
- For the cross-method generalization experiments, we use the single forge method as the training set and the other forge method as the testing set. To ensure that the test set does not overlap with the training set, we choose the same test video names in the cross-method experiments as in the intra-method experiments. In other words, the test set here has the same test individuals as the above test set, and the difference is the synthesis method. For example, we choose the same 750 Deepfakes videos, which are the same in intra-method experiments, as the training set, and 250 Face2Face videos, where the video name is the same as the 250 Deepfakes videos in the intra-method experiments, as the testing set. The rest of the dataset is arranged in the same way.

In addition, for each video, we use the face detector MTCNN for face cropping to obtain 32 frames of face images. After the dataset preprocessing operation, we have 24000 images (750×32) in each experiment as the training set and 8000 images (250×32) as the test set. For the acquisition of facial landmarks and masks, we use the dlib package. All experiments are based on the Pytorch framework and implemented on NVIDIA GeForce RTX 2080.

All experiments in this paper use the same parameter settings for a fair comparison. Specifically, we use the stochastic gradient descent (SGD) optimizer when training the model. The learning rate is 0.001, each model is trained for ten epochs, and the batch size is 4.

In order to compare our method with SOTA, we selected four methods. The first one, Xception [3], is an architecture based on Inception, where the traditional modules have been replaced with depthwise separable convolutions. DSP-FWA [9], where convolutional neural networks (CNNs) are used to capture the artifacts originating from the warping process necessary to adapt the new face to the source image. EfficientNetB4, taken from a wider study, is currently one of the best networks in the field of deepfake detection. Moreover, EfficientNetB4ATTST [2] integrates attention mechanism and siamese training on the basis of EfficientNetB4 and is committed to improving the generalization ability of the model.

Ablation Experiments on Parameter. To correctly set the parameters of the proposed method, we conducted comparative experiments on the threshold τ to explore which value is the most reasonable (Table 3). For the parametric ablation experiments, we consider all fake methods in the FF++ dataset. In particular, we extract the five deepfake techniques in the FF++ dataset at the same proportion. Precisely, we extract 150 videos for each forgery method (a total of 750 videos) and 750 real videos as the training set and 50 videos for each forgery method (whole 150 videos) and 150 real videos as the test set. We adopted the accuracy rate to measure the performance. The experimental results show that when $\tau = 0.5$, the average accuracy rate is maximum, equal to 62.73%.

4.3 Results

We first verify the intra-method performance of the proposed approach, which means training and testing on the same deepfake technique. Related experimental results are shown in Table 4, highlighted in blue.

The intra-method performance of the proposed method is not competitive with the analyzed SOTA. It never performs better than the best one, i.e. EfficientNetB4, and outperforms DSP-FWA only in the tests on Deepfakes and Face2Face. However, the performance is generally good on face swap generation techniques, especially on FaceShifter and Deepfakes (Table 4). A dramatic performance drop occurs when training and testing on FaceSwap images. However, this drop could be explained by the fact that FaceSwap is based on 3D template synthesis instead of 2D, as in the case of FaceShifter and Deepfakes.

Table 4. Intra-method and Cross-method accuracy based on different training sets in FF++.

Train Set	Model	Test Set (ACC %)					cross-method
		Deepfakes	Face2Face	FaceSwap	FaceShifter	NeuralTextures	Average
Deepfakes	Xception[3]	98.38	51.25	49.87	52.03	52.92	51.52
	DSP-FWA[9]	89.79	51.73	52.03	54.94	53.17	52.97
	EfficientNetB4[2]	99.06	52.83	49.77	54.53	55.69	53.21
	EfficientNetB4ATTST[2]	73.42	52.57	48.51	53.74	57.29	53.03
	Ours(τ=0.5)	93.97	54.81	48.04	57.61	64.91	**56.34**
Face2Face	Xception	56.28	98.17	50.73	49.74	51.43	52.04
	DSP-FWA	55.70	82.98	51.63	50.76	51.80	52.47
	EfficientNetB4	58.98	98.91	51.28	50.02	51.75	53.01
	EfficientNetB4ATTST	54.26	65.04	53.66	47.98	53.56	52.37
	Ours(τ=0.5)	56.57	85.49	53.96	52.00	53.78	**54.08**
FaceSwap	Xception	51.22	52.67	97.72	50.14	49.42	50.86
	DSP-FWA	61.13	51.52	70.56	52.94	50.59	**54.05**
	EfficientNetB4	50.65	53.16	98.03	50.27	50.09	51.04
	EfficientNetB4ATTST	47.53	53.02	69.71	48.08	47.33	48.99
	Ours(τ=0.5)	55.38	52.42	58.77	53.72	50.89	53.10
FaceShifter	Xception	51.51	48.73	49.16	96.06	51.18	50.14
	DSP-FWA	51.54	49.88	49.69	98.06	50.34	50.36
	EfficientNetB4	50.52	49.75	49.94	98.23	50.09	50.08
	EfficientNetB4ATTST	53.71	48.48	49.33	69.41	52.57	51.02
	Ours(τ=0.5)	60.67	50.40	50.01	91.28	51.82	**53.22**
NeuralTextures	Xception	67.06	61.21	48.16	54.81	89.50	57.88
	DSP-FWA	62.36	59.18	50.18	55.79	87.61	56.88
	EfficientNetB4	72.10	59.49	48.61	58.99	93.80	59.80
	EfficientNetB4ATTST	59.22	53.14	48.79	52.33	62.08	53.37
	Ours(τ=0.5)	82.43	61.17	45.20	57.94	83.13	**61.69**

The general good behavior on face swap techniques is because the proposed method bases its functioning on the inconsistencies between the inner and outer face. This observation agrees with the reported results, where it is worth remarking that the face swap technique modifies the inner face, thus increasing the differences with the outer face.

We believe the low intra-method performance is due to the global description obtained from the difference between the target component and the source component. In other words, the proposed method pays more attention to global information that allows for increasing the cross-methods accuracy.

Cross-method experiments allow us to evaluate whether the proposed method can correctly classify unknown types of deepfakes. As shown in Table 4, we conduct cross-method testing based on different training sets. Specifically, we use each forgery method in the FF++ dataset as the training set and the non-overlapping data based on the remaining methods as test sets for cross-validation. The specific data construction method is shown in Sect. 4.2.

As seen from Table 4, the proposed method obtains the highest average accuracy compared to the SOTA methods. This aspect is crucial because deepfake detection is an arms race problem, and new manipulations are created frequently.

Furthermore, we notice that the training method with a greater generalization is NeuralTexture. This is expected as this reenactment method is recent and produces realistic deepfakes. The most significant increase over SOTA is obtained by training on Deepfakes. In particular, from Table 4, we can see that it reaches 56.34% accuracy, which is 3.13% higher than that of EfficientNetB4. Furthermore, for the FaceShifter method, the average cross-method performance of our approach is 53.22%, which is 2.86% better than DSP-FWA, and 2.14% better than EfficientNetB4ATTST.

The cross-method findings show that the proposed method's main contribution is connected to the classification of face swap-type samples, but it also helps the classification of reenacted images. For example, we have reported in Fig. 3 an input image of low quality. It can be observed by looking at each column that the deep fake probability is high in most cases (columns 4–8), thus independent of the training set. On the other hand, the probability related to the real image leads to the correct classification in 4 cases out of 5.

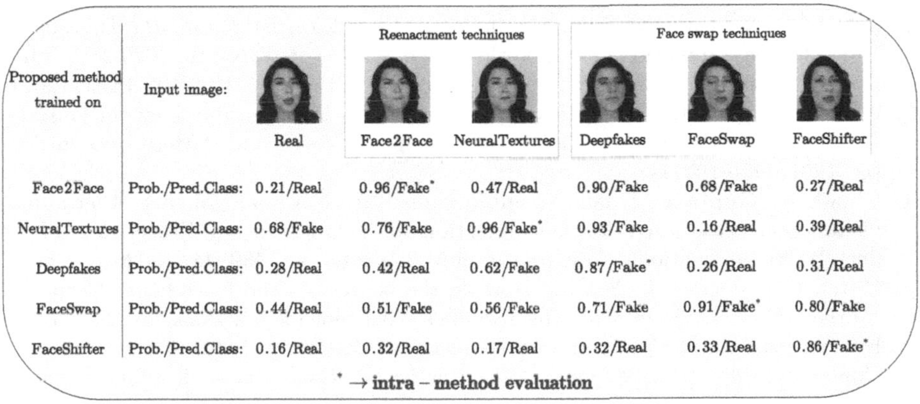

Proposed method trained on	Input image:	Real	Reenactment techniques: Face2Face	Reenactment techniques: NeuralTextures	Face swap techniques: Deepfakes	Face swap techniques: FaceSwap	Face swap techniques: FaceShifter
Face2Face	Prob./Pred.Class:	0.21/Real	0.96/Fake*	0.47/Real	0.90/Fake	0.68/Fake	0.27/Real
NeuralTextures	Prob./Pred.Class:	0.68/Fake	0.76/Fake	0.96/Fake*	0.93/Fake	0.16/Real	0.39/Real
Deepfakes	Prob./Pred.Class:	0.28/Real	0.42/Real	0.62/Fake	0.87/Fake*	0.26/Real	0.31/Real
FaceSwap	Prob./Pred.Class:	0.44/Real	0.51/Fake	0.56/Fake	0.71/Fake	0.91/Fake*	0.80/Fake
FaceShifter	Prob./Pred.Class:	0.16/Real	0.32/Real	0.17/Real	0.32/Real	0.33/Real	0.86/Fake*

* → intra – method evaluation

Fig. 3. Classification of input samples obtained with various types of manipulation with models trained on FF++ deepfake types.

5 Conclusions

In this paper, we described a novel general-purpose deepfake detection framework. Specifically, we exploited facial landmark detection algorithms to extract masks by assuming that the input image combines source and target images. Furthermore, we segment the face image into inner and outer faces and construct different encoders for feature extraction. The inconsistency is used to discriminate real from fake, and a reconstruction loss is proposed to force the two encoders to obtain richer detailed information.

The experimental findings demonstrate that the proposed technique is promising regarding generalization ability. However, the loss function must be improved to get more details on the differences between inner and outer faces. This will be addressed in future works.

Acknowledgment. This work is partially supported by China Scholarship Council (No.202206290093), by SERICS (PE00000014) under the Italian Ministry of University and Research (MUR) National Recovery and Resilience Plan funded by the European Union - NextGenerationEU and within the PRIN2017 - BullyBuster - A framework for bullying and cyberbullying action detection by computer vision and artificial intelligence methods and algorithms (CUP: F74I19000370001). The project has been included in the Global Top 100 list of AI projects addressing the 17 United Nations Strategic Development Goals by the International Research Center for Artificial Intelligence under the auspices of UNESCO.

References

1. Bekci, B., Akhtar, Z., Ekenel, H.K.: Cross-dataset face manipulation detection. In: 2020 28th Signal Processing and Communications Applications Conference (SIU), pp. 1–4. IEEE (2020)
2. Bonettini, N., Cannas, E.D., Mandelli, S., Bondi, L., Bestagini, P., Tubaro, S.: Video face manipulation detection through ensemble of CNNs. In: 2020 25th International Conference on Pattern Recognition (ICPR), pp. 5012–5019. IEEE (2021)
3. Chollet, F.: Xception: deep learning with depthwise separable convolutions. In: Proceedings of the IEEE Conference on Computer Vision and Pattern Recognition, pp. 1251–1258 (2017)
4. Chugh, K., Gupta, P., Dhall, A., Subramanian, R.: Not made for each other-audio-visual dissonance-based deepfake detection and localization. In: Proceedings of the 28th ACM International Conference on Multimedia, pp. 439–447 (2020)
5. Ciftci, U.A., Demir, I., Yin, L.: How do the hearts of deep fakes beat? Deep fake source detection via interpreting residuals with biological signals. In: 2020 IEEE International Joint Conference on Biometrics (IJCB), pp. 1–10. IEEE (2020)
6. Concas, S., Perelli, G., Marcialis, G.L., Puglisi, G.: Tensor-based deepfake detection in scaled and compressed images. In: 2022 IEEE International Conference on Image Processing (ICIP), pp. 3121–3125. IEEE (2022)
7. Kim, M., Tariq, S., Woo, S.S.: FReTAL: generalizing deepfake detection using knowledge distillation and representation learning. In: Proceedings of the IEEE/CVF Conference on Computer Vision and Pattern Recognition, pp. 1001–1012 (2021)
8. Li, L., Bao, J., Zhang, T., Yang, H., Chen, D., Wen, F., Guo, B.: Face x-ray for more general face forgery detection. In: Proceedings of the IEEE/CVF Conference on Computer Vision and Pattern Recognition, pp. 5001–5010 (2020)
9. Li, Y., Lyu, S.: Exposing deepfake videos by detecting face warping artifacts. In: IEEE Conference on Computer Vision and Pattern Recognition Workshops (CVPRW) (2019)
10. Lyu, S.: Deepfake detection: current challenges and next steps. In: 2020 IEEE International Conference on Multimedia & Expo Workshops (ICMEW), pp. 1–6. IEEE (2020)
11. Nadimpalli, A.V., Rattani, A.: On improving cross-dataset generalization of deepfake detectors. In: Proceedings of the IEEE/CVF Conference on Computer Vision and Pattern Recognition, pp. 91–99 (2022)
12. Ramachandran, S., Nadimpalli, A.V., Rattani, A.: An experimental evaluation on deepfake detection using deep face recognition. In: 2021 International Carnahan Conference on Security Technology (ICCST), pp. 1–6. IEEE (2021)

13. Rossler, A., Cozzolino, D., Verdoliva, L., Riess, C., Thies, J., Nießner, M.: FaceForensics++: learning to detect manipulated facial images. In: Proceedings of the IEEE/CVF International Conference on Computer Vision, pp. 1–11 (2019)
14. Shiohara, K., Yamasaki, T.: Detecting deepfakes with self-blended images. In: Proceedings of the IEEE/CVF Conference on Computer Vision and Pattern Recognition, pp. 18720–18729 (2022)
15. Taeb, M., Chi, H.: Comparison of deepfake detection techniques through deep learning. J. Cybersecurity Priv. **2**(1), 89–106 (2022)
16. Zhao, H., Zhou, W., Chen, D., Wei, T., Zhang, W., Yu, N.: Multi-attentional deepfake detection. In: Proceedings of the IEEE/CVF Conference on Computer Vision and Pattern Recognition, pp. 2185–2194 (2021)

Real-Time Multiclass Face Spoofing Recognition Through Spatiotemporal Convolutional 3D Features

Salvatore Giurato(✉), Alessandro Ortis, and Sebastiano Battiato

Image Processing Laboratory, Dipartimento di Matematica e Informatica, Universita' degli studi di Catania, Viale A. Doria 6, 95125 Catania, Italy
salvatore.giurato@phd.unict.it, {ortis,battiato}@dmi.unict.it

Abstract. Face recognition is used in numerous authentication applications, unfortunately they are susceptible to spoofing attacks such as paper and screen attacks. In this paper, we propose a method that is able to recognise if a face detected in a video is not real and the type of attack performed on the fake video. We propose to learn the temporal features exploiting a 3D Convolution Network that is more suitable for temporal information. The 3D ConvNet, other than summarizing temporal information, allows us to build a real-time method since it is so much more efficient to analyse clips instead of analyzing single frames. The learned features are classified using a binary classifier to distinguish if the person in the clip video is real (i.e. live) or not, multi class classifier recognises if the person is real or the type of attack (screen, paper, ect.). We performed our test on 5 public datasets: Replay Attack, Replay Mobile, MSU-MSFD, Rose-Youtu, RECOD-MPAD.

Keywords: Antispoofing Attack · 3D Features · Multi-Class detection · liveness

1 Introduction

Face Recognition is a biometric system that has been deployed in real life applications such as recognizing people's identity. Face Recognition made significant progress during the past years with DeepFace [13], DeepIDs [14], VGG Face [15], FaceNet [16], SphereFace [17] and ArcFace [18]. However the more Face Recognition is popular the newer spoofing attacks appears. Common attacks can be categorized as video replay attacks, photo attacks, and 3D mask attacks. Common Face Recognition methods are unable to detect the difference between real faces and attack faces. In literature there are different approaches on how to distinguish real face or attack face. One approach is to detect the motion in a video in order to prevent photo attack, Li et al. [19] used the Fourier spectra to estimate the temporal changes due to the motion. In contrast to Li et al. [19], Kollreider et al. [20] and Kollreider et al. [21] worked on RGB space instead of frequency domain to estimate respectively the 2D and 3D motion. The work in

G. L. Foresti et al. (Eds.): ICIAP 2023 Workshops, LNCS 14365, pp. 356–367, 2024.
https://doi.org/10.1007/978-3-031-51023-6_30

Pan et al. [22] and Sun et al. [23] exploited the human physiological behaviour to detect anomalies in eye blinking, in case of anomalies the video is considered a spoofing attack. Bao et al. [24] used optical flow to distinguish between 3D photo and planar photo attack. A different approach was proposed by Li et al. [25] and Nowara et al. [26] that exploit remote PhotoPlethysmoGraphy (rPPG) that detect blow flood using RGB images, this method is able to detect photo based and 3D mask attack. The first work that used a machine learning approach through Convolutional Neural Networks (CNN) was Yang et al. [27] where an AlexNet [9] architecture with the last layer replaced by an SVM binary classifier was employed. Patel et al. [28] also used an AlexNet architecture replacing the 1000-way SoftMax with a binary classifier, in this case the network is pretrained on ImageNet [11] and WebFace [12] and fine tuned on the facial spoofing dataset. Li et al. [29] proposed a CNN based on VGG-Face [15] pretrained on massive dataset and fine-tuned. George et al. [30] proposed Deep Pixelwise Binary Supervision (DeepPixBiS) based on DenseNet [10] that uses two losses during the training. While the motions method works on the video the convolutional methods works on the single frame of the video, it means that they are so much more time consuming. To reduce the complexity of the video analysis Tran et al. [7] proposed a 3D Convolutional Network to solve the action recognition problem. Sultani et al. [8] exploited the model of Tran et al. [7] to build a new model for anomaly detection. Sultani et al. [8] using the 3D Convolutional neural network trained by Tran et al. [7] obtained optimal results on the anomaly detection problem. Our proposed approach is to build a novel method to detect anti-spoofing attacks exploiting 3D Convolutional Network, and then we will present our studies of a novel method that is able to distinguish the type of attack. To the best of our knowledge this is the first work that recognise whether the video is real or a print attack or replay attack or mask attack. To exploit well the 3D Convolutional Network we divide the videos in clips, from now on we will refer as segment when we talk about clips. The problem of spoofing detection still does not have a benchmark dataset, some dataset that has been used in previous works are not available, some others are new so they have not been tested in newer works. There is not an anti-spoofing dataset in the wild which means that all the videos in all datasets are collected in a protected environment, this means that each dataset has different characteristics. For this reason, we decided to work on 5 different datasets: Idiap Replay Attack [1], Idiap Replay Mobile [2], Recod-MPAD [3], MSU-MFSD [4], Rose-Youtu Dataset [5] [6].

2 Proposed Method

This section presents the keystone of our approach. First, the videos are segmented in temporal video clips of a fixed duration, to properly analyse the temporal changes in the scenes. As previously mentioned, face detection and face recognition are deployed in different real-life applications. Once the face is detected, a recognition algorithm is applied to the portion of image depicting the face, discarding the background information. In order to get the same

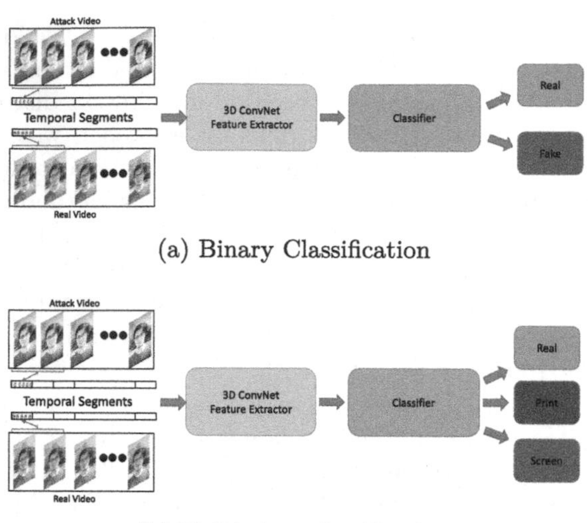

(a) Binary Classification

(b) Multi-class classification

Fig. 1. Considered binary and multiclass pipelines.

information of the face recognition algorithm, the Face Detection procedure is performed on each frame of each segment. The proposed method exploits two neural network or machine learning models: the first network extracts the feature for each segment (i.e. short video clip), the second model is a binary classifier or a multiclass classifier. The first network is a pretrained 3D Convolutional Network [7], by which the features of the segments are extracted, in this way it is generated a 3D array of features that contains a single feature representing a segment of video. The 3D Convolutional Network is fed with a segment in order to get the temporal features, which help reducing the computational complexity of the method, since a single segment is able to summarise the information of a certain number of frames. Such features of segments represent instances in a bag of features, which will be fed to a classifier. Once all the features of the segments are extracted the second network is trained for spoofing attack detection in a binary or multiclass classification setting. In the binary classifier we consider the features bag as negative for real videos or as positive for attack videos, regardless the type of attack. In the multiclass classifier we consider the features bag as negative (0) for real videos, and as positive screen (1) for screen attack videos and as positive print (2) for print attack videos, and as positive mask (3) for mask attack videos. It is also possible to compute an interpolation to the feature bag to double the features dimensions and try to get more information from them. The sequence of features extracted from a video segment can be eventually interpolated. The result is a sequence of features doubled in number. Such a technique is successfully applied in [8] to detect anomalies in videos. The underlying idea is that the interpolation (i.e., average) of two temporally subsequent features results in a more smooth feature, in which anomaly signals may emerge. In the context of our work, we expect that feature anomalies in attack attempts videos may be detected in a similar way. After the classification of the interpolated network to have only one result as in the other classifiers, or a different number of results, for each feature bag, it is computed the extrapolation process. In Fig. 1a and 1b it is possible to observe the pipeline of how our method works.

Table 1. Information of the dataset (Attack: Print (P), Screen (S), Mask (M))

Dataset	Train Real/Fake	Segment Train Real/Fake	Test Real/Fake	Segment Test Real/Fake	Attack
Replay Attack	60/300	1380/4200	80/400	1840/5600	P/S
Replay Mobile	120/192	2108/3413	110/192	1931/3399	P/
Recod-MPAD	250/1000	824/1185	200/800	670/952	P/S
MSU-MFSD	30/90	516/1492	40/120	671/1986	P/S
Rose-Youtu	147/398	1081/4989	171/468	1521/5229	P/S/M

3 Evaluation

3.1 Dataset and Evaluation Metrics

We evaluated the performance of our model with 5 known datasets: Idiap Replay Attack [1], Idiap Replay Mobile [2], Recod-MPAD [3], MSU-MFSD [4], Rose-Youtu Dataset [5,6]. In Table 1 a summary of the details about the datasets. The videos are divided in train/test sets as recommended by the authors of each dataset. It is worth highlighting that there is no person that appears in both train and test sets. Note that the chosen extraction process forces the frames of the video to be resized in 112×112 format before the feature extraction. In both binary and multiclass classifiers we build a confusion matrix from the classification. From the confusion matrix we extract the value of True Positive (TP), True Negative (TN), False Positive (FP), False Negative (FN) Therefore, we consider:

- True Positive (TP) the elements correctly classified as "Attack" in the binary classifier and as "Print" and "Screen" in the multiclass classifier.
- True Negative (TN) the elements correctly classified as "Real" in both binary classifier and multiclass classifier.
- False Positive (FP) the elements incorrectly classified as "Attack" in the binary classifier and as "Print" and Screen" in the multiclass classifier.
- False Negative (FN) the elements incorrectly classified as "Real" in both binary classifier and multiclass classifier.

The metrics we used to evaluate our classification are:

- Accuracy represents the percentage of correct predictions in the classification:

$$Accuracy = \frac{TP + TN}{TP + FP + FN + TN}. \quad (1)$$

- False Acceptance Rate is the ratio of False Positive predictions:

$$FAR = \frac{FP}{FP + TP}. \quad (2)$$

- False Rejection Rate is the ratio of False Negative predictions:

$$FRR = \frac{FN}{FN + TN}. \quad (3)$$

- HTER (Half Total Error Rate) [38] is a measure of the error, it is one of the most common metrics in biometric system to evaluate the system performance:

$$HTER = \frac{FAR + FRR}{2}. \quad (4)$$

Another metric similar to HTER is EER (Equal Error Rate) that is the threshold where FAR = FRR, however this metric can be used only to evaluate a training, but cannot be used to measure the performance of a model [39].

Table 2. Results of binary classifiers on All Datasets

Method	Binary		Multiclass	
	Acc	HTER	Acc	HTER
K-NN	75.7	29.1	65.9	36.7
LDA	80.3	24.7	**76.7**	**26.4**
SVM linear	78.8	27.0	73.6	30.0
Random Forest	**82.8**	**19.8**	74.2	27.7
Decision Tree	69.7	37.8	59.8	43.7
GBoost	79.5	26.1	75.6	27.0
3L-NN	79.8	25.7	/	/

3.2 Implementation Details

To start the pre-processing of the videos, we detect the face in each frame using FaceNet [16]. After the detection, the face is cropped from the image, and we store in the memory only the face for each frame. To continue the faces are resized in 112×112 and stored in segments containing 16 frames each. We extract features from the fully connected (FC) layer FC6 of the C3D network [7], which have a feature of dimensions of 4096. As a classifier we used different methods:

- K-NN with K = 3
- LDA
- SVM using the linear kernel
- Random Forest
- Decision Tree
- GBoost
- 3L-NN: A 3 Fully Connected Neural Network, each fully connected layer is followed by a dropout layer. The output of the 3 Fully Connected layers is respectively 512, 32 and 1. The last output is passed through a sigmoid function to convert to 0 and 1. The optimizer used is Adam with a learning rate = 0.001.

Table 3. Binary Classifier intra-dataset analysis:Replay Attack (I-RA), Replay-Mobile (I-RM), MSU-MFSD (MFSD), Recod-MPAD (MPAD), Rose-Youtu (Y)

Method	I-RA		I-RM		MFSD		MPAD		Y	
	Acc	HTER	Acc	HTER	Acc	HTER	Acc	HTER	Acc	HTER
K-NN	87.8	16.5	86.6	14.6	72.9	35.0	68.3	30.7	84.3	22.3
LDA	**91.4**	11.0	**93.7**	**6.3**	67.5	36.0	68.9	31.8	83.0	24.2
SVM linear	90.6	16.5	91.8	7.1	80.8	25.3	77.3	23.1	**84.5**	**21.3**
Random Forest	87.0	**9.0**	89.2	11.1	75.5	32.3	76.1	24.2	84.3	21.5
Decision Tree	83.9	19.4	78.0	23.6	67.1	40.0	64.5	36.2	78.9	30.4
GBoost	87.5	14.9	88.6	12.3	72.5	35.2	**77.8**	**22.6**	82.5	24.4
3L-NN	88.29	11.1	88.5	12.2	78.2	29.0	66.2	35.0	84.0	21.8
3L-NN$_{interp}$	91.2	11.7	88.3	8.7	**81.2**	**22.4**	72.7	26.2	83.1	**21.3**

3.3 Binary Classifier

In this section we summarise the experiments made with the binary classifier. For the sake of comparison, we performed the classification on the features of the video with different classifiers (K-NN, Support Vector Machine, Random Forest, Decision Tree, GBoost, 3L-NN). In Table 2 there is the comparison between these methods and a simple binary classifier. To perform a fair comparison the features are extracted from C3D Network, all these methods are trained with the train set of all the dataset, to get the results shown in the table the test set is used on the trained models. However, considering all the dataset together is not a good idea because they are from different protected environments, and they contain different information so this may lead a model to be more error prone. In addition, the presence of video attack in Dataset is higher than the presence of real video, this may cause overfit in models and networks. In Table 3 there are the results of the intra-dataset analysis using different datasets and different classifiers. As we can see for each dataset the classifiers have a different behaviour, this is because of the difference between the datasets. From the results we notice that some dataset are "easier" to generalise with the classifier such as Replay-Attack and Replay Mobile and more difficult to generalise such as MSU-MFSD and Recod-MPAD. In bold are highlighted the best results for each dataset.

Table 4. Multi Class Classifier: Replay Attack (I-RA), Replay-Mobile (I-RM), MSU-MFSD (MFSD), Recod-MPAD (MPAD), Rose-Youtu (Y)

Method	I-RA		I-RM		MFSD		MPAD		Y	
	Acc	HTER	Acc	HTER	Acc	HTER	Acc	HTER	Acc	HTER
K-NN	70.5	28.7	80.9	19.2	65.5	40.1	62.6	36.3	85.3	15.3
LDA	**87.0**	**13.1**	**92.1**	**6.8**	66.5	38.2	61.6	36.8	87.6	14.5
SVM linear	83.6	15.7	91.5	7.7	**74.1**	**31.4**	71.7	28.4	**87.9**	**13.2**
Random Forest	77.3	16.4	87.2	13.8	69.3	36.4	68.7	31.4	87.0	14.2
Decision Tree	69.5	28.9	67.9	32.5	61.3	45.1	58.3	40.2	87.0	14.2
GBoost	81.0	17.2	88.7	11.8	70.4	35.0	**71.9**	**28.2**	79.3	23.8

3.4 Multiclass Classifier

In this section we summarise the experiments made with the multiclass classifier. We performed the classification on the features of the video with different classifiers (i.e. K-NN, Support Vector Machine, Random Forest, Decision Tree, GBoost). This is the first time that a method is able to distinguish the type of attacks. In Table 2 there is the comparison between different models, to perform a fair comparison the features are extracted from C3D Network, all these methods are trained on with the train set of all Dataset, to get the results we use the test set. In Table 4 there are the results on intra-dataset analysis on different models using different multiclass classifiers, the aim is to distinguish different types of attack. As we can see the classifiers work in a different way, for each dataset. However, we observe from Tables 3 and 4 that the behaviour of binary and multiclass classifiers is coherent.

3.5 Comparison with the State-of-the-Art

To evaluate our method, we decided to use more than one classifier. In order to obtain a fair comparison, we performed intra dataset and cross dataset tests on binary classifiers. As mentioned above, the datasets contain videos that have different characteristics and different conditions because they have been recorded in "under a controlled environment" so the cross-dataset, so the expectation is to have higher HTER than in intra-dataset. In Table 5 there is the comparison of our method with other state of art methods, in intra-dataset analysis, existing methods outperform our method. Although these methods outperform our method they analyse the videos frame by frame, on the other hand we analyse our video using segments of 16 frames, this means that our method has a less computation complexity. In Table 6 there is the comparison of our method

Table 5. Comparison with the state of the art: Replay Attack (I-RA), Replay-Mobile (I-RM), MSU-MFSD (M), Rose-Youtu (Y)

Method	I-RA	I-RM	M	Y
	HTER	HTER	HTER	HTER
LBP + SVM [1]	13.87	N/A	N/A	N/A
SVM RBF [2]	5.28	7.8	N/A	N/A
SURF (Gray) [31]	21.2	N/A	N/A	N/A
CSURF (RGB) [31]	13.5	N/A	N/A	N/A
CSURF (HSV) [31]	11.5	N/A	N/A	N/A
CSURF (YCbCr) [31]	8.9	N/A	N/A	N/A
CSURF (HSV + YCbCr) [31]	8.2	N/A	N/A	N/A
MRCNN [32]	1.6	N/A	N/A	N/A
Patch-Based CNN [33]	1.25	N/A	N/A	N/A
Depth-Based CNN [33]	0.75	N/A	N/A	N/A
Patch-Depth CNN [33]	**0.72**	N/A	N/A	N/A
DeepPixBiS [30]	N/A	**0.0**	N/A	N/A
CoALBP (HSV) [4,31]	3.7	N/A	9.8	26.6
CoALBP (YCbCr) [4,31]	1.4	N/A	8.1	17.1
LPQ (HSV) [4,31]	7.9	N/A	12.2	30.4
LPQ (YCbCr) [4,31]	6.3	N/A	7.4	27.6
Deep Learning Features [4,27]	2.1	N/A	**5.8**	**8.0**
Our_{LDA}	11.0	6.3	36.0	24.2
Our_{SVM}	16.5	7.1	25.3	21.3
$Our_{RandomForest}$	9.0	11.1	32.3	21.5
$Our_{3L-NN_{interp}}$	11.7	8.7	22.4	21.3

with other state of art methods in cross-dataset analysis. As mentioned for the intra-dataset analysis all methods work in a different way for each dataset. In both Tables 5 and 6 we choose methods that to the best of our knowledge are the best state of art methods, we reported the results that they presented in their papers, so in case of missing information we reported "N/A". We observe that Our K-NN method is competitive with most of other state of art models when trained on Replay-Attack and tested on MSU-MFSD, our SVM method is competitive when trained on Rose-Youtu and tested on Replay Attack and our Random Forest Method is competitive when trained on Rose-Youtu and tested on MSU-MFSD.

Table 6. Cross Analysis: Replay Attack (I), MSU-MFSD (M), Rose-Youtu (Y)

Method	I -> M	I -> Y	M -> I	M -> Y	Y -> I	Y -> M
	HTER	HTER	HTER	HTER	HTER	HTER
SURF (Gray) [31]	43.8	N/A	48.2	N/A	N/A	N/A
CSURF (RGB) [31]	44.1	N/A	47.8	N/A	N/A	N/A
CSURF (HSV) [31]	44.3	N/A	54.6	N/A	N/A	N/A
CSURF (YCbCr) [31]	31.8	N/A	53.8	N/A	N/A	N/A
CSURF (HSV + YCbCr) [31]	33.0	N/A	50.6	N/A	N/A	N/A
SA [4,27]	33.2	42.8	33.3	30.0	36.2	24.9
KSA [4,27]	33.3	40.1	34.9	30.4	38.8	26.1
ADA [34]	30.5	N/A	5.1	N/A	N/A	N/A
PAD-GAN [35]	23.2	N/A	8.7	N/A	N/A	N/A
ML-Net [36]	35.3	42.8	11.5	34.6	30.7	32.6
UDA [36]	29.0	**39.8**	3.0	**29.7**	**23.7**	**24.4**
Li,Zi et al. [37]	**20.8**	N/A	**2.9**	N/A	N/A	N/A
Our$_{K-NN}$	32.1	47.3	46.4	40.2	52.0	40.7
Our$_{SVM}$	35.0	45.5	36.9	42.5	25.3	39.1
Our$_{RandomForest}$	37.9	55.3	37.1	42.0	51.1	29.2

3.6 Ablation Study

In our experiment we investigated the effectiveness of using temporal features instead of using single features for each frame. This study is performed on the dataset Idiap Replay-Attack. In Table 7 are reported the results of this study considering the classes "real" and "attack" for the binary classifier and the class "real", "print attack" and "screen attack" for the multiclass classifier. We observe from the Table 7 that the single features have slightly better results; however, the time of training and classification are so much higher than the model based on temporal features.

Table 7. Ablation Study

Method	Binary				Muliclass			
	Temporal		Single Frame		Temporal		Single Frame	
	Accuracy	HTER	Accuracy	HTER	Accuracy	HTER	Accuracy	HTER
K-NN	87.8	14.9	88.7	14.7	70.5	25.7	75.4	23.5
LDA	**91.4**	**11.6**	**92.7**	**9.8**	**87.0**	**13.3**	**90.3**	**10.9**
SVM linear	90.6	0.132	91.6	11.7	83.6	17.8	87.4	13.9
Random Forest	87.4	17.5	87.9	49.0	/	/	/	/
Decision Tree	83.0	22.6	83.9	22.0	/	/	/	/
GBoost	87.5	17.0	88.0	16.6	/	/	/	/

4 Conclusion

In this study we proposed different ways to recognise face spoofing attacks, taking into account either the classic binary task (i.e., real vs attack) and the multiclass task concerning different types of spoofing attack. This method is able to execute the real time analysis in both binary and multiclass classifiers thanks to the 3D features. We performed a big series of test on different classifiers (K-NN, Support Vector Machine, Random Forest, Decision Tree, GBoost, 3L-NN) and on different datasets (Replay-Attack, Replay-Mobile, MSU-MFSD, RECOD-MPAD, Rose-Youtu) As far as we know this is the first work able to distinguish between the type of attack and obtaining closer results to the binary classifier. In addition this is the first time that an anti-spoofing method has used the 3D Convolutional Neural Network to extract the features. Even though our method already has a good computational complexity, the aim in the future is to optimize the classification and to improve the results of our model using a more complex classifier and a new 3D features extractor. In future work, a comprehensive study will be presented on the complexity of the anti-spoofing methods.

Acknowledgements. This work is partially funded by TIM S.p.A. through its UniversiTIM granting program.

Portions of the research in this paper used the Replay-Attack Dataset made available by the Idiap Research Institute, Martigny, Switzerland.

Portions of the research in this paper used the Replay-Mobile Dataset made available by the Idiap Research Institute, Martigny, Switzerland. Such Corpus was captured in collaboration with the Galician R and D Center on Advanced Telecommunications (GRADIANT), Vigo, Spain.

References

1. Chingovska, I., André, A., Sébastien, M.: On the effectiveness of local binary patterns in face anti-spoofing. In: 2012 BIOSIG-Proceedings of the International Conference of Biometrics Special Interest Group (BIOSIG). IEEE (2012)
2. Costa-Pazo, A., et al.: The replay-mobile face presentation-attack database. In: 2016 International Conference of the Biometrics Special Interest Group (BIOSIG). IEEE (2016)
3. Almeida, W.R., et al.: Detecting face presentation attacks in mobile devices with a patch-based CNN and a sensor-aware loss function. PloS one **15**(9), e0238058 (2020)
4. Wen, D., Han, H., Jain, A.K.: Face spoof detection with image distortion analysis. IEEE Trans. Inf. Forensics Secur. **10**(4), 746–761 (2015)
5. Li, H., et al.: Unsupervised domain adaptation for face anti-spoofing. IEEE Trans. Inf. Forensics Secur. **13**(7), 1794–1809 (2018)
6. Li, Z., et al.: One-class knowledge distillation for face presentation attack detection. IEEE Trans. Inf. Forensics Secur. **17**, 2137–2150 (2022)
7. Tran, D., et al.: Learning spatiotemporal features with 3D convolutional networks. In: Proceedings of the IEEE International Conference on Computer Vision (2015)

8. Sultani, W., Chen, C., Mubarak, S.: Real-world anomaly detection in surveillance videos. In: Proceedings of the IEEE Conference on Computer Vision and pattern recognition (2018)
9. Krizhevsky, A., Sutskever, I., Hinton, G.E.: ImageNet classification with deep convolutional neural networks. Commun. ACM **60**(6), 84–90 (2017)
10. Huang, G., et al.: Densely connected convolutional networks. In: Proceedings of the IEEE Conference on Computer Vision and Pattern Recognition (2017)
11. Deng, J., et al.: ImageNet: a large-scale hierarchical image database. In: 2009 IEEE Conference on Computer Vision and Pattern Recognition. IEEE (2009)
12. Yi, D., et al.: Learning face representation from scratch. arXiv preprint arXiv:1411.7923 (2014)
13. Taigman, Y., et al.: DeepFace: closing the gap to human-level performance in face verification. In: Proceedings of the IEEE Conference on Computer Vision and Pattern Recognition (2014)
14. Sun, Y., Xiaogang, W., Xiaoou, T.: Deeply learned face representations are sparse, selective, and robust. In: Proceedings of the IEEE Conference on Computer Vision and Pattern Recognition (2015)
15. Parkhi, O.M., Andrea, V., Andrew, Z.: Deep face recognition. In: BMVC 2015-Proceedings of the British Machine Vision Conference 2015 (2015)
16. Schroff, F., Dmitry, K., James, P.: FaceNet: a unified embedding for face recognition and clustering. In: Proceedings of the IEEE Conference on Computer Vision and Pattern Recognition (2015)
17. Liu, W., et al.: SphereFace: deep hypersphere embedding for face recognition. In: Proceedings of the IEEE Conference on Computer Vision and Pattern Recognition (2017)
18. Deng, J., et al.: ArcFace: additive angular margin loss for deep face recognition. In: Proceedings of the IEEE/CVF Conference on Computer Vision and Pattern Recognition (2019)
19. Li, J., et al.: Live face detection based on the analysis of Fourier spectra. Biometric Technology for Human Identification, vol. 5404. SPIE (2004)
20. Kollreider, K., Hartwig, F., Josef, B.: Evaluating liveness by face images and the structure tensor. In: Fourth IEEE Workshop on Automatic Identification Advanced Technologies (AutoID 2005). IEEE (2005)
21. Kollreider, K., Fronthaler, H., Bigun, J.: Non-intrusive liveness detection by face images. Image Vis. Comput. **27**(3), 233–244 (2009)
22. Pan, G., et al.: Eyeblink-based anti-spoofing in face recognition from a generic webcamera. In: 2007 IEEE 11th International Conference on Computer Vision. IEEE (2007)
23. Sun, L., Pan, G., Wu, Z., Lao, S.: Blinking-based live face detection using conditional random fields. In: Lee, S.-W., Li, S.Z. (eds.) ICB 2007. LNCS, vol. 4642, pp. 252–260. Springer, Heidelberg (2007). https://doi.org/10.1007/978-3-540-74549-5_27
24. Bao, W., et al.: A liveness detection method for face recognition based on optical flow field. In: 2009 International Conference on Image Analysis and Signal Processing. IEEE (2009)
25. Li, X., et al.: Generalized face anti-spoofing by detecting pulse from face videos. In: 2016 23rd International Conference on Pattern Recognition (ICPR). IEEE (2016)
26. Nowara, E.M., Ashutosh, S., Ashok, V.: PPGSecure: biometric presentation attack detection using photopletysmograms. In: 2017 12th IEEE International Conference on Automatic Face and Gesture Recognition (FG 2017). IEEE (2017)

27. Yang, J., Zhen, L., Stan, Z.L.: Learn convolutional neural network for face anti-spoofing. arXiv preprint arXiv:1408.5601 (2014)
28. Patel, K., Han, H., Jain, A.K.: Cross-database face antispoofing with robust feature representation. In: You, Z., Zhou, J., Wang, Y., Sun, Z., Shan, S., Zheng, W., Feng, J., Zhao, Q. (eds.) CCBR 2016. LNCS, vol. 9967, pp. 611–619. Springer, Cham (2016). https://doi.org/10.1007/978-3-319-46654-5_67
29. Li, L., et al.: An original face anti-spoofing approach using partial convolutional neural network. In: 2016 Sixth International Conference on Image Processing Theory, Tools and Applications (IPTA). IEEE (2016)
30. George, A., Sébastien, M.: Deep pixel-wise binary supervision for face presentation attack detection. In: 2019 International Conference on Biometrics (ICB). IEEE (2019)
31. Boulkenafet, Z., Komulainen, J., Hadid, A.: Face antispoofing using speeded-up robust features and fisher vector encoding. IEEE Signal Process. Lett. **24**(2), 141–145 (2016)
32. Ma, Y., Lifang, W., Li, Z.: A novel face presentation attack detection scheme based on multi-regional convolutional neural networks. Pattern Recogn. Lett. **131**, 261–267 (2020)
33. Atoum, Y., et al.: Face anti-spoofing using patch and depth-based CNNs. In: 2017 IEEE International Joint Conference on Biometrics (IJCB). IEEE (2017)
34. Wang, G., et al.: Improving cross-database face presentation attack detection via adversarial domain adaptation. In: 2019 International Conference on Biometrics (ICB). IEEE (2019)
35. Wang, G., et al.: Cross-domain face presentation attack detection via multi-domain disentangled representation learning. In: Proceedings of the IEEE/CVF Conference on Computer Vision and Pattern Recognition (2020)
36. Wang, G., et al.: Unsupervised adversarial domain adaptation for cross-domain face presentation attack detection. IEEE Trans. Inf. Forensics Secur. **16**, 56–69 (2020)
37. Li, Z., et al.: One-class knowledge distillation for face presentation attack detection. IEEE Trans. Inf. Forensics Secur. **17**, 2137–2150 (2022)
38. Hong, Y.: Performance evaluation metrics for biometrics-based authentication systems, Diss. (2021)
39. Bengio, S., et al.: Confidence measures for multimodal identity verification. Inf. Fusion **3**(4), 267–276 (2002)

Computer Vision for Environment Monitoring and Preservation (CVEMP) Enhancing Air Quality Forecasting Through

Enhancing Air Quality Forecasting Through Deep Learning and Continuous Wavelet Transform

Pietro Manganelli Conforti(✉), Andrea Fanti, Pietro Nardelli, and Paolo Russo

DIAG Department, Sapienza University of Rome, Via Ariosto 25, 00185 Rome, Italy
{manganelliconforti,fanti,russo}@diag.uniroma1.it

Abstract. Air quality forecasting plays a crucial role in environmental management and public health. In this paper, we propose a novel approach that combines deep learning techniques with the Continuous Wavelet Transform (CWT) for air quality forecasting based on sensor data. The proposed methodology is agnostic to the target pollutant and can be applied to estimate any available pollutant without loss of generality. The pipeline consists of two main steps: the generation of stacked samples from raw sensor signals using CWT, and the prediction through a custom deep neural network based on the ResNet18 architecture.

We compare our approach with traditional one-dimensional signal processing models. The results show that our 2D pipeline, employing the Morlet mother wavelet, outperforms the baselines significantly. The localized time-frequency representations obtained through CWT highlight hidden dynamics and relationships within the parameter behavior and external factors, leading to more accurate predictions. Overall, our approach demonstrates the potential to advance air quality forecasting and environmental management for healthier living environments worldwide.

Keywords: Air Quality Forecasting · Deep Learning · Continuous Wavelet Transform (CWT)

1 Introduction

The quality of the air we breathe is a critical aspect of our daily lives, directly impacting human health and the overall well-being of our environment. With the continuous rise in urbanization and industrialization, several contaminants have been introduced into the atmosphere that cause adverse changes either directly, by releasing toxic or harmful chemicals into the air, or indirectly, through the disruption of the delicate natural equilibrium, reflected in the composition of the air. Monitoring, forecasting, and preservation of air quality are of uttermost importance, in order to manage the chemical human footprint and adapt to the constant development of technologies. Moreover, accurate analysis through new technologies offers the potential to reduce costs associated with air sensing

G. L. Foresti et al. (Eds.): ICIAP 2023 Workshops, LNCS 14365, pp. 371–382, 2024.
https://doi.org/10.1007/978-3-031-51023-6_31

and monitoring. By leveraging advanced processing techniques and predictive models, more precise forecasts can be generated, enabling optimized deployment of sensing resources and targeted monitoring efforts.

In recent years, advancements in deep learning techniques combined with signal processing methods have emerged as powerful processing tools for addressing these challenges. In fact, deep neural networks demonstrated exceptional performance in capturing intricate interactions among numerous environmental scenarios, including meteorological conditions predictions [17], pollutant emissions monitoring [5], and environment feature analysis [10]. In addition, transform techniques such as the short-time Fourier transform and the wavelet transform have demonstrated substantial advantages for extracting valuable information about the time-varying frequency properties of one-dimensional signals in several fields of application, such as speech recognition [6], anomaly detection [14], and biomedical signal processing [11]. By transforming these signals into two-dimensional, time-frequency representations, it is possible to conduct a more thorough analysis of the underlying data, exploiting two-dimensional, state-of-the-art deep models with their accumulated knowledge via transfer learning [18].

In this research, we specifically focus on the Continuous Wavelet Transform (CWT) and explore the impact of different mother wavelets on enhancing the time-frequency analysis of air chemical concentration and other environmental measurements. Moreover, we investigate the potential synergies between deep learning techniques and transform-based methods in order to improve air quality forecasting. Therefore, the aim of this research is threefold, with a specific focus on the subsequent key areas:

- Utilizing a deep learning pipeline along with a proposed convolutional compression module to process the transformed data. We stack multiple time-frequency representations and fed them as input to the network, enabling the estimation of an air pollutant, (e.g., CO), starting from the other measured elements (e.g., C6H6, CO, NO2, NOx, NMHC, O3, AH, RH, T).
- Exploring the methodology for generating informative two-dimensional representations from several one-dimensional data segments and the effects of different mother wavelets employed by the CWT to enhance the time-frequency analysis. By comparing the images produced by different wavelets, we aim to identify the one that captures a broader range of temporal features and thus improving the final accuracy of air quality predictions.
- We demonstrate the possibility of accurately predicting one pollutant starting from a subset of the available measurements. This could enable the production of cheaper air sensors that can measure fewer chemical concentrations to obtain the same final results

The rest of the paper is organized as follows: firstly Sect. 2 will provide a description of the current state of the art, discussing the related works concerning air quality analysis and prediction and the use of mathematical transforms to analyze one-dimensional signals. Thereafter, in Sect. 3 we present the work methodology employed and the full model pipeline. Following will be described the experimental setup in Sect. 4, where are shown the hyperparameters employed in the 1D and 2D cases, the dataset composition, and the accuracy measures used. Finally, in Sect. 5 a comparison table of all the research findings is provided. The last section Sect. 6 will conclude the paper with future works and additional considerations.

2 Related Work

In recent years, deep learning has witnessed remarkable success across various domains. However, the performance of these models heavily relies on the quality of data representation, and this is where the power of transformations comes into play. Akansu et al. [1] provide a comprehensive assessment of the diverse and evolving applications of wavelet transforms across various fields. In order to identify complex mixtures in Raman spectroscopy, [12] propose a novel scheme based on a CWT and a deep neural network. The multi-label network model is used for classifying complex mixtures with improved performance thanks to CWT that decomposes the desired molecular information and noise from the Raman spectrum. Another approach [13] utilizes the Short-Time-Fourier transform (STFT) and a new convolutional neural network for lung cancer diagnosis from spectrochemical analysis. Similarly but in a different domain, in [16] authors utilize a deep neural network to detect audio operations using two types of transform techniques: STFT and Modified Discrete Transform. They improve audio authentication for forensics, successfully recognizing spoofed voices with high accuracy.

Recently, the field of air quality assessment has witnessed significant advancements through the application of machine learning techniques. We highlight the key works and contributions in this domain, outlining the methodologies and outcomes of these studies. Liu et al. [7] focused on predicting both the Air Quality Index and air pollutant concentrations through the application of various machine learning algorithms such as random forest regression and support vector regression. The work of Ly et al. [8] investigates the predictions of two important compounds in air pollution: NO_2 and CO. The authors address the regression problem by combining fuzzy logic with neural networks and a metaheuristic optimization chosen between simulated annealing and particle swarm optimization algorithms. Tackling the problem of noisy data, Sabzekar et al. [15] increase the robustness of the support vector regression, by converting constraints into fuzzy inequalities. In the work of Ma et al. [9] recurrent neural networks (RNNs) are exploited for time-series regression to tackle air quality estimation. They introduced a new RNN family, called Particle Filter RNNs, that is able to model the uncertainty by approximating a set of weighted particles, updated through to a particle filtering algorithm. Finally, Chen et al. [2] exploit a Gaussian process for multi-output prediction proposing a framework with a novel multivariate Student-t process regression model.

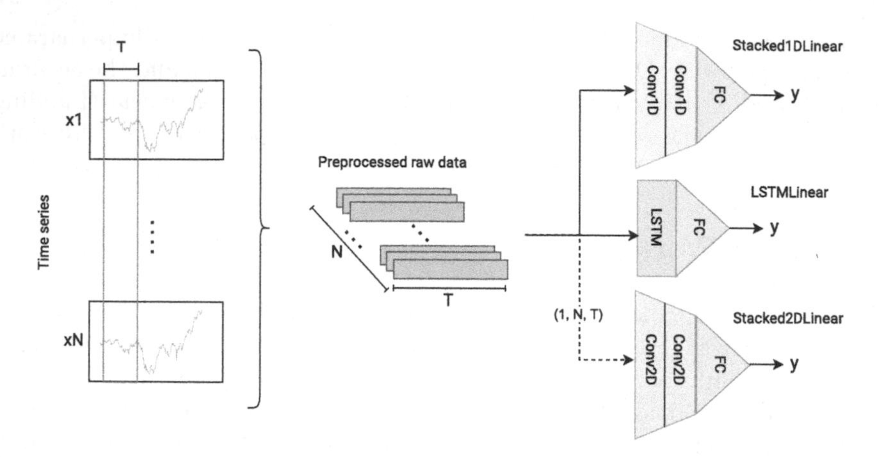

Fig. 1. Schema of the three baseline models, Stacked1DLinear, LSTMLinear, and Stacked2DLinear, employed in the 1D analysis.

3 Methodology and Data

The task we address through our pipeline is the accurate estimation of the mean future values of a single pollutant, accordingly to specific time windows. More in detail, given hourly measurements of the input parameters over a certain week, our task is to predict the average concentration of a target pollutant on the day which follows that week. Starting from the list of chemical elements inside the analyzed dataset, we have chosen CO to be predicted from a selection of these elements. However, the proposed methodology is agnostic with respect to the target pollutant and thus can be applied to estimate any of the other available pollutants without loss of generality. We compared our approach (which is described in 2D analysis subsection) with some traditional approaches (1D analysis subsection) which work directly on the one-dimensional signals. A detailed comparison in Sect. 4 will show the performance differences between the different techniques.

3.1 1D Analysis

In the field of air quality estimation, the analysis of raw time-series data has been widely employed by researchers as a standard approach. For this reason, we outline the development of three baseline models for air quality forecasting, emphasizing the usage of raw time-series data. The visual representation of the three baseline models can be seen in Fig. 1.

As a first step, the raw data undergoes a preprocessing stage, detailed in Sect. 4.1. The resulting output naturally forms a time series due to the sequential nature of the data. To effectively model the time-dependent patterns, we adopt a first baseline approach, called Stacked1DLinear, that utilizes a sequence of one-dimensional convolutional layers. Another model suitable for this type of data is the Long Short-Term Memory (LSTM) architecture, which is able to capture features from the sequential information contained in the data. This architecture (LSTMLinear) utilizes relevant information from earlier time steps, thereby enhancing its capacity to make accurate predictions for air quality values. In our third baseline approach, called Stacked2DLinear, we explored a different strategy by utilizing 2D convolutions to process the time series data. This approach benefits from the ability of convolutional layers to detect spatial patterns across different parameters and temporal patterns along the time axis. In our pursuit of a robust and comprehensive baseline comparison, we intentionally adopted the 2D convolution approach to closely resemble the Conv2D part of our proposed network architecture. This will highlight the contribution of the 2D transformation which we propose in our 2D pipeline.

3.2 2D Data Generation

To generate 2D images from the 1D time series, we use the Continuous Wavelet Transform (CWT). It is a mathematical technique used for analyzing and processing signals or time series data in various scientific and engineering applications, providing a multidimensional analysis of both time and frequency components. This can be fundamental to highlight specific spectral features of air chemical concentrations and meteorological parameters hidden in the input data, showing the time-frequency representations of the signals dynamics and their relationship. Furthermore, by employing these two-dimensional representations, the CWT can be synergistically integrated with a 2D deep neural network. This integration allows us to harness both the informative time-frequency representation and the robust generalization capabilities of visual features present in pre-trained deep neural networks.

Given a positive scale parameter a, a translation value b, a mother wavelet function $\psi(t)$, and an input signal $x(t)$, the mathematical formulation [4] used by this technique is defined as:

$$X_{\psi}(a,b) = \langle x(t), \psi_{\text{ab}}(t) \rangle = \int_{-\infty}^{+\infty} x(t)\psi_{\text{ab}}(t)dt$$

Where ψ_{ab} is the complex conjugate of the mother wavelet ψ and $X_{\psi}(a,b)$ is the time-frequency representation of the signal. Usually, the wavelets are brief, low-energy oscillatory signals. The mother wavelet function produces daughter wavelets based on the a and b parameters, which can be exploited to examine various frequency components at different levels [19]. Depending on input data, mother wavelets may create quite diverse time-frequency representations, revealing different trends. To analyze this behavior, in Sect. 5 we compare the different performances obtained by employing three popular mother wavelets: `morlet`, `morlet2`, and `ricker`. We exploit the CWT to produce a so-called *scalogram* of a signal, i.e. a heat map where the axes are time t and frequency f, and the color represents the magnitude or phase of the CWT $X_{\psi}(f,t)$. In

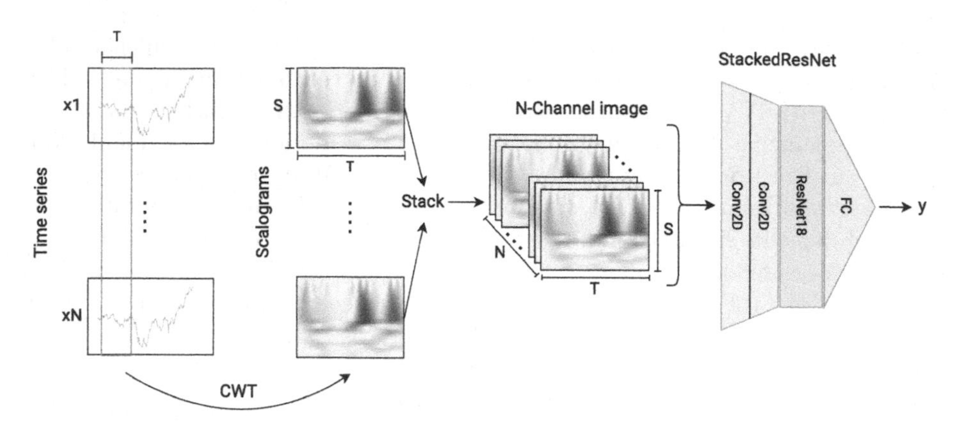

Fig. 2. The full 2D analysis pipeline. T is the window size, S is the number of scales, and N is the number of input signals.

our case, we chose to only use the magnitude, to allow us to generate a single gray-scale image from a single signal. The domain of the f axis is also referred to as *scales*, and its size as the *number of scales*. To increase the number of samples from a single time series, we split the time series into signal windows of fixed size, which are allowed to overlap for a fixed number of time steps in order to further increase the dataset size. Each window is then passed independently through the CWT, and the resulting magnitude is then used to produce the gray-scale image associated with that window by plotting its scalogram. Note that the resulting images are composed of a single channel. This means that, given a window size of N and the number of scales M, the resulting image will have dimensions (N, M). To split the time series into signal windows we use a sliding window, meaning that subsequent windows are partially overlapping. We call the temporal distance between the beginning of a signal window and the beginning of the next one the *stride* of the sliding window. More formally, let us assume the time series has length $\hat{T}$, the window size is T and the stride is S. Then, if $\hat{x}_t$ is the t-th value in the time series, the j-th element of the i-th window x_{ij} is computed as follows: $x_{ij} = \hat{x}_{iS+j}$ and thus the i-th signal window x_i will be $x_i = (\hat{x}_{iS}, \hat{x}_{iS+1}, \ldots, \hat{x}_{iS+N})$. If $\hat{N}$ is not a multiple of N, then the last signal window will be discarded since its size would be smaller than N. Finally, we would like to remark that the previous procedure is applied along the number of pollutant signals employed as input (e.g. if 11 different pollutant measurements are provided as input, for a single signal window 11 different scalograms are produced).

3.3 2D Analysis

Convolutional Neural Networks pre-trained on ImageNet provide powerful visual features which can be further specialized on the task of interest; for this reason, their usage is able to increase the accuracy of our regression task. However, those deep models require the input image to be represented as a three-channels image (usually in the RGB format). In our case, for a given signal window we have a number of scalograms equal to the number of pollutants measured and provided as input. For this reason, as a first step we compose each input sample by stacking the images corresponding to the Wavelet transform of the same window of the different input time series. More specifically, given the images I_j corresponding to the scalograms of the magnitudes of the CWTs of the i-th temporal windows of the parameters $j = 1, \ldots, N_j$, each of dimensions (N, M), the corresponding input sample will be a 3D tensor T of dimensions (N_j, N, M), such that $(T_j) = I_j$, or equivalently, a (N, M) image with N_j channels. In order to be able to feed this tensor to our pre-trained ResNet18 model, we provide the tensor as input to a sequence of two convolutional layers, with the last layer producing as output a $(3, N, M)$ tensor which has the correct shape in order to be provided to the backbone model for producing the output. Figure 2 reports the architecture of the proposed 2D pipeline.

4 Experimental Setup

4.1 Dataset

The dataset used to assess the validity of our method is the "Air Quality Dataset" provided by De Vito et al. [3]. It contains measurements of air pollutant concentration parameters and environmental parameters on a main road of an Italian city. More specifically, it consists of a collection of time series, one for each air pollutant or meteorological parameter. For some parameters, two types of measurements are available: one from a sensor tower (which we refer to as GT), and the other representing raw readings from cheaper, smaller embedded sensors (which we refer to as PT08); this is because the dataset was originally designed to calibrate the smaller sensor using the measurements from the GT sensors. All time series are aligned, i.e. the t-th samples of all time series correspond to measurements taken at the same real-time. There are 13 parameters in total: 5 from the GT sensors (CO, NMHC, C6H6, NOx, NO2), 5 from the PT08 sensors (CO, NMHC, NOx, NO2, O3), and 3 meteorological measurements (T, AH, RH). The measurements were taken hourly, aside from the first few samples. Some measurements are missing in the original dataset, and are marked with a -200 value. We split the dataset into training and test sets with a ratio of 80/20. We further split the former into training and validations sets with a ratio of 90/10. Thus, after splitting the time series into windows, we obtained 865 samples divided into 623 training samples, 69 validation samples and 173 test samples.

4.2 Implementation Details

Stacked1DLinear (6.2M parameters) and Stacked2DLinear (11M parameters) consist of 2 blocks made of convolutional layers, batch normalizations, and ReLU activations, followed by 2 fully connected layers. The blocks in the

Stacked2DLinear are exactly the same as the ones used in the StackedResnet in order to ensure a fair comparison. The LSTMLinear model (13.7M) has 2 LSTM layers as the main architecture with 512 hidden sizes. The LSTM output is fed as input into two fully connected linear layers. StackedResNet (11M trainable parameters) is composed of two 3×3 Conv2D blocks, a ResNet18 pre-trained model used as backbone and a Fully Connected (FC) block. The last layer of the original ResNet18 model is substituted with a new FC layer to allow for the correct output size. (a single floating point number in our case). The first Conv2D layer has 6 output channels, with the second having 3 output channels; both Conv2D layers have a 3×3 kernel with stride 1 and padding 1. We train all the learnable parameters of the resulting architecture, including those of the pre-trained on ImageNet ResNet18 block. To generate the dataset used in all experiments, we used a sliding window size of 168 data points (which correspond to the hours in a week) with a stride of 6. To plot each scalogram, we used $\{1, \ldots, 126\}$ as the scales, resulting in a number of scales of 126. For model optimization, we used the Adam optimizer with a batch size of 8 and an initial learning rate of 1e-5. All models were optimized for 100 epochs, with the weights of the 2 convolutional layers, the ResNet18 and the FC block jointly training (with the ResNet18 weights pre-trained on ImageNet, while the other layers were trained from scratch). Additional implementation details and source code can be found in the GitHub repository[1].

4.3 Performance Measures

Two performance measures are employed to evaluate the considered models in the chosen task: mean relative error (MRE), and mean squared error (MSE). More specifically, we computed the mean relative error $\text{RE}(y, \hat{y})$ and mean squared error $\text{MSE}(y, \hat{y})$ over predictions $y = (y_1, \ldots, y_N)$ and targets $\hat{y} = (\hat{y}_1, \ldots, \hat{y}_N)$ as:

$$\text{MRE}(y, \hat{y}) = \frac{1}{N} \sum_{i=1}^{N} \frac{|y_i - \hat{y}_i|}{|\hat{y}_i|} \qquad \text{MSE}(y, \hat{y}) = \frac{1}{N} \sum_{i=1}^{N} (y_i - \hat{y}_i)^2$$

The MSE is also used as the loss function to train all models, which is a common choice in Deep Learning when directly predicting real-valued quantities.

5 Results

We experimented with several combinations of input parameters, both including and excluding CO; results are reported in Table 1. For the sake of lighter notation, we group some parameters and refer to them as follows. GT stands for all parameters measured with the GT sensor, i.e. CO, C6H6, NOx, NO2. PT

[1] https://github.com/PietroManganelliConforti/Deep-Learning-and-Wavelet-Transform-for-Air-Quality-Forecasting.

Table 1. 2D pipeline results using different combinations of input parameters. The table is split vertically to separate the configurations that included CO among input parameters from those that didn't include it.

Input variables	Test MRE	Train MRE	Test MSE	Train MSE
GT*	**0.0621**	0.0281	**0.0327**	0.0053
PT*	0.0699	0.0327	0.0367	0.0075
PT*, HT	0.0650	0.0323	0.0372	0.0067
GT*, HT	0.0661	0.0315	0.0351	0.0065
GT*, PT*, HT	0.0806	**0.0275**	0.0658	**0.0049**
HT	0.1023	0.0472	0.0764	0.0145
PT	0.0755	0.0319	0.0416	0.0069
GT	**0.0522**	0.0277	**0.0221**	0.0056
PT*,CO(GT)	0.0577	0.0345	0.0259	0.0075
GT*,CO(PT08)	0.0575	0.0305	0.0281	0.0059
CO(GT)	0.0655	**0.0213**	0.0267	**0.0031**
CO(PT08)	0.0770	0.0268	0.0515	0.0045

Table 2. Performances of our 2D pipeline using different mother wavelet functions. All experiments use the GT* combination of input parameters.

Mother wavelet	Test MRE	Train MRE	Test MSE	Train MSE
Morlet	**0.0621**	**0.0281**	**0.0327**	**0.0053**
Morlet2	0.0762	0.0349	0.0405	0.0073
Ricker	0.0792	0.0337	0.0521	0.0070

Table 3. Performances of our 2D pipeline (from Table 1) against the three baseline models. All experiments use the GT* combination of input parameters.

Model	Test MRE	Train MRE	Test MSE	Train MSE
LSTMLinear	0.086	0.0549	0.0466	0.0217
Stacked1DLinear	0.121	0.0340	0.1367	0.0070
Stacked2DLinear	0.137	0.0385	0.1411	0.0103
StackedResNet (Morlet)	**0.0621**	**0.0281**	**0.0327**	**0.0053**

stands for all parameters measures with the PT08 sensor, i.e. CO, O3, NOx, NO2, NMHC. HT stands for the 3 atmospheric parameters T, AH, and RH. GT* and PT* indicate the same sets as GT and PT, but excluding their respective CO measurements. As expected, most of the combinations which include CO perform better than those that don't; the exceptions are PT, CO(GT) and CO(PT08). Note, however, that two of these combinations actually only include one parameter as input, which is expected to make the 2D pipeline less effective.

This is also supported by the GT* and CO(GT) combinations obtaining similar performances, even though the former doesn't contain the target parameter at all. Moreover, since the measurements from the PT08 sensor are significantly less reliable than those from the GT sensor, using only those measurements instead of CO(GT) is also expected to result in worse performances. These results also confirm that the GT signals are the most informative, as the best-performing combinations with and without CO are GT and GT*, while as expected the meteorological parameters (HT) seem to be the less informative ones. We also compared different mother wavelet functions, namely Morlet, Morlet2 and Ricker. We used GT* as the set of input parameters, since this was the best-performing combination of inputs that didn't include CO as an input. This still allows us to obtain good performances while not requiring any information on the target parameter itself. Results are reported in Table 2. It is clear from these outcomes that the Morlet mother wavelet seems to give the best performances, while the other two give similar performances, which further validates the choice of the Morlet wavelet for all other experiments.

Finally, we evaluated the three 1D baseline approaches described above, to compare them against the results obtained with the 2D pipeline. Again, we used GT* as the set of input parameters. The comparison is reported in Table 3. In particular, our 2D pipeline with the Morlet mother wavelet significantly outperforms all 3 baselines, which further validates the usefulness of this methodology for this kind of task. The baseline is mostly outperformed even when using the other wavelets, as seen by comparing the results in Tables 2 and 3. Our method performs better than 2 out of the 3 baselines even when only using the meteorological parameters as inputs, as seen by comparing results in Tables 1 and 3.

6 Conclusion

In this study, we explored the application of the Continuous Wavelet Transform in combination with different types of deep neural networks for air quality forecasting using sensor data. Through its employment, we obtained localized time-frequency representations of pollutant and environmental values, thereby highlighting concealed dynamics and relationships within the parameter behavior and the other possible external factors. Exploiting the time-frequency features extracted from three different mother wavelets and leveraging the knowledge acquired by a pre-trained deep neural network, our approach surpassed the performance of conventional one-dimensional signal processing methods.

In our future work, we aim to enhance the functionalities of our system by exploring and testing new pipeline architectures. One potential avenue is to apply an ensemble learning approach using multiple networks, which can lead to improved predictive performance and more robust results. In addition to the CWT, we will also explore other time-frequency signal representation methods, such as the Short-Time Fourier Transform. Moreover, we plan to expand the scope of our predictions beyond the average concentration values for the day following the considered time window. Ultimately, the future work outlined will significantly advance our understanding of the proposed task and improve the accuracy and applicability of our prediction system in order to reduce the cost of air quality management and to promote healthier living environments for communities worldwide.

References

1. Akansu, A.N., Serdijn, W.A., Selesnick, I.W.: Emerging applications of wavelets: a review. Phys. Commun. **3**(1), 1–18 (2010). https://doi.org/10.1016/j.phycom.2009.07.001, https://linkinghub.elsevier.com/retrieve/pii/S1874490709000482
2. Chen, Z., Wang, B., Gorban, A.N.: Multivariate Gaussian and student-t process regression for multi-output prediction. Neural Comput.Appl. **32**(8), 3005–3028 (2020). https://doi.org/10.1007/s00521-019-04687-8
3. De Vito, S., Massera, E., Piga, M., Martinotto, L., Di Francia, G.: On field calibration of an electronic nose for benzene estimation in an urban pollution monitoring scenario. Sens. Actuators B **129**(2), 750–757 (2008). https://doi.org/10.1016/j.snb.2007.09.060, https://www.sciencedirect.com/science/article/pii/S0925400507007691
4. Gargour, C., Gabrea, M., Ramachandran, V., Lina, J.M.: A short introduction to wavelets and their applications. IEEE Circuits Syst. Mag. **9**(2), 57–68 (2009). https://doi.org/10.1109/MCAS.2009.932556
5. Huang, L., et al.: Exploring deep learning for air pollutant emission estimation. Geoscientific Model Dev. **14**(7), 4641–4654 (2021). https://doi.org/10.5194/gmd-14-4641-2021, https://gmd.copernicus.org/articles/14/4641/2021/, publisher: Copernicus GmbH
6. Krawczyk, M., Gerkmann, T.: STFT phase reconstruction in voiced speech for an improved single-channel speech enhancement. IEEE/ACM Trans. Audio Speech Lang. Process. **22**(12), 1931–1940 (2014). https://doi.org/10.1109/TASLP.2014.2354236
7. Liu, H., Li, Q., Yu, D., Gu, Y.: Air quality index and air pollutant concentration prediction based on machine learning algorithms. Appl. Sci. **9**(19), 4069 (2019). https://doi.org/10.3390/app9194069, https://www.mdpi.com/2076-3417/9/19/4069
8. Ly, H.B., et al.: Development of an AI model to measure traffic air pollution from Multisensor and Weather data. Sensors **19**(22), 4941 (2019). https://doi.org/10.3390/s19224941, https://www.mdpi.com/1424-8220/19/22/4941
9. Ma, X., Karkus, P., Hsu, D., Lee, W.S.: Particle filter recurrent neural networks. Proc. AAAI Conf. Artif. Intell. **34**(04), 5101–5108 (2020). https://doi.org/10.1609/aaai.v34i04.5952, https://ojs.aaai.org/index.php/AAAI/article/view/5952
10. Ma, Z., Mei, G.: Deep learning for geological hazards analysis: data, models, applications, and opportunities. Earth Sci. Rev. **223**, 103858 (2021). https://doi.org/10.1016/j.earscirev.2021.103858, https://www.sciencedirect.com/science/article/pii/S0012825221003597
11. Manganelli Conforti, P., D'Acunto, M., Russo, P.: Deep learning for chondrogenic tumor classification through wavelet transform of Raman spectra. Sensors 22(19), 7492 (2022). https://doi.org/10.3390/s22197492, https://www.mdpi.com/1424-8220/22/19/7492

12. Pan, L., Pipitsunthonsan, P., Daengngam, C., Chongcheawchamnan, M.: Identification of complex mixtures for Raman spectroscopy using a novel scheme based on a new multi-label deep neural network. IEEE Sens. J. **21**(9), 10834–10843 (2020)
13. Qi, Y., et al.: Accurate diagnosis of lung tissues for 2D Raman spectrogram by deep learning based on short-time Fourier transform. Anal. Chim. Acta **1179**, 338821 (2021). https://doi.org/10.1016/j.aca.2021.338821, https://linkinghub.elsevier.com/retrieve/pii/S0003267021006474
14. Russo, P., Schaerf, M.: Anomaly detection in railway bridges using imaging techniques. Sci. Rep. **13**(1), 3916 (2023)
15. Sabzekar, M., Hasheminejad, S.M.H.: Robust regression using support vector regressions. Chaos, Solitons Fractals **144**, 110738 (2021). https://doi.org/10.1016/j.chaos.2021.110738, https://www.sciencedirect.com/science/article/pii/S0960077921000916
16. Saleem, S., Dilawari, A., Khan, U.G.: Spoofed voice detection using dense features of STFT and MDCT spectrograms. In: 2021 International Conference on Artificial Intelligence (ICAI), pp. 56–61 (2021). https://doi.org/10.1109/ICAI52203.2021.9445259
17. Salman, A.G., Kanigoro, B., Heryadi, Y.: Weather forecasting using deep learning techniques. In: 2015 International Conference on Advanced Computer Science and Information Systems (ICACSIS), pp. 281–285 (2015). https://doi.org/10.1109/ICACSIS.2015.7415154
18. Tan, C., Sun, F., Kong, T., Zhang, W., Yang, C., Liu, C.: A survey on deep transfer learning. In: Kůrková, V., Manolopoulos, Y., Hammer, B., Iliadis, L., Maglogiannis, I. (eds.) ICANN 2018. LNCS, vol. 11141, pp. 270–279. Springer, Cham (2018). https://doi.org/10.1007/978-3-030-01424-7_27
19. Tary, J.B., Herrera, R.H., Van Der Baan, M.: Analysis of time-varying signals using continuous wavelet and synchrosqueezed transforms. Philos. Trans. R. Soc. A Math. Phys. Eng. Sci. **376**(2126), 20170254 (2018)

Optimize Vision Transformer Architecture via Efficient Attention Modules: A Study on the Monocular Depth Estimation Task

Claudio Schiavella(✉), Lorenzo Cirillo, Lorenzo Papa, Paolo Russo, and Irene Amerini

Department of Computer, Control and Management Engineering (DIAG), Sapienza University of Rome, Rome, Italy
schiavella.1884561@studenti.uniroma1.it

Abstract. IoT and edge devices, capable of capturing data from their surroundings, are becoming increasingly popular. However, the onboard analysis of the acquired data is usually limited by their computational capabilities. Consequently, the most recent and accurate deep learning technologies, such as Vision Transformers (ViT) and their hybrid (hViT) versions, are typically too cumbersome to be exploited for onboard inferences. Therefore, the purpose of this work is to analyze and investigate the impact of efficient ViT methodologies applied to the monocular depth estimation (MDE) task, which computes the depth map from an RGB image. This task is a critical feature for autonomous and robotic systems in order to perceive the surrounding environment. More in detail, this work leverages innovative solutions designed to reduce the computational cost of self-attention, the fundamental element on which ViTs are based, applying this modification to METER architecture, a lightweight model designed to tackle the MDE task which can be further enhanced. The proposed efficient variants, namely Meta-METER and Pyra-METER, are capable of achieving an average speed boost of 41.4% and 34.4% respectively, over a variety of edge devices when compared with the original model, while keeping a limited degradation of the estimation capabilities when tested on the indoor NYU dataset.

Keywords: Computer vision · Edge device · Efficient vision transformer · Monocular depth estimation

1 Introduction

The last years have seen the wide exploitation of edge devices for various applications that involve IoT and data analytics, such as robotics, autonomous systems, and augmented reality.

These devices can collect real-time data from their surroundings through connections with sensors and detectors and are able to computer onboard analysis thanks to the integrated specific hardware. One of the main application scenarios involves computer vision (CV) paradigms, which are adopted to understand

G. L. Foresti et al. (Eds.): ICIAP 2023 Workshops, LNCS 14365, pp. 383–394, 2024.
https://doi.org/10.1007/978-3-031-51023-6_32

and interact with the surrounding environments. This process always involves the solution of different tasks based on the considered use case. In this work, we focus on the monocular depth estimation (MDE) task, where a dense depth map is estimated given a single input RGB image. When applied in real-world contexts, this task should often be performed in real-time. [24]. Indeed, the perception of the surroundings has to be fast but at the same time accurate and precise. These constraints are particularly true in applications like augmented reality and robotics, where the lack of these features would certainly lead to unsatisfactory and unsuitable results in terms of quality. In terms of estimation quality, ViTs and Hybrid ViTs (hViT) are able to achieve high accuracy performances at the cost of slow inference speed, which makes them challenging to utilize on edge devices [4].

However, in the current research literature, a number of methods have been proposed that aim to optimize the speed performance of vision transformers [11,13,20]. In addition, in the MDE task, recent works have been developed in order to design ad hoc models specifically designed to infer on embedded devices, such as [11,13,20]. More in detail, Papa et al. [11] propose METER, an hViT model particularly successful in achieving accurate estimation, taking advantage of the ViT global processing, while keeping fast inference performances. However, as commonly happen in ViT architectures, the self-attention [18] blocks (such as the ones used in METER architecture) exhibit a quadratic cost with respect to the input features, requiring a large amount of computation that is difficult to achieve on edge devices, which are typically low-powered, computationally constrained hardware.

Therefore, in this paper, we propose two techniques to reduce METER attention module computational cost, analyzing their application in order to define the best trade-off between inference performance and estimation capabilities. More in detail, inspired by Metaformer [22] technique and Pyramid ViT [19] technique, we introduce Meta-METER and Pyra-METER, two METER architecture modifications that are able to reduce its computational time by acting on its attention module.

The proposed solutions will take advantage of the optimization technique previously introduced while making METER network more accessible by resource-constrained devices. In this work, a special emphasis will be placed on how the inclusion of efficient variants of the attention module affects the non-optimized model. In particular, the effects of this change on both speed and accuracy will be analyzed. Summarizing, the contributions of our work are following reported:

- We propose two ways to reduce the computation time of METER architecture, namely Meta-METER and Pyra-METER.
- We perform some extensive experiments on the NYU Depth v2 dataset [17] and the inference time on the Intel Xeon CPU, Coral DevBoard[1], and NVIDIA Jetson TX1[2].

[1] https://coral.ai/products/dev-board/.
[2] https://developer.nvidia.com/embedded/jetson-tx1.

The rest of the paper is organized as follows: Sect. 2 reports the most related works regarding our method. Section 3 describes our proposed techniques in detail. Section 4 analyzes the obtained results with the used metrics and finally Sect. 5 reports our conclusions and reasonings on the possible choices and the corresponding trade-offs.

2 Related Works

In this section, we report state-of-the-art related works. In particular, Sect. 2.1 regards depth estimation, reporting how to approach that task as well as some real applications. We first discuss the general concept of depth estimation, on which monocular depth estimation is based. In Sect. 2.2, we focus on the methods that have been proposed to enhance the efficiency of Vision Transformers. We also discuss the specific techniques that we use in this paper, to achieve faster inferences with METER [11]. In particular, we will focus on those novel implementations that strongly affect the attention module performances.

2.1 Depth Estimation

Depth estimation involves calculating the distance of an object from the center of the camera. Measuring the depth of a scene is a very important task with relevant practical applications, such as in robotics, augmented reality, biometrics and many others field [2,9,12]. Depth estimation techniques can be grouped into two main categories:

- **Active depth sensing**: depth is obtained by perturbing the environment with suitable devices (e.g. lidar).
- **Passive depth sensing**: depth is estimated from one or more RGB images of the target scene.

In our application context, the focus is on monocular depth estimation [24]. This approach is classified as a passive depth-sensing technique which, due to the single input image, cannot make use of triangulation algorithms. Recently some ViT architectures [7–15] tackled the monocular depth estimation task reaching high accuracy, but the computational requirement of such networks makes them unsuitable for devices with hardware constraints. On the contrary, convolutional neural network architectures [13–20] are able to reach some good results on embedded devices regarding the monocular depth estimation, but they don't have the same low estimation error of ViTs.

Some lightweight architectures have been introduced in order to solve the monocular depth estimation task on hardware-constrained devices. A first architecture is described in [14] called PyD-Net, with a pyramidal structure to be able to get the features of the input image by using different levels of the pyramid. Another fully convolutional network called CReaM has been introduced in [16], while the encoder-decoder FastDepth [20] can also obtain very good results in this context. Finally, a network with high performances on embedded devices has been proposed in [23].

Another model with good performance on embedded devices is METER, a lightweight vision transformer architecture proposed by Papa et al. [11]. Its structure consists of an encoder-decoder architecture that can achieve state-of-the-art estimation and low-latency inference performance on the same embedded hardware taken in consideration in our work: NVIDIA Jetson TX1 and NVIDIA Jetson Nano. It has been proposed in three variants (S, XS, XXS) that will be the subject of our tests.

2.2 Efficient Vision Transformers

Vision Transformers are the equivalent of the transformer technology [18] (originally designed to work with text input) applied on images. The attention [18] module of a Vision Transformer is often very heavy in terms of computational time due to the quadratic complexities with respect to the input features. Therefore, some efficient ViTs have been introduced in the last years in order to reduce the computational time and also the number of operations of the attention [18] module.

Two of those novel techniques are the Softmax-free Transformer with Linear Complexity (SOFT) [8], in which the Softmax operation exploited in self-attention modules is substituted by a Gaussian kernel function to perform a similar dot-product function, and SimA [6] where the Softmax layer is replaced by a l_1-norm in order to normalize the query and key matrices. Another technique is Flowformer [21] in which the transformer is linearized by following the flow network theory and taking into consideration the conservation flows. An additional attempt has been performed in the MobileViT [10] work, which introduces a lightweight architecture for mobile devices. A further solution is proposed by Metaformer [22], that presents a revolutionary new baseline in the world of transformers. In fact, as per common thought, the role of the attention [18] module in the context of transformer architecture seems to be very important, if not the most important, for the quality of the result and the efficiency and performance of the model. However, the authors of the paper [22] showed how the latter concept may not actually be true. In a very drastic approach, the entire attention module was removed and replaced with a very simple, almost trivial, token mixer. In this way, the entire conceptual structure of the attention module is dramatically reduced. Particular focus should be given to this token mixer block, as the latter retains a very general nature, being referred to as a simple tool that can mix information, as is similarly the case with the query, values and key fundamental matrices of the transformer block that are in the standard attention module of [18].

Finally, Pyramid Vision Transformers [19] are characterized by the application of a spatial reduction function to the input. In this way, the output of the attention module is smaller and so it can be processed faster by the rest of the architecture. This is due to the so-called spatial-reduction attention layer (SRA) [19] that is substituted to the classical multi-head attention layer [18].

As Metafomers and Pyramid ViTs demonstrate to be two of the most successful method to speed up the inference phase of ViT networks, we have chosen to analyze both of them in order to study their application to a depth estimation architecture.

3 Proposed Method

This section outlines how METER architecture [11] is modified to improve its time performances. In particular, in Sects. 3.1 and 3.2 we describe the proposed optimizations and the new METER structures, where the attention modules of the transformer blocks will be modified as the methods shown in Sect. 2.

More in detail, we will focus on the traditional self-attention mechanism described in [18]:

$$Attention(Q, K, V) = Softmax(\frac{QK^T}{\sqrt{d_{head}}})V \tag{1}$$

where d_{head} is the dimension of each head, Q is the query, K is the key, V is the value and $Softmax$ is the well-known activation function.

The choice to optimize METER's attention module derives mainly from the fact that its computational cost, being a classical attention mechanism, is quadratic with respect to the input. Optimizations proposed in the literature such as [19,22], as seen in 2.2, are some of the most promising solutions to deal with this issue.

In Fig. 1 we can see how the traditional attention module is changed with Meta-METER and Pyra-METER modules, with a detailed description of the modifications reported in the following sections.

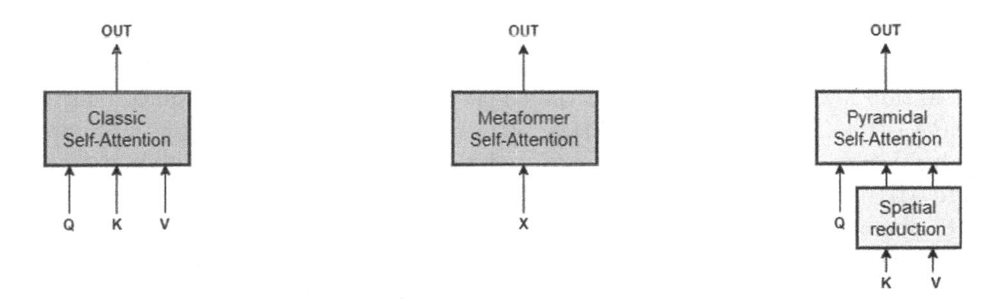

Fig. 1. Differences between attention modules: Classic Self-Attention, Metaformer Self-Attention, Pyramidal Self-Attention with spatial reduction

3.1 Meta-METER

As introduced in Sect. 2.2, thanks to the work presented in Metaformer [22], we have at our disposal a general technique able to create an attention module with reduced computational complexity. Initially, in the Metaformer [22], the nature of the token mixer is left general. Subsequently, the use of spatial pooling as a token mixer is proposed in an in-depth discussion of the baseline [22], due to its simplicity. By doing so, we can specify our new attention module as follows

$$Attention(X) = Pooling(Norm(X)) + X \tag{2}$$

where the terms X indicate the embedded input, $Norm(...)$ refers to the normalization technique chosen. In particular, we use a Layer Normalization technique [1], while *Pooling* is a standard spatial pooling of a suitable size to fit the other METER blocks.

We can assume that the pooling operator is the most appropriate one for the attention mechanism simplification, due to how the pooling context is closely linked to tasks related to convolutional neural networks, so computer vision and image processing in general. Moreover, as emphasized by the authors of [22], the choice of pooling avoids the introduction of further learnable parameters, reducing not only the computational complexity but also the chance of overfitting. This reinforces the choice of this particular token mixer, as it is a further approach to lightening the network, completely in tune with one of the intentions stated in Sect. 1. The modification applied to the transformer blocks of METER [11] is completed by the addition of a small convolutional network, as described in Sect. 2.2, composed of two convolutional layers followed by ReLU activation functions. This last step slightly increases the number of learnable parameters, but this increase is balanced by the nature of the new attention module and is necessary to avoid a reduction in accuracy performances.

Finally, the two attention blocks employed by METER architecture are equipped with a residual connection.

3.2 Pyra-METER

In the pyramid ViT paper [19] the authors propose to substitute the multi-head attention layer (MHA) [18] with the so-called spatial-reduction attention layer (SRA) [19]. The SRA, as the MHA, takes as input a query Q, a key K, and a value V and returns as output a refined feature. The difference between SRA and MHA is that SRA reduces the spatial scale of K and V before the attention [18] operation. This reduces the computational operation of the network.

Pyramid ViTs can be successfully employed also on high-resolution images by splitting and processing the data in patches, as suggested in [19]. As a consequence, we can use them for dense prediction tasks. Furthermore, they are characterized by a shrinking pyramid that allows us to cut the computational time of feature maps. The spatial scale of K and V have done as follows [19]:

$$SpatialReduction(X) = Norm(Reshape(X, R)W) \tag{3}$$

where X is the input sequence, R is the reduction ratio of the attention layers, C is the number of channels of the output, $Reshape(X, R)$ is an operation of reshaping X to size $\frac{HW}{R^2} \times (R^2C)$, $W \in \mathbb{R}^{(R^2C)\times C}$ is a linear projection that reduces the dimension of X to C and $Norm(...)$ refers to Layer Normalization. The attention operation is computed as reported in (1).

By looking at these Formulas (1)–(3), we can notice that the computational cost of the pyramid ViT attention operation is reduced by R^2 times with respect to the classical multi-head attention operation [18].

To build the Pyra-METER architecture, we have substituted METER attention module with the pyramid ViT attention module. In particular, we reshaped the input image only if $R > 1$, while the reshaping of K and V is performed even if $R < 1$ by dividing the number of channels by the number of heads. In the attention layers, the kernel has a shape equal to $R \times R$. The greater that ratio is, the more the image is reduced with the spatial reduction. After that, we put a simple normalization layer and we applied also the dropout mechanism to reduce overfitting. Finally, the projection mentioned in (3) is performed before the last layer which is the classical Softmax operation layer described in [18].

4 Results

In this section, we report the results obtained by Meta METER and Pyra-METER structures described in Sect. 3. The rest of the section is organized as follows: Sect. 4.1 describes the implementation details and the evaluation metrics used. Section 4.2 analyzes the obtained results by comparing proposed models with the original METER architecture when tested over the NYU dataset. Finally, Sect. 4.3 highlights how the optimization strategy applied to the self-attention and respective transformer blocks impacts the inference performances.

4.1 Experimental Setup

All the algorithms have been implemented in Python by leveraging the PyTorch deep learning framework. We use the same training procedure, loss function, and data augmentation strategy described in the reference METER paper with ADAM [5] optimizer and a batch size of 64. Moreover, we evaluate the performance of compared models by exploiting some common metrics used in monocular depth estimation tasks: the root-mean-square error (RMSE), the mean absolute error (MAE), respectively measured in meters [m], the relative error (REL), and the accuracy values δ_1, δ_2 and δ_3 [3]. Furthermore, we also evaluate the inference time, namely, the time measured in milliseconds [ms] that a model takes to perform a single inference over different devices: the Intel Xeon CPU 2.20GHz, which is representative of the x86 architecture, a Coral DevBoard equipped with an ARM Cortex-A53 and the NVIDIA Jetson TX1 equipped with a Cortex-A57. In the next Section they will be reported as CPU, DEV, and TX1 respectively. In particular, we also specifically evaluate the inference timings of the three transformer blocks of METER architecture, in order to highlight the impact of the proposed architectural changes on the transformer modules speed. Finally, we tested the proposed attention modifications on three METER configurations, namely S, XS, and XXS, which represent the same model structure but with a different number of layers and trainable parameters, in order to create S, XS, and XXS variants of Meta-METER and Pyra-METER.

4.2 Performance Analysis

The overall estimation performances of introduced models, tested over the three configurations (S, XS, and XXS) are compared in this section. The experiments were carried out on the Xeon CPU 2.20GHz, exploiting a single core, and then on the Coral DevBoard and NVIDIA Jetson TX1 boards, all while avoiding task parallelization. Ee report in Table 1 the obtained results over the evaluation metrics introduced in Sect. 4.1.

Table 1. Error metrics, accuracy values and total inference time of S, XS, XXS METER, Meta-METER and Pyra-METER measured on Intel Xeon CPU (T_{CPU}), Coral DevBoard (T_{DEV}) and NVIDIA Jetson TX1 (T_{TX1}).

Model	RMSE [cm]	MAE [cm]	Abs_{Rel}	δ_1	δ_2	δ_3	T_{CPU} [ms]	T_{DEV} [ms]	T_{TX1} [ms]
METER S	**47.035**	**35.151**	**0.137**	0.816	**0.962**	**0.990**	68.576	154.880	105.837
Meta-METER S	49.129	36.650	0.144	0.811	0.959	0.988	**40.164**	**147.608**	**102.167**
Pyra-METER S	48.424	36.350	0.142	**0.817**	0.960	0.989	46.586	155.584	112.521
METER XS	49.565	37.301	0.151	0.811	0.953	0.987	49.616	128.613	80.824
Meta-METER XS	51.499	38.710	0.154	0.801	0.951	0.986	**31.373**	**120.406**	**77.077**
Pyra-METER XS	**49.389**	**36.843**	**0.145**	**0.812**	**0.956**	**0.989**	36.246	124.221	81.743
METER XXS	54.665	41.505	0.172	0.770	0.944	0.983	29.734	68.906	45.343
Meta-METER XXS	57.111	43.610	0.182	0.754	0.936	0.981	**15.927**	**61.084**	**31.183**
Pyra-METER XXS	**53.684**	**40.832**	**0.162**	**0.775**	**0.946**	**0.984**	16.567	64.744	48.675

Based on the reported values, we can notice that on the S-configuration, the non-optimized model is slightly more accurate, while for the XS and XXS variants, the Pyra-METER is able to achieve the most accurate estimations. In contrast, by looking at the inference timings, Meta-METER always achieves the fastest inference performances (+20.7%) with a limited average degradation of −4.3%, and −1.3% for the RMSE e δ_1 accuracy respectively, when compared with the original METER model.

More in detail, when tested over different devices, i.e., CPU, DEV, and TX1, Meta-METER is able to achieve an average performance boost of +41.5%, +7.5%, and +13.1% respectively, while, Pyra-METER improves the non-optimized version by +35.1% and +3.0% respectively for the CPU and DEV. In contrast, on TX1 it obtains a decrement with respect to the original METER inference time of −4.9%; this fact could be due to the usage of the GPU by the TX1 which performs in parallel some processes that improve the performances with respect to the CPU. Moreover, by comparing the two introduced models, we can observe that the averaged timing boost equal to +10.6% of Meta-METER with respect to Pyra-METER comes at the expense of worse (−2.9%) estimation performances. The latter results are due to the fact that Meta-METER exploits only one single pooling layer that is able to considerably speed up the inference time, at the expense of some details loss, while Pyra-METER still employs a lot of computational time on its attention module, because it gives its main contribution with large resolution features, decreasing it as the features get smaller.

4.3 Inference Time Analysis

Once the general model's performances have been analyzed, we report in this section an in-depth analysis of the inference times of the different self-attention blocks by measuring the inference time of the three METER transformer blocks only. In fact, by isolating the transformer blocks with respect to the other model layers, the impact of attention modifications is better highlighted.

We report the results obtained over the three configurations (S, XS, and XXS) in Table 2.

Table 2. Transformer blocks inference time of S, XS, XXS configurations of METER, Meta-METER and Pyra-METER measured on Intel Xeon CPU, Coral DevBoard and NVIDIA Jetson TX1.

Model	CPU			DEV			TX1		
	T_1 [ms]	T_2 [ms]	T_3 [ms]	T_1 [ms]	T_2 [ms]	T_3 [ms]	T_1 [ms]	T_2 [ms]	T_3 [ms]
METER S	0.011	0.006	0.004	0.028	0.011	0.008	0.016	**0.0075**	**0.005**
Meta-METER S	**0.004**	**0.002**	**0.001**	**0.013**	**0.008**	**0.005**	**0.011**	0.010	0.006
Pyra-METER S	0.005	0.004	0.003	0.022	0.017	0.018	0.017	0.017	0.019
METER XS	0.006	0.003	0.002	0.023	0.008	0.005	0.013	**0.006**	0.004
Meta-METER XS	**0.002**	**0.001**	**0.0006**	**0.008**	**0.005**	**0.0026**	**0.007**	**0.006**	**0.003**
Pyra-METER XS	0.003	0.002	0.002	0.012	0.011	0.009	0.010	0.009	0.009
METER XXS	0.006	0.002	0.001	0.020	0.006	0.004	0.012	0.004	0.003
Meta-METER XXS	**0.002**	**0.001**	**0.0004**	**0.005**	**0.003**	**0.0017**	**0.005**	**0.003**	**0.001**
Pyra-METER XXS	**0.002**	**0.001**	0.001	0.008	0.006	0.005	0.007	0.006	0.005

Based on the reported values, it can be noticed that Meta-METER always achieves the lowest inference timings. Although the latter result could have been inferred from the general results given in Table 1, in this analysis, we report in detail the individual contribution of the optimized method in each transformer block, i.e., T_1, T_2, and T_3. Therefore, when Meta-METER and Pyra-METER are compared with the original structure, the inference times over each transformer block are equal to −58.5%, −32.2%, and −46.6% for the first model and −39.9%, +14.1%, and +75.2% for the second one. Moreover, when respectively tested over the CPU, DEV, and TX1 edge devices, the overall performance of the

Meta-METER and Pyra-METER self-attention over the non-optimized variant are equal to −65.0%, −50.2%, and −22.1% for the first model and −34.8%, +13.1%, and +71.1% for the second one. From these results, we can derive that Meta-METER attention is able to effectively guarantee an inference boost over both transformer blocks with various input shapes and devices. In contrast, the spatial-reduction self-attention layer of Pyra-METER better fits scenarios where the resolution of input features is high.

5 Conclusions

In this paper, we propose two optimized versions of METER [11] architecture, Meta-METER and Pyra-METER. The introduced models can achieve comparable estimation performances over the non-optimized model with lower inference timings. The latter results are due to the usage of a pooling layer for the Meta-METER self-attention block and to the spatial reduction for the Pyra-METER self-attention block.

We can assess that the proposed models are promising optimized alternatives of METER architecture which enable us to tackle the MDE task on resource-constrained devices. Based on the reported results, we can conclude that the optimization of self-attention quadratic cost is a valuable starting point to close the gap between high-accurate deep learning models and devices with limited computational power. In particular, Meta-METER is able to sensibly improve both the total and the transformer blocks inference time, even if some reduction in the estimation metrics due to the usage of a single pooling layer that discards some features. On the other side, Pyra-METER improves all the estimation metrics, especially for the XS and XXS sizes, and also the inference times on the CPU, whereas, on the other two boards, is not able to reach excellent inference times on the second and third transformer blocks because the employed structure works well when the reduction in resolution is large, so as one goes towards the small end features of the encoder, its contribution is less.

Given the above, if is possible to take some losses on estimation metrics in favor of large gains on inference time, then Meta-METER is a good application choice. On the other hand, Pyra-METER is a more balanced choice because it has a smaller boost on inference time in favor, however, of some gains on estimation metrics as well.

Given the positive results of such approaches on hybrid ViTs architectures, we plan in the future to extend such optimizations to more complex architectures and test over different benchmark datasets. Furthermore, we will investigate other ways to make ViTs efficient by using pruning/knowledge distillation or quantization strategies.

Acknowledgments. This study has been partially supported by SERICS (PE00000014) under the MUR National Recovery and Resilience Plan funded by the European Union - NextGenerationEU, Sapienza University of Rome project 2022–2024 "EV2" (003_009_22), and project 2022–2023 "RobFastMDE".

References

1. Ba, J.L., Kiros, J.R., Hinton, G.E.: Layer normalization. arXiv preprint arXiv:1607.06450 (2016)
2. Dong, X., et al.: Towards real-time monocular depth estimation for robotics: a survey. IEEE Trans. Intell. Transport. Syst. **23**(10), 16940–16961 (2022)
3. Eigen, D., Puhrsch, C., Fergus, R.: Depth map prediction from a single image using a multi-scale deep network. Adv. Neural Inf. Process. Syst. **27** (2014)
4. Han, K., et al.: A survey on vision transformer. IEEE Trans. Pattern Anal. Mach. Intell. **45**(1), 87–110 (2022)
5. Kingma, D.P., Ba, J.: Adam: a method for stochastic optimization. arXiv preprint arXiv:1412.6980 (2014)
6. Koohpayegani, S.A., Pirsiavash, H.: Sima: simple softmax-free attention for vision transformers. arXiv preprint arXiv:2206.08898 (2022)
7. Li, Z., et al.: Binsformer: revisiting adaptive bins for monocular depth estimation. arXiv preprint arXiv:2204.00987 (2022)
8. Jiachen, L., et al.: Soft: Softmax-free transformer with linear complexity. Adv. Neural. Inf. Process. Syst. **34**, 21297–21309 (2021)
9. Makarov, I., Borisenko, G.: Depth inpainting via vision transformer. In: 2021 IEEE International Symposium on Mixed and Augmented Reality Adjunct (ISMAR-Adjunct), pp. 286–291. IEEE (2021)
10. Mehta, S., Rastegari, M.: Mobilevit: light-weight, generalpurpose, and mobile-friendly vision transformer. arXiv preprint arXiv:2110.02178 (2021)
11. Papa, L., Russo, P., Amerini, I.: METER: a mobile vision transformer architecture for monocular depth estimation. IEEE Trans. Circuits Syst. Video Technol. (2023)
12. Papa, L., et al.: Lightweight and energy-aware monocular depth estimation models for IoT embedded devices: challenges and performances in terrestrial and underwater scenarios. Sensors **23**(4), 2223 (2023)
13. Papa, L., et al.: Speed: separable pyramidal pooling encoder-decoder for real-time monocular depth estimation on low-resource settings. IEEE Access **10**, 44881–44890 (2022)
14. Poggi, M., et al.: Towards real-time unsupervised monocular depth estimation on CPU. In: 2018 IEEE/RSJ International Conference on Intelligent Robots and Systems (IROS), pp. 5848–5854. IEEE (2018)
15. Ranftl, R., Bochkovskiy, A., Koltun, V.: Vision transformers for dense prediction. In: Proceedings of the IEEE/CVF International Conference on Computer Vision, pp. 12179–12188 (2021)
16. Sandler, M., et al.: Mobilenetv2: inverted residuals and linear bottlenecks. In: Proceedings of the IEEE Conference on Computer Vision and Pattern Recognition, pp. 4510–4520 (2018)
17. Silberman, N., Hoiem, D., Kohli, P., Fergus, R.: Indoor segmentation and support inference from RGBD images. In: Fitzgibbon, A., Lazebnik, S., Perona, P., Sato, Y., Schmid, C. (eds.) Computer Vision – ECCV 2012. LNCS, vol. 7576, pp. 746–760. Springer, Heidelberg (2012). https://doi.org/10.1007/978-3-642-33715-4_54
18. Vaswani, A., et al.: Attention is all you need. In: Advances in Neural Information Processing Systems, pp. 5998–6008 (2017)
19. Wang, W., et al.: Pyramid vision transformer: a versatile backbone for dense prediction without convolutions. In: Proceedings of the IEEE/CVF International Conference on Computer Vision, pp. 568–578 (2021)

20. Wofk, D., et al.: Fastdepth: fast monocular depth estimation on embedded systems. In: 2019 International Conference on Robotics and Automation (ICRA), pp. 6101–6108. IEEE (2019)
21. Wu, H., et al.: Flowformer: linearizing transformers with conservation flows. arXiv preprint arXiv:2202.06258 (2022)
22. Yu, W., et al.: Metaformer is actually what you need for vision. In: Proceedings of the IEEE/CVF Conference on Computer Vision and Pattern Recognition, pp. 10819–10829 (2022)
23. Yucel, M.K., et al.: Real-time monocular depth estimation with sparse supervision on mobile. In: Proceedings of the IEEE/CVF Conference on Computer Vision and Pattern Recognition, pp. 2428–2437 (2021)
24. Zhao, C.Q., Sun, Q.Y., Zhang, C.Z., Tang, Y., Qian, F.: Monocular depth estimation based on deep learning: an overview. Sci. China Technol. Sci. **63**(9), 1612–1627 (2020)

Assessing Machine Learning Algorithms for Land Use and Land Cover Classification in Morocco Using Google Earth Engine

Hafsa Ouchra[1](✉), Abdessamad Belangour[1], Allae Erraissi[2], and Mouad Banane[3]

[1] Laboratory of Information Technology and Modeling LTIM, Faculty of Sciences Ben M'sik, Hassan II University, Casablanca, Morocco
ouchra.hafsa@gmail.com

[2] Polydisciplinary Faculty of Sidi Bennour, Chouaib Doukkali University, El Jadida, Morocco
erraissi.a@ucd.ac.ma

[3] Laboratory of Artificial Intelligence and Complex Systems Engineering (AICSE), Hassan II University, ENSAM, Casablanca, Morocco
mouad.banane-etu@etu.univh2c.ma

Abstract. Google Earth Engine constitutes a cloud-based geospatial data processing platform. It grants free access to vast volumes of satellite data along with unlimited computational power, enabling the monitoring, visualization, and analysis of environmental features on a petabyte scale. The platform's capacity to accommodate various land use and land cover (LULC) classification approaches, utilizing both pixel-based and object-oriented methods, has been facilitated by providing an array of machine learning algorithms. Earth observation data has emerged as a valuable resource, offering temporally and spatially consistent quantitative information compared to traditional ground surveys. It presents numerous opportunities for urban mapping, monitoring, and a wide array of physical, climatic, and socio-economic data to support urban planning and decision-making. In this study, Landsat 8 satellite data was harnessed for supervised classification. Three advanced machine learning techniques—Support Vector Machine (SVM), Random Forest (RF), and Minimum Distance (MD)—were employed to categorize areas within Morocco, encompassing water bodies, built-up regions, cultivated land, sandy areas, barren zones, and forests. The classification outcomes are presented using a set of accuracy indicators, including Overall Accuracy (OA) and the Kappa coefficient.

Keywords: Remote sensing · satellite image classification · Machine learning · Google Earth Engine

1 Introduction

Using GEE for satellite image classification can be a complex and time-consuming process, but it can also be a powerful way to extract valuable information from geospatial data, which involves training machine learning algorithms to recognize satellite

G. L. Foresti et al. (Eds.): ICIAP 2023 Workshops, LNCS 14365, pp. 395–405, 2024.
https://doi.org/10.1007/978-3-031-51023-6_33

imagery patterns and classify the image into different land cover or land use categories [1-3]. Google Earth Engine is indeed a powerful platform that allows for spectral and spatial analysis of satellite images. With GEE, you can perform various operational functions using both simple mathematical operations and more advanced techniques like supervised and unsupervised machine learning algorithms.

The platform provides a library of application programming interfaces (APIs) that support popular programming languages such as Python and JavaScript [1, 3].

This platform includes supervised classification algorithms like RF [4], SVM [5], and Classification and Regression Tree (CART) [6], as well as clustering algorithms like K-Means, and Simple Non-Iterative Clustering (SNIC) [3, 7]. The limitations of this platform are that it does not support hierarchical algorithms [8], few spatial functions are provided due to parallel implementation issues, and it is difficult to implement filtering techniques such as Gaussian, Laplace, Sobel and Haun transforms. is to be [3].

This paper provides an overview of the GEE platform's capabilities and describes the findings of a study that focuses on the classification of Landsat 8 OLI satellite images applied to Moroccan territory. In our literature review on satellite image classification, we used the most used supervised machine learning classifiers [9, 10] to achieve our goal.

The following is the structure of this research paper. The second section describes related work in satellite image classification using the GEE platform. Section 3 presents our methodology. Section 4 develops our experimental study and discusses the results. Section 5 discusses the conclusion.

2 Related Work

Remote sensing data can provide a large amount of information about the entire planet. Remote sensing data can be used to describe and model the urban environment, allowing for environmental management and planning, as well as spatial data analysis [11, 12]. There are different applications of remote sensing that include several fields such as urban planning, agriculture, resource management, mineralogy, etc. Many research papers have explored the use of GEE for satellite image classification [13, 14] using machine learning algorithms [15-17].

Pérez-Cutillas et al. [3] demonstrated that Landsat 8 images, captured by the OLI and TIRS, are indeed widely used in various applications provided by satellite platforms, including those within the Google Earth Engine. Landsat 8 provides valuable multispectral and thermal data, making it suitable for a range of remote sensing analyses.These algorithms, such as Support Vector Machines (SVM), Random Forests, Decision Trees, and Artificial Neural Networks, can be utilized within the GEE platform for processing and analysis tasks.

Tassi et al. [2] created and tested an object combining the SNIC algorithm for spatial cluster identification and GLMC for calculating cluster texture indices. Further evidence of the reliability of the overall methodology was provided in their study. Nevertheless, the object-oriented type tends to be computationally intensive and slows down the execution of GEE code, particularly when high-resolution data are used.

From the literature, we found that all researchers demonstrate the potential of GEE for satellite image classification using machine learning algorithms and they develop accurate and efficient methods to classify satellite images to obtain land cover information for better planning such as urban planning.

3 Methodology

The method we proposed, presented in Fig. 1 and applied to the Moroccan territory, was entirely based on the use of the Google Earth Engine platform environment.

3.1 Study Area and Dataset

Morocco is a North African country that is highly vulnerable to climate change. It is located between 31° 47′ 30.127" N and 7° 5′ 33.432" W and its surface area covers 710,850 km^2. Figure 1 shows the Moroccan territory.

Fig. 1. Image composite from Landsat 8 OLI of Morocco case study

Landsats sensors record the energy reflected from the planet in different wavelengths of the electromagnetic spectrum because each object has different reflective strengths. The spectrum consists of many different wavelengths [5, 16].

Today, the Landsat 7 and 8 satellites record visible blue, green, and red light, as well as near, mid, and thermal infrared light that the human eye cannot see. They digitally record this data and transmit it to ground stations, where it is processed and stored in a data archive [19, 20].

We used the Landsat 8 surface reflectance collection 1 Tier 1 in this study. The atmospherically corrected surface reflectance of Landsat 8 OLI/TIRS sensors is represented in this dataset. These satellite images contain five visible and near-infrared (VNIR) bands, two shortwave infrared (SWIR) bands, and two thermal infrared (TIR) bands processed at orthorectified brightness temperature. Their spatial resolution is 30 m [21]. We used the Large-Scale International Boundary (LSIB) dataset [18] provided by the United States Office of the Geographer to define our study area on a geographical map.

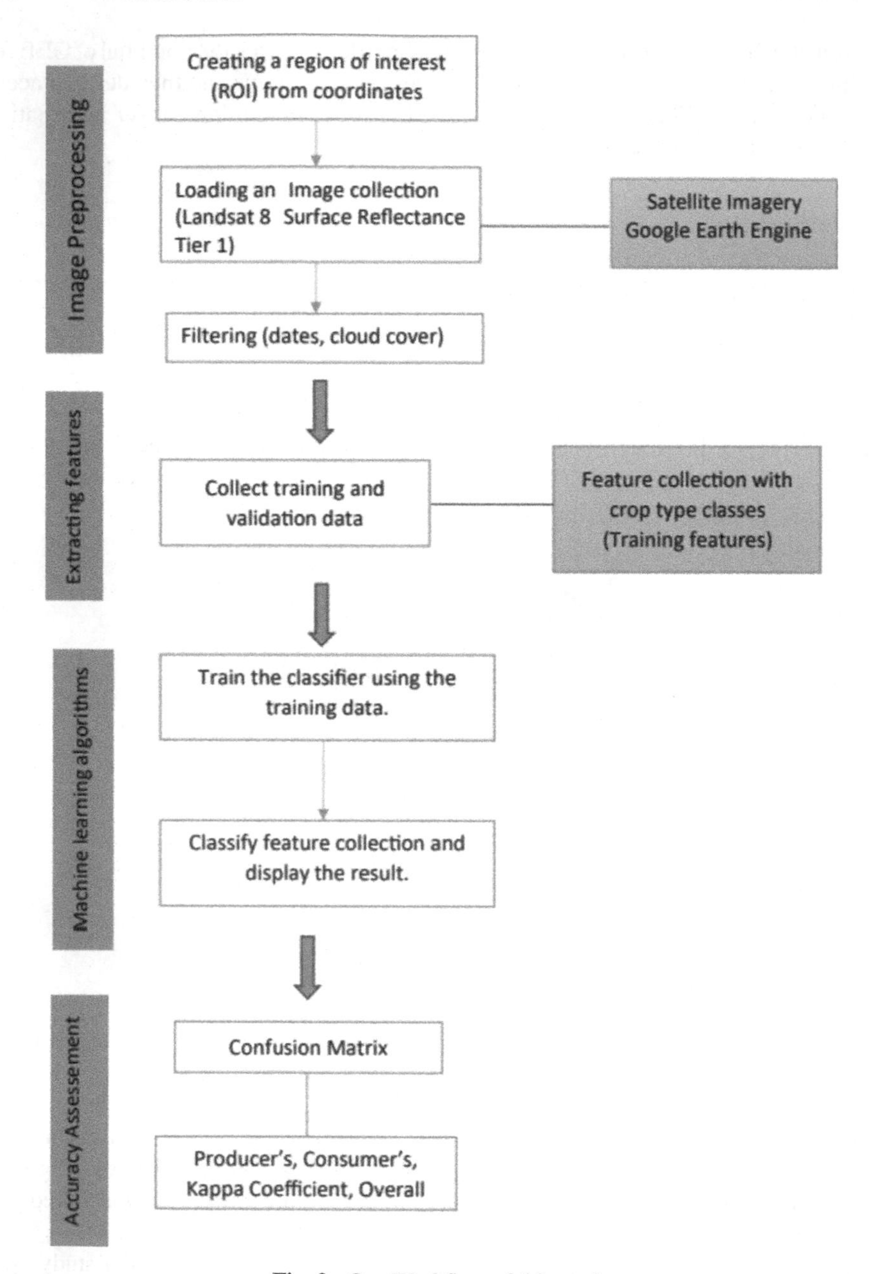

Fig. 2. Our Workflow of this study

It is derived from two other datasets: an LSIB line vector file and the World Vector Shorelines dataset from the National Geospatial-Intelligence Agency (NGA) (WVS) [22]. In this research, we are interested in our Moroccan country from Tangier city to Geuira city. In our study, we merged two codes "MO" and "WI" to extract the Moroccan territory, our study area (pixel on the Moroccan territory). They codes defined a mask of such a territory.

3.2 Data Preparation

Image Preprocessing

The image preprocessing step is an essential step that allows users to create noise-free (cloud) and topographically corrected image composites for a user-defined extent and period in Google Earth Engine (GEE) [23]. In this work, we obtained Landsat 8 collection 1 tier 1 satellite composite images [21] of pixel size 30 m of our study area by filtering them by date to get just the images of the following period of the year 2021 from the month of January until the month of December (2021-01-01, 2021-12-31). Image preprocessing can also reduce the model learning time and increase the inference speed of the algorithm. The composite image we obtained can then be exported to the user's Google Drive for offline use.

Table 1. Land cover classes

Class	Numbers of points	Description
Water	103	River, lakes, reservoirs, etc.
Forest	102	Mixed and uncultivated forest areas: land dominated by trees
Barren	107	Barren Land typically has thin soil, sand, or rock. Deserts, dry salt flats, beaches, sand dunes, exposed rock, strip mines, quarries, and gravel pits are all examples of Barren land
Built-up	107	Residential, industrial, commercial, or mixed-use areas: including road and transportation networks, etc.
Sand	102	A large area of sand, for example a beach, shore, and strand, etc....
Cropped	102	Agricultural land refers to the portion of the land area that is arable, under permanent crops and under permanent pasture: Crop fields and fallow lands, for example, are examples of organized or unorganized agricultural practices
Total	623	

Features Extraction

Feature extraction is a critical step in the selection of features to be used as input data for classification algorithms. Large datasets contain many variables, the processing of which requires a great deal of computing resources. This step reduces the number of resources required while retaining relevant information [24]. In this article, we have collected a set of geometric points (features) for each category: aquatic areas, built-up areas, cultivated areas, sandy areas, barren areas and forest areas as shown in Table 1 from the composite image of our study area and merging them with the spectral indices NDVI [25, 26], NDBI [27], BSI [28], and MNDWI [29] these indices are mentioned in Table 2, then we created the training data using this collection of features to randomly divide these training data 80% for training and 20% for validation.

Table 2. Spectacles indices Equations

Index	Equation using Landsat 8 OLI
Normalized Difference Vegetation Index (NDVI) [25, 26]	$NDVI = \frac{NIR-RED}{NIR+RED}$
Modified Normalized Difference Water Index (MNDWI) [29]	$MNDWI = \frac{Green-SWIR1}{Green+SWIR1}$
Bare Soil Index (BSI) [28]	$BSI = \frac{Green+NIR}{Green-NIR}$
Normalized Difference Built-up Index (NDBI) [27]	$NDBI = \frac{SWIR-NIR}{SWIR+NIR}$

3.3 Machine Learning Algorithms

After collecting training data and assembling properties that store the known class label and numerical values for the predictors. The next step involves creating classifier instances and, if needed, specifying their parameters. In the GEE platform, there is the "Classifier" package that manages supervised classification by Machine Learning algorithms. In this step, we trained 3 classifiers RF, MD, and SVM algorithms using the training data we constructed, then classified our composite image collection with specific bands to finally find a classified map of the study area. Table 3 shows the description of those algorithms.

Table 3. Description of each algorithm

Machine learning algorithms	Description
Support vector machine (SVM) [18]	It is a family of machine learning algorithms that allow for solving classification, regression, or anomaly detection problems. They are known for their strong theoretical guarantees, their great flexibility, and their simplicity of use even without much knowledge of data mining [18]. Their goal is to divide the data into classes using the simplest boundary possible, so that the distance between the different groups of data and the boundary that separates them is as small as possible. This is also known as the "margin". [1]

(*continued*)

Table 3. (*continued*)

Machine learning algorithms	Description
Random Forest (RF) [18]	It is a popular machine learning algorithm known for its accuracy, simplicity, and versatility. It is widely used for both classification and regression tasks. The name "Random Forest" comes from the fact that it builds a collection or ensemble of decision trees, forming a forest-like structure. Each decision tree in the Random Forest is constructed by using a random subset of the available features from the training data. This random selection of features helps to introduce diversity among the trees and reduces the chances of overfitting
Minimum distance classification (MD) [18]	It is a special case of maximum likelihood classification. The classifier calculates the distance of each pixel to each class mean, in several standard deviations in the direction of the pixel and assigns the pixel to the class with the smallest value according to these Mahalonobis distance units [30]

3.4 Evaluation of Accuracy

This step is an essential component of modeling and mapping to test the effectiveness and scientific significance of the models. This step compares the classified image to another data source that is thought to be field reality data. To evaluate and calculate the accuracy of each classification method (the accuracy of a classified map), we used a confusion matrix to compare a set of random points from field reality data sets to the classified data (classified map). The confusion matrix is a technique for summarizing prediction results for a specific classification problem. It compares actual data for a target variable to predictions from a model. Correct and false predictions are revealed and classified, allowing them to be compared to predefined values.

4 Results and Discussion

In this research, satellite imagery Landsat 8 OLI was used to map the land cover of our study area.

Using spectral features, we made a land-use classification. An operation was performed to obtain reflectance data with bands of Landsat 8 OLI image from January 1, 2021, to December 31, 2021, and resample all image bands from band B1 to band B7 in a single spatial resolution 30 m using the GEE cloud platform, which allows us to solve problems related to the availability and storage of free computer resources, as well as data preprocessing. Furthermore, the study area is characterized by a cartographic

Table 4. Overall accuracy and kappa coefficient of each classifier

Classifier	Overall Accuracy %	Kappa Coefficient
RF	91.22	0.89
MD	93.85	0.92
SVM	74.56	0.69

complexity marked by the existence of numerous land-use categories, made up of six classes that can be found in Table 1. This feature is time-consuming in local calculation mode.

To map our study area (Morocco), we applied the following classification methods RF, SVM, and MD. The land cover maps produced by each approach were presented in Fig. 3 and were evaluated using the confusion matrix resulting in the overall accuracy, and the Kappa coefficient. The results obtained in this study used LANDSAT 8 OLI datasets and processing tools such as the Google Earth Engine platform and classifiers that are used to map the territory of Morocco into categories (Water Zone, Forest, Built-up, Sand, Barren, and Cropped) and to evaluate the performance of each classifier used in this research. According to our study on previous works [8, 31] and from this work, we found that the methodology we used in this experiment could be adapted to map and evaluate other areas in different countries or cities. The most important contribution in this research was the use of the following classifiers: RF, SVM, and MD to map the Moroccan territory into the following zones: water zone, Forest, Built-up, Sand, Barren, and Cropped using the Landsat 8 OLI dataset and the GEE platform.

Our analysis showed that the Minimum Distance classifier is the best-performing classifier because its classification accuracy is very high compared to the other classifiers used. The accuracy obtained by this classifier is 93.85% and for its kappa coefficient, we obtained 0.93 as shown in Table 4. The classified maps of the territory of Morocco from the following period 1-1-2021 to 31-12-2021 are presented in Fig. 2. The figure presents the distribution of land use categories for this year, a total of 6 classes were produced for each of the images (Water, Built-up areas, Cropped, Sand, Barren, and Forest). The image classification results were compared in terms of the accuracy of each classifier as shown in Table 4.

From the confusion matrix of each classifier used, we deduced that the best classification is the Minimum Distance classifier because its Kappa coefficient is 0.92 which is a value close to 1 which indicates that the classification is significantly better than random and it has an accuracy of 93.85%, this is a high accuracy compared to the other accuracy 91.22% and 74.56% of RF and SVM, respectively. The SVM classifier is a classifier that has an accuracy of 74.56%, it is a very low accuracy compared to the other classifiers, and for its Kappa coefficient is 0.69.

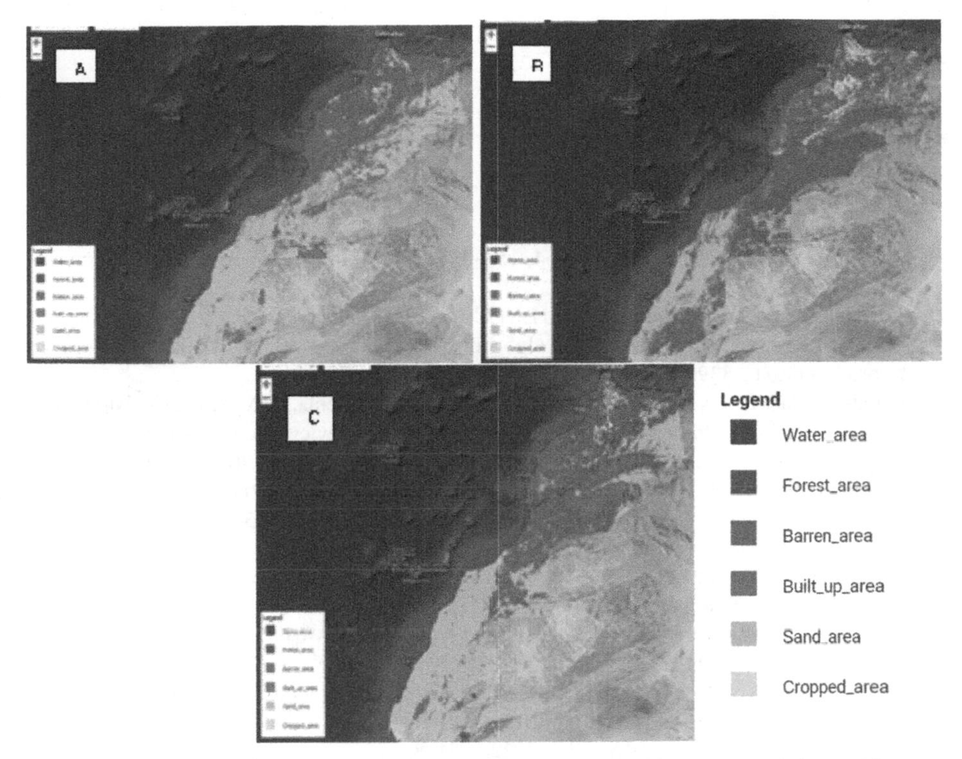

Fig. 3. Classification results of Morocco land cover map. A) Classification by Minimum Distance; B) Classification by Random Forest; C) Classification by Support Vector Machine.

5 Conclusion

The primary goal of this study was to evaluate and compare the performance of different classification algorithms in terms of classification accuracy and Kappa coefficient, as well as to test the ability of the GEE platform and Landsat 8 OLI data to classify areas in Moroccan territory (LC classes) (Water, Cropped, Forest, Barren, Sand, and Built-up). The plan was to use the GEE platform to collect a large amount of Landsat 8 OLI satellite data, then filter it to only include images from 2021.

Based on the results presented in this paper, we can conclude that the GEE platform's dependability and versatility, thanks to its cloud architecture, allow us to avoid the integration of other external, often commercial software. In this paper, a high accuracy has been achieved by the Minimum Distance supervised classifier thanks to the machine learning methodology which is based on the collection of training and validation points of the study area to be used during the application of this methodology to obtain a better classification of the study area.

In general, GEE offers substantial opportunities for Earth observation and geospatial applications, and it has provided very good performance in enabling access to remote sensing products via the cloud platform and allowing seamless execution of complex workflows for processing satellite data. It opened a new mega data paradigm for storing and analyzing remote sensing data at a scale that was not feasible with desktop processing machines.

References

1. Tamiminia, H., Salehi, B., Mahdianpari, M., Quackenbush, L., Adeli, S., Brisco, B.: Google earth engine for geo-big data applications: a meta-analysis and systematic review. ISPRS J. Photogrammetry Remote Sens. **164**, 152–170 (2020). https://doi.org/10.1016/j.isprsjprs.2020.04.001
2. Tassi, A., Vizzari, M.: Object-oriented LULC classification in google earth engine combining SNIC, GLCM, and machine learning algorithms. Remote Sens (Basel) **12**(22), 1–17 (2020). https://doi.org/10.3390/rs12223776
3. Pérez-Cutillas, P., Pérez-Navarro, A., Conesa-García, C., Zema, D.A., Amado-Álvarez, J.P.: What is going on within google earth engine? A systematic review and meta-analysis. Remote Sens. Appl. Soc. Environ. **29** (2023). https://doi.org/10.1016/j.rsase.2022.100907
4. Magidi, J., Nhamo, L., Mpandeli, S., Mabhaudhi, T.: Application of the random forest classifier to map irrigated areas using google earth engine. Remote Sens. **13**(5), 876 (2021). https://doi.org/10.3390/RS13050876
5. Awad, M.: Google earth engine (GEE) cloud computing based crop classification using radar , optical images and support vector machine algorithm (SVM). In: 2021 IEEE 3rd International Multidisciplinary Conference on Engineering Technology, IMCET 2021, pp. 71–76 (2021). https://doi.org/10.1109/IMCET53404.2021.9665519
6. Chen, H., Yunus, A.P., Nukapothula, S., Avtar, R.: Modelling arctic coastal plain lake depths using machine learning and google earth engine. Phys. Chem. Earth, Parts A/B/C **126**, 103138 (2022). https://doi.org/10.1016/J.PCE.2022.103138
7. Gorelick, N., Hancher, M., Dixon, M., Ilyushchenko, S., Thau, D., Moore, R.: Google earth engine: planetary-scale geospatial analysis for everyone. Remote Sens. Environ. **202**, 18–27 (2017). https://doi.org/10.1016/J.RSE.2017.06.031
8. Amani, M., et al.: Google earth engine cloud computing platform for remote sensing big data applications: a comprehensive review. IEEE J. Sel. Top. Appl. Earth. Obs. Remote Sens **13**, 5326–5350 (2020). https://doi.org/10.1109/JSTARS.2020.3021052
9. Ouchra, H., Belangour, A.: Satellite image classification methods and techniques: a survey. In: Proceedings of IEEE International Conference on Imaging Systems and Techniques, IST 2021 (2021). https://doi.org/10.1109/IST50367.2021.9651454
10. Ouchra, H., Belangour, A., Erraissi, A.: Machine learning for satellite image classification: a comprehensive review. In: 2022 International Conference on Data Analytics for Business and Industry (ICDABI), pp. 1–5, October 2022. https://doi.org/10.1109/ICDABI56818.2022.10041606
11. Nelson, P.R., et al.: Satellite remote sensing. An introduction. J. Geophys. Res. Biogeosci. **127**(2) (1987). https://doi.org/10.1029/2021JG006697
12. Ouchra, H., Belangour, A., Erraissi, A.: Spatial data mining technology for GIS: a review. In: 2022 International Conference on Data Analytics for Business and Industry (ICDABI), pp. 655–659, October 2022. https://doi.org/10.1109/ICDABI56818.2022.10041574
13. Ouchra, H., Belangour, A., Erraissi, A.: A comparative study on pixel-based classification and object-oriented classification of satellite image. Int. J. Eng. Trends Technol. **70**, 206–215 (2022). https://doi.org/10.14445/22315381/IJETT-V70I8P221
14. Ouchra, H., Belangour, A., Erraissi, A.: Satellite data analysis and geographic information system for urban planning: a systematic review. In: 2022 International Conference on Data Analytics for Business and Industry (ICDABI), pp. 558–564, October 2022. https://doi.org/10.1109/ICDABI56818.2022.10041487

15. Ouchra, H., Belangour, A.: Object detection approaches in images: a survey. vol. 11878, pp. 132–141, June 2021, https://doi.org/10.1117/12.2601452
16. Ouchra, H., Belangour, A.: Object detection approaches in images: a weighted scoring model based comparative study. www.ijacsa.thesai.org
17. Ouchra, H., Belangour, A., Erraissi, A.: An overview of GeoSpatial artificial intelligence technologies for city planning and development. In: 2023 Fifth International Conference on Electrical, Computer and Communication Technologies (ICECCT), pp. 1–7, February 2023, https://doi.org/10.1109/ICECCT56650.2023.10179796
18. Borra, S., Thanki, R., Dey, N.: Satellite image analysis : clustering and classification (2019)
19. Venkatappa, M., Sasaki, N., Shrestha, R.P., Tripathi, N.K., Ma, H.O.: Determination of vegetation thresholds for assessing land use and land use changes in Cambodia using the google earth engine cloud-computing platform. Remote Sens (Basel) **11**(13) (2019). https://doi.org/10.3390/rs11131514
20. Bouzekri, S., Lasbet, A.A., Lachehab, A.: A new spectral index for extraction of built-up area using landsat-8 data. J. Indian Soc. Remote Sens. **43**(4), 867–873 (2015). https://doi.org/10.1007/S12524-015-0460-6
21. Landsat 8 | Landsat Science. https://landsat.gsfc.nasa.gov/satellites/landsat-8/. Accessed 30 Jan 2023
22. LSIB 2017: large scale international boundary polygons, Simplified | Earth Engine Data Catalog | Google for Developers. https://developers.google.com/earth-engine/datasets/catalog/USDOS_LSIB_SIMPLE_2017. Accessed 24 Aug 2023
23. Yang, L., Driscol, J., Sarigai, S., Wu, Q., Chen, H., Lippitt, C.D.: Google earth engine and artificial intelligence (AI): a comprehensive review. Remote Sens. **14**(14) MDPI (2022). https://doi.org/10.3390/rs14143253
24. Ouchra, H., Belangour, A., Erraissi, A.: Machine learning algorithms for satellite image classification using google earth engine and landsat satellite data: Morocco case study. IEEE Access (2023). https://doi.org/10.1109/ACCESS.2023.3293828
25. Yengoh, G.T., Dent, D., Olsson, L., Tengberg, A.E., Tucker III, C.J.: Use of the normalized difference vegetation index (NDVI) to assess land degradation at multiple scales, Springer. in SpringerBriefs in Environmental Science. Cham: Springer International Publishing (2016). https://doi.org/10.1007/978-3-319-24112-8
26. Gascon, M., et al.: Normalized difference vegetation index (NDVI) as a marker of surrounding greenness in epidemiological studies: the case of Barcelona city. Urban For Urban Green **19**, 88–94 (2016). https://doi.org/10.1016/J.UFUG.2016.07.001
27. NDBI—ArcGIS Pro | Documentation. https://pro.arcgis.com/en/pro-app/latest/arcpy/spatial-analyst/ndbi.htm. Accessed 19 May 2023
28. Abutaleb, K., et al.: Assessment of urban heat island using remotely sensed imagery over greater Cairo, Egypt. Adv. Remote Sens. **4**(1), 35–47 (2015). https://doi.org/10.4236/ARS.2015.41004
29. Ngandam Mfondoum, A.H., Etouna, J., Nongsi, B.K., Mvogo Moto, F.A., Noulaquape Deussieu, F.G.: Assessment of land degradation status and its impact in arid and semi-arid areas by correlating spectral and principal component analysis neo-bands. Int. J. Adv. Remote Sens. GIS **5**(1), 1539–1560 (2016). https://doi.org/10.23953/CLOUD.IJARSG.77
30. Abburu, S., Golla, S.B.: Satellite image classification methods and techniques: a review (2015)
31. Ouchra, H., Belangour, A., Erraissi, A.: A comprehensive study of using remote sensing and geographical information systems for urban planning. Internetworking Indonesia J. **14**(1), 15–20 (2022)

An Application of Artificial Intelligence and Genetic Algorithm to Support the Discovering of Roman Centuriation Remains

Pietro Fusco(✉) and Salvatore Venticinque

Department of Engineering, University of Campania "Luigi Vanvitelli", Aversa, Italy
pietro.fusco@unicampania.it

Abstract. Methodologies and technologies for uncovering archaeological ruins and objects have been advanced by the utilization of Artificial Intelligence (AI). AI supports archaeologists to discover remains that are difficult to be identified manually as they must be sought on a vast territory or because they are hidden from direct observation. Here we present a methodology that integrates deep learning, computer vision and genetic algorithm techniques to identify, from large aerial pictures, remains of the Centuriation, which is an ancient Roman system for the division of lands.

Keywords: Image Processing · Convolutional Networks · Roman Centuriation · Smart Archaeology · Image segmentation · Genetic Algorithm

1 Introduction

The investigation of the archaeological landscape of ancient urban structures that are now mostly hidden by crops, vegetation, and modern settlements is usually expensive and time-consuming [11]. It is also a difficult issue for automatic algorithms due to the traces similarity to other image artifacts or to their poor boundary information and discontinuities [6]. In the last years, methodologies and technologies for uncovering archaeological ruins and objects have been advanced by the utilization of Artificial Intelligence.

Centuriation was a system used in the Roman culture for the division of the territory assigned to citizens [15]. It is founded on the layout of parallel and perpendicular streets called *decumani* and *cardines*. The decumans were usually arranged along the most developed axis of the territory or were parallel to a major communication route. A divided territory assignment (to more owners) was made up of squares, whose limits were 20 *actus* distant from each other. An *actus* is the agricultural unit of measurement, equivalent to about 35 m. In Fig. 1 it is shown how the main streets and pathways of the Italian city of Marcianise, in the province of Caserta, match the same organization of Roman Centuriation. The automatic recognition of these remains is a relevant and challenging activity in the field of smart archaeology [16]. In fact, nowadays the study of such

G. L. Foresti et al. (Eds.): ICIAP 2023 Workshops, LNCS 14365, pp. 406–417, 2024.
https://doi.org/10.1007/978-3-031-51023-6_34

Fig. 1. Example of ancient Roman centuriation system in the City of Marcianise

archaeological remains is still made manually and it is not trivial because original divisions are partially lost. Many streets have been covered or moved and pathways are lost. Moreover, there exist variations to the original schema that use, for example, rectangular patterns whose edge measures half an actus (or its multiple). In this paper, we advanced the innovative methodology presented in [8] for the automatic identification of Roman Centuriated systems through the elaboration of aerial images. The original contribution of this work is a methodology to align a grid, which is compliant with the Centuriation schema, with the traces detected by the segmentation and processing of aerial images retrieved from the Google Map service. A genetic algorithm is used to find the best rotation and translation of the grid that allows estimating the degree of likelihood for that area to be divided according to the Centuriation schema in ancient times. In the next Section, we discuss related work. In Sect. 3 the original methodology and the last advance proposed in this paper are described. We provide the implementation details in Sect. 4. Experimental results are discussed in 5 and at the end conclusions are drawn.

2 Related Work

The usage of satellite imagery and digital image processing for the detection and surveying of ancient land-use patterns, and in particular for the identification of possible remains is investigated in several works [1,6,12]. Deep learning models, such as U-Net and Mask R-CNN, are used in many visual recognition tasks [9,10], and especially in [3] for semantic segmentation of airborne LiDAR data related to Maya archaeology. Although these techniques have been around for a long time [4], their success was limited due to the size of the available training sets and the size of the considered networks. In [8] we exploited an improved version of such a method that is known as the U-Net neural network [14]. The U-Net architecture uses skip connections to combine low-level feature maps with higher-level ones, which enables precise pixel-level localization. In this paper we focus on the next phase that processes the results of deep learning elaboration to estimate the degree of Centuriation trough the identification of land-use patterns and/or regular shapes. In [16] authors developed an approach for the detection of ruins of livestock enclosures in alpine areas estimating a rectangularity feature that quantifies the degree of alignment of an optimal subset of extracted linear segments with a contour of rectangular shape. In [1] an image analysis procedure utilizing Radon transforms has been used to disclose landscape divisions relating to the Roman settlement and land use obliterated by agricultural intensification. In [6] a multiphase active contour model has been used to improve the visibility of buried remains showing in the image as crop marks (i.e. centuriations, agricultural allocations, ancient roads, etc.). An improvement to such kind of technique is proposed in [5] for archaeological trace identification.

3 Methodology

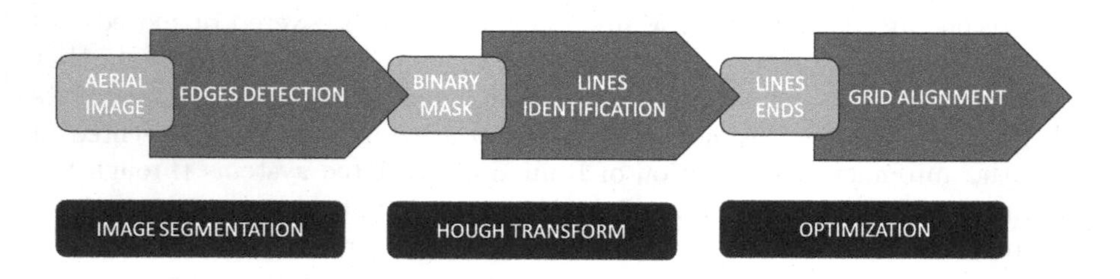

Fig. 2. Methodology

The proposed methodology consists of the data processing pipeline shown in Fig. 2. It exploits the integrated utilization of different technologies in three sequential steps. It aims at extracting as the set of parallel and perpendicular lines $\mathbb{C}$ from an aerial image whose size is $W \times H$ pixels, which correspond to the remains of an ancient Roman Centuriation.

3.1 Edges Detection

In this phase a deep learning technique is used to extract from an aerial image like the one shown in Fig. 3 (a), a mask image shown in Fig. 3 (b), which contains roads and footpaths, or any other edges, which emerge from the ground.

The aerial image I, which is the input of this phase, is defined as a 2D array of $W \times H$ rgb pixels.

Let's define a first function ψ which crops the I image into a set of $n \times m$ smaller images $C_{i,j}$ of 1500×1500 pixels, with $n = ceil[W/1500]$ and $m = ceil[H/1500]$

$$\psi\colon \mathbf{I}^{W \times H} \longrightarrow \{\mathbf{C}_{i,j}^{1500 \times 1500}\, \forall\, i \in \{0..n\}\; and\, j \in \{0...m\}\}$$

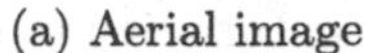

(a) Aerial image

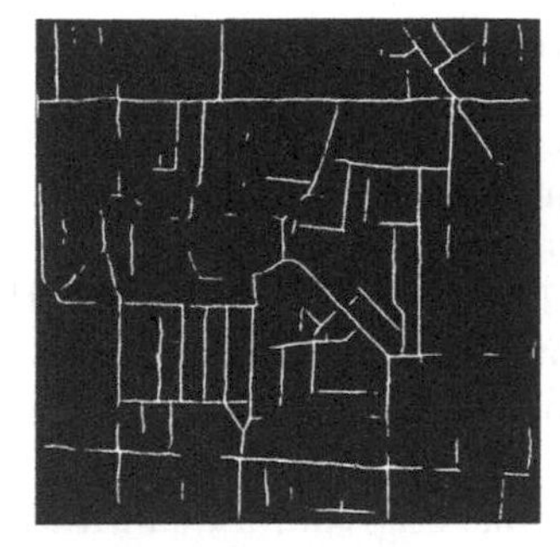

(b) Binary mask

Fig. 3. Trace detection

A CNN network is used to classify each pixel of each cropped image $C_{i,j}$ as a *background* pixel or as a foreground pixel belonging to any identified *paths*. For each $C_{i,j}$ input image, the CNN network generates as output a mask image $M_{i,j}$, that is a 2D array with the same width and height as the input one but with a depth size of 1, where each value is the classified class key for the corresponding pixel.

$cnn\colon \mathbf{C}_{i,j} \longrightarrow \mathbf{M}_{i,j} \in K^{1500\times1500} \quad \forall i,j$ with $K = \{'background','path'\}$.

Let's define the ϕ function that merges the mask images $\mathbf{M}_{i,j}$ into one image $\mathbf{O}$ with the original dimensions.

$$\phi\colon \{\mathbf{M}_{i,j}^{1500\times1500}\, \forall\, i \in \{0..n\}\, and\, j \in \{0...m\}\} \longrightarrow \mathbf{O}^{W\times H}$$

The visualization of the image encoding the two class labels of k with the RGB values $\{[0,0,0],[255,255,255]\}$ is shown in Fig. 3 (b).

3.2 Lines Identification

In this phase, segments are identified from the mask image.

The $\mathbf{O}$ image represents the input of this phase, where the classified pixels are grouped into k segments by the probabilistic Hough transformation $\mathbf{H}$.

$$H\colon \mathbf{O}^{\mathbf{W}\times\mathbf{H}} \longrightarrow \{S_k \forall k \in \{1,...,n_s\}\}$$

Each segment S_k corresponds to a set of l_k pixels $(x_i,y_i)\colon x_i cos(\theta) + y_i \sin(\theta) = r \quad \forall i \in \{1,...,l_k\}$. In Fig. 4 line segments, in green, are identified in the maps image by the probabilistic Hough transform [13].

3.3 Grid Alignment

The solution implemented in [8] for the last phase of the propose methodology used a dense grid to extends vertical and horizontal segments identified at the previous step to identify rectangular shape that could be compliant with Centuriation. However, the problem of the optimal grid rotation according to the direction of Centuriation scheme was neglected and the estimation of the degree of likelihood of Centuriation was still manually delegated to experts of the field. Here we look for an optimal alignment of a Centuriation grid with the lines found in the previous step and a quantitative estimate of the same alignment is used to estimate the compliance of the traces with a Centuriation scheme.

Fig. 4. Lines identified by Hough transform.

To achieve this goal we first compute the ends of the lines, as it is shown in Fig. 5a. Then, a fixed-size grid, shown in Fig. 5b, is generated by the user defining the number of rows and columns in order to maximize the number of lines' ends within the vertical and horizontal strips of the above grid.

In a xy plane the grid has 3° of freedom (DOFs) only, a x and y displacement, and a rotation around a point contained in the xy plane itself. We have to find the optimal solution, in terms of rotation and displacement, for the grid positioning.

A Genetic Algorithm has been used to solve the positioning problem. A Genetic Algorithm (GA) is a metaheuristic inspired by the process of natural selection that belongs to the larger class of evolutionary algorithms (EA). Genetic algorithms are commonly used to generate high-quality solutions to optimization and search problems by relying on biologically inspired operators such as mutation, crossover and selection. In the GA formulation DOFs are considered as genes of an individual and the fitness function returns the number of lines fitting the strips of the grid.

4 Implementation

The proposed methodology has been implemented integrating open source software, the code developed in [8] and an original implementation of the GA based grid alignment algorithm.

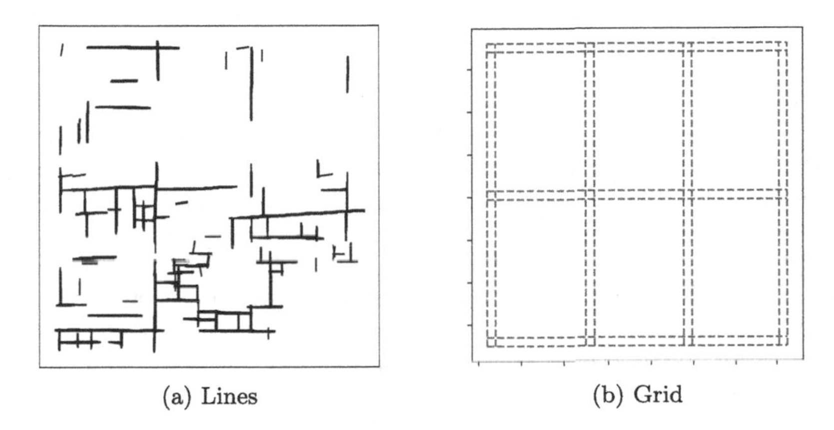

(a) Lines (b) Grid

Fig. 5. Starting point

4.1 The U-Net Convolutional Neural Network

A Pytorch implementation of a U-Net CNN has been used for image segmentation, to identify all the main streets and footpaths within the input image. It has been trained by the Massachusetts Roads Dataset[1], which consists of 1171 aerial images of the state of Massachusetts as, to the author's knowledge, no similar datasets are available for this kind of problem. Each image is 1500×1500 pixels in size, covering an area of 2.25 square kilometers. The U-Net provides a mask

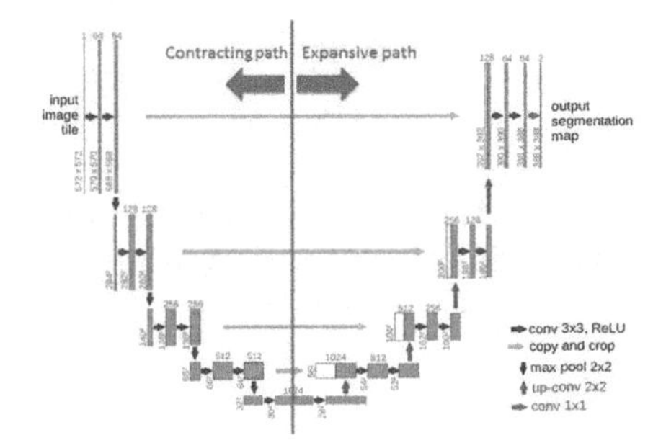

Fig. 6. A generalized U-Net architecture

[1] https://www.cs.toronto.edu/~vmnih/data/.

image as output that contains detected segments, which usually match roads and footpaths eventually included in the input image.

Figure 6 shows a generalized U-Net architecture. Its architecture can be broadly considered an encoder network followed by a decoder network. Unlike classification where the end result of the deep network is the only important thing, semantic segmentation not only requires discrimination at the pixel level, but also a mechanism to project the discriminative features learned at different stages of the encoder onto the pixel space.

- The encoder (contracting path) is the first half of the architecture diagram (Fig. 6). It usually is a pre-trained classification network such as VGG or ResNet where convolution blocks are applied which are followed by a maxpool downsampling layer to encode the input image into feature representations at multiple different levels.
- The decoder (expansive path) is the second half of the architecture. The goal is to semantically project the discriminative features (lower resolution) learned by the encoder onto the pixel space (higher resolution) to get a dense classification. The decoder consists of upsampling and concatenation layers followed by regular convolution operations.

To evaluate the behavior of the above network it is possible to compute two metrics that are commonly used for semantic segmentation problems. The Dice Loss is defined in Eq. 1, while the *Dice* coefficient is defined in Eq. 2.

$$Dice\ Loss = 1 - Dice \tag{1}$$

$$Dice = \frac{2\sum_{i=1}^{N} p_i g_i}{\sum_{i=1}^{N} p_i^2 + \sum_{i=1}^{N} g_i^2} \tag{2}$$

Equation (2) shows the formula of *Dice* coefficient, in which p_i and g_i represent pairs of corresponding pixel values of prediction and ground truth, respectively. The values of p_i and p_i are either 0 or 1, representing whether the pixel belongs to the foreground (value of 1) or not (value of 0). Therefore, the denominator is the sum of total foreground pixels of both prediction and ground truth, and the numerator is the sum of correctly predicted foreground pixels because the sum increments only when p_i and g_i match (both of value 1). Figure 7a shows the curves of *Dice Loss* which approaches almost one, indicating that in the training step, the model is capable to generalize enough. The Intersection over Union (IoU) metric, shown in Eq. 3, is essentially another method to quantify the percent overlap between the target mask and our prediction output. This metric is closely related to the Dice coefficient.

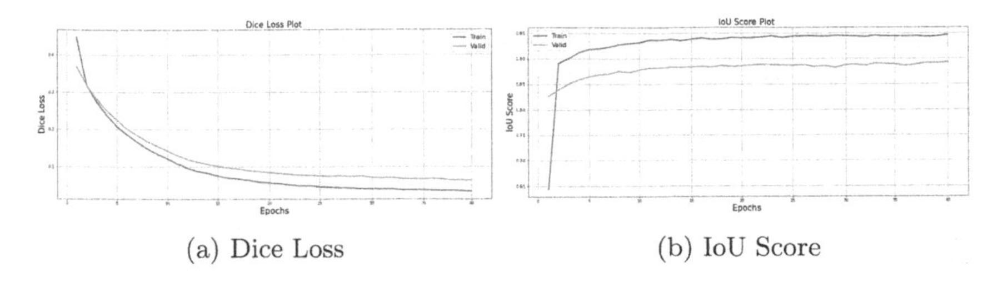

(a) Dice Loss (b) IoU Score

Fig. 7. Semantic segmentation training metrics

Figure 7b shows the *Intersection over Union* variation during the training steps.

$$IoU = \frac{target \cap prediction}{target \cup prediction} \tag{3}$$

Finally, the model, after the training and validation steps over the first dataset, was tested over the DeepGlobe Road Extraction Dataset which consists of 6226 images of the same kind of aerial images as the Massachusetts Roads Dataset. The final results are reported in Table 1.

Table 1. U-Net validation metrics

metric	*training*	*validation I*	*validation II*
IoU	0.95	0.89	0.88
Dice Loss	0.02	0.05	0.07

4.2 Image Processing

For the image processing of the image mask, we used an optimized version of the standard probabilistic Hough transform provided by the OpenCV library [2], and in this case, it is less computationally intensive and executes faster. Moreover, it was done a pre-processing step on the mask image in order to select lines as vertical and horizontal as possible. In other words, vertical and horizontal thresholds slope were defined to select all lines which did not exhibit a slope greater than the above threshold both in vertical and horizontal directions.

4.3 Genetic Algorithm

The system implementation was made possible by using DEAP framework [7] which seeks to make algorithms explicit and data structures transparent. DEAP is compatible out of the box with Python 3. The implementation consists of two main parts. The first one is represented by a *System_Manager_Grid* class which contains several methods among which there are methods for creating a grid, rotating and translating data points and grid lines and lastly there is a method for running the genetic algorithm simulation. The second part of the system is a Python class that contains a function for loading data and an instance of the *System_Manager_Grid* class.

5 Experimental Results

Our test bed is composed of 4 aerial images downloaded by the Google Map application. They correspond to four cities in the province listed in Table 2. Each geographic area is identified by the latitude and longitude of its upper left corner (lat1, long1) and lower right corner (lat2, long2) by archaeologists who had previously investigated this area tracing Centuriation remains with classic techniques.

Table 2. Test cases

Use Case	*(lat1, long1) (lat2, long2)*
Marcianise	(41.03699966153493, 14.282332207120104)
	(41.02356625889453, 14.300298406435248)
Casagiove	(41.08261546935178, 14.303388824263887)
	(41.07064562316231, 14.324267139938394)
Casapulla	(41.0825150404040, 14.268917099445803)
	(41.0663241840082, 14.302277371886024)
Capodrise	(41.05162612435181, 14.288554711937884)
	(41.03859863125092, 14.311235472131052)

For each geographic area to be processed, the corresponding aerial image is downloaded with the same zoom level.

To make convergence easy, pre-setting has been done. Pre-setting means that initially, the user could have placed the grid at any distance from the lines and with any rotation angle in the xy plane.

As an example, Fig. 8 shows the best individual of an intermediate generation of the GA algorithm with only 323 lines contained in the grid, out of the 354 which would be fit by the optimal solution. It is straightforward to observe that the regular schema of the detected lines is not aligned with the grid yet.

By choosing a random initial position for the grid could have obtained a sub-optimal result. At worst, convergence could have not been achieved. By reducing the initial distance between the grid and the lines themselves made it easier to converge towards a good solution avoiding falling in any local minimum.

Making pre-setting step, the optimization simulation showed fewer numerical problems and the user had to test fewer GA algorithm hyper-parameters to find the convergence. It must be emphasized that the GA-based method for the right

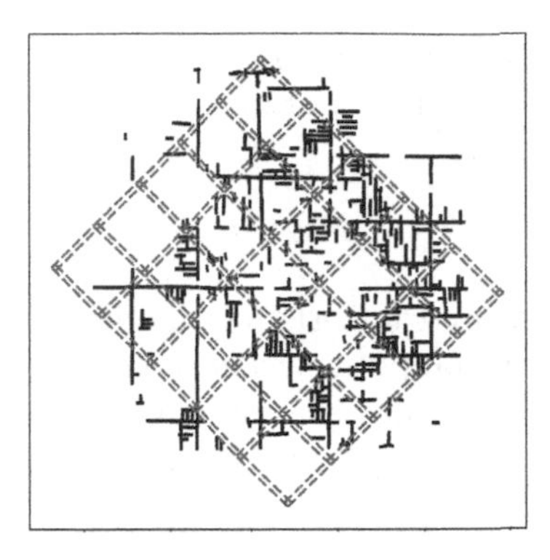

Fig. 8. An intermediate generation of the GA algorithm

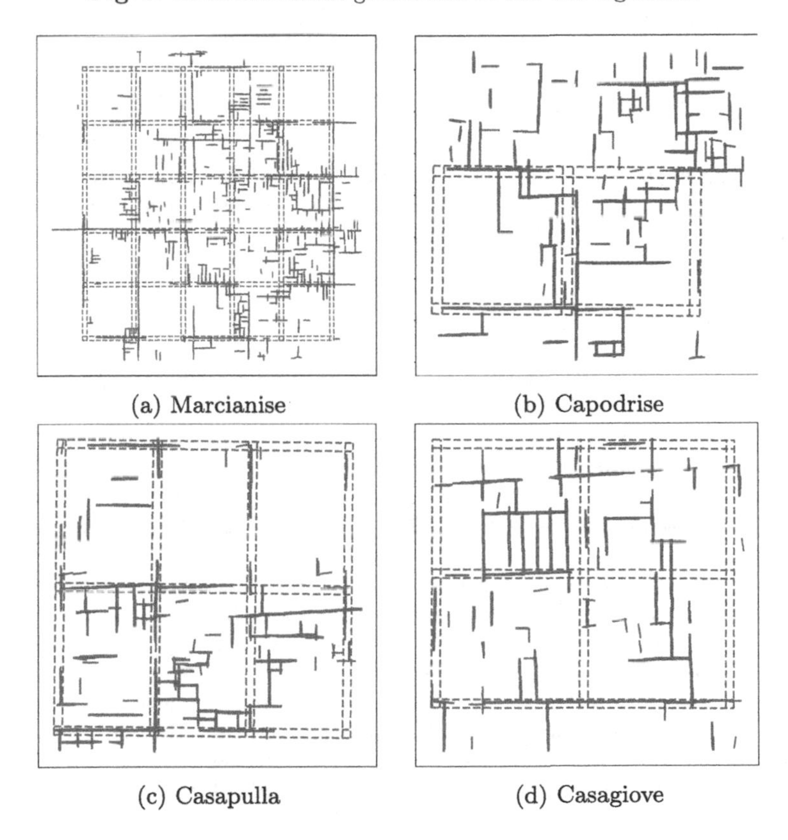

(a) Marcianise

(b) Capodrise

(c) Casapulla

(d) Casagiove

Fig. 9. Final detection

positioning of the grid in the plane w.t.r. the lines is quite immature and it needs to be tweaked.

Lastly, Fig. 9 shows the final results provided by the implemented system where it is possible to observe the position of the grid w.r.t. the lines. The final position of the grid is the one that contains as many lines as possible.

6 Conclusions

In this paper, it was presented an original methodology that is based on the integration of deep learning, computer vision and genetic algorithm techniques to automatically support the identification of undiscovered remains of the ancient Roman Centuriation system from aerial images. A prototype implementation demonstrated the feasibility of the approach. Future research direction aims at automatically scanning wide areas to estimate the accuracy of results. More accurate analytical models will be investigated to provide a quantitative index for evaluating the probability that the selected area includes remains of Roman Centuriation, and in particular for the fitness function. In a second phase the methodology will be tested to support new studies in geographical areas not yet investigated.

Acknowledgments. We would like to acknowledge Prof. Giuseppina Renda and Dott. Sabrina Mataluna for their support as experts in the application domain for the requirements analysis and the evaluation of results.

References

1. Bescoby, D.J.: Detecting roman land boundaries in aerial photographs using radon transforms. J. Archaeol. Sci. **33**(5), 735–743 (2006). ISSN: 0305-4403. https://doi.org/10.1016/j.jas.2005.10.012
2. Bradski, G.: The OpenCV library. Dr. Dobb's J. Softw. Tools **25**, 120–123 (2000)
3. Bundzel, M., et al.: Semantic segmentation of airborne LiDAR data in Maya archaeology. Remote Sens. **12**(22) (2020). ISSN: 2072-4292. https://doi.org/10.3390/rs12223685
4. Cireşan, D.C. et al.: Deep neural networks segment neuronal membranes in electron microscopy images. In: Proceedings of the 25th International Conference on Neural Information Processing Systems - Vol 2, NIPS'12, pp. 2843–2851. Curran Associates Inc., Lake Tahoe, Nevada (2012)
5. D'Orazio, T., Palumbo, F., Guaragnella, C.: Archaeological trace extraction by a local directional active contour approach. Pattern Recogn. **45**(9) (2012). Best Papers of Iberian Conference on Pattern Recognition and Image Analysis (IbPRIA'2011), pp. 3427–3438. ISSN: 0031-3203. https://doi.org/10.1016/j.patcog.2012.03.003
6. Figorito, B., Tarantino, E.: Semi-automatic detection of linear archaeological traces from orthorectified aerial images. Int. J. Appl. Earth Obs. Geoinf. **26**, 458–463 (2014). ISSN: 1569-8432. https://doi.org/10.1016/j.jag.2013.04.005
7. Fortin, F.-A., et al.: DEAP: evolutionary algorithms made easy. J. Mach. Learn. Res. **13**, 2171–2175 (2012)
8. Fusco, P., Venticinque, S., Aversa, R.: An application of artificial intelligence to support the discovering of roman centuriation remains. IEEE Access **10**, 79192–79200 (2022). https://doi.org/10.1109/ACCESS.2022.3194147

9. Girshick, R., et al.: Rich feature hierarchies for accurate object detection and semantic segmentation. In: Proceedings of the IEEE Computer Society Conference on Computer Vision and Pattern Recognition (2013). https://doi.org/10.1109/CVPR.2014.81
10. Krizhevsky, A., Sutskever, I., Hinton, G.E.: ImageNet classification with deep convolutional neural networks. Commun. ACM **60**(6), 84–90 (2017). ISSN: 0001–0782. https://doi.org/10.1145/3065386
11. Magli, G., et al.: Uncovering a masterpiece of roman engineering: the project of via Appia between Colle Pardo and Terracina. J. Cult. Heritage **15**(6), 665–669 (2014). ISSN: 1296–2074. https://doi.org/10.1016/j.culher.2013.11.014
12. Montufo, A.M.: The use of satellite imagery and digital image processing in landscape archaeology. a case study from the island of Mallorca, Spain. Geoarchaeology **12**(1), 71–85 (1997). https://doi.org/10.1002/(SICI)1520-6548(199701)12:1⟨71::AID-GEA4⟩3.0.CO;2-6
13. Mukhopadhyay, P., Chaudhuri, B.B.: A survey of Hough transform. Pattern Recogn. **48**(3), 993–1010 (2015). ISSN: 0031–3203. https://doi.org/10.1016/j.patcog.2014.08.027
14. Ronneberger, O., Fischer, P., Brox, T.: U-Net: convolutional networks for biomedical image segmentation. CoRR abs/1505.04597 (2015). arXiv: 1505.04597
15. Sparavigna, A.C.: Roman centuriation in satellite images. PHILICA (2015). https://doi.org/10.5281/zenodo.3361974
16. Zingman, I., Saupe, D., Lambers, K.: Detection of incomplete enclosures of rectangular shape in remotely sensed images. In: 2015 IEEE Conference on Computer Vision and Pattern Recognition Workshops (CVPRW), pp. 87–96 (2015). https://doi.org/10.1109/CVPRW.2015.7301387

Convolutional Neural Networks for the Detection of Esca Disease Complex in Asymptomatic Grapevine Leaves

Alberto Carraro[1(✉)], Gaetano Saurio[2], Ainara López-Maestresalas[3], Simone Scardapane[4], and Francesco Marinello[1]

[1] TESAF, Università di Padova, Viale dell'Università 16, 35020 Legnaro, PD, Italy
alberto.carraro.11@studenti.unipd.it

[2] DIAG Department, Sapienza University of Rome, Via Ariosto 25, 00185 Rome, Italy
gaetano.saurio@uniroma1.it

[3] Department of Engineering, ETSIAB-ISFOOD, Universidad Pública de Navarra, Campus de Arrosadia, 31006 Pamplona, Spain
ainara.lopez@unavarra.es

[4] DIET Department, Sapienza University of Rome, Via Eudossiana 18, 00184 Rome, Italy
simone.scardapane@uniroma1.it

Abstract. The Esca complex is a grapevine trunk disease that significantly threatens modern viticulture. The lack of effective control strategies and the intricacy of Esca disease manifestation render essential the identification of affected plants before symptoms become evident to the naked eye. This study applies Convolutional Neural Networks (CNNs) to distinguish, at the pixel level, between healthy, asymptomatic and symptomatic grapevine leaves of a Tempranillo red-berried cultivar using Hyperspectral imaging (HSI) in the 900–1700 nm spectral range. We show that a 1D CNN performs semantic image segmentation (SiS) with higher accuracy than PLS-DA, one of HSI data's most widely used classification algorithms.

Keywords: Esca complex · Convolutional Neural Networks · Partial Least Squares Discriminant Analysis · Hyperspectral imaging

1 Introduction

HSI has emerged as a powerful technology in various fields, including remote sensing, agriculture, environmental monitoring, and medical imaging. The

This study was carried out within the Agritech National Research Centre and received funding from the European Union Next-GenerationEU (PIANO NAZIONALE DI RIPRESA E RESILIENZA (PNRR)-MISSIONE 4 COMPONENTE 2, INVESTIMENTO 1.4-D.D. 1032 17/06/2022, CN00000022). This manuscript reflects only the authors' views and opinions, neither the European Union nor the European Commission can be considered responsible for them.

G. L. Foresti et al. (Eds.): ICIAP 2023 Workshops, LNCS 14365, pp. 418–429, 2024.
https://doi.org/10.1007/978-3-031-51023-6_35

ability to capture detailed spectral information over a wide range of wavelengths offers significant advantages for material identification and classification tasks. However, HSI data has inherent challenges, such as high dimensionality and noise, which can hinder efficient and accurate data analysis.

Partial Least Squares Discriminant Analysis (PLS-DA) [7] is considered a valuable and effective classification technique for hyperspectral data analysis. HSI datasets are characterized by high dimensionality, where each pixel in an image is associated with a vast number of spectral bands, making traditional classification methods challenging to implement. PLS-DA addresses this issue by projecting the original spectral data onto a lower-dimensional subspace that captures the most discriminative information related to class separability. By identifying latent variables that maximize the covariance between the input data and their corresponding class labels, PLS-DA reduces the complexity of the HSI data while preserving its essential spectral characteristics. The resulting lower-dimensional representation facilitates efficient and accurate classification, enabling PLS-DA to be particularly well-suited for land cover classification, disease detection, and material identification using HSI. Moreover, the interpretability of the latent variables in PLS-DA provides valuable insights into the spectral features contributing significantly to classification, making it a helpful tool in understanding complex HSI datasets. This technique has been successfully used in [14] to perform SiS of hyperspectral images of grapevine leaves affected by Esca disease.

In recent years, various Neural Networks (NN) emerged among Machine Learning (ML) techniques with promising results in HSI data analysis.

In [4], the authors apply 1-Dimensional Convolutional Neural Networks (1D CNN) to solve the pixel classification problem on the Pavia [1] and Indian Pines [2] datasets. They also use contribution maps to spot the most relevant wavelengths in the classification of each sample; the experiments carried out in [9] demonstrate that 1D CNNs can achieve better classification performance than some traditional ML methods, such as Support Vector Machines on Indian Pines and Salinas dataset [3].

This study compares the performance of CNNs *versus* Partial Least Squares Discriminant Analysis in identifying asymptomatic grapevine leaves using hyperspectral images. We obtained encouraging results with an increase in terms of accuracy of almost five percentage points. The literature contains many successful applications of Deep Learning (DL) on medical HSI images [17] as well as on RGB tomato images [5]. Moreover 1D CNN have been also applied in [10] to classify leaves and in [12] to tackle identification of grapevines inoculated with the Grapevine Vein-Clearing Virus. This work adds one more approach to the problem of Esca disease detection with DL.

2 Materials and Methods

2.1 Leaf Samples

This study was carried out on 72 grapevine leaves collected close to the 2018 harvest (September 20), at stage 89 according to the BBCH scale [11], from an experimental rainfed vineyard naturally infected with Esca located in Olite, Spain (42°26'19.06" N, 1°38'52.57" W), which belongs to the Viticulture and Enology Station of Navarra (EVENA). This vineyard consisted of *Vitis vinifera* L. cv. 'Tempranillo'. The leaves were first subject of a previous investigation in [14]. This is a monitored vineyard where Esca foliar symptoms have been recorded since 2014, which makes it possible to know if Esca has appeared on a specific plant. Taking this historical record of the last five consecutive seasons into account, it was possible to collect leaves from three categories: control (class CO), asymptomatic leaves from vines that had never exhibited Esca symptoms and could therefore be considered to be healthy; asymptomatic leaves from plants that were showing Esca symptoms (class E1); and symptomatic leaves from the same Esca-affected vines (class E2). Twenty-four leaves of each category, 72 in total, were taken from no more than six different vines, always considering mature but not senescent leaves. Leaves were kept refrigerated (3°C) until the measurement started, around 24 h after the sample gathering.

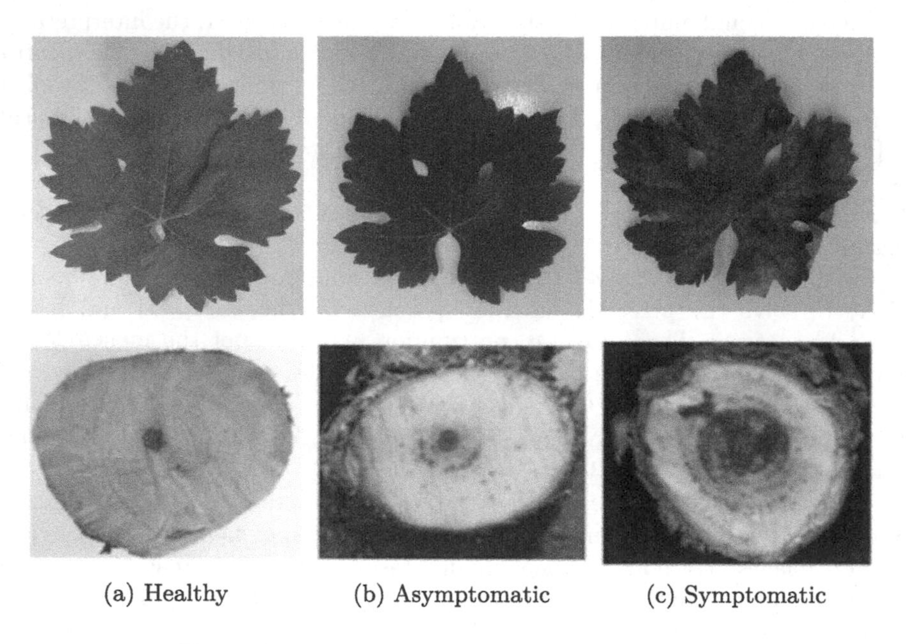

(a) Healthy (b) Asymptomatic (c) Symptomatic

Fig. 1. State of the leaf and state of the trunk [8].

Esca complex is a grapevine trunk disease caused by multiple fungal species, including Phaeoacremonium oleophilum, Phaeomoniella chlamydospora, and Fomitiporia mediterranea, which colonize the grapevine's woody tissues. As these fungi spread through the vascular system, they disrupt the flow of water and nutrients within the plant. Figure 1a contains a healthy leaf (class CO) and a section of a healthy trunk. Until the leaf receives the essential nutrients, its normal chlorophyll production gives it its green colour. The disruption of water flow within the trunk impairs the leaves' ability to receive essential nutrients before the chlorosis becomes visible to the human eye. Figure 1b displays an asymptomatic leaf (class E1) and a section of an infected trunk. Symptomatic grapevine leaves affected by Esca disease exhibit characteristic yellow and brown stripes due to the impact of the pathogenic fungi on the vascular system and chlorophyll production. Figure 1c shows a symptomatic leaf (class E2) and a section of an infected trunk.

2.2 Hyperspectral Image Acquisition and Processing

Hyperspectral Image Acquisition. The hyperspectral system used in this study comprised the following devices: (1) an InGaAs camera (Xeva 1.7-320 Xenics, Leuven, Belgium), with a 320 × 256-pixel resolution sensitive in the 900–1700 nm spectral range (3.14 nm spectral resolution), coupled to a spectrograph (ImSpector N17E, Specim, Spectral Imaging Ltd., Oulu, Finland) with a 30 μm slit and a 16 mm C-mount lens (f/1.4); (2) a linear translation stage driven by a stepper motor (LEFS25, SMC Corporation, Tokyo, Japan) attached to a black sample holder plate to move the sample during image acquisition; (3) four 46 W halogen lamps pointing at the sample holder; (4) a black cover enclosing the entire setup to avoid possible external light interference; (5) a computer equipped with Xeneth 2.5 and ACT Controller software to adjust configuration settings and control the camera and translation stage during scanning, as well as to store the leaf images. Samples were placed at a distance of 400 mm from the lens resulting in an image spatial resolution of 0.75 mm/pixel along the scanning line (320 pixels). The vendor's calibration package was applied for image correction, and a 2 ms integration time was set to avoid detector saturation.

A hyperspectral image of the adaxial side of each leaf was recorded. Moreover, an RGB image was also obtained for each sample by a Lumix DMC-TZ25 digital camera (Panasonic, Japan) and used as a reference.

Spectral Data Extraction and Labelling. Every pixel of each hyperspectral image has been annotated with an integer from 0 to 3 (0 = Background, 1 = CO, 2 = E1, and 3 = E2) as a result of a multi-step process illustrated by the five images of Fig. 2 using sample Y20328H5 as a running example.

(1) Fig. 2a shows an RGB depiction of the leaf Y20328H5; (2) Fig. 2b: using RGB images as a reference, the visually symptomatic leaf zones (avoiding nerves) were identified and manually selected using a graphical user-friendly interface; (3) all pixels of all 72 images are reunited in a single dataset on which the unsupervised K-Means clustering algorithm is run. In Fig. 2c, the white pixels are those whose assigned cluster has the most significant overlap with the white pixels of Fig. 2b; (4) The background is isolated through a binary mask using Otsu thresholding [13]. In Fig. 2d, the white pixels correspond to the leaf surface computed from the hyperspectral file, not the RGB image; (5) the white pixels of Fig. 2e are those that are white both in Fig. 2d, Fig. 2c, and Fig. 2b, i.e. the binary mask in Fig. 2e is the intersection of the other three binary masks. Finally, the pixels selected by this last mask were assigned the label E2 (symptomatic).

(a) (b) (c)

(d) (e)

Fig. 2. Ground truth labelling of pixels.

The stages depicted in Fig. 2 have been adopted for the ground truth labelling of all 72 hyperspectral images, with the suitable adaptation of step 2. For the leaves of class CO and E1, respectively, a large part of the leaf surface has been selected and attributed to the respective label with the only care of avoiding leaf nerves.

The raw hyperspectral data contained 256 absolute reflectance values (non-negative integers) for each pixel. Moreover, a low signal-to-noise ratio was observed at the beginning of the spectrum, resulting in noisy images in this region. Therefore, bands in the 900–1000 nm range were excluded entirely for further analysis, reducing the spectral range to 1000–1700 nm with a total of 224 bands.

As a result of the ground truth label attribution, not all pixels were included in the training and testing of the classification algorithms. Table 1 shows how pixels are subdivided in the dataset.

Table 1. Distribution of pixels in the dataset according to the labels (ground truth annotation).

Pixel type		Amount
Leaf surface	CO	639, 344
	E1	259, 346
	E2	71, 854
	not annotated	898, 508
Background		3, 653, 188
Total		5, 522, 240

Among all reflectance values recorded on bands from 33 to 256 (i.e. from 1000 nm to 1700 nm), the smallest and largest reflectance values recorded were 0 and 31680, respectively. The final pre-processing step was applying a global min-max scaling to all reflectance values transforming x into $\frac{x}{31680}$.

3 Experiments and Results

The scaled dataset consisting only of the 639,344 spectra of CO pixels + 259,346 spectra of E1 pixels + 71,854 spectra of E2 pixels (see Table 1) was then randomly divided into train (80%) and test (20%) subsets.

The same train and test sets were used to train and test the PLS-DA algorithm and our model.

It is a deep neural network consisting of five 1D convolutional layers with a kernel size of three and one padding. We adopt max-pooling layers with a kernel size of two, batch normalization, and a dropout layer with a 20% probability after the fourth convolutional layer. The network employs CrossEntropyLoss, Adam, with a learning rate of 0.001 over 30 epochs.

Table 2 reports the basic evaluation metrics for the multi-class classifier implemented through PLS-DA: the Confusion Matrix, precision and recall for each class, together with the overall accuracy reached by the algorithm on the test set.

Table 2. Confusion matrix for PLSDA classification, with precision and recall for each class.

		Predicted class				
		CO	E1	E2	Precision %	Recall %
Actual class	CO	112,213	15,459	197	99.044	87.756
	E1	1,083	50,641	145	75.955	97.632
	E2	0	572	13,799	97.582	96.020
					Overall accuracy %	
					91.007	

Table 3 reports the same evaluation metrics for the multi-class classifier implemented through our 1D CNN. All values are relative to the same test set used to evaluate the PLS-DA classifier.

Finally Fig. 3a and Fig. 3b show, respectively, the ROC-AUC curves relative to our 1D CNN and to the PLS-DA classifier.

Table 3. Confusion matrix for our 1DCNN classification, with precision and recall for each class.

		Predicted class				
		CO	E1	E2	Precision %	Recall %
Actual class	CO	122,989	4,880	0	97.395	96.184
	E1	3,287	48,582	0	90.736	93.663
	E2	2	80	14,289	100.000	99.429
					Overall accuracy %	
					95.750	

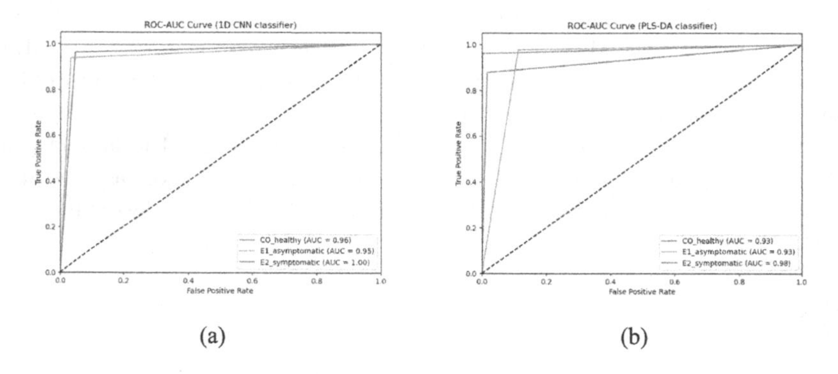

Fig. 3. ROC-AUC curves for our 1D CNN and for the PLS-DA classifier.

The most common error type is the misclassification of E1 and CO pixels, predicting one class for the other. This is probably because asymptomatic leaves are not all equal. The amount of nutrients reaching them has not been measured, so the section of the different trunks transporting water to those leaves may have different levels of internal damage. In some cases, the damage may be light so that the algorithms do not notice its effect on the reflectance spectrum of leaf tissue. This is also visible from the ROC-AUC curves. While class E2 achieves perfect discrimination (AUC = 1), the classes CO (AUC = 0.96) and E1 (AUC = 0.95) have excellent but lower scores. The necrotic tissue (E2 pixels) has a very distinguishable hyperspectral signature. In contrast, asymptomatic pixels have a range of signatures according to the severity of the damage to the trunk that supports each leaf.

Table 4. Juxtaposition of RGB images of leaves and the semantic segmentation produced by our 1D CNN on the corresponding hyperspectral images.

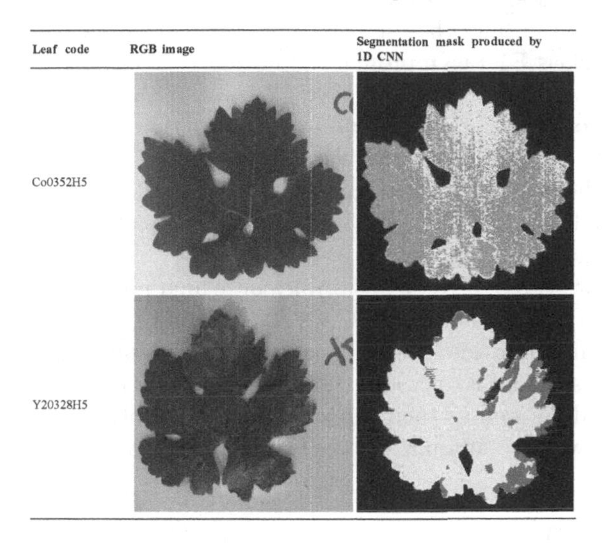

Leaf code	RGB image	Segmentation mask produced by 1D CNN
Co0352H5		
Y20328H5		

Table 5. Examples of classification performances by PLS-DA and our 1D CNN.

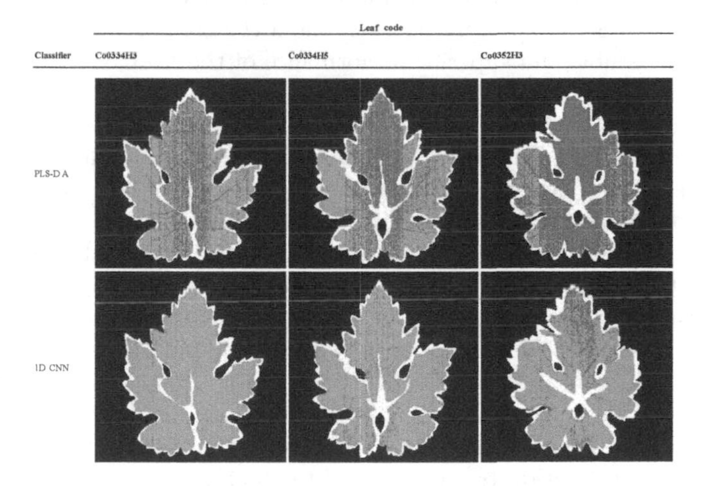

	Leaf code		
Classifier	Co0334H3	Co0334H5	Co0352H3
PLS-DA			
1D CNN			

In Table 4 the right-hand part of each row illustrates the segmentation outputs from the hyperspectral images. On the left-hand side the corresponding RGB image gives a chance to visualize the leaves, although the RGBs are not perfectly coincident with hyperspectral images (due to their acquisition at separate time instances). In that table, one can see the class predicted by our 1D CNN for each pixel of the leaves Co0352H5 and Y20328H5. In the computed segmentation mask, green stands for pixels predicted of class CO, yellow stands for pixels predicted of class E1, and brown stands for pixels predicted of class

E2. Note that inside a given leaf image of class E2, as Y20328H5, no annotated pixel was ever attributed a ground truth class different from E2: the same protocol was followed for all other classes. This means that the CNN quite sensibly attributes the class E1 (asymptomatic) outside the portions of leaf tissue with localized necrosis. Moreover, in sample Co0352H5, the CNN predicts the presence of asymptomatic (E1) pixels, mainly concentrated on the upper end of the leaf. This might be because nutrients are less supplied by peripheral areas of leaf tissue, even in healthy leaves. On the other hand the 1D CNN predicts the presence of E1 pixels on leaf Co0352H5: a thorough explanation of this fact needs further investigation.

Table 5 shows three of the five samples on which almost all prediction errors occur, both for PLSDA and our 1D CNN. In each image: black is the background, white is non-annotated leaf surface pixels, and red and green are misclassified and correctly classified. At present time, these concentration of errors is not yet fully understood, as well as their placement on the central portion of the leaves. On all other leaves the amount of misclassified pixels is much lower.

4 Discussion

Throughout our evaluation, it became evident that our 1D CNN consistently outperformed PLS-DA on every leaf sample in the context of Esca disease classification. The stable and superior performance of the 1D CNN model across all leaf samples reaffirms the robustness and effectiveness of DL techniques for HSI data analysis. The 1D CNN's ability to automatically extract and learn intricate features from the HSI data proved advantageous in discriminating healthy portions from asymptomatic and symptomatic regions. It is possible that PLS-DA, as a linear-based method, exhibited limitations in fully capturing the complexity and non-linear relationships in this kind of data, leading to sub-optimal classification results on various leaf samples.

Remarkably, the vast majority of prediction errors in our study, both for the PLS-DA algorithm and our 1D CNN, were found to be concentrated on just five leaf samples of class CO (healthy leaves). This observation is particularly intriguing, as it highlights the potential influence of these specific samples on the overall classification performance. Identifying these critical samples warrants further investigation to comprehend the factors contributing to their misclassification and their potential impact on the model's generalizability. Possible sources of misclassification could include anomalies in the HSI data, artefacts during data acquisition, or unique variations in the leaf tissue affected by Esca disease. By analyzing and addressing the underlying causes of these errors, we can refine our model and develop more robust strategies for future prediction tasks. Additionally, this finding underscores the importance of carefully curating and balancing the training dataset to ensure representative coverage of all classes, as the over-reliance on a few challenging samples might lead to biased predictions in real-world applications.

5 Conclusions and Future Work

In conclusion, this study successfully applies 1D CNNs as a powerful SiS technique for discriminating healthy portions of leaf tissue from asymptomatic and symptomatic regions affected by Esca disease. The achieved results exhibit high accuracy, precision, and recall rates, underscoring the potential of this approach for early detection and monitoring of Esca disease in vineyards. By automating the identification process, the developed model offers a time-efficient and reliable solution for vineyard management, contributing to better disease control and overall crop yield improvement. The success of 1D CNNs in this context aligns with the broader utilization of DL techniques in precision agriculture and plant pathology, with the potential to revolutionize disease detection and crop management practices. As further research continues to refine the model's performance and adapt it to various crop diseases, the integration of 1D CNNs in agricultural practices holds great promise for enhancing food security and sustainability on a global scale.

The superior accuracy achieved by 1D Convolutional Neural Networks (CNNs) over Partial Least Squares Discriminant Analysis (PLS-DA) may be attributed to several key factors not yet fully understood. 1D CNNs may be able to capture more intricate patterns and relationships within the data, enabling more discriminative representations for classification. Additionally, 1D CNNs are more robust to noise and variations in the data thanks to their ability to learn dependencies and patterns from local receptive fields. This resilience to noise enhances the overall classification accuracy, especially in complex and noisy hyperspectral images. Overall, the adaptive learning capacity, ability to handle high-dimensionality, and robustness to noise make 1D CNNs a compelling choice for SiS and classification tasks in HSI data analysis, leading to superior accuracy compared to traditional PLS-DA approaches.

In light of the promising results obtained from this study, our future research directions will focus on advancing feature selection techniques for hyperspectral data in classification problems. Particularly exploring and implementing intrinsic supervised methods for feature selection (in classification and SiS). Comparative investigations could be conducted to assess the effectiveness of traditional transformation-based feature selection methods, such as PLS-DA and PCA, against other approaches which are directly integrable into the framework of NN like Group Lasso [16,18], contribution maps [4] and gradient tracking [15]. These comparative analyses would shed light on the strengths and weaknesses of each method, providing valuable insights into their respective capabilities in extracting the most relevant spectral features for accurate classification.

Another of our future interests is exploring Conv2D (in [17] is applied DL SiS for tumour detection in HSI data accounting for more spatial context) and advanced architectures like Vision Transformers [6]. We aim to capture spectral and spatial correlations between neighbouring pixel areas and long-range interactions by doing so. These advancements hold great potential to enhance further the accuracy and robustness of disease detection and classification in vineyard environments.

References

1. Indian Pavia university dataset. https://paperswithcode.com/dataset/pavia-university. Accessed 28 July 2023
2. Indian pines dataset. https://paperswithcode.com/dataset/indian-pines. Accessed 28 July 2023
3. Salinas dataset. https://paperswithcode.com/dataset/salinas. Accessed 28 July 2023
4. Cai, R., Yuan, Y., Lu, X.: Hyperspectral band selection with convolutional neural network. In: Lai, J.-H., et al. (eds.) PRCV 2018. LNCS, vol. 11259, pp. 396–408. Springer, Cham (2018). https://doi.org/10.1007/978-3-030-03341-5_33
5. Chowdhury, M.E.H., et al.: Automatic and reliable leaf disease detection using deep learning techniques. AgriEngineering **3**(2), 294–312 (2021). https://doi.org/10.3390/agriengineering3020020
6. Dosovitskiy, A., et al.: An image is worth 16x16 words: transformers for image recognition at scale. In: International Conference on Learning Representations (2020)
7. Fordellone, M., Bellincontro, A., Mencarelli, F.: Partial least squares discriminant analysis: a dimensionality reduction method to classify hyperspectral data. Statistica Applicata - Ital. J. Appl. Stat. **31**(2), 181–200 (2020). https://doi.org/10.26398/IJAS.0031-010, https://www.sa-ijas.org/ojs/index.php/sa-ijas/article/view/31-10
8. Goufo, P., Singh, R.K., Cortez, I.: Metabolites differentiating asymptomatic and symptomatic grapevine plants (vitis vinifera "malvasia-fina") infected with esca complex disease-associated fungi. Biol. Life Sci. Forum **11**(1) (2022). https://doi.org/10.3390/IECPS2021-11923, https://www.mdpi.com/2673-9976/11/1/87
9. Hu, W., Huang, Y., Wei, L., Zhang, F., Li, H.: Deep convolutional neural networks for hyperspectral image classification. J. Sens. **2015** (2015). https://doi.org/10.1155/2015/258619. publisher: Hindawi Publishing Corporation
10. Kuang, D.: A 1D convolutional network for leaf and time series classification. CoRR abs/1907.00069 (2019). http://arxiv.org/abs/1907.00069
11. Meier, U.: Growth Stages of Mono-and Dicotyledonous Plants. Blackwell Wissenschafts-Verlag, Hoboken (1997)
12. Nguyen, C., Sagan, V., Maimaitiyiming, M., Maimaitijiang, M., Bhadra, S., Kwasniewski, M.T.: Early detection of plant viral disease using hyperspectral imaging and deep learning. Sensors **21**(3), 742 (2021). https://doi.org/10.3390/s21030742
13. Otsu, N.: A threshold selection method from gray-level histograms. IEEE Trans. Syst. Man Cybern. **9**(1), 62–66 (1979). https://doi.org/10.1109/TSMC.1979.4310076
14. Pérez-Roncal, C., Arazuri, S., Lopez-Molina, C., Jarén, C., Santesteban, L.G., López-Maestresalas, A.: Exploring the potential of hyperspectral imaging to detect esca disease complex in asymptomatic grapevine leaves. Comput. Electr. Agric. **196**, 106863 (2022). https://doi.org/10.1016/j.compag.2022.106863, https://www.sciencedirect.com/science/article/pii/S0168169922001806
15. Pruthi, G., Liu, F., Sundararajan, M., Kale, S.: Estimating training data influence by tracking gradient descent. CoRR abs/2002.08484 (2020). https://arxiv.org/abs/2002.08484
16. Scardapane, S., Comminiello, D., Hussain, A., Uncini, A.: Group sparse regularization for deep neural networks. Neurocomputing **241**, 81–89 (2017). https://doi.org/10.1016/j.neucom.2017.02.029, https://www.sciencedirect.com/science/article/pii/S0925231217302990

17. Trajanovski, S., Shan, C., Weijtmans, P.J.C., de Koning, S.G.B., Ruers, T.J.M.: Tongue tumor detection in hyperspectral images using deep learning semantic segmentation. IEEE Trans. Biomed. Eng. **68**(4), 1330–1340 (2021). https://doi.org/10.1109/TBME.2020.3026683
18. Yuan, M., Lin, Y.: Model selection and estimation in regression with grouped variables. J. Roy. Stat. Soc. Ser. B: Stat. Methodol. **68**(1), 49–67 (2005). https://doi.org/10.1111/j.1467-9868.2005.00532.x

ArcheoWeedNet: Weed Classification in the Parco archeologico del Colosseo

Gaetano Saurio[1(✉)], Marco Muscas[1], Indro Spinelli[1], Valerio Rughetti[3], Irma Della Giovampaola[2], and Simone Scardapane[1]

[1] Sapienza University of Rome, Rome, Italy
gaetano.saurio@uniroma1.it
[2] Ministry of Culture, Rome, Italy
[3] UniNettuno University, Rome, Italy

Abstract. This paper summarizes the development of a weed monitoring system in the Parco archeologico del Colosseo (hereinafter, Parco) using Deep Learning (DL) techniques to recognize forty-one species of plants now present in the area. The project is part of SyPEAH (System for the Protection and Education of Archaeological Heritage), a platform designed to safeguard the Parco by its Authority. This study emanates from an extended phase of the photographic collection spanning ten months. This endeavour facilitated the compilation of a dataset comprising nearly 5,000 photographs depicting the flora of pertinent significance. In the paper, we detail the first version of the system, consisting of a neural network trained to predict the species of plants and the materials on which they grow. We also describe transfer learning techniques aimed at improving performance. The present system attains recognition accuracy exceeding 90% for common species, enabling near real-time monitoring of the entire Park's flora through image analysis using supplied fixed and mobile devices. It will support proactive interventions for maintenance. The paper details data analysis and neural network design and envisions future developments.

Keywords: Weed classification · Deep Learning · CNN · PlantNet · Archaeological site conservation

1 Introduction

The Colosseum, a symbolic representation of ancient Roman engineering and architecture, is a testament to Italy's wealthy historical and cultural heritage. As a preeminent historical site with global recognition, the Parco holds profound import, drawing millions of tourists and scholars annually. Yet, within its grandeur resides a persistent challenge-the relentless proliferation of weeds imperilling the integrity and conservation of this cultural gem. Effectively managing weed proliferation in historical locales is pivotal for safeguarding their intrinsic worth and sustainable protection. Conventional manual methodologies for identifying and categorizing weeds have demonstrated protracted timelines and susceptibility to inaccuracies. With the evolution of technology, Artificial

G. L. Foresti et al. (Eds.): ICIAP 2023 Workshops, LNCS 14365, pp. 430–441, 2024.
https://doi.org/10.1007/978-3-031-51023-6_36

Intelligence (AI) emerges as a promising avenue for grappling with this difficulty, presenting accelerated and more precise weed classification procedures. To our knowledge, this approach has yet to be investigated in the existing literature.

Our work is part of the Decision Support System SyPEAH project for protecting the Parco through better forecasting, planning, and reducing the risk of deterioration of the condition of the materials. SyPEAH was developed in the framework of the project of monitoring and maintenance of the Parco as a preliminary tool for the effective activity of programmed conservation of cultural heritage with particular regard to the archaeological structures [4,5,18,19]. For some specific technical aspects of the project the Parco has involved the Italian Space Agency, INGV, ISPRA, as well as other universities such as the Milan Polytechnic and the Federico II of Naples, along with numerous professionals, including archaeologists, engineers, architects, geologists, and restorers. Within this ambitious project, obtaining an accurate classification of plant images is a challenging task, even when we are interested in a delimited area characterized by a contained number of species [17], such as the Forum Romanum.

In this context, epistemic uncertainty [9] leads to a much more significant overlap between classes than in other domains [14]. It is due to the speed and frequency with which the appearance of a plant varies through the different stages of growth and other dynamics. Observations of the same species can appear completely different, and, at the same time, images of specimens even of different genera can be indistinguishable. The spectral characteristics of plants change with growth and health conditions, but also with the climate and weather situation, precipitation and light, angle and intensity of the sun's rays [23]. The elevated level of uncertainty might also arise from the compact arrangement of vegetation within confined spatial boundaries. This juxtaposition and resultant shading can present complexities in image-based classification endeavours. The co-occurrence of intertwined plant structures may obscure the discernible attributes of individual specimens and impede the accurate demarcation of inter-species boundaries. Additionally, the resultant shading effects can engender fluctuations in image illumination, thereby introducing supplementary intricacies.

Almost all weed classification problems in the literature can be traced back to agricultural contexts with substantial differences compared to an archaeological park. On the one hand, there's the goal to identify the species that compete with the crops in the field, threatening the harvest (it is often a binary weed-crop classification, which involves a relatively small number of botanical species [16]); on the other hand, we have to understand the species that attack the integrity of historical materials (in a multi-classification in which weeds can belong to any species, with no limits related to edibility or human consumption [15]). The critical difference is in the substrate on which plants grow. The agricultural land is always more uniform and precisely prepared for growing crops. Agricultural datasets tend to be more unbalanced, with underrepresented classes and much greater pixel counts for the soil than weeds. The terrain can be highly varied in the archaeological context, with differences in soil composition, humidity, and other parameters. It can strongly influence the species present and require adaptation of classification models [1]. Weed management in agriculture always involves using control techniques (herbicides, mechanical or biological tools); conversely, in the archaeological context, the approach is more cautious and aimed at preserving the historical characteristics of the site. It requires an even more accurate identification of weeds, tending to avoid damaging the surrounding materials or soil. These factors influence the approach and priorities in weed classification and require specific adaptations in the archaeological context, different from most of the literature [8]. While post-germination recognition is a significant challenge in general, it is crucial in safeguarding archaeological structures.

Machine Learning (ML), particularly DL, provides techniques capable of addressing many of these difficulties [2]. Patterns can be learned automatically from data without needing to be explicitly pre-programmed to recognise specific plant features [13]. They exhibit the capacity to discern weeds based on intricate and subtle attributes that could elude the scrutiny of human experts. Moreover, this framework facilitates the transference of acquired knowledge from training to novel data instances. Thus, once suitably instructed, the model may proficiently discern weeds even in contexts and settings distinct from those it has learned from. Following the training phase, the neural network can extrapolate insights from previously unseen images, thereby automating the identification and surveillance of infestations. This system engenders a heightened efficiency and efficacy in the holistic management of the archaeological park, consequently curtailing the necessity for manual intervention and minimizing the duration required for species identification. Additionally, as elaborated subsequently, training neural networks to recognize multiple facets of interest from singular photographs, such as plants and their respective growth substrates, emerges as a comparatively straightforward endeavour.

In this paper, after focusing on the dataset collection and analysis, we describe how a preliminary model, a Deep Neural Network (DNN), is trained and evaluated to obtain strong performance in the classification of weeds in the context of the Parco. We present initial classification results to assess these ideas and future possible developments.

2 Dataset Collection and Analysis

Supervised DL excels at processing large amounts of labelled data, and consequently, their availability is critical to practical DNN training. Within limits, the greater the data, the more likely the trained model will identify weeds correctly [20] and generalize to new lighting or distribution situations not seen in the training phase. As anticipated, we use a set of 4,771 photographs of plants taken at the Park between January and December 2022. The temporal amplitude of the collection tries to intercept as many possible variations to allow the trained models to be more robust against them and to generalize [22]. The original images have a resolution of $5{,}456 \times 3{,}632$ pixels. For each image in the dataset, information is available regarding the species (in a set of 41 possible species) and the material on which the plant grows (in a group of 13 possibilities). Some examples are shown in Fig. 1, while the species list can be found later in Fig. 6. In Table 1, we find species with dataset tags organised by botanical taxonomy. Utilizing taxonomic analysis for weed species management has the potential to enhance the safeguarding of the archaeological ecosystem and its historical significance. This approach facilitates nuanced risk assessment, resource allocation, and intervention prioritization, enabling tailored strategies for distinct species, while concurrently facilitating the preservation of potentially endangered or uncommon species coexisting within the weed population.

Table 1. Species in our dataset grouped into botanical Orders and Families according to the APG IV genomic plant taxonomy [10] and class tag.

Order	Family	Genus-Species	Dataset tag
Polypodiales	Pteridaceae	Adiantum capillus-veneris	1
Sapindales	Simaroubaceae	Ailanthus altissima	35
Lamiales	Plantaginaceae	Antirrhinum majus	37
		Cymbalaria muralis	8
		Veronica hederifolia	30
	Lamiaceae	Clinodopium nepeta	6
		Micromeria graeca	15
		Salvia verbenaca	32
		Thymus vulgaris	39
	Oleaceae	Olea europaea	17
	Verbenaceae	Verbena officinalis	34
Brassicales	Brassicaceae	Brassica nigra	33
	Resedaceae	Reseda phyteuma	38
Capparales	Capparaceae	Capparis spinosa	2
Rosales	Cannabaceae	Celtis australis	3
	Moraceae	Ficus carica	10
	Urticaceae	Parietaria officinalis	19
	Rhamnaceae	Rhamnus alaternus	21
	Rosaceae	Rubus ulmifolius	22
		Sanguisorba minor	23
	Ulmaceae	Ulmus minor	28
Asterales	Asteraceae	Chamaemelum nobile	4
		Crepis bursifolia	7
		Sonchus tenerrimus	25
		Dittrichia viscosa	9
	Campanulaceae	Trachelium caeruleum	26
Caryophyllales	Amaranthaceae	Chenopodium album	5
Malpighiales	Euphorbiaceae	Euphorbia maculata	36
Ranunculales	Papaveraceae	Fumaria capreolata	11
Geraniales	Geraniaceae	Geranium rotundifolium	31
Apiales	Araliaceae	Hedera helix	12
Laurales	Lauraceae	Laurus nobilis	13
Malvales	Malvaceae	Malva sylvestris	14
Solanales	Solanaceae	Nicotiana glauca	16
Oxalidales	Oxalidaceae	Oxalis dillenii	18
Arecales	Arecaceae	Phoenix canariensis	20
Zygophyllales	Zygophyllaceae	Tribulus terrestris	27
Gentianales	Rubiaceae	Valantia muralis	41
Dipsacales	Viburnaceae	Viburnum tinus	40
Saxifragales	Crassulaceae	Sedum reflexum	24
		Umbilicus rupestris	29

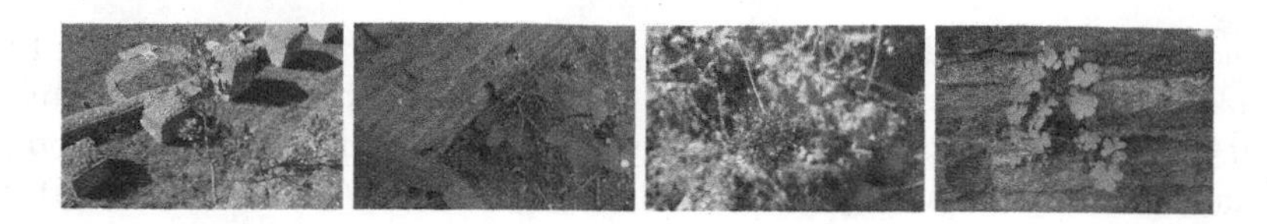

Fig. 1. Four examples from the dataset. From the left: Laurus nobilis, Rubus ulmifolius, Trachelium caeruleum and Oxalis dillenii. We note a wide variability concerning the shape of the plants, the angle, and the materials on which they reside.

The starting dataset was divided into two parts covering 80% (training set) and 20% (evaluation set), to be used respectively to train the models and to evaluate their performance. For training, two scaled-down versions of the photographs are used (one with a medium resolution of $1{,}024 \times 680$ pixels as input for the main models, and the other, with a low resolution of 512×340 for some preliminary experiments). Reducing the resolution simplifies image processing and diminishes the computational load when training and using models.

Figure 2 shows the subdivision into species of the two datasets (training and test). As can be seen, about half of the species have an average of 80 photos in the training and 20 in the test set (excluding Valantia muralis, which has fewer), up to a maximum of over 150 overall images in the two sets (e.g., Dittrichia viscosa). These aspects of the dataset make it appears well-balanced. In Fig. 3, a similar graph is reported concerning the materials on which the plants grow. In this case, the situation is more critical, with some extremely common substrates (e.g. conglomerate, mortar and brick) and others almost absent (e.g. plaster, lime).

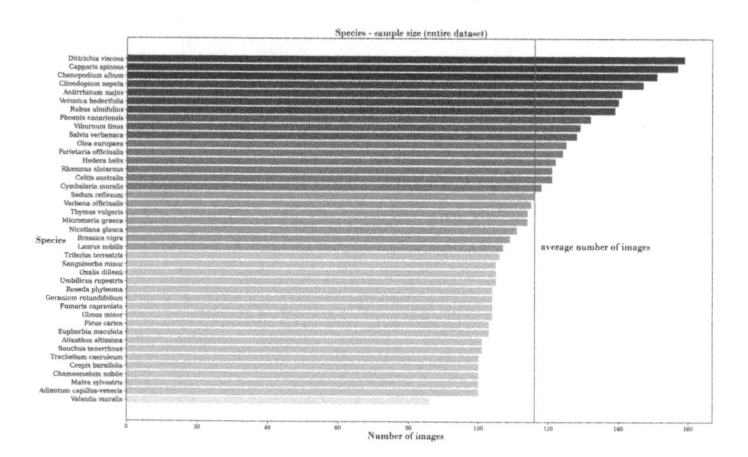

Fig. 2. The number of samples grouped by species. The average number of photos per species is shown with a vertical bar.

Our dataset has several images whose classification can be challenging even for an experienced human observer (as seen in Fig. 4). Many images have intricate

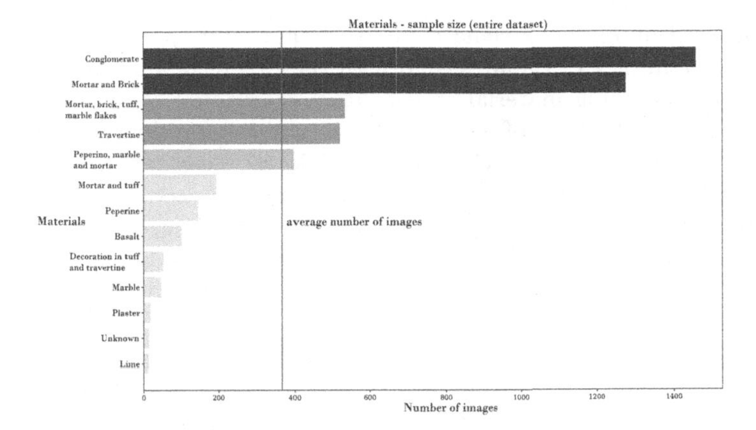

Fig. 3. The number of samples grouped by materials. Note the substantial imbalance between classes in this case. The average number of photos per material is shown with a vertical bar.

visual characteristics requiring detailed analysis and careful evaluation for proper classification.

Fig. 4. Example of similarity between Laurus nobilis and Rhamnus alaternus, even in different conditions.

3 Methods

The application of DL models offers advanced tools to identify and classify weed species accurately, making it possible to face the associated challenges, even in archaeological contexts [11]. This section explores strategies to overcome the limitations of traditional weed recognition systems, revisiting models in literature [24] and opening up new possibilities to improve the sustainability and efficiency of weed management processes-preservation of historical sites. Our DL model, called ArcheoWeedNet, has to classify both the weeds and the materials on which these plants grow within the archaeological park. If, on the one hand, this integrated approach aims to provide a complete understanding of the plant ecosystem present in the specific context of the study, on the other hand, it increases the complexities addressable with the ML techniques available in the literature [3]. Including material classification is essential to identify unwanted plant species and the type of environment in which they thrive. This aspect is crucial in the context of the archaeological park, as weeds can directly influence the conservation and integrity of the archaeological structure [8]. Recognizing the different materials allows for a deeper understanding of the environmental dynamics and the factors that favour the growth and spread of unwanted species. It contributes to more effective pest monitoring and allows for targeted management strategies [21].

At the base of our prototype, we consider a classic strategy of transfer learning. This technique involves using a pre-trained model on a large dataset as a starting point for training on a new dataset of interest. This approach allows us to address the size and limited representativeness of the available training data. As a dataset for pre-training, we consider Pl@ntNet-300k [9] (or more simply, from now on, PlantNet), a dataset developed by a group of researchers and developers at the artificial intelligence laboratory of CIRAD (Centre de coopération internationale en recherche agronomique pour le développement), in collaboration with the Tela Botanica network and INRIA (National Research Institute in Computer Science and Automation). It comprises 300,000 images from diverse sources and geographies and is a reference for classifying and recognising over 1000 plant species [9]. This diverse dataset offers a wide range of plant species, including flowering plants, trees, shrubs, grasses, and other forms of vegetation [9]. Employing fine-tuning using a PlantNet-trained ResNet network gave us more significant benefits than using an ImageNet-trained ResNet, as we will see later. This difference derives from the peculiar distributions of the images present in the respective datasets. PlantNet, as mentioned, represents a dataset specifically focused on plants. Training a ResNet network on PlantNet allows the model to understand better plant-specific visual characteristics, such as shape, texture, and leaf characteristics. It results in a more remarkable ability of the model to discriminate and classify weeds in the specific context of the problem [9]. On the other hand, ImageNet is a more general dataset encompassing a wide range of categories, including animals, objects, people, and other entities [6]. While a ResNet trained on ImageNet can learn visual characteristics of a general nature [12], they may need to be more specialised in the specific features of plants. As a result, it shows less ability to discriminate and less precision in classifying weeds within a particular context. As we will see later, ArcheoWeedNet already achieves an extremely high accuracy, higher than 75%, considering all classes, and higher than 90%, focusing on the most common species.

The adapted ResNet-50 backbone has a final fully connected layer with 1024 units, and the model includes separate linear layers for species and materials prediction, each having two hidden layers (512 and 128 units) with ReLU activation. These new blocks enable the model to distinguish the specific characteristics of weeds and archaeological materials. We trained the model using a sum of a cross-entropy loss on the species and a separate cross-entropy loss on the materials. Optimization is performed with Adam, using a learning rate of 0.001 for 15 epochs and a batch size of 32. The general scheme of ArcheoWeedNet is illustrated in Fig. 5.

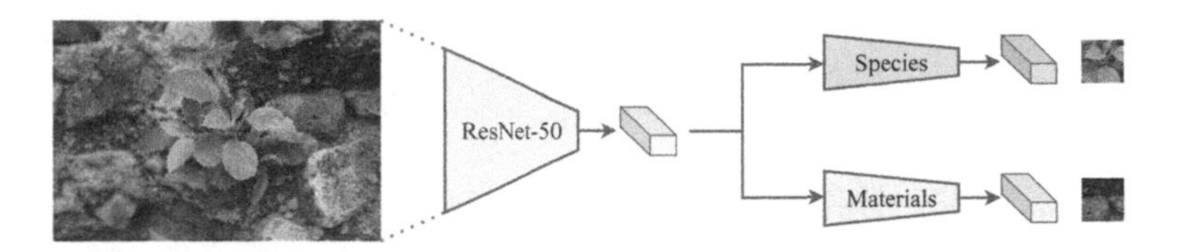

Fig. 5. Scheme of ArcheoWeedNet, trained to recognize the plant's species and the materials on which it grows, starting from a photo taken by an operator or a camera.

The architecture thus defined allows to address both tasks simultaneously, improving the ability of classification in the context of the archaeological park. The accuracy of our solution depends both on the representativeness and quality of the weed dataset and on the correct configuration of the initial parameters of the model. The appropriate selection of evaluation metrics and the iterative process of model optimization led to accurate and reliable results for most pest classes in the context of the archaeological park.

4 Results

The trained model simultaneously predicts botanical species (41 possible classes) and archaeological materials (13 possible classes). Considering the tremendous seasonal variability of the plants in the images and the imbalance of the materials, as previously described, we obtain a variable performance between the different classes. As said, we consider two distinct scenarios as a starting point: a pre-trained ResNet-50 on ImageNet, used as a comparison, and a pre-trained ResNet-50 on PlantNet. The results obtained on the original set of classes and without data augmentation showed that the network pre-trained on PlantNet exceeded the network pre-trained on ImageNet by more than 15% points. In Fig. 6, we present the confusion matrix relating to the test set of this second version of the model, in which each row represents a species, and the columns describe the distribution of the model's predictions for that species. The average accuracy of the model is 76.65% and, as evidenced by the matrix, the accuracy is balanced between the different species, as well as the errors (with some exceptions, for example, several confusions between Celtis australis and Rubus ulmifolius).

The situation of the materials, as highlighted in the confusion matrix shown in Fig. 7, presents criticalities. As expected, due to the imbalance of the photos, the average accuracy of 54.76% is split between some materials that are recognized very well (such as mortar and brick) and others that are almost completely ignored due to the lack of examples in the training set. This point will be the

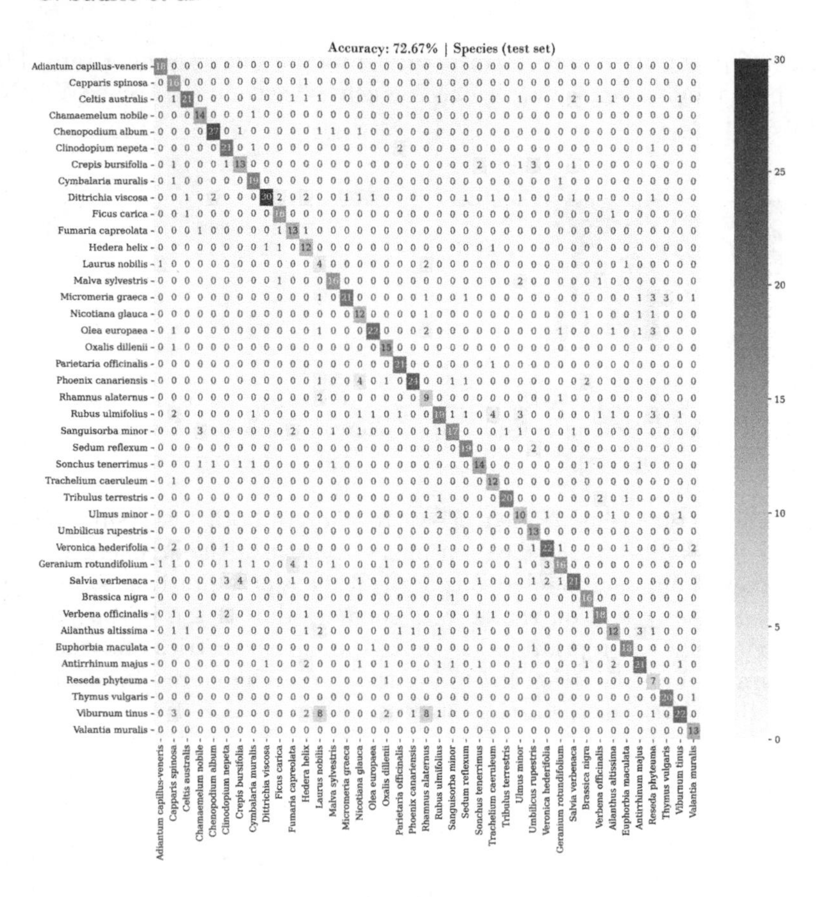

Fig. 6. Species confusion matrix (test set).

subject of further investigations in the future, in particular by increasing the dataset available regarding the materials.

Table 2. The accuracies achieved with the two versions of ArcheoWeedNet, pre-trained on ImageNet and Pl@ntNet-300k.

	Materials (accuracy)	Species (accuracy)
ImageNet pre-training	56.44%	60.42%
Pl@ntNet-300k pre-training	54.76%	76.65%

Table 2 shows the accuracies obtained with the two versions of ArcheoWeedNet: pre-trained on ImageNet and pre-trained on PlantNet, as described in the section on the methods.

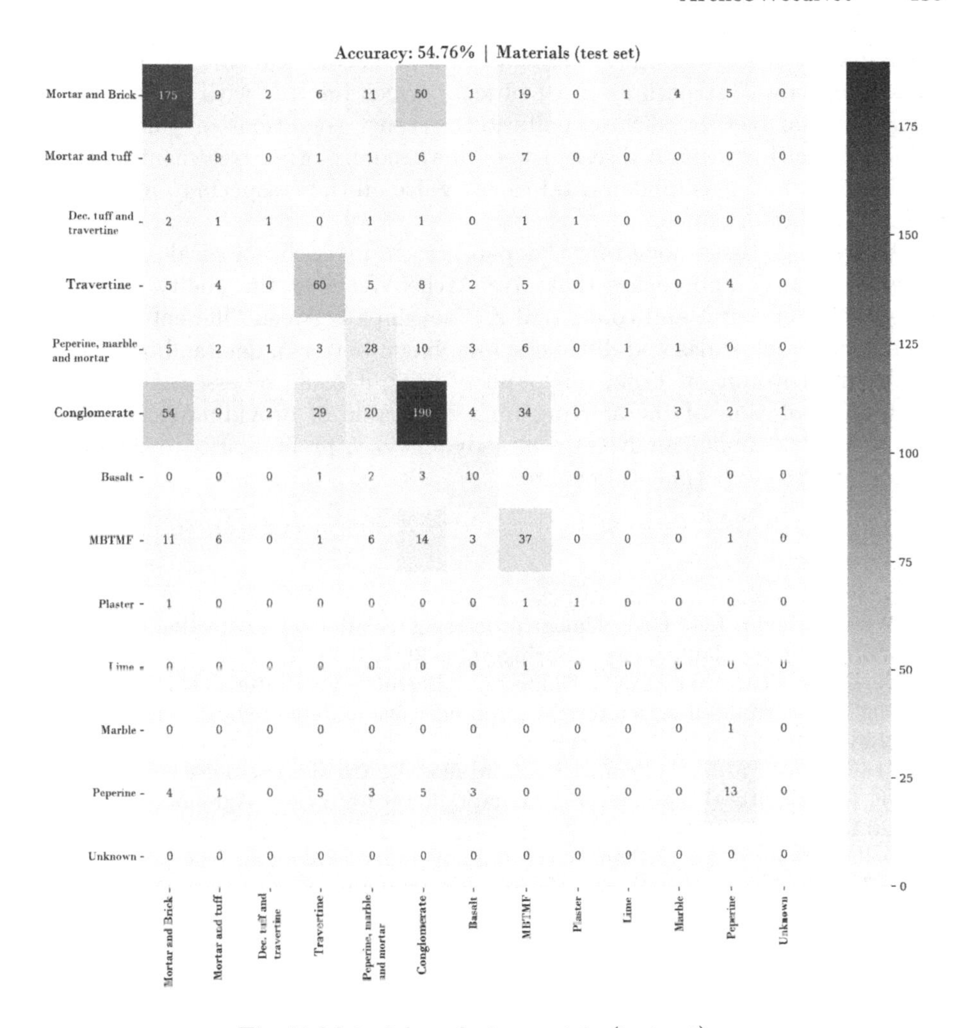

Fig. 7. Material confusion matrix (test set).

5 Conclusions

This paper presented preliminary experiments on a novel dataset collected to detect and recognize weed infestations in the Parco. On the current dataset version, transfer learning strategies from neural models pre-trained on botanic datasets already show good performance on most species. Future work will consider significantly increasing the size of the dataset (especially concerning seasonality aspects) and designing more sophisticated models going beyond standard Convolutional Neural Networks (CNNs), such as Vision Transformers (ViT) [7]. In particular, we plan to develop ad-hoc models to address the high variability of the dataset. Almost all species undergo significant variations during the different seasons (leaf shape, flowering, and others), and the features of the surrounding habitat are also fundamental for plant life, such as the type and conditions of soil and exposure to the sun. In the context of the archaeological park, there are artefacts, such as wall structures or art objects, which interact with plants. Finally, environmental factors, such as pollution, weather conditions or parasites, can influence health status. A diseased weed may require other treatments, perhaps less invasive, and it is fundamental to recognize all these aspects to improve the recognition performance.

Addressing these long-range dependencies in archaeological park plant imagery requires approaches that give ArcheoWeedNet the ability to capture the spatial, temporal and contextual relationships between different image features at different scales and distances, enabling a better understanding of plants and their environment. From this point of view, it could be essential to exploit the high resolution of the starting images capable of providing rich and precious details for image analysis, as already analyzed preliminarily in the dataset section and the experimental results.

References

1. Arnal Barbedo, J.G.: Digital image processing techniques for detecting, quantifying and classifying plant diseases. Springerplus **2**(1), 1–12 (2013)
2. Benos, L., Tagarakis, A.C., Dolias, G., Berruto, R., Kateris, D., Bochtis, D.: Machine learning in agriculture: a comprehensive updated review. Sensors **21**(11), 3758 (2021)
3. Cravero, A., Pardo, S., Sepúlveda, S., Muñoz, L.: Challenges to use machine learning in agricultural big data: a systematic literature review. Agronomy **12**(3), 748 (2022)
4. Della Giovampaola, I.: Piano sostenibile di tutela e valorizzazione del patrimonio archeologico e di educazione continua al patrimonio culturale: SyPEAH (A platform Systemfor the Protection and Education of Archaeological Heritage). Bullettino della Commissione Archeologica Comunale CXII, pp. 61–76 (2021)
5. Della Giovampaola, I.: SyPEAH: the WebAPP system for protection and education to archaeological heritage in the parco archeologico del colosseo. Geosciences **11**(6), 246 (2021)
6. Deng, J., Dong, W., Socher, R., Li, L.J., Li, K., Fei-Fei, L.: ImageNet: a large-scale hierarchical image database. In: 2009 IEEE Conference on Computer Vision and Pattern Recognition, pp. 248–255. IEEE (2009)
7. Dosovitskiy, A., et al.: An image is worth 16×16 words: transformers for image recognition at scale. In: International Conference on Learning Representations (2020)
8. Fracchiolla, M., Lasorella, C., Cazzato, E., Vurro, M.: Weeds in non-agricultural areas: how to evaluate the impact? A preliminary case study in archaeological sites. Agronomy **12**(5), 1079 (2022)
9. Garcin, C., et al.: Pl@ntNet-300K: a plant image dataset with high label ambiguity and a long-tailed distribution. In: NeurIPS 2021–35th Conference on Neural Information Processing Systems (2021)
10. Angiosperm Phylogeny Group: An update of the angiosperm phylogeny group classification for the orders and families of flowering plants: APG iv. Bot. J. Linn. Soc. **181**(1), 1–20 (2016)

11. Hasan, A.M., Sohel, F., Diepeveen, D., Laga, H., Jones, M.G.: A survey of deep learning techniques for weed detection from images. Comput. Electron. Agric. **184**, 106067 (2021)
12. He, K., Zhang, X., Ren, S., Sun, J.: Deep residual learning for image recognition. In: Proceedings of the IEEE Conference on Computer Vision and Pattern Recognition, pp. 770–778 (2016)
13. LeCun, Y., Bengio, Y., Hinton, G.: Deep learning. Nature **521**(7553), 436–444 (2015)
14. Nilsback, M.E., Zisserman, A.: Automated flower classification over a large number of classes. In: 2008 Sixth Indian Conference on Computer Vision, Graphics and Image Processing, pp. 722–729. IEEE (2008)
15. Olsen, A., et al.: DeepWeeds: a multiclass weed species image dataset for deep learning. Sci. Rep. **9**(1), 2058 (2019)
16. Panda, B., Mishra, M.K., Mishra, B.S.P., Tiwari, A.K.: An extensive review on crop/weed classification models. In: Web Intelligence, pp. 1–16. No. Preprint, IOS Press (2023)
17. Ricotta, C., Grapow, L.C., Avena, G., Blasi, C.: Topological analysis of the spatial distribution of plant species richness across the city of Rome (Italy) with the echelon approach. Landsc. Urban Plan. **57**(2), 69–76 (2001)
18. Russo, A., Della Giovampaola, I.: Il monitoraggio e la manutenzione delle aree archeologiche. Il piano per il futuro del Parco archeologico del Colosseo, pp. 13–31. "L'Erma" di Bretschneider (2020)
19. Russo, A., Giovampaola, I.D., Spizzichino, D., Leoni, G., Coletta, A., Virelli, M.: The project of parco archeologico del colosseo and the Italian network of archaeological parks: from satellite monitoring to conservation and preventive maintenance policies. In: El-Qady, G.M., Margottini, C. (eds.) Sustainable Conservation of UNESCO and Other Heritage Sites Through Proactive Geosciences, pp. 659–678. Springer, Cham(2023). https://doi.org/10.1007/978-3-031-13810-2_34
20. Sudars, K., Jasko, J., Namatevs, I., Ozola, L., Badaukis, N.: Dataset of annotated food crops and weed images for robotic computer vision control. Data Brief **31**, 105833 (2020)
21. Sun, Y., Liu, Y., Wang, G., Zhang, H., et al.: Deep learning for plant identification in natural environment. Comput. Intell. Neurosci. **2017**, 7361042 (2017)
22. Therrien, R., Doyle, S.: Role of training data variability on classifier performance and generalizability. In: Medical Imaging 2018: Digital Pathology, vol. 10581, pp. 58–70. SPIE (2018)
23. Wäldchen, J., Mäder, P.: Plant species identification using computer vision techniques: a systematic literature review. Arch. Comput. Methods Eng. **25**, 507–543 (2018)
24. Wu, Z., Chen, Y., Zhao, B., Kang, X., Ding, Y.: Review of weed detection methods based on computer vision. Sensors **21**(11), 3647 (2021)

Automatic Alignment of Multi-scale Aerial and Underwater Photogrammetric Point Clouds: A Case Study in the Maldivian Coral Reef

Federica Di Lauro[1(✉)], Luca Fallati[2], Simone Fontana[3], Alessandra Savini[2], and Domenico G. Sorrenti[1]

[1] Department Informatica, Sistemistica e Communicazione, Milano, Italy
{federica.dilauro,domenico.sorrenti}@unimib.it
[2] Department Scienze dell'Ambiente e della Terra, Lombardy, Italy
{luca.fallati,alessandra.savini}@unimib.it
[3] School of Law, Università degli Studi di Milano, Bicocca, Italy
simone.fontana@unimib.it

Abstract. The research question that the paper investigates is whether the usage of state of the art algorithms for point clouds registration solves the problem of multi-scale vision-based point clouds registration in mixed aerial and underwater environments. This paper reports very preliminary results on the data we have been able to procure, in the context of a coral reef restoration project nearby Magoodhoo Island (Maldives). The results obtained by exploiting state of the art algorithms are promising, considering that those data presents hard samples, in particular for their multi-scale nature (noise in captured 3D points increases with depth). However, further investigation on larger data-sets is needed to confirm the overall applicability of the current algorithms to this problem.

1 Introduction

Coral reefs are essential to tropical marine ecosystems, providing various benefits, such as being a habitat for a quarter of all marine species, protecting shorelines, reducing flood risks, producing fisheries, and regulating climate [4,22]. Unfortunately, coral reefs globally have declined significantly due to human activity and climate change impacts [15,16]. In response, marine scientists and ecologists have started coral restoration activities to improve the resilience of vulnerable reefs. Researchers are particularly concerned about optimizing the site selection of restoration sites, and measuring coral growth. However, quantitative monitoring during the post-outplanting and post-nursery stages is often limited, and there are concerns about a lack of standardized methodology for assessing the success of restoration programs [10,20]. Gaining a complete understanding of the alterations that occur pre- and post-transplanting each year is crucial. Doing so will help determine the success of the restoration program, facilitate adaptive management, and enhance decision-making by providing specific spatial

G. L. Foresti et al. (Eds.): ICIAP 2023 Workshops, LNCS 14365, pp. 442–453, 2024.
https://doi.org/10.1007/978-3-031-51023-6_37

information. Some historical monitoring methods include single colony growth measurements, point intercept transects, and benthic quadrants. An increasingly popular method for quantitatively measuring coral growth changes, from above and below the water, relies on the computer vision technique *Structure from Motion* (SfM) [8,9,18,19]. Data can be collected through snorkeling or diving using an underwater camera, or by an aerial drone flying above the sea surface. Integrating technology in ecological research requires a comprehensive understanding of its potential, to enhance predictions of biological processes. SfM is a valuable tool in that it generates detailed models of landscapes and organisms from images, enabling extensive analysis of habitat structure [3,10]. Moreover, the combination of various multi-scale data (both above and underwater) can aid in monitoring the restoration site across multiple levels: from the individual coral fragments level to the entire restoration area, enabling us to comprehend and evaluate the spatial patterns that could affect the outcome of restoration.

In this work, we investigate the application of registration algorithms to the problem of multi-scale point cloud registration, that is, finding the best alignment of two point clouds that sport a different level of detail. A point cloud is an unordered list of 3D points, all referred to the same reference frame, possibly complemented by an intensity channel or 3 color channels, depending on the sensor used for gathering the data. Different levels of detail relate to the spacing between the points making up the point cloud. This characteristic can typically be obtained by the usage of different sensors or by the same sensor positioned at different distances from the scene. As an example, images from a stereo rig with high-resolution cameras, held by a diver in a few meters deep water, can easily allow reconstructing points from the single stripes of *Posidonia Oceanica*. The very same rig, mounted on a drone flying at 20m high above water, would only allow to roughly reconstruct the 3D envelope of the entire *Posidonia Oceanica* bush. Given that registration algorithms rely on matching points in the point clouds to be registered, a different level of detail entails difficulties in finding such matches. The registration aims to build a larger map of the environment, w.r.t. what can be observed with just one single activation of the sensor. In this work, we specifically focus on aligning point clouds created with monocular SfM technique, using images collected by an Unmanned Aerial Vehicle (UAV), and by a human moving underwater in a natural environment.

Point cloud registration can be classified as "local" if there is an initial guess about the alignment, otherwise it is called"global". In our case, since only the UAV point cloud is georeferenced and we have no spatial information about the pose of the observer for the underwater point clouds, we need to solve a global registration problem.

Differently from point clouds built with LiDARs, vision-based point clouds are affected by a not stationary noise; the noise actually grows quadratically [17] with the depth, i.e. the distance between the sensor and the point in the scene, and is reduced by an increase in the baseline, i.e. the distance in space between the two sensors' acquisitions. In the monocular approach the baseline depends on the observer motion. Should one consider modelling such uncertainty on the reconstructed points, such modeling is made difficult because the observer motion is unknown, altogether with the 3D reconstruction, which is the SfM problem. Because of the dependency on depth, the uncertainty would also be specific to each reconstructed 3D point. State-of-the-art point cloud registration algorithms, to the best of our knowledge, do not explicitly take into account the uncertainty affecting the points in a pair of matched points in the two point clouds to register. One might argue that an implicit and partial handling of such uncertainty has been taken into consideration by the proposal in [1], for data association only, i.e. during the process of finding correspondences between the two clouds. However, this approach only deals with local registration and thus is not suitable for our goals. From these considerations, we drew our research question: are global state-of-the-art algorithms capable to deal with point clouds coming from vision-based 3D reconstruction? In the application domain at hand there is another aspect, that makes this issue more delicate: the observer is at greatly different distances from the 3D points, which originates the differences in the level of detail mentioned before.

This work is structured as follows: in Sect. 2, we describe the data acquisition and point cloud generation methods, in Sect. 3 we present the method used to derive the ground truth pose for the evaluation of the algorithms, in Sect. 4 we present the automatic registration algorithms used and the results obtained, and finally we draw our conclusions in Sect. 5.

2 Data Acquisition and Monocular Photogrammetric Point Clouds Generation

The data we used were acquired in Magoodhoo, an island in the southern part of Faafu Atoll (Republic of the Maldives). This island is the base-station of the MaRHE Center, a research outpost of the University of Milano - Bicocca, active in the country since 2011.

In the last decade, the researchers of MaRHE have conducted long term monitoring activities on the coral reef environments around the island, and recently they have started an ambitious coral reef restoration program. Part of this program involves the transplantation of coral fragments on portions of the reef flat on existing natural structures and on metal frames called "spiders". A view of the area where the data was acquired is shown in Fig. 1. To assess the restoration efforts' success, aerial drone and underwater photogrammetry surveys were performed on the restored area.

The aerial photos were acquired with a DJI Phantom 4 quad-copter equipped with one 12Mpx camera, flying at 22.9m over the sea surface following parallel transects and maintaining high frontal and lateral overlap levels. In total, we acquired 77 photos, and the area covered by the flight is 1370 square meters. Moreover, the drone is equipped with a GPS system, which allowed a rough geolocalization of the pictures (non RTK-level GPS for the position, and gyros for the orientation).

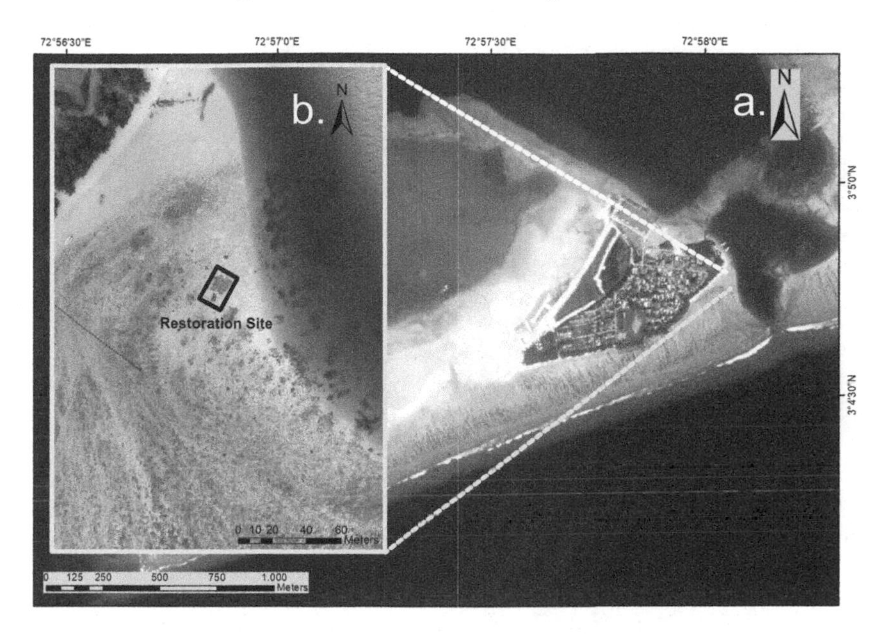

Fig. 1. Study area: a. represents Magoodhoo island (Faafu Atoll, Rep. of the Maldives). b. represent a zoom on the study area on the eastern part of the island, facing a small lagoon. The "restoration site" is on the shallow reef flat close to the shore.

The underwater photos were acquired in snorkeling with a Canon PowerShot G7 X camera (20Mpx) lodged in an underwater housing. The survey area spans approximately 87 square meters, and a total of 813 images were captured while swimming 1 m above the seafloor. To accurately capture the complexity of the area, photos were taken both orthogonal and at a 45° angle to the seafloor.

The areas surveyed are shown, together with each camera location, in Fig. 2.

Each of two sets of images was processed using Agisoft MetaShape [2], following its standard workflow: photo alignment, sparse cloud building and finally dense cloud building, each resulting in a point cloud. The clouds are composed of 3930461 and 136016717 points, respectively for the UAV and the underwater cloud. Both point clouds were then used as input to the registration pipeline.

Since the point clouds were created using a monocular photogrammetric technique, they need to be properly scaled. The UAV's point cloud was scaled according to the GPS positions. The underwater point cloud was scaled by placing objects of known size (a metric tape) underwater and using them as reference.

3 Manual Registration and Its Use as Ground Truth

To properly evaluate an automatic registration pipeline, we need to obtain the ground truth transformation between the two clouds. Since no ground truth is available for our case study, we decided to perform a manual alignment to use as reference. We first roughly align the clouds by manually selecting corresponding

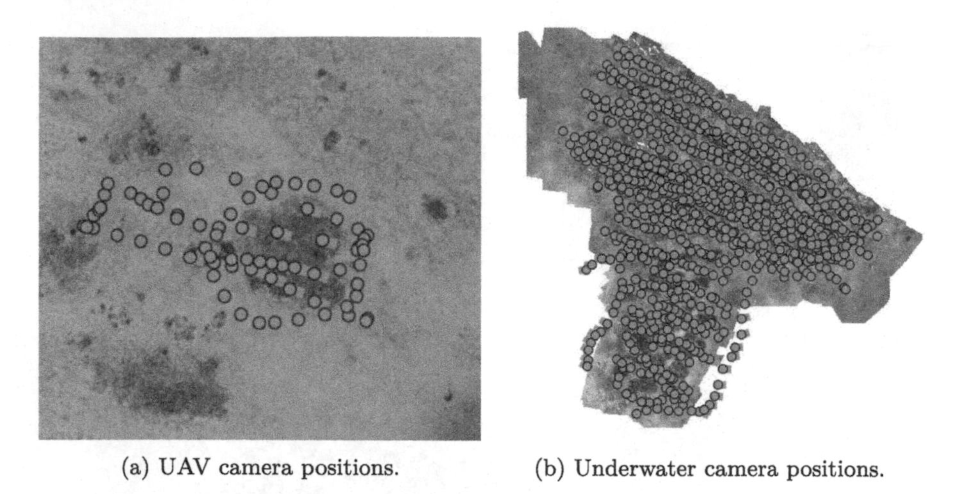

(a) UAV camera positions. (b) Underwater camera positions.

Fig. 2. Camera locations, shown as grey circles, overlapped on their respective point cloud. Figures exported from Agisoft MetaShape. (Color figure online)

points and estimating the transformation that matches these pairs, and then applying the well-known local registration algorithm ICP [5,6,25].

We used the CloudCompare software [7] to estimate an initial roto-translation between the source and target point cloud: using the "Align (point pairs picking)" tool we selected 3 pairs of correspondences in the clouds, and estimated the transformation which best aligns them. Note that this step requires an expert to identify the correspondences. Next, we applied the ICP algorithm implemented in CloudCompare, to refine the coarse initial guess provided by the semi-manual alignment. We set the parameters as follows:

- Final expected overlap between the two clouds: 100%. Since the underwater point cloud is a sub-map of the drone point cloud we expect a full overlap between the clouds.
- Final error to reach convergence: $1e - 05$. This small value is to allow a large number of iterations, since we aim to find a very accurate transform.
- Random sampling limit: 50000. Given that the photogrammetric point clouds are very dense, the random sampling allows to improve the computation times.

The results obtained with the manual pipeline are shown in Fig. 3. As experts, we can confirm the attainment of a high-quality result, suitable as ground truth for the evaluation of algorithms.

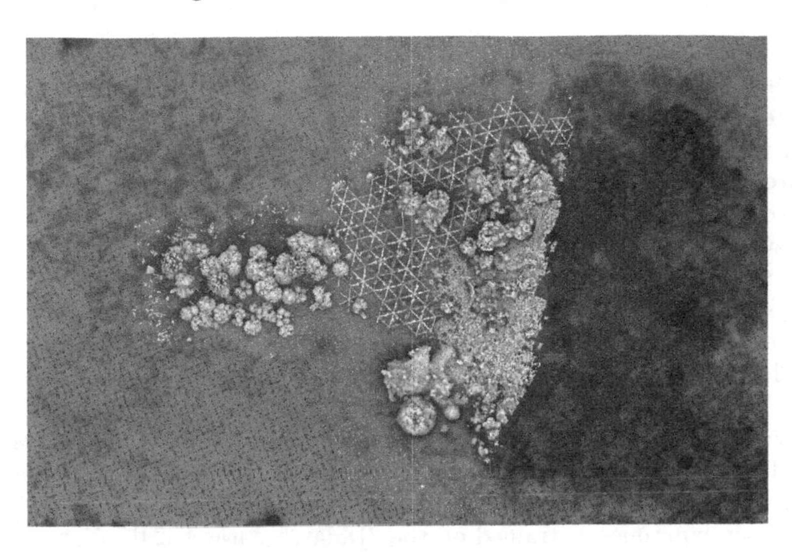

Fig. 3. Results obtained with the manual registration pipeline. The UAV point cloud is shown in RGB color, the underwater point cloud is shown in white, with the CloudCompare ShadeVis shader applied, to highlight the details. The background of CloudCompare is visible where the darker spots are. The triangle-shaped structures are the so-called "spiders".

4 The Automatic Registration Pipeline

To select the global registration algorithms, we take a previous comparison [13] as reference. This comparison exploits the Point Cloud Registration Benchmark [12], which is a collection of registration problems under different environments, including plains, woods and a planetary emulation environment, and also introduces a quality metric to evaluate the results. This metric takes into account both rotation and translation errors together, and thus allows a direct comparison between algorithms. In this comparison, FPFH [21], a purely geometric hand-crafted descriptor, and 3DSmoothNet [14], a deep learning based descriptor, showed the best performance among different feature extractors, especially in natural environments. For this reason, even though the sensors of the point clouds used in the benchmark were not from photogrammetry, we decided to rely on them for a comparison on our data. However, these are only feature extractors. Therefore, we also need a technique that estimates a roto-translation from a set of correspondences. Among the different available techniques, TEASER++ [23], according to [12,13], showed the best performance, especially in the presence of the large number of outliers we expect in natural environments, which usually have low structure and produce noisy point clouds. An approach based on RANSAC [11] came close to the results of TEASER++, and thus we also decided to compare these two roto-translation estimation techniques, to investigate how much of an impact the roto-translation step estimate has, compared to the feature extractor step.

The preprocessing and FPFH features extraction have been performed using the Open3D library [26]. First, the clouds are downsampled by applying a voxel grid filter with a voxel leaf size equal to 10cm. Then, we estimate the normals, needed for computing the FPFH features, with a radius of 20cm, and finally estimate the FPFH features with a radius of 50cm. These parameters were chosen to try to capture the details of the coral reef at a sufficient level to characterize the relevant spots to align.

To extract the features using the 3DSmoothNet neural network, some preprocessing steps are needed. We first apply a voxel grid downsampling filter with a voxel leaf size equal to 2cm to the underwater point cloud, to reduce the computational burden, while maintaining a high level of detail in the cloud. The neural network requires a smoothed density value representation of the point clouds, which we calculated using the parameters supplied by the authors. Finally, the point clouds are fed to the neural network, generating a feature vector of 64 elements. The authors provide the pre-trained weights of the network, which we used for our experiments, trained on the 3DMatch dataset [24].

We select 10000 random points with their associated descriptor to find the correspondences between the two clouds, using the euclidean distance between the feature vectors. Finally, we apply TEASER++ or RANSAC to find the rigid transformation.

We then quantify the quality of the registration by measuring the error between the source point cloud in the estimated ground truth pose G, and the source point cloud in pose T, determined using the chosen algorithm. A first error measure, proposed in the Point Cloud Registration Benchmark [12], is the weighted distance between the same points of point cloud S, composed of n points s_i, in the two poses, and is defined in Eq. 1.

$$WeigthedDistance = \frac{\sum_i \frac{\|G \cdot s_i - T \cdot s_i\|}{\|s_i - \overline{S}\|}}{n} \tag{1}$$

where $\overline{S}$ is the centroid of S, $G \cdot s_i$ and $T \cdot s_i$ are the application of the ground truth and the T roto-translations to s_i, and $\|x\|$ is the L_2 norm of the vector x. The weighted distance represents the average distance between the ground truth and the estimated positions of a point, normalized by the average distance to the centroid of the point cloud. While this measure is useful to provide a good comparison between algorithms, it is a pure number and does not provide an intuitive grasp of the absolute quality. For this reason, we also show the average unweighted distance (in meters) between the two clouds, shown in Eq. 2

$$AverageDistance = \sum_i \frac{\|G \cdot s_i - T \cdot s_i\|}{n} \quad (2)$$

We made the code for registration and evaluation of the results available at the following repository: https://github.com/iralabdisco/cvemp-2023

Table 1. Errors for the different registration algorithms used.

Registration algorithm	Weighted distance	Avg. distance [m]
TEASER++ with 3DSmoothNet	0.025	0.068
TEASER++ with FPFH	8.428	20.259
RANSAC with 3DSmoothNet	0.025	0.085
RANSAC with FPFH	6.609	15.991

The results of the different algorithms using the presented metrics are shown in Table 1. We also provide a visual representation of the results in Fig. 4 and Fig. 5. Both TEASER++ and RANSAC obtained very good results while using 3DSmoothNet as feature extractor, with TEASER++ slightly performing better than RANSAC. However, they both completely failed when using FPFH. This agrees with our previous experience in a quite extensive set of scenarios, where in some scenarios 3DSmoothNet was performing better then FPFH. We can affirm that in our case the feature extractor has a much stronger impact on the result with respect to the roto-translation estimation technique. When providing a significant descriptor of the points, both RANSAC and TEASER++ were able to estimate the correct transformation, while the FPFH features were not useful in identifying the correct reef patch. It can be seen in Fig. 5 that using FPFH features resulted in the alignment of the underwater point cloud to the wrong reef patches.

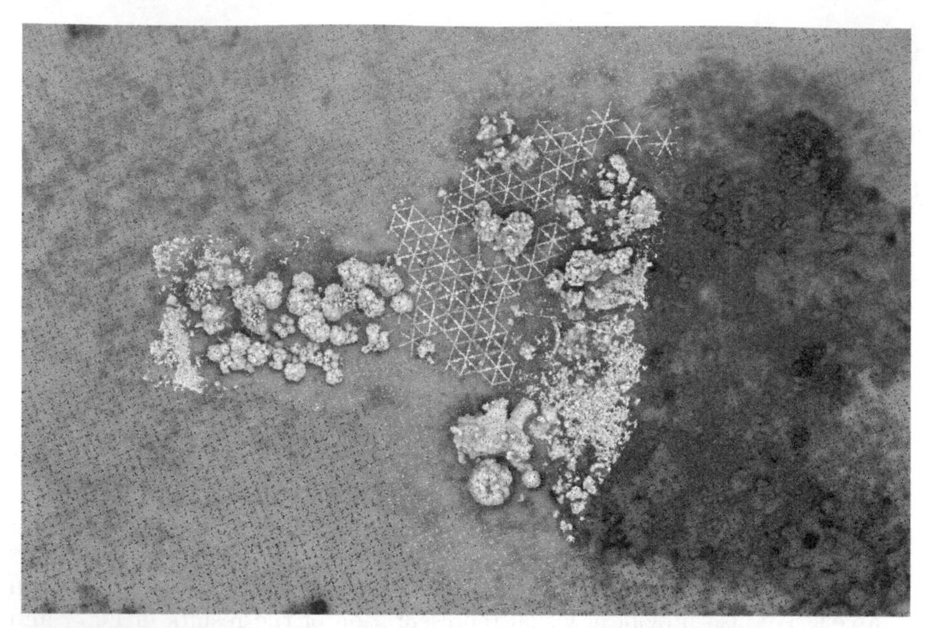

(a) Results using 3DSmoothNet features and TEASER++

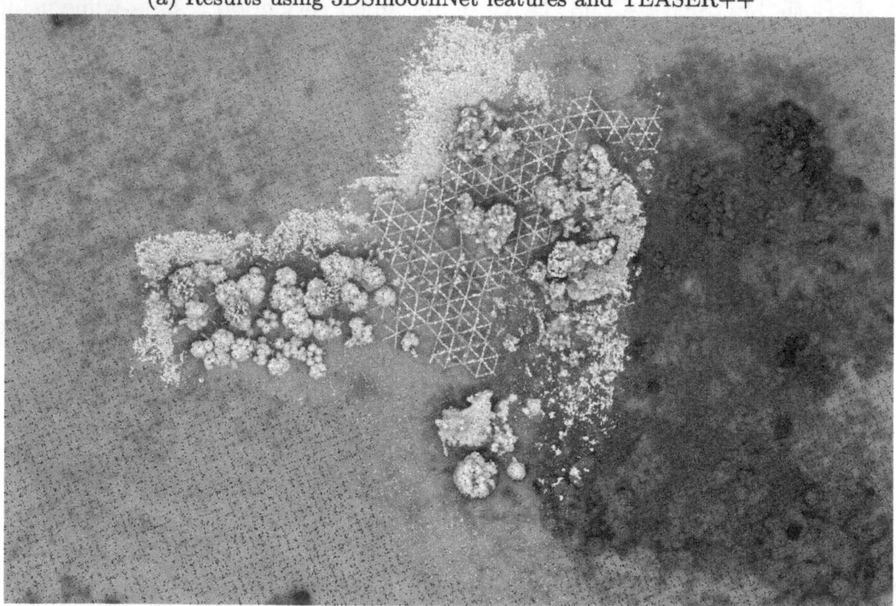

(b) Results using 3DSmoothNet features and RANSAC based approach

Fig. 4. Registration results using 3DSmoothNet as feature extractor. The UAV point cloud is shown in color, the underwater point cloud is shown in white.

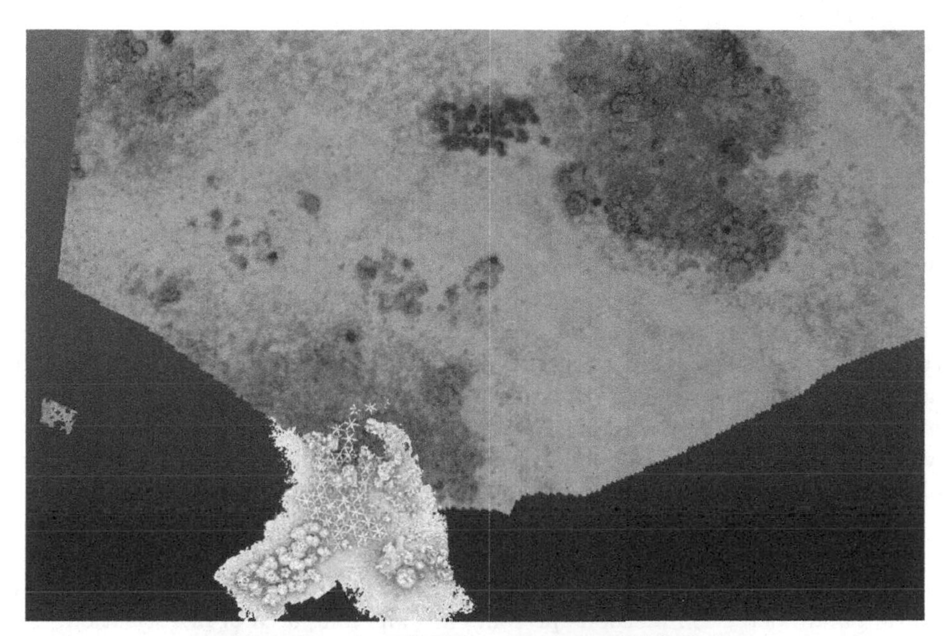

(a) Results using FPFH features and TEASER++

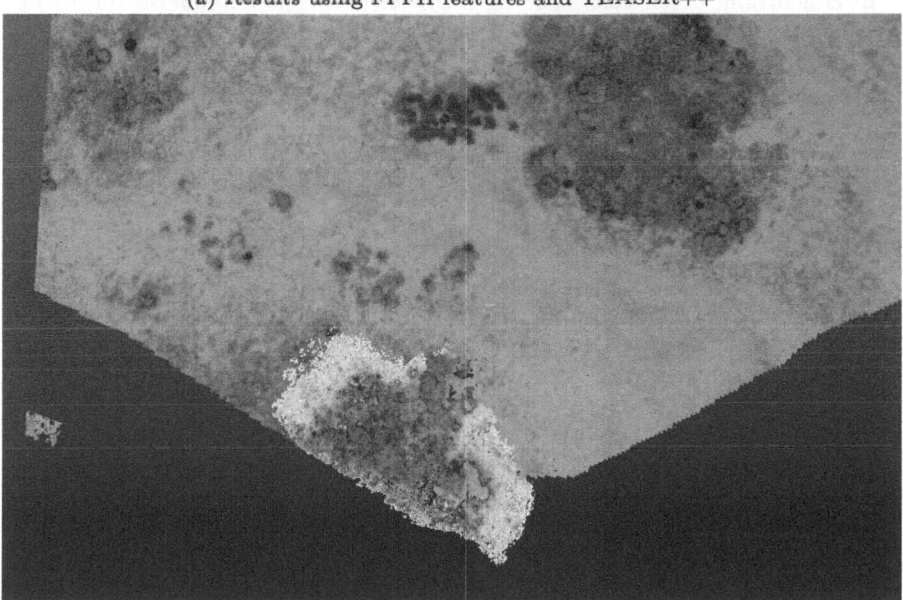

(b) Results using FPFH features and RANSAC based approach

Fig. 5. Registration results using FPFH as feature extractor. The UAV point cloud is shown in color, the underwater point cloud is shown in white.

5 Conclusions

In this paper, we analyzed the applicability of state-of-the-art registration algorithms in a specific case study in a natural environment. In particular we tried to align point clouds, coming from a monocular photogrammetric pipeline, which may vary by level of detail. Our findings emphasize the greater relevance of the feature extraction step over the roto-translation estimation technique, highlighting its crucial role in accurate point cloud alignment.

Despite the progress made, several limitations persist. While we obtained a very good result using global registration, these approaches can fail and thus the result must be manually checked. Moreover, evaluation of registration in aerial-underwater still requires a manual ground truth estimation since the usual methods, such as using an RTK GPS or a total station, cannot be used underwater. The scarcity of data available for evaluating algorithms, and the absence of a ground truth, make it difficult to draw definitive conclusions.

Looking ahead, there are promising avenues for research. One direction involves conducting a more comprehensive comparison of different approaches, thereby paving the way for improved techniques. Moreover, the pursuit of automatic failure detection methods presents a compelling area of investigation.

In conclusion, given that the alignment of point clouds has proven to be a valuable tool in advancing our understanding of complex ecosystems, such as coral reefs, we are pleased with the outcomes of this tentative analysis, as it serves as a bridge between computer science and environmental sciences.

References

1. Agamennoni, G., Fontana, S., Siegwart, R.Y., Sorrenti, D.G.: Point clouds registration with probabilistic data association. In: 2016 IEEE/RSJ International Conference on Intelligent Robots and Systems (IROS), pp. 4092–4098 (2016). https://doi.org/10.1109/IROS.2016.7759602
2. Agisoft MetaShape, Version 2.0.14. http://www.agisoft.com, software; (Accessed 31 July 2023)
3. Bayley, D.T.I., Mogg, A.O.M.: A protocol for the large-scale analysis of reefs using structure from motion photogrammetry. Methods Ecol. Evolut. **11**(11), 1410–1420 (2020). https://doi.org/10.1111/2041-210X.13476, https://besjournals.onlinelibrary.wiley.com/doi/abs/10.1111/2041-210X.13476
4. Bellwood, D.R., Hughes, T.P., Folke, C., Nyström, M.: Confronting the coral reef crisis. Nature **429**(6994), 827–833 (2004). https://doi.org/10.1038/nature02691,
5. Besl, P.J., McKay, N.D.: Method for registration of 3-d shapes. In: Sensor fusion IV: Control Paradigms and Data Structures, vol. 1611, pp. 586–606. SPIE (1992)
6. Chen, Y., Medioni, G.: Object modelling by registration of multiple range images. Image Vis. Comput. **10**(3), 145–155 (1992)
7. CloudCompare, Version 2.12.4. http://www.cloudcompare.org, software; (Accessed 31 July 2023)
8. Curtis, J.S., Galvan, J.W., Primo, A., Osenberg, C.W., Stier, A.C.: 3d photogrammetry improves measurement of growth and biodiversity patterns in branching corals. Coral Reefs **42**(3), 623–627 (Mar 2023). https://doi.org/10.1007/s00338-023-02367-7

9. Fallati, L., Saponari, L., Savini, A., Marchese, F., Corselli, C., Galli, P.: Multi-temporal UAV data and object-based image analysis (OBIA) for estimation of substrate changes in a post-bleaching scenario on a maldivian reef. Remote Sensing **12**(13), 2093 (2020). https://doi.org/10.3390/rs12132093,
10. Ferrari, R., et al.: Photogrammetry as a tool to improve ecosystem restoration. Trends Ecol. Evolu. **36**(12), 1093–1101 (2021). https://doi.org/10.1016/j.tree.2021.07.004
11. Fischler, M.A., Bolles, R.C.: Random sample consensus: a paradigm for model fitting with applications to image analysis and automated cartography. Commun. ACM **24**(6), 381–395 (1981). https://doi.org/10.1145/358669.358692
12. Fontana, S., Cattaneo, D., Ballardini, A.L., Vaghi, M., Sorrenti, D.G.: A benchmark for point clouds registration algorithms. Rob. Auton. Syst. **140**, 103734 (2021). https://doi.org/10.1016/j.robot.2021.103734, https://www.sciencedirect.com/science/article/pii/S0921889021000191
13. Fontana, S., Di Lauro, F., Sorrenti, D.G.: Assessing the practical applicability of neural-based point clouds registration algorithms: A comparative analysis. Authorea, Inc. (Jul 2023). https://doi.org/10.22541/au.168908592.24833908/v1
14. Gojcic, Z., Zhou, C., Wegner, J.D., Andreas, W.: The perfect match: 3d point cloud matching with smoothed densities. In: International Conference on Computer Vision and Pattern Recognition (CVPR) (2019)
15. Hughes, T.P., et al.: Spatial and temporal patterns of mass bleaching of corals in the anthropocene. Science **359**(6371), 80–83 (2018). https://doi.org/10.1126/science.aan8048
16. Hughes, T.P., et al.: Global warming transforms coral reef assemblages. Nature **556**(7702), 492–496 (2018). https://doi.org/10.1038/s41586-018-0041-2
17. Matthies, L., Shafer, S.: Error modeling in stereo navigation. IEEE J. Rob. Autom. **3**(3), 239–248 (1987). https://doi.org/10.1109/JRA.1987.1087097
18. Montalbetti, E., et al.: Reef complexity influences distribution and habitat choice of the corallivorous seastar culcita schmideliana in the maldives. Coral Reefs **41**(2), 253–264 (2022). https://doi.org/10.1007/s00338-022-02230-1
19. Peterson, E., Carne, L., Balderamos, J., Faux, V., Gleason, A., Schill, S.: The use of unoccupied aerial systems (UASs) for quantifying shallow coral reef restoration success in belize. Drones **7**(4), 221 (2023). https://doi.org/10.3390/drones7040221
20. Ridge, J.T., Johnston, D.W.: Unoccupied aircraft systems (UAS) for marine ecosystem restoration. Front. Marine Sci. **7** (2020). https://doi.org/10.3389/fmars.2020.00438
21. Rusu, R.B., Blodow, N., Beetz, M.: Fast point feature histograms (fpfh) for 3d registration. In: 2009 IEEE International Conference on Robotics and Automation, pp. 3212–3217 (2009). https://doi.org/10.1109/ROBOT.2009.5152473
22. Woodhead, A.J., Hicks, C.C., Norström, A.V., Williams, G.J., Graham, N.A.J.: Coral reef ecosystem services in the anthropocene. Functional Ecol. **33**(6), 1023–1034 (2019). https://doi.org/10.1111/1365-2435.13331, https://besjournals.onlinelibrary.wiley.com/doi/abs/10.1111/1365-2435.13331
23. Yang, H., Shi, J., Carlone, L.: TEASER: fast and certifiable point cloud registration. IEEE Trans. Robotics (2020)
24. Zeng, A., Song, S., Nießner, M., Fisher, M., Xiao, J., Funkhouser, T.: 3dmatch: learning local geometric descriptors from rgb-d reconstructions. In: CVPR (2017)
25. Zhang, Z.: Iterative point matching for registration of free-form curves and surfaces. Int. J. Comput. Vision **13**(2), 119–152 (1994)
26. Zhou, Q.Y., Park, J., Koltun, V.: Open3D: A modern library for 3D data processing. arXiv:1801.09847 (2018)

Generation of Human Face and Body Behavior (GHB)

Upsampling 4D Point Clouds of Human Body via Adversarial Generation

Lorenzo Berlincioni, Stefano Berretti(✉), Marco Bertini, and Alberto Del Bimbo

Media Integration and Communication Center (MICC), University of Florence, Florence, Italy
{lorenzo.berlincioni,stefano.berretti, marco.bertini,alberto.delbimbo}@unifi.it

Abstract. Time varying sequences of 3D point clouds, or *4D* point clouds, are acquired at an increasing pace in several applications (*e.g.*, LiDAR in autonomous or assisted driving). In many cases, such volume of data is transmitted, thus requiring that proper compression tools are applied to either reduce the resolution or the bandwidth. In this paper, we propose a new solution for upscaling of time-varying 3D video point clouds. Our model consists of a specifically designed Graph Convolutional Network that combines Dynamic Edge Convolution and Graph Attention Networks for feature aggregation in a Generative Adversarial setting. To make these modules work in synergy, we present a specific way to sample dense point clouds and provide each node with enough features of its neighbourhood to generate new vertices. Compared to other solutions in the literature that address the same task, our proposed model is capable of obtaining similar results in terms of quality of reconstruction, while using a substantially lower number of parameters (≃ 300KB).

Keywords: Time varying 3D point clouds · 3D upscaling · Graph Attention Network · Generative Adversarial setting · Super Resolution

1 Introduction

There is a rising interest in capturing the real world in 3D at high-resolution. For dynamic real time settings, such as 3D sensing for robotics, telepresence, automated driving using LiDAR [8], this technology might need high-resolution point clouds with up to millions of points per frame. Taking into consideration the average point-cloud video, under some constraints such as keeping the identity of a human subject recognizable, the size of a single frame can be approximated as ∼10 Mbytes, which translates to a bitrate of ∼300 Mbytes per second without compression for a 30 fps dynamic point cloud. The high data rate is one of the main problems faced by dynamic point clouds, and efficient compression technologies allowing for the distribution of such content are widely investigated. One result in this direction is represented by the Point Cloud Compression standard

G. L. Foresti et al. (Eds.): ICIAP 2023 Workshops, LNCS 14365, pp. 457–469, 2024.
https://doi.org/10.1007/978-3-031-51023-6_38

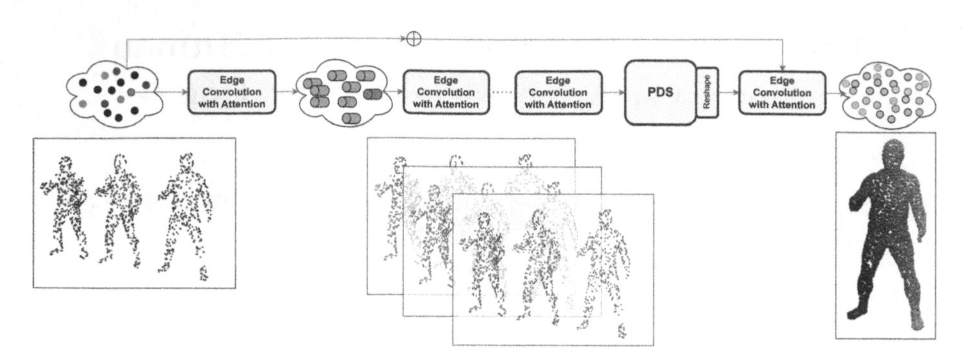

Fig. 1. Overview of our GCN generator architecture: Given a set of input point clouds, a set of Edge Convolution with Attention layers (in the middle) are first cascaded to generate an embedding; the Parallel Double Sampling (PDS) layer is then used for upsampling; last, the output is summed with the input following a residual-like schema to generate the upsampled point cloud.

specifications that include Video-based Point Cloud Compression (V-PCC) and Geometry-based PCC (G-PCC) [5] as released in 2020 by the Moving Picture Expert Group (MPEG).

Given these premises, our task is to perform *upscaling* and *artifact removal* of sparsely populated 3D point cloud sequences. Here, we will use the term *upscale* to indicate the operation by which the total number of vertices of an input point cloud is increased; instead, we will use *artifact removal* to imply the reconstruction process after some sort of compression or subsampling has been performed on an input point cloud. A high compression rate can achieve acceptable bandwidth requirements with a huge decrease in fidelity. For some applications, for example where the user experience is important, the identity of the subject must be maintained and such a low-resolution is not acceptable. In recent approaches that tackled this task, such as [22], long sequences of input frames and a large encoder-decoder model were used. As detailed below, we followed a different approach.

In this paper, we pose the upscaling problem in a Generative Adversarial setting using two architectural modules: the Edge Convolution [24] and the Graph Attention Network (GAT) [21]. In particular, the input point clouds are modeled as graphs and processed by a Graph Convolutional Network (GCN) that performs Edge Convolution. This operation incorporates local neighborhood information and is stacked to learn global shape properties. In addition, affinity in the feature space captures semantic characteristics over potentially long distances in the original embedding. While this module was used for CNN-based high-level tasks on point clouds, including classification and segmentation, the GAT was used for feature aggregation performing an attentioned learned mean of the neighbourhood features instead of a simple average. Figure 1 provides an overview of the proposed architecture. Our experiments on the FAUST 4D dataset [2] shown upscaling reconstructions that are comparable with those reported in the literature, while using a lower number of input frames and an architecture with a lower number of parameters.

The main contributions of our work can be summarized as follows:

- We propose a new GAN-like architecture for time varying point cloud upscaling that, to our knowledge, was not tried before for this task. It combines a *discriminator* network and a *generator* in an adversarial game. The generator makes use of Edge Convolution on the input graphs derived from the point clouds and a graph attention mechanism for aggregating the features of the local neighbourhood;
- The proposed solution demonstrated a clear advantage over the existing methods in the capability of producing upscaled 3D point clouds with comparable accuracy to state-of-the-art methods but using a way lower number of parameters in the architecture. The inference time resulted compatible with an online application handling a stream of input frames.

2 Related Work

Point cloud upsampling was first approached using optimization based solutions [1,6,7,14,26], while deep learning based methods were applied only more recently. In the following, we focus on method in the latter category, restricting our analysis to the 4D case since this is the case we will consider in this work.

Upsampling of static 3D point clouds was studied in [12,13,23,27]. Recently, more works shifted the attention to *4D reconstruction*, where a sequence of 3D objects is reconstructed from time-varying point clouds given as inputs [11,17].

In the Occupancy Network (ONet) proposed by Mescheder *et al.* [16], a 3D object was described using a continuous function that indicates which sub-sets of the 3D space the object occupies, and an iso-surface retrieved by employing the Marching Cube algorithm. Tang *et al.* [20] learned a temporal evolution of the 3D human shape through spatially continuous transformation functions among cross-frame occupancy fields. To this end, they established, in parallel, the dense correspondence between predicted occupancy fields at different time steps via explicitly learning continuous displacement vector fields from spatio-temporal shape representations. Niemeyer *et al.* [18] introduced a learning-based framework for object reconstruction directly from 4D data without predefined templates. The OFlow method they proposed calculates the integral of a motion field of 3D points in a 3D point cloud specified in space and time to implicitly represent trajectories of all the points in dense correspondences between occupancy fields. Vu *et al.* [22] proposed a network architecture, called RFNet-4D, that jointly reconstructs objects and their motion flows from 4D point clouds. It is shown that jointly learning spatial and temporal features from a sequence of point clouds can leverage individual tasks, leading to improved overall performance. To this end, a temporal vector field learning module using unsupervised learning approach for flow estimation was designed that, in turn, leveraged by supervised learning of spatial structures for object reconstruction. Jiang *et al.* [9]

introduced a compositional representation that disentangles shape, initial state, and motion for a 3D object that deforms over a temporal interval. Each component is represented by a latent code via a trained encoder. A neural Ordinary Differential Equation (ODE) was used to model the motion: it is trained to update the initial state conditioned on the learned motion code, while a decoder takes the shape code and the updated state code to reconstruct the 3D model at each time stamp. An Identity Exchange Training (IET) strategy is also proposed to encourage the network to learn decoupling each component.

With respect to the above solutions, our approach is characterized by a specific design that combines two GCNs to work in an adversarial setting. The resulting architecture proved to be flexible in the number of frames used as inputs and conjugated effective reconstructions with inference times that are compatible with online execution.

3 Upsampling of Time Varying Point Clouds

Let us consider a sequence of point clouds in the 3D space, where each point cloud can be regarded as a *frame* of a 4D video at time t. In the following, we consider n frames *fused* together forming a time varying point cloud as an unordered list of points in $\mathbb{R}^4$ with $\{x, y, z, t\}$ coordinates. Our task is to *upscale* each frame of the input sequence p_t, and get a more detailed one by leveraging the information of the previous $n-1$ low-resolution point cloud frames (*i.e.*, $p_{t-1}, \ldots, p_{t-n+1}$). More in detail, assuming the frame to upscale is p_0 and given a buffer of n previous frames, the input point cloud P_{in} is defined as:

$$P_{in} = \{p_{-n+1}, p_{-n+2}, ..., p_0\}, \;\; P_{in} \in \mathbb{R}^{4\times L\times n}, \tag{1}$$

where each low-resolution point cloud $p_i \in \mathbb{R}^{4\times L}$ is composed of L points.

We are interested in learning a map $f(P_{in}, \theta)$ from $P_{in} \rightarrow P_{out}$, where P_{out} is the target point cloud representing the initial frame at $t = 0$ upscaled to have $H = S \times L \times n$ points, with S being the *scale factor*, *i.e.*, $P_{out} \in \mathbb{R}^{3\times H}$ (we note the target frame has just the three spatial components $\{x, y, z\}$).

3.1 Proposed Architecture

We model the input point cloud as a graph, using message passing based convolution. In particular, we propose to perform an adversarial training using a new architecture, which is composed of two GCNs. Details for the *discriminator* and the *generator* are given below (see Fig. 2(a) for an overview).

Discriminator. It targets a classification task on a point cloud, so its architecture is inspired by PointNet++ [19]. We propose a structure that progressively reduces the number of points with *max-pooling* operations, then uses a sequence of linear layers before the output (see the bottom part of Fig. 2(a)).

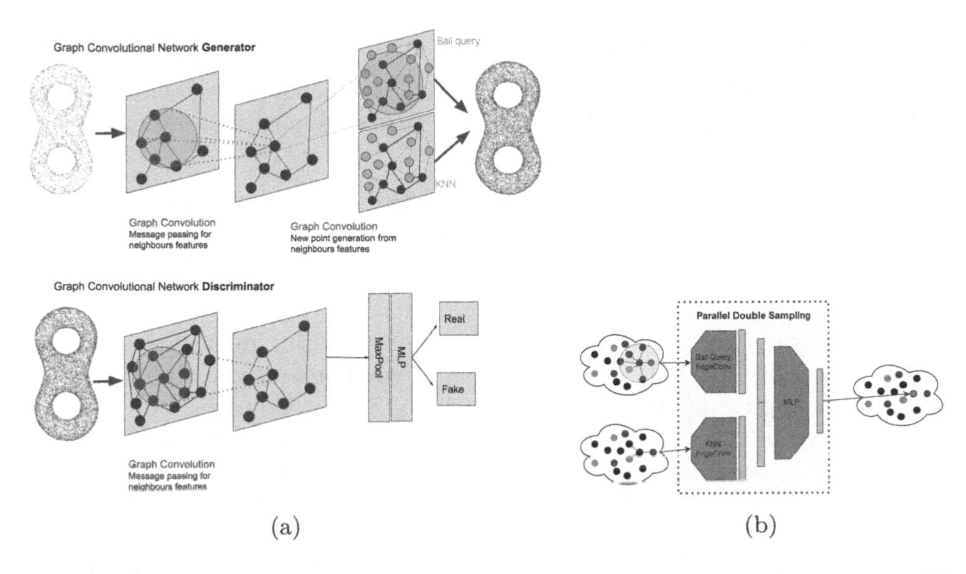

Fig. 2. (a) Proposed GCN architecture. *Top*: Generator; *Bottom*: Discriminator. (b) Details of the proposed Parallel Double Sampling (PDS) module.

Generator. The basic module of our generator network combines Edge Convolution [24] and GAT [21]. The Edge Convolution performs message passing over a dynamic graph in which the edges are updated as the point cloud changes. The GAT, instead, performs an attentional aggregation over the features collected from the dynamic local neighbourhood. This is in contrast with more common choices for aggregation such as *max* or *average*. We refer to this module as *Edge Convolution with Attention*.

This is followed by our Parallel Double Sampling (PDS) module that collects various features for each node using different neighborhood sampling techniques. Once the original node has been enriched with the local features, the PDS module uses them to generate multiple new vertices according to the scale factor. Finally, the position of these new vertices is summed with the position of the closest point that originated it, resembling a residual way of operation (see Fig. 1).

Parallel Double Sampling. The PDS module is the core of the generator. For each point, it performs two sets of operations in parallel (a simplified illustration of this module is presented in Fig. 2(b)). The first set is composed of:

- **Radius filtering**: For each vertex, only those vertices that belong to a sphere of radius r, centered on the vertex, are accounted as neighbors with the capability of passing messages;
- **Furthest Point Subsampling (FPS)**: The FPS algorithm [19] is used to temporarily sample a fraction s of the original points that are the farthest away, inside the radius, from a starting point;
- **Convolution**: Graph convolution is applied over the remaining vertices, independently of their number, and their features are aggregated.

The second set of operations, performed in parallel to the first one, consists of:

- **K-NN**: A fixed number of k closest vertices is selected as neighbors;
- **Convolution**: Graph convolution is applied over the vertices, and their aggregated features.

Finally, the two sets of features are concatenated and fed to a linear layer that maps $2 \times Channels_{in} \rightarrow Channels_{out}$ (values for $Channels_{in}$ and $Channels_{out}$ change across internal layers).

Loss Function. The model was trained end-to-end using multiple losses. The generator loss $L_G = \lambda_1 L_{Rec} + \lambda_2 L_{Dnt} + \lambda_3 L_{Adv}$ combines three components with values of λ_i empirically determined as $\lambda_1 = 1.0$, $\lambda_2 = 0.5$, and $\lambda_3 = 0.1$.

The *reconstruction loss* L_{Rec} computes the point-to-set Chamfer distance between the reconstructed (output) and the ground truth (GT) point clouds. Similar to [25], we take into account the *neighbourhood* of each reconstructed point $p_r \in P_{out}$, by finding the closest point $p_t \in P_{GT}$ in the GT point cloud, and compute both the distance between them and the difference in terms of local neighbors:

$$L_{Rec}(P_{out}, P_{GT}) = \sum_{p_r \in P_{out}} \min_{p_t \in P_{GT}} ||p_r - p_t||_2^2 + \sum_{p_t \in P_{GT}} \min_{p_r \in P_{out}} ||p_r - p_t||_2^2. \quad (2)$$

The *density loss* L_{Dnt} was defined as:

$$L_{Dnt}(P_{out}, P_{GT}) = \sum_{p_r \in P_{out}} \min_{p_t \in P_{GT}} ||Dnt(p_r) - Dnt(p_t)||_2^2 + \sum_{p_t \in P_{GT}} \min_{p_r \in P_{out}} ||Dnt(p_r) - Dnt(p_t)||_2^2. \quad (3)$$

where a vertex p *neighbourhood density* $Dnt(p)$ is given as the normalized sum of its neighbours in a given radius: $Dnt(p \in P) = \frac{1}{N_{max}} \sum_{n \in Ball_p} 1$.

For the *adversarial* component L_{Adv}, we used the LSGAN loss [15] as:

$$L_{Adv} = \min_G L(G) = \frac{1}{2} \mathbb{E}_{\mathbf{z} \sim p_{\mathbf{z}}(\mathbf{z})} \left[(D(G(\mathbf{z})) - c)^2 \right], \quad (4)$$

for the generator G (being $\mathbf{z}$ the union of input point clouds), and:

$$\min_D L(D) = \frac{1}{2} \mathbb{E}_{\mathbf{x} \sim p_{data}(\mathbf{x})} \left[(D(\mathbf{x}) - b)^2 \right] + \frac{1}{2} \mathbb{E}_{\mathbf{z} \sim p_{\mathbf{z}}(\mathbf{z})} \left[(D(G(\mathbf{z})) - a)^2 \right], \quad (5)$$

for the discriminator D, where a and b are the labels for fake and real data, respectively, and c denotes the value that G wants D to believe for fake data.

4 Experiments

The proposed solution for point clouds upscaling was evaluated in a comprehensive set of experiments, both qualitative and quantitative (see Sect. 4.2). An ablation study showing the relevance of the different components of our architecture is also reported in Sect. 4.3.

4.1 Implementation Details and Used Dataset

Our model was implemented using the PyTorch Geometric (PyG) library [4], which is specifically designed for Graph Neural Networks. The two networks were implemented using the Message Passing paradigm posed in an adversarial setting. Both the Discriminator and the Generator were optimized with Adam, using the standard learning rate $lr = 1e^{-4}$, with $\beta_1 = 0.9$, and $\beta_2 = 0.999$. A linear decaying scheduler that drops the learning rate to 1/10th every 10 epochs was also applied. Other hyperparameters, such as the radii for the *Ball Query* for the FPS sampling ($r_{small} = 0.06$, $r_{large} = 0.1$) and the number of neighbours for the *KNN* sampling ($n_{neighbours} = 9$) were determined using grid search.

The training data was augmented using different operations. Each sequence of input point clouds and its relative ground truth point cloud was randomly flipped along any of its axes, per point random noise was added, and a random scale along any axes was applied in the range $[0.9, 1.1]$. We also exploited the fully convolutional nature of the generator architecture; similar to the case of 2D image super-resolution, where patches of the target high-resolution image are used in the training, we randomly feed a 3D slice of the video instead of the full body. A final form of augmentation is a time inversion inside the sequence.

To evaluate our proposed solution, we used the Dynamic FAUST (D-FAUST) dataset [2]. It contains animated meshes of 129 sequences of 10 subjects (5 females and 5 males) with various motions such as "shake hips", "punching", running on "spot", or "one leg jump". In order to compare with other methods, we used the train/test split proposed in [18]. For each sequence, at training time, we randomly picked an index, then subsampled the following frames according to the model's frame rate. We trained multiple models at different frame rates.

We also followed the evaluation setup used in [18]. Specifically, for each evaluation, we carried out two case studies: *seen individuals but unseen motions* (*i.e.*, test subjects were included in the training data but their motions were not given in the training set); and *unseen individuals but seen motions* (*i.e.*, test subjects were found only in the test data but their motions were seen in the training set).

4.2 Results

Qualitative. Some qualitative results of the proposed upscaling are given in Figs. 3(a) and (b). In Fig. 3(a), the input low-resolution frame, our reconstruction point cloud and the ground truth are given from left to right. A second example is shown in Fig. 3(b), where the input frame, our reconstruction and the ground truth are compared both in terms of point clouds (top) and in terms of mesh reconstruction using the Poisson algorithm (bottom).

Quantitative. To measure the reconstruction quality, we applied the Chamfer point-to-set metric following the same protocol as reported in [22].

We compared our approach with respect to six state-of-the-art solutions in the literature of 4D reconstruction from point cloud sequences, namely PSGN 4D, ONet 4D, OFlow, LPDC, 4DCR, and RFNet-4D. The PSGN 4D extended

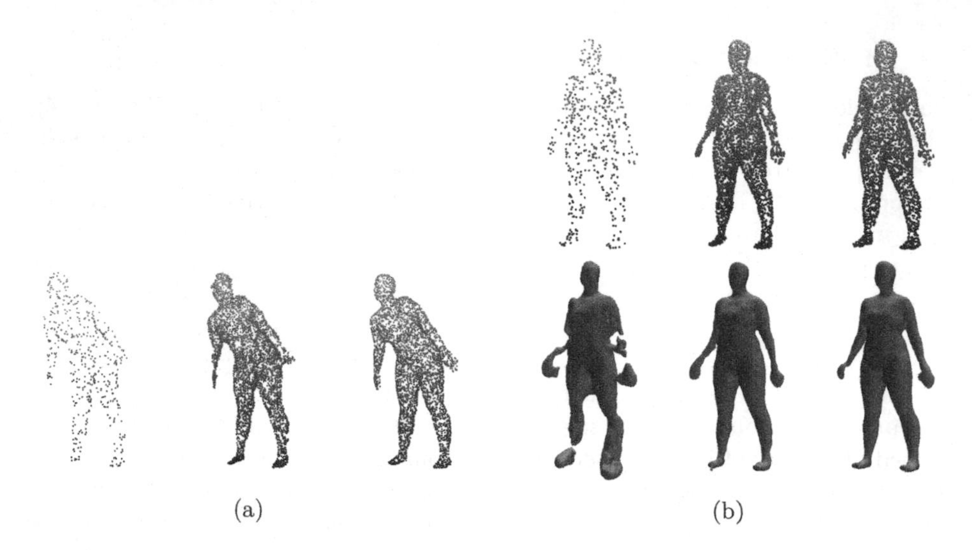

Fig. 3. (a) A single frame from an input low-resolution point cloud with ~512 vertices, our reconstruction and the GT are shown from left to right. (b) Point clouds are shown on the top, while the corresponding meshes obtained using Poisson surface reconstruction [10] are shown on the bottom.

the PSGN approach [3] to predict the point cloud trajectory (4D point cloud) instead of a single point set. The ONet 4D network extended ONet [16] to define the occupancy field in the spatio-temporal domain by predicting occupancy values for points sample in space and time. The OFlow network [18] assigned each 4D point an occupancy value and a motion velocity vector relying on a differential equation to calculate the trajectory. The LPDC [20] learned a temporal evolution of the 3D human shape through spatially continuous transformation functions among cross-frame occupancy fields. The 4DCR solution [9] used a compositional representation that disentangles shape, initial state, and motion for a 3D object that deforms over a temporal interval. Finally, RFNet-4D [22] jointly reconstructed objects and their motion flows from 4D point clouds.

Table 1 reports results for our solution and for the other methods as given in [22]. For our method (last line in the table), we used 3 frames for upscaling at 60fps with a ×4 scale factor starting from low-resolution point clouds composed of 1024 vertices. For the *unseen individual and seen motion* protocol (second column in the table), our approach achieved the second best score. In the third column of the table, it can be observed that our method reached a reconstruction error of similar magnitude with respect to the two best performing methods, *i.e.*, RFNet-4D and LPDC. It is worth noting that RFNet-4D obtained the reported error using 17 input frames against 3 to 8 as used in our tests.

In Table 2, we report the inference time, in seconds, for different configurations of our model. All the measurements correspond to experiments executed on an Nvidia 2080Ti GPU. The values reported in the table evidence that our approach can open the way to real-time upscaling. As reported in [22], their

Table 1. Reconstruction error using the Chamfer distance for the *unseen individuals and seen motions* (UISM), and *seen individuals and unseen motions* (SIUM) protocols. Results for the best and second best performing methods are in bold and underlined, respectively. Our approach scored the second best error for the first protocol and the third best for the second one.

Method	(UISM) Chamfer dist. $\times 10^{-3} \downarrow$	(SIUM) Chamfer dist. $\times 10^{-3} \downarrow$
PSGN-4D [3]	0.6877	0.6189
ONet-4D [16]	0.7007	0.5921
OFlow [18]	0.2741	0.1773
4DCR [9]	0.2220	0.1667
LPDC [20]	0.2188	<u>0.1526</u>
RFNet-4D [22]	**0.1594**	**0.1504**
Ours	<u>0.1758</u>	0.1638

method used 17 input frames to reconstruct an output frame, while our range of frames is between 3 (for models using larger input point clouds) and 8 (for smaller inputs) due to memory constraints at training time.

Table 2. Inference time for different configurations of our model using a three-frame buffer. Every test was performed on an Nvidia2080Ti. For the other models it must be noted that they used a 17 frame input sequence to output a frame.

Method	Input size	Upscale $\times$	Inference time (s) $\downarrow$
Ours	1024	3	0.103
Ours	1024	2	0.089
Ours	512	4	0.046
Ours	512	2	0.039
Ours	256	8	0.034
Ours	256	4	0.030
Oflow [16]	–	–	0.95
LDPC [20]	–	–	0.44
RFNet-4D [22]	–	–	0.24

4.3 Ablation Study

We verified different aspects involved in the design of our model in a set of ablation studies. In particular, we investigated every component of our proposed architecture for 4D point clouds reconstruction by comparing the percentage decrease of the reconstruction, when specific features are removed.

We performed a first set of experiments by using a stream of input point clouds at 60fps and with 256 points per frame; on this stream, we performed upscaling from subsets of consecutive 3 frames, using an upscale factor of $\times 2$. From Table 3 (column two and three), we can notice that by removing individual components of our architecture, the performance of the model significantly and consistently decreases. In particular, we removed the attention aggregation module and we substituted it with a *mean* aggregation. We also ablated the impact of the Density Loss and the adversarial component.

In column four and five of the same table, we repeated the ablation experiments as above but using a different setup: the frame rate was changed to 30fps, the input resolution to 512 points per frame, and we performed upscaling using a $\times 4$ factor. Also in this case, ablating the density loss term results into the most significant decrease in the accuracy of the upscaled model. It is also interesting to observe that, while the percentage increment in the Chamfer distance when removing the attention layer and the adversarial loss shows small differences between the two settings, this is not the case for the density loss: removing this term has a much larger impact on the results in column four and five ($\sim$+13%) than in column two and three ($\sim$+6%).

Table 3. Ablation results: (column two/three) 256 input points, 3 frames, 60fps, and upscale factor $\times 2$; (column four/five) 512 input points, 3 frames, 30fps, and upscale factor $\times 4$. (FF stands for our architecture fully featured).

Variant	Chamfer $\times 10^{-3} \downarrow$	% wrt FF	Chamfer $\times 10^{-3} \downarrow$	% wrt FF
No Attention	1.193	+3.11%	0.5856	+2.82%
No Density Loss	1.226	+5.96%	0.6433	+12.95%
No Adversarial Loss	1.213	+4.84%	0.5930	+4.13%
Ours FF	<u>1.157</u>	–	<u>0.5695</u>	–

Importance of the Temporal Information. A question that arises with the proposed solution is the actual impact of having the time buffer compared to using just the last point cloud as an input. To compare these two solutions, we feed our model with the same frame repeated n times. In this way, we keep the comparison fair by not changing the input size and the amount of starting points but only the *information* contained within it. We refer to this setup as *static* sequence, whilst we use the term *dynamic* to refer the proposed procedure that uses n different frames. In Table 4, we report some comparative results between the two ways of using the frames. It can be observed that there is a useful information in the time and movement of the cloud: as in a 2D video, the same frame repeated n times does not contain the same amount of useful data for reconstruction as n different subsequent frames do.

Table 4. Result for our method using dynamic and *static* sequences (*i.e.*, the final frame is repeated n times) at different resolutions.

Sequence	Input size	Frames	$\times$	Chamfer $\times 10^{-3} \downarrow$
Static	256	3	4	2.876
Dynamic	256	3	4	1.109
Static	256	4	3	2.825
Dynamic	256	4	3	0.745
Static	512	3	2	1.851
Dynamic	512	3	2	0.677

5 Conclusions

In this paper, we presented a fully convolutional graph-based approach for time-varying point clouds upscaling using a novel approach with respect to most of the state-of-the-art models. Our proposed method is comparable with state-of-the-art solutions in terms of upsampling performance but it has a lighter architectural complexity that allows the implementation on devices with limited computational capabilities. However, our method still shows some limitations and drawbacks: *(i)* The whole point cloud at every stage of the network is kept in memory, thus slowing down the training operation and posing some limitations in the number of input frames; *(ii)* Results for the reconstruction accuracy are comparable with those reported in the state-of-the-art, though a bit lower.

Acknowledgments. This work was supported by the European Commission under European Horizon 2020 Programme, grant number 951911-AI4Media.

References

1. Alexa, M., Behr, J., Cohen-Or, D., Fleishman, S., Levin, D., Silva, C.: Computing and rendering point set surfaces. IEEE Trans. Visualization Comput. Graph. **9**(1), 3–15 (2003). https://doi.org/10.1109/TVCG.2003.1175093
2. Bogo, F., Romero, J., Pons-Moll, G., Black, M.J.: Dynamic FAUST: registering human bodies in motion. In: IEEE Conference on Computer Vision and Pattern Recognition (CVPR) (2017)
3. Fan, H., Su, H., Guibas, L.J.: A point set generation network for 3D object reconstruction from a single image. In: IEEE Conference on Computer Vision and Pattern Recognition (CVPR) (2017)
4. Fey, M., Lenssen, J.E.: Fast graph representation learning with pytorch geometric. arXiv preprint arXiv:1903.02428 (2019)
5. Graziosi, D., Nakagami, O., Kuma, S., Zaghetto, A., Suzuki, T., Tabatabai, A.: An overview of ongoing point cloud compression standardization activities: video-based (v-pcc) and geometry-based (g-pcc). APSIPA Trans. Signal Inf. Process. **9**, e13 (2020)

6. Huang, H., Li, D., Zhang, H., Ascher, U., Cohen-Or, D.: Consolidation of unorganized point clouds for surface reconstruction. ACM Trans. Graph. **28**(5), 1–7 (2009)
7. Huang, H., Wu, S., Gong, M., Cohen-Or, D., Ascher, U., Zhang, H.R.: Edge-aware point set resampling. ACM Trans. Graph. **32**(1), 1–12 (2013)
8. Jang, E.S., et al.: Video-based point-cloud-compression standard in MPEG: from evidence collection to committee draft [standards in a nutshell]. IEEE Signal Process. Mag. **36**(3), 118–123 (2019)
9. Jiang, B., Zhang, Y., Wei, X., Xue, X., Fu, Y.: Learning compositional representation for 4D captures with neural ode. In: IEEE/CVF Conference on Computer Vision and Pattern Recognition (CVPR), pp. 5340–5350 (2021)
10. Kazhdan, M., Bolitho, M., Hoppe, H.: Poisson surface reconstruction. In: Eurographics Symposium on Geometry Processing, SGP 2006, pp. 61–70. Eurographics Association, Goslar (2006)
11. Leroy, V., Franco, J.S., Boyer, E.: Multi-view dynamic shape refinement using local temporal integration. In: IEEE International Conference on Computer Vision (ICCV) (2017)
12. Li, R., Li, X., Fu, C., Cohen-Or, D., Heng, P.: PU-GAN: a point cloud upsampling adversarial network. CoRR arxiv:1907.10844 (2019)
13. Li, R., Li, X., Heng, P., Fu, C.: Point cloud upsampling via disentangled refinement. CoRR arxiv:2106.04779 (2021)
14. Lipman, Y., Cohen-Or, D., Levin, D., Tal-Ezer, H.: Parameterization-free projection for geometry reconstruction. ACM Trans. Graph. **26**(3), 22-es (2007)
15. Mao, X., Li, Q., Xie, H., Lau, R.Y.K., Wang, Z., Smolley, S.P.: Least squares generative adversarial networks. In: IEEE International Conference on Computer Vision (ICCV), pp. 2813–2821 (2016)
16. Mescheder, L., Oechsle, M., Niemeyer, M., Nowozin, S., Geiger, A.: Occupancy networks: learning 3D reconstruction in function space. In: IEEE/CVF Conference on Computer Vision and Pattern Recognition (CVPR), pp. 4460–4470 (2019)
17. Mustafa, A., Kim, H., Guillemaut, J.Y., Hilton, A.: Temporally coherent 4D reconstruction of complex dynamic scenes. In: IEEE Conference on Computer Vision and Pattern Recognition (CVPR) (2016)
18. Niemeyer, M., Mescheder, L., Oechsle, M., Geiger, A.: Occupancy flow: 4D reconstruction by learning particle dynamics. In: IEEE/CVF International Conference on Computer Vision (ICCV) (2019)
19. Qi, C.R., Yi, L., Su, H., Guibas, L.J.: PointNet++: deep hierarchical feature learning on point sets in a metric space. Adv. Neural Inf. Process. Syst. (NeurIPS) **30**, 1–10 (2017)
20. Tang, J., Xu, D., Jia, K., Zhang, L.: Learning parallel dense correspondence from spatio-temporal descriptors for efficient and robust 4D reconstruction. IEEE/CVF Conference on Computer Vision and Pattern Recognition (CVPR), pp. 6018–6027 (2021)
21. Veličković, P., Cucurull, G., Casanova, A., Romero, A., Lio, P., Bengio, Y.: Graph attention networks. arXiv preprint arXiv:1710.10903 (2017)
22. Vu, T.A., Nguyen, D.T., Hua, B.S., Pham, Q.H., Yeung, S.K.: Rfnet-4D: joint object reconstruction and flow estimation from 4D point clouds. In: European Conference on Computer Vision (ECCV), pp. 36–52. Springer, Heidelberg (2022). https://doi.org/10.1007/978-3-031-20050-2_3
23. Wang, Y., Wu, S., Huang, H., Cohen-Or, D., Sorkine-Hornung, O.: Patch-based progressive 3D point set upsampling. CoRR arxiv:1811.11286 (2018)

24. Wang, Y., Sun, Y., Liu, Z., Sarma, S.E., Bronstein, M.M., Solomon, J.M.: Dynamic graph CNN for learning on point clouds. ACM Trans. Graph. **38**(5), 1–12 (2019)
25. Wu, T., Pan, L., Zhang, J., Wang, T., Liu, Z., Lin, D.: Density-aware chamfer distance as a comprehensive metric for point cloud completion. arXiv preprint arXiv:2111.12702 (2021)
26. Wu, Z., et al.: 3D shapenets: a deep representation for volumetric shapes. In: IEEE Conference on Computer Vision and Pattern Recognition (CVPR) (2015)
27. Yu, L., Li, X., Fu, C., Cohen-Or, D., Heng, P.: Pu-net: point cloud upsampling network. CoRR arxiv:1801.06761 (2018)

Decoding Deception: Understanding Human Discrimination Ability in Differentiating Authentic Faces from Deepfake Deceits

Shelina Khalid Jilani[1,2](✉), Zeno Geradts[3,5], and Aliyu Abubakar[4]

[1] University of Bradford, Bradford, UK
[2] INTERPOL, Lyon, France
shelinajilani63@gmail.com
[3] Informatics Institute, University of Amsterdam, Sciencepark 900, 1098 XH Amsterdam, The Netherlands
[4] Department of Electrical Engineering and Electronics, University of Liverpool, Liverpool, UK
[5] Netherlands Forensic Institute, Laan Van Ypenburg 6, 2497 GB Den Haag, The Netherlands

Abstract. Advances in innovative digital technologies present a maturing challenge in differentiating between authentic and manipulated media. The evolution of automated technology has specifically exacerbated this issue, with the emergence of DeepFake content. The degree of sophistication poses potential risks and raise concerns across multiple domains including forensic imagery analysis, especially for Facial Image Comparison (FIC) practitioners. It remains unclear as to whether DeepFake videos can be accurately distinguished from their authentic counterparts, when analysed by domain experts. In response, we present our study where two participant cohorts (FIC practitioners and novice subjects) were shown eleven videos (6 authentic videos and 5 DeepFake videos) and asked to make judgments about the authenticity of the faces. The research findings indicate that when distinguishing between DeepFake and authentic faces, FIC practitioners perform at a similar level to the untrained, novice cohort. Though, statistically, the novice cohort outperformed the practitioners with an overall performance surpassing 70%, relative to the FIC practitioners. This research is still in its infancy stage, yet it is already making significant contributions to the field by facilitating a deeper understanding of how DeepFake content could potentially influence the domain of Forensic Image Identification.

Keywords: DeepFake Detection · Face Identification · Artificial Intelligence · Forensic Practitioners and Deep Learning

1 Introduction

In recent years the proliferation of DeepFake media has emerged as a formidable challenge that poses a significant risk to multiple industries including human society, politics, democracy and forensic science [1, 2, 3]. The term came into existence when an individual known as "deepfakes," posting on Reddit, asserted in late 2017 that they had

G. L. Foresti et al. (Eds.): ICIAP 2023 Workshops, LNCS 14365, pp. 470–481, 2024.
https://doi.org/10.1007/978-3-031-51023-6_39

created a machine learning algorithm capable of superimposing celebrity faces onto adult content videos [4]. Ever since then, the domain of non-existent identities driven by artificial intelligence, has captivated the attention of researchers and the public alike. DeepFake is an umbrella term used for a broad range of synthetic, computer-generated media wherein the features of a target person in an original image or video are altered to resemble the facial characteristics of another individual. Such advanced technology typically produces media that is exceptionally lifelike in appearance, with many researchers reporting 'seeing is not believing' [5, 6, 7]. Concurrently, the use of biometric technology has propelled the use of physiological and/or behavioral attributes of an individual, to assist with person verification. In particular, the growth and ubiquitous nature of facial recognition-based technology has been multifarious, given that faces hold a pivotal position in human communication. A human face can share both verbal and non-verbal cues [8, 9], and the acquisition of face related material from a digital perspective, enables this external structure to hold prime position in the field of computer vision research.

The transformative aspect of DeepFake lies in its extensive reach, complexity, and magnitude of the underlying technology, which qualifies anyone with access to a computer, to create counterfeit videos that are indistinguishable from genuine media [10]. The issue is further fuelled with the availability of open-source software, which allows the public to test the latest technology; introducing them to the world of artificial intelligence, with a 'try before you buy' enticement. Further, the ease by which artificial faces are generated in images and videos is because of the: (*i*) availability of large-scale datasets [11, 12] and, (*ii*) the advancement of deep learning methods which reduce the need for manual editing, streamlining the process [13, 14].

In response to the increasingly sophisticated media, substantial endeavours are being carried out by researchers to understand the underlying processes and levels of accuracy associated with human discrimination ability. In the forensic sector, image falsification is not a novel challenge initiated exclusively by DeepFakes. The act of image manipulation through the means of editing software such as Photoshop, remains prevalent even today and the field of digital forensics has long been tackling this challenge [15]. An underdeveloped domain of study relates to the impact of DeepFake media on human facial perception. Furthermore, the question of whether forensic practitioners outperform inexperienced individuals in discerning manipulated media from genuine content remains an open inquiry.

It is already well documented that DeepFake media has the power to be misconstrued and accepted as authentic by human observers [16, 17, 18]. Human judgment is influenced by a range of factors inclusive of emotions. Recent studies in social psychology suggest that negative emotions have the potential to lower susceptibility to deception [19, 20], which may improve an individual's sensitivity. Anger is also reported to diminish cognitive processing depth by encouraging individuals to rely on stereotypes and pre-existing beliefs [21].

Nevertheless, existing scholarly work in the field of perceptual psychology and visual neuroscience indicates that the human visual system possesses specialised mechanisms designed for the perception of faces [22]. For example, within the Fusiform Gyrus of the human brain, there is a distinct region known as the Fusiform Face Area (FFA), which is dedicated to the processing of facial information. It has been reported that the FFA exhibits selective activation to faces as opposed to other control stimuli [23]. Regardless of one's stance in the debate surrounding whether facial recognition is an inherent ability, or a skill acquired through experience the consensus is that the processing of face-related information tends to take place holistically for many [24, 25].

To investigate human capabilities in detecting DeepFakes, we created a survey named "Decoding Deception" using the Google Forms. The survey was accessible for anyone with an internet connection and featured the DeepFake videos with the original images sampled from the FakeAVCeleb dataset. Each participant had the opportunity to evaluate the level of difficulty or ease involved in distinguishing between the media. Each participant was asked to rate their level of confidence on a three-point scale (50% likened to someone being *unsure,* 75% suggested a *more than likely* response and 100% equated to *extremely confident),* when assigning their response.

Considering the research signifying the visual processing abilities of humans, it is reasonable to anticipate that the participants would exhibit proficient performance in identifying artificial face manipulations. Our hypothesis suggests that the group of forensic practitioners are expected to outperform, if not significantly outperform the group of novice participants.

1.1 Forms of Facial Manipulation

A photograph of a human face can be divided into two independent attributes as outlined in [26]. Firstly, there is the two-dimensional shape which encompasses the arrangement and contours of face features such as the eyes, nose, and mouth. Secondly, there is the representation of the facial surface which covers coloration, skin, hair, and luminosity and provide indicators of the three-dimensional face shape which can be influenced by lighting conditions. Facial manipulations can be classified into four distinct groups:

- **Entire Face Synthesis** [27]: This form of manipulation technique involves the creation of entirely fabricated facial images which are often achieved through Advanced Generative Adversarial Networks (GANs) architecture, such as StyleGAN [28] and StyleGAN2 [29]. This method of manipulation has reported remarkable outcomes, producing facial images of exceptional quality and realism.
- **Identity Swap** [30]: This manipulation involves the substitution of one person's face in a video with the face of another subject. In general, two approaches are considered, (i) conventional techniques such as FaceSwap [31] and (ii) newer deep learning practices commonly referred to a DeepFakes [32].
- **Face Editing/Retouching:** This involves the modification of facial characteristics such as hair, skin colour, age and the addition of accessories such as eyewear [33].
- **Face Reenactment:** Involves the seamless process of replacing a face in a video sequence whilst keeping the gestures and facial expressions of the target. A popular technique associated this form of face manipulation is NeutralTextures [34].

2 Methodology

DeepFakes have become a dominant form of deception in the realm of digital technology and fabricated media. Utilising advanced deep learning algorithms, particularly Generative Adversarial Networks (GANs), these manipulations generate astonishingly realistic content that can effectively mislead human observers.

2.1 Generative Adversarial Networks (GANs)

The field of Artificial Intelligence has been revolutionised by Generative Adversarial Networks (GANs), which have paved the way for generating highly realistic synthetic data. GANs operate through a competitive framework using a generator and a discriminator neural network, as depicted in Fig. 1. The generator's task is to produce synthetic data, while the discriminator's role is to evaluate and differentiate between authentic and generated samples. The ultimate objective of the generator is to create synthetic data that is indistinguishable from real data, challenging the discriminator's ability to discern between the two [35].

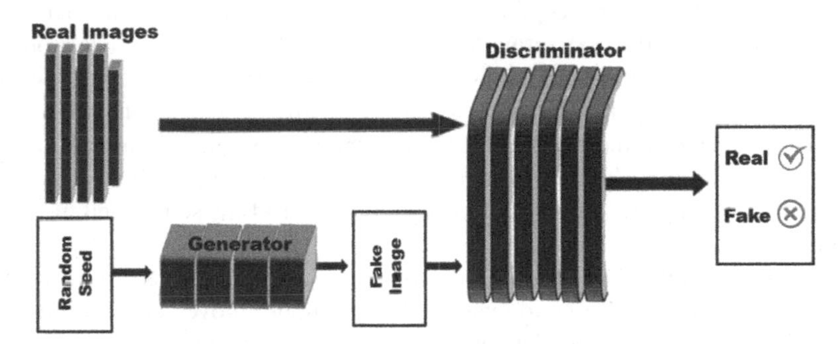

Fig. 1. Schematic representation of a Generative Adversarial Network (GAN) framework.

The generator in a GAN learns a distribution, $Dist_g$ over the data x. This is achieved by creating a mapping function from a prior noise distribution, $Dist_z(z)$ to the data space. The function is defined as $G(z; \theta_g)$. On the other hand, the discriminator, $D(x; \theta_d)$, provides a scalar output signifying the likelihood that x is derived from the training data instead of $Dist_g$. Both the generator and the discriminator are trained simultaneously. The parameters for G are adjusted to minimise $log(1 - D(G(z)))$, while parameters for D are adjusted to minimize $logD(X)$. This training procedure can be likened to a two-player min-max game, where the value function $V(G, D)$ is being optimized. The objective function is defined as:

$$min_G max_D V(D, G) = E_{xp_{data}(x)}\left[logD(x)\right] + E_{zp_z(z)}\left[log(1 - D(G(z)))\right] \quad (1)$$

2.2 Conditional GANs (CGANs)

In a conventional GAN framework (Fig. 1), both the generator and the discriminator operate without any constraints, allowing for unrestricted data generation. However, the lack of specific conditions can lead to inefficiency if the generated data is not required within a specific framework or context. In contrast, the architectural variant, CGANs, presents an option for conditionality in both the generator and the discriminator [36]. These conditions correspond to the class labels of images or other specified properties.

Thus, a traditional GAN model can be transitioned into a CGAN by introducing supplementary conditions to both the generator and the discriminator. For both the generator and the discriminator extra information y is added to the input x.

$$\begin{aligned} & min_G max_D V(D, G) \\ & = E_{xp_{data}(x)}\left[logD(x \vee y)\right] \\ & + E_{zp_z(z)}\left[log(1 - D(G(z \vee y)))\right] \end{aligned} \tag{2}$$

DeepFake videos employ various techniques such as lip-sync and faceswap to manipulate specific facial areas and create authentic, non-existent identities. Lip-sync entails synchronising mouth movements with an audio clip, whereas faceswap involves altering the entire face. Faceswap DeepFakes, employ a combination of two encoder-decoder pairs. The process involves extracting facial features such as the eyes, nose mouth, and ears using an encoder, and then reconstructing the face using a decoder. Typically, to accomplish faceswap, a pair of encoders and decoders are trained on both the source and target images or videos; the duration of the training process directly impacts the level of detail and specificity achieved in the final deepfake video. Once trained, the encoders and decoders are swapped, allowing the original encoder of the source and the decoder of the target to generate a manipulated video.

For lip-sync DeepFakes, a generator coupled with a lip-sync discriminator is employed. The generator learns to synchronise the mouth movements with the audio by using the target individual's data as a reference. By training on the target individual's data, the generator learns to produce realistic lip movements that align with audio. Figure 2 shows an illustrative example.

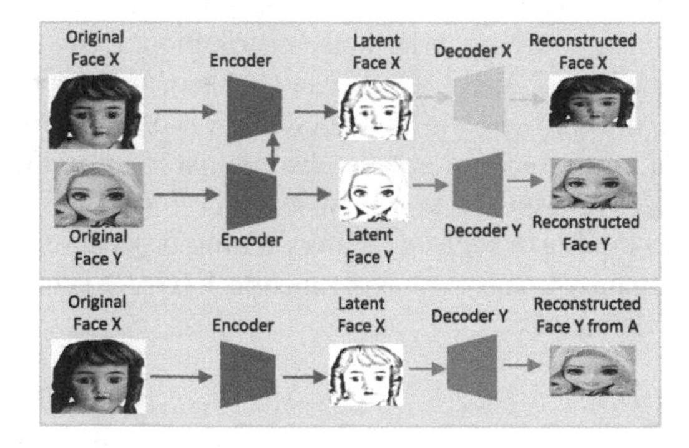

Fig. 2. Graphic to show the process of DeepFake development with encoder-decoder pairs.

The lip-sync DeepFake videos in this study were created using the Wav2lip application. The application employs a specialised model that focuses on synchronising lip movements in a video with an audio clip. The generator component of the model is trained on a range of audio samples and learns to align an individual's mouth movements with the corresponding audio content. To enhance accuracy, a lip-sync discriminator is utilised, assisting the generator in refining its output and rectifying any inconsistencies. By incorporating the lip-sync discriminator during training, the occurrence of artefacts in the final deepfake videos is minimised.

2.3 Faceswap

Faceswap encompasses the substitution of the source's face with another person's face (the target), while maintaining the target's facial expressions. Typically, this procedure entails a sequence of stages. During face detection and alignment, facial landmarks are identified for precise face alignment. Face encoding transforms features of the source and target faces into a numerical format, using deep neural networks. During the face swapping phase, the source face is overlaid onto the target face using techniques such as image warping and/or blending, the goal being to ensure the swapped face blends naturally. Texture blending involves the matching of colours and textures of the source face to the target face. Typically, using techniques such as colour correction and texture mapping. Lastly, during facial expression transfer phase, the swapped face exhibits the target's facial expressions.

2.4 Lip-Sync

Lip-sync techniques strive to achieve harmonisation between the lip movements of an individual in a video and the accompanying audio file. The objective is to ensure precise synchronisation between the spoken words and the corresponding lip movements. Again, similar to the process of faceswap, a series of sequential steps are required for lip-syncing. During audio analysis, the file containing sound information is processed to extract phonetic or timing information. This element of analysis helps to identify the specific sounds that need to be synchronised with the lip movements. Lip Motion Extraction analyses extracted lip shape and movements over time. This is achieved by tracking the movement of specific lip landmarks of the lips. For alignment, the extracted lip movements along with the phonetic/ timing information (from the audio file) are combined. Once the alignment is achieved, the lip movements are animated in a way that corresponds to the audio file. This typically involves warping or morphing the target person's lips to match the desired phonetic shapes or timing.

2.5 Dataset

To support the development of detection software, researchers have curated diverse datasets that serve as valuable resources for research. For this study, the FakeAVCeleb dataset [37] was utilised. FakeAVCeleb is one of the most recent dataset releases; a novel audio-visual DeepFake database which also includes synthesised lip-sync DeepFake audios.

A total of 11 authentic videos of varying image quality, duration and facial viewpoint were selected. Consideration was given to include videos that represented a range of racial backgrounds and maintain equal representation of genders. For each genuine video, a corresponding deepfake version was created utilising both the faceswap and lip-sync techniques, resulting in a total of 11 deepfake videos. The reason for creating a small data sample was to ensure manageability over the quality of the video files, over quantity. In addition, feasibility, to ensure the process wasn't time intensive, especially since our research serves purpose as an exploratory study which we endeavour broadens the discussion of identification abilities across a cohort of individuals.

2.6 Experimental Procedure

The experimental procedure involved presenting a series of videos (authentic and DeepFake), to two participant samples, (i) a group of facial image comparison (FIC) practitioners, from European forensic laboratories, and (ii) novice participants with no experience of working in the field of facial image comparison. The FIC practitioners were selected based on their expertise in the domain of facial image comparison. Each participant was tasked with carefully examining each video and deciding whether it was a DeepFake or not. Participants marked each video, accordingly, providing a clear indication of their judgment. To further evaluate the confidence of their assessments, participants were also asked to rate their level of confidence using a three-point scale.

The confidence scale included three categories: "likely" (50% confidence level), "very likely" (75% confidence level), and "extremely likely" (nearly 100% confidence level). By providing these confidence ratings, the experts were able to express the degree of certainty they had in their judgments regarding the authenticity of the videos. Such an experimental procedure ensured the objectivity and integrity of the assessment process.

3 Results and Discussion

In our experiment, a total of 51 participants were instructed to watch a series of vidoes and identify those of authentic nature and DeepFake. The survey results were analysed with the aim to determine the core elements of this study. Initially, it was hypothesised that given the level of expertise in unfamilair facial identification, Facial Image Comparison (FIC) practitioners would perform exceedingly better compared to the novice cohort. However considering the overall performance between both the participant cohorts, the novice participants marginally outperformed the FIC practitioners when identifying authentic faces in the survey amongst the DeepFakes (Fig. 3).

Upon closer inspection, for the *correct authentic* category, the median score is >65%, and inclined towards the upper quartile of the data distribution which indicates that many of the FIC practitioners performed highly. The results are promising considering that only a small population sample were tested and that the practitioners only work with material consisting of true, authentic identities. Additionally, some if not all the FIC practitioners will not have had the opportunity (prior to this study), to test their discrimination ability using DeepFake material. In contrast, for the *incorrect authentic* category the median is <35% and closer to the lower quartile of the box. This suggests that several FIC practitioners struggled to make judgements about the authenticity of the videos.

Shifting our attention to the novice population cohort, the median score for the *correct authentic category* reached the boundary of the upper quartile (>85%), indicating a higher discrimination ability relative to FIC practitioners. This may be reflective of the participants who work in the domain of digital forensics, but not facial image comparison.

In Fig. 4, the data suggests that both the FIC practitioners and the novices exhibit a comparable, average performance level in correctly identifying DeepFakes with a median score hovering around 60% for both groups. Likewise, a similar pattern is observed in the incorrect responses for DeepFake identities, with both cohorts exhibiting an average performance level of approximately 40%, except for a few responses. Amongst the FIC practitioners, the highest *correct DF* (DeepFake) distribution is 80%, with the

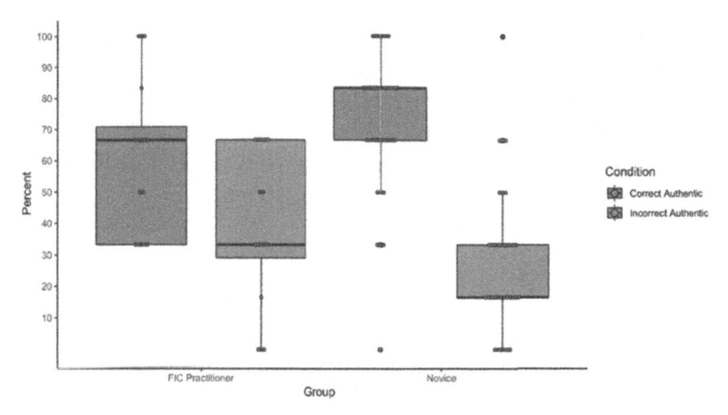

Fig. 3. A Boxplot to illustrate the discrimination abilities between Facial Image Comparison (FIC) Practitioners and novice subjects, when viewing authentic (non-computer-generated) videos with true, human identities.

maximum value depicted by the end of the 'whisker' in the box plot, reaching 100%. Conversely, the upper quartile for the *incorrect DF* distribution is at least 60%, with the 'whisker' extending to 80%. In summary, FIC practitioners generally perform highly in correctly identifying DeepFakes compared to incorrectly identifying them, with the majority performing above chance-level, at 60% for both cases.

In comparison, the novice cohort performance follows a similar pattern. The novice data reports that three-quarter of the incorrect responses is below 45%, indicating a low error rate.

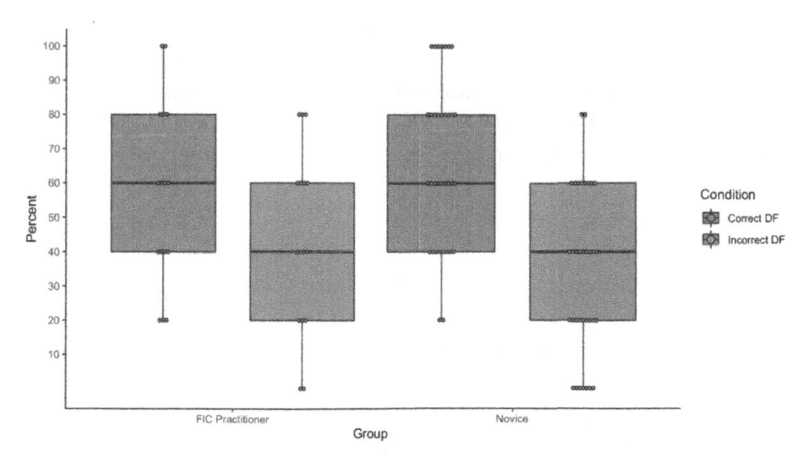

Fig. 4. A Boxplot to illustrate the discrimination abilities between Facial Image Comparison (FIC) Practitioners and novice subjects, when viewing computer-generated DeepFake videos.

Figure 5 (below) depicts the overall performance of both cohorts, and the findings reveal that FIC practitioners do not a perform as highly as their novice counterparts. Generally, the novice participants perform significantly better with a median score reaching

>70%. For the FIC participants, the incorrect distribution is positively skewed with an upper quartile of 55%. This suggests that at least 75% of their incorrect answers fall below the 55% mark. Instead, the incorrect distribution for novice participants shows an upper quartile value of approximately 45%. This suggests that at least three-quarters of the incorrect results from novice participants are lower than 45%. This indicates less accuracy in their results as compared to the FIC participants.

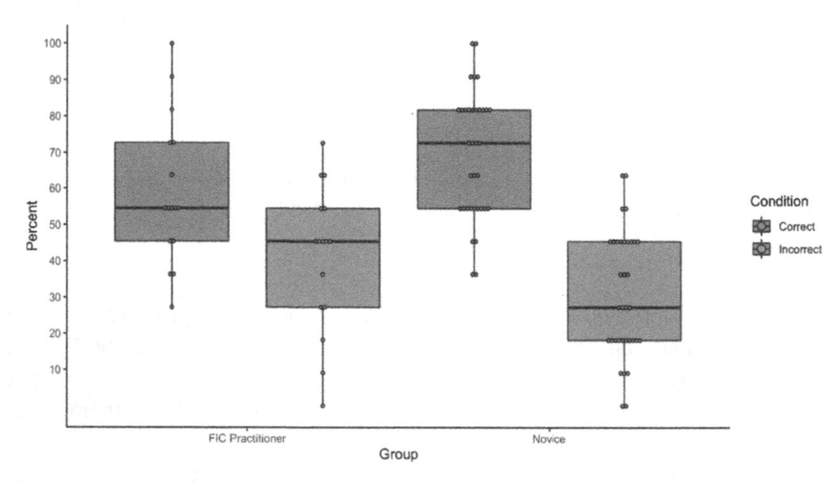

Fig. 5. A Boxplot to illustrate the overall performance between Facial Image Comparison (FIC) Practitioners and novice subjects, when viewing computer-generated DeepFake videos with authentic (non-computer-generated) videos with true, human identities.

Our experimental results appear to contradict current literature, which suggest that novice participants generally have limited ability to identify media manipulations. Our research has demonstrated that novice participants are better at identifying DeepFake faces, compared to FIC practitioners, especially when shown with a short viewing window. Such exceptional performance is to be considered with a caveat that there was a limitation to the number of times the videos could be shown and that too, without any voice information. In addition, there were only two participants in the novice cohort and one in the FIC practitioners' cohort, who performed exceptionally, achieving 100% accuracy across all eleven tested videos. Hence, it is evident that there are underlying variables which affect human performance, and our research highlights the requirement for further study.

Importantly, our research is not free from limitations. The greatest limitation is the participant size, especially for the practitioner cohort. The analysed data came from 51 participants, (16 FIC practitioners and 35 novices). This sample size is not particularly representative, although the results can sufficiently provide exploratory insights. Another potential limitation may be in the process of data collection itself, whilst the participants were asked to rate their level of confidence when making judgements on authenticity, they were not probed about what they were looking for within the videos, when determining whether the face was an authentic or a DeepFake. The judgement rating is not a core focus of this paper, it was only included to ensure participants responded as honestly and confidently as possible. Such form of qualitative information may have provided a deeper insight into the similarities and/or differences between perception for the tested cohorts.

4 Conclusion

Detecting DeepFakes in this modern world is an increasingly challenging problem on various fronts. Such innovations have a significant impact on online safety, crime, forensic science, and society. In this paper, we have provided preliminary data to show the distinctions in human discrimination ability between an expert (facial image comparison practitioners) and a (non-expert) novice cohort. Our aspiration is that these discoveries will trigger more in-depth studies within the forensic science realm and explain the effects that DeepFakes have on facial image identification.

Acknowledgement. The authors would like to thank Bas Roosenstein, (Forensics Educational Institution, University of Applied Science, Amsterdam) and Dr. Reuben Morton (Open University, UK), for their valuable contributions to this article.

References

1. Borges, L., Martins, B., Calado, P.: Combining similarity features and deep representation learning for stance detection in the context of checking fake news. J. Data Inf. Q. (JDIQ) **11**(3), 1–26 (2019)
2. Dack, S.: Deep fakes, fake news, and what comes next. The Henry M. Jackson School of International Studies (2019)
3. Mansoor, N., Iliev, A.: Artificial intelligence in forensic science. In: Arai, K. (eds.) Advances in Information and Communication. FICC 2023. LNNS, vol. 652, pp. 155–163. Springer, Cham (2023). https://doi.org/10.1007/978-3-031-28073-3_11
4. Bitesize, B.B.C.: deepfakes: what are they and why would I make one? (2019)
5. Maras, M.H., Alexandrou, A.: Determining authenticity of video evidence in the age of artificial intelligence and in the wake of Deepfake videos. The International Journal of Evidence & Proof **23**(3), 255–262 (2019)
6. Cochran, J.D., Napshin, S.A.: Deepfakes: awareness, concerns, and platform accountability. Cyberpsychol. Behav. Soc. Netw. **24**(3), 164–172 (2021)
7. Hancock, J.T., Bailenson, J.N.: The social impact of deepfakes. Cyberpsychol. Behav. Soc. Netw. **24**(3), 149–152 (2021)
8. Jilani, S.K., Ugail, H., Logan, A.: Man vs machine: the ethnic verification of Pakistani and non-Pakistani mouth features. In: 41st ISTANBUL International Conference on “Advances in Science, Engineering & Technology” (IASET-22) (2022)
9. Adyapady, R.R., Annappa, B.: A comprehensive review of facial expression recognition techniques. Multimedia Syst. **29**(1), 73–103 (2023)
10. Fletcher, J.: Deepfakes, artificial intelligence, and some kind of dystopia: the new faces of online post-fact performance. Theatr. J. **70**(4), 455–471 (2018)
11. Narayan, K., Agarwal, H., Thakral, K., Mittal, S., Vatsa, M., Singh, R.: DF-Platter: multi-face heterogeneous Deepfake dataset. In: Proceedings of the IEEE/CVF Conference on Computer Vision and Pattern Recognition, pp. 9739–9748 (2023)

12. Korshunov, P., Marcel, S.: Vulnerability assessment and detection of deepfake videos. In: 2019 International Conference on Biometrics (ICB), pp. 1–6. IEEE, June 2019
13. Goodfellow, I., et al.: Generative adversarial networks. Commun. ACM **63**(11), 139–144 (2020)
14. Brophy, E., Wang, Z., She, Q., Ward, T.: Generative adversarial networks in time series: a systematic literature review. ACM Comput. Surv. **55**(10), 1–31 (2023)
15. Battiato, S., Giudice. O., Paratore, A.: Multimedia forensics: discovering the history of multimedia contents. In: Proceedings of the 17th International Conference on Computer Systems and Technologies 2016, pp. 5–16 (2016)
16. Rössler, A., Cozzolino, D., Verdoliva, L., Riess, C., Thies, J., Nießner, M.: Faceforensics: A large-scale video dataset for forgery detection in human faces (2018). arXiv preprint arXiv: 1803.09179
17. Vaccari, C., Chadwick, A.: Deepfakes and disinformation: exploring the impact of synthetic political video on deception, uncertainty, and trust in news. Soc. Media+ Soc. **6**(1), 2056305120903408 (2020)
18. Groh, M., Epstein, Z., Firestone, C., Picard, R.: Deepfake detection by human crowds, machines, and machine-informed crowds. Proc. Natl. Acad. Sci. **119**(1), e2110013119 (2022)
19. Forgas, J.P., East, R.: On being happy and gullible: mood effects on skepticism and the detection of deception. J. Exp. Soc. Psychol. **44**(5), 1362–1367 (2008)
20. Brashier, N.M., Marsh, E.J.: Judging truth. Annu. Rev. Psychol. **71**, 499–515 (2020)
21. Clore, G., et al.: Affective feelings as feedback: some cognitive consequences. In: Martin, L.L., Clore, G.L. (eds.) Theories of Mood and Cognition: A User's Handbook. pp. 27–62, L. Erlbaum, 2001
22. Sinha, P., Balas, B., Ostrovsky, Y., Russell, R.: Face recognition by humans: nineteen results all computer vision researchers should know about. Proc. IEEE **94**(11), 1948–1962 (2006)
23. Kanwisher, N., McDermott, J., Chun, M.M.: The fusiform face area: a module in human extrastriate cortex specialized for face perception. J. Neurosci. **17**(11), 4302–4311 (1997)
24. Richler, J.J., Gauthier, I.: A meta-analysis and review of holistic face processing. Psychol. Bull. **140**(5), 1281 (2014)
25. Young, A.W., Burton, A.M.: Are we face experts? Trends Cogn. Sci. **22**(2), 100–110 (2018)
26. Bruce, V., Young, A.W.: Face perception. Psychology Press, Milton Park (2012)
27. Sabel, J., Johansson, F.: On the robustness and generalizability of face synthesis detection methods. In: Proceedings of the IEEE/CVF Conference on Computer Vision and Pattern Recognition, pp. 962–971 (2021)
28. Karras, T., Laine, S., Aila, T.: A style-based generator architecture for generative adversarial networks. In: Proceedings of the IEEE/CVF Conference on Computer Vision and Pattern Recognition, pp. 4401–4410 (2019)
29. Karras, T., Laine, S., Aittala, M., Hellsten, J., Lehtinen, J., Aila, T.: Analyzing and improving the image quality of stylegan. In: Proceedings of the IEEE/CVF Conference on Computer Vision and Pattern Recognition, pp. 8110–8119 (2020)
30. Chen, R., Chen, X., Ni, B., Ge, Y.: SimSwap: an efficient framework for high fidelity face swapping. In: ACM Multimedia (2020)
31. Bitouk, D., Kumar, N., Dhillon, S., Belhumeur, P., Nayar, S.K.: Face swapping: automatically replacing faces in photographs. ACM Trans. Graph. **27**(3), 1–8 (2008)
32. Pu, J., et al.: Deepfake videos in the wild: analysis and detection. In: Proceedings of the Web Conference 2021, pp. 981–992, April 2021
33. Gonzalez-Sosa, E., Fierrez, J., Vera-Rodriguez, R., Alonso-Fernandez, F.: Facial soft biometrics for recognition in the wild: recent works, annotation, and COTS evaluation. IEEE Trans. Inf. Forensics Secur. **13**(8), 2001–2014 (2018)
34. Thies, J., Zollhöfer, M., Nießner, M.: Deferred neural rendering: image synthesis using neural textures. ACM Trans. Graph. (TOG) **38**(4), 1–12 (2019)

35. Soni, R., Arora, T.: A review of the techniques of images using GAN. In: Generative Adversarial Networks for Image-to-Image Translation, pp. 99–123 (2021)
36. Mirza, M., Osindero, S.: Conditional generative adversarial nets. arXiv preprint arXiv:1411.1784 (2014)
37. Khalid, H., Tariq, S., Kim, M., Woo, S.S.: FakeAVCeleb: a novel audio-video multimodal deepfake dataset (2021). arXiv preprint arXiv:2108.05080

Generative Data Augmentation of Human Biomechanics

Halldór Kárason[1(✉)], Pierluigi Ritrovato[1], Nicola Maffulli[2], and Francesco Tortorella[1]

[1] Department of Information and Electrical Engineering and Applied Mathematics, University of Salerno, Salerno, Italy
hkarason@unisa.it

[2] Department of Medicine, Surgery and Dentistry, University of Salerno, Salerno, Italy

Abstract. Wearable sensors are miniature and affordable devices used for monitoring human motion in daily life. Data-driven models applied to wearable sensor data can enhance the accuracy of movement analysis outside of controlled settings. However, obtaining a large and representative database for training these models is challenging due to the specialised motion laboratories and expensive equipment required. To address this limitation, this study proposes a data augmentation approach using generative deep learning to enhance biomechanical datasets. A novel conditional generative adversarial network (GAN) was developed to synthesise biomechanical data during gait. The GAN takes into account the subject's anthropometric measures to generate data that represents specific body types as well as information about the gait cycle for reconstruction back into the time domain. The proposed model was evaluated for generating biomechanical data of unseen subjects and fine-tuning the model with small percentages (1%, 2% and 5%) of the test dataset. Researchers and practitioners can overcome the limitations of obtaining large training datasets from human participants by synthesising realistic and diverse synthetic data. This paper outlines the methodology and experimental setup for developing and evaluating the GAN and discusses its potential impact on the field of biomechanics and human motion analysis.

Keywords: biomechanics · deep learning · wearable sensors · generative adversarial networks

1 Introduction

Wearable sensors are highly portable and cheap devices with low power consumption that can be used to monitor human motion patterns in daily life. They generally lack the accuracy and precision of expensive laboratory instruments, but deploying data-driven models on wearable sensors' time-series data could further facilitate the accuracy of movement and biomechanical analysis outside controlled laboratory settings and learn non-linear relationships between easily collected data and complex human biomechanics.

G. L. Foresti et al. (Eds.): ICIAP 2023 Workshops, LNCS 14365, pp. 482–493, 2024.
https://doi.org/10.1007/978-3-031-51023-6_40

Different models have been proposed to predict kinematic and kinetic variables from wearable sensors during running and walking. The most commonly used architectures are Long Short-Term Memory (LSTM) [15,19,20] and Convolutional Neural Networks (CNNs) [6,7]. A prerequisite for such modalities is a large and representative database. Training data is typically captured inside specialised motion laboratories with expensive equipment, which requires expert knowledge to operate and is time-consuming. Thus, a sufficiently large number of samples from human participants can be challenging to achieve.

Biomechanical datasets have been enhanced using simulations of musculoskeletal models created by solving optimal control problems. Simulation-enhanced datasets have been shown to improve wearable sensor predictions of kinematics. Still, they show limited improvement in predicting more complex mechanisms like joint forces/torques and ground reaction forces [6,15].

In a study by Renani et al. [20], authors used data augmentation techniques (by warping joint angle trajectories) on measured joint angles to generate further training samples and enhance their dataset. The new joint angles were then used to simulate a musculoskeletal model. From virtual sensors placed on the musculoskeletal models, a significant decrease in RMSE (54% for hip angles and 45% for knee angles) was achieved when training a Bidirectional LSTM model on the enhanced dataset, compared to only training on measured kinematic data. In another study [19], data from virtual IMU sensors was augmented to account for the variability of sensor orientation and limb alignment by assigning random rotations to a virtual sensor whose original orientations were aligned with a cluster of reflective markers. Both studies were limited to kinematic data, neglecting the augmentation of kinetic data.

Recently, generative networks have been proposed to generate synthetic but realistic data for various applications, including wearable sensor data for human activity recognition [1,16,23]. In the context of biomechanics data, limited work has been proposed using neural networks for generating data that represent realistic patterns of human movement and the associated forces applied. In a recent paper by Bicer et al. [3], an autoencoder-based network was proposed to generate synthetic motion capture data of marker trajectories and ground reaction forces (GRFs). They found the dataset enhanced with synthetic data improved the accuracy of deep learning models predicting joint angles and ground reaction forces from data of a single simulated pelvic inertial measurement unit. Their approach did not offer a solution to influence the anthropometrics of the subject or to reconstruct the time-normalised data to the time domain.

Here, a generative data augmentation technique is proposed using a GAN network to synthesise biomechanical data. Anthropometric measures of human subjects will be used to condition the model and generate data from specific body types, allowing for the reconstruction of the generated gait cycles to the time domain. Unconstrained generative models could yield physically implausible samples. The main hypothesis is that a generative network can be trained to generate realistic movement patterns and forces using only anthropometric measures of the subject.

2 Methods

2.1 Experimental Dataset

An open-source dataset [4] of multimodal lower limb biomechanics was collected from twenty-two healthy individuals (age 21 ± 3.4 yr, height 1.70 ± 0.07 m, mass 68.3 ± 10.83 kg). The dataset offers individual trials containing kinematics and kinetics together with wearable sensor signals (EMG, IMU, and goniometer) during human locomotion. Different walking conditions, including walking on a treadmill, level ground, stairs and ramps, and different contexts of walking speed, stair height, and ramp inclination. Specifically, kinematic marker and force platform data from treadmill walking were used.

2.2 Data Pre-processing

On the kinematic marker data, inverse kinematic was performed in OpenSim [21] to derive joint angle coordinate data. The gait trials were cut into gait cycles by detecting a 'toe-off' event in force platform data. In total, 37687 gait cycles were derived from the 22 subjects. From each stride, gait parameters were calculated. Each gait cycle was time-normalised to 0–100% to standardise the size of the samples during training. Eighteen coordinates describing joint angles (18 degree-of-freedom) of the lower body were extracted from each gait cycle along with ten features of force platform data (GRFs in x-, y-, and z-directions and COP in x- and z- directions; COP in the vertical y-direction is always zero). Information about the subjects' age, height, leg height, weight and gender were added to the data along with the side of the foot (left or right) and time of gait cycle (for reconstruction). Finally, each training sample consisted of a 28×100 array of 28 feature columns and 100-time points, along with a 1×7 conditional vector of auxiliary information. The conditional vector consists of age, height, leg height, weight, gender, side, and duration of the gait cycle. Data extraction and processing previously listed were performed in MATLAB (R2021b, The MathWorks Inc).

Then, the data was imported to Python for further processing before training. The joint angle coordinate data was scaled using a MinMaxScaler (Scikit-learn library) with a feature range from -1 to 1. Force platform data were scaled using the MaxAbsScaler, which scales each feature to its maximum absolute value. The auxiliary data vectors were also scaled using the MinMaxScaler with a feature range from -1 to 1.

2.3 Model Architecture

The model used here is a type of Generative Adversarial Network (GAN) [8], where two neural networks (Generator and Discriminator) compete with each other in the form of a zero-sum game, where one agent's gain is another agent's loss. The Discriminator tries to identify whether a specific sample is fake (comes from the Generator) or real (comes from the actual data domain). The Generator model generates new synthetic data similar to the problem domain and tries to fool the Discriminator. The Generator takes a vector drawn randomly from the latent space to ensure that each generated data is unique. Given a training set, this technique learns to create new data within the same statistical distribution as the training set. Conditional GANs (cGANs) [14] allow for conditioning the network with additional information to generate samples from a specific distribution. This conditional information is usually class labels. Then, along with the random vector, a one-hot encoded vector of class labels is passed to the Generator to generate data for the given class. During training, this label (usually as one-hot encoded image channels) is also passed to the Discriminator. With this technique, we gain the ability to ask our model to generate samples of a specific type (Fig. 1).

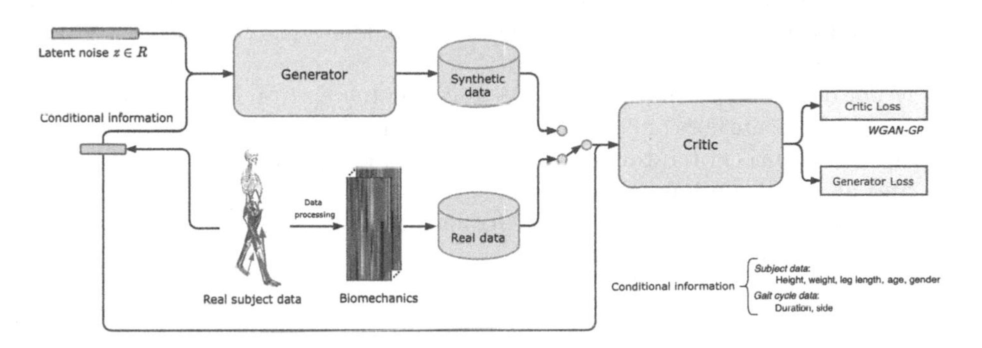

Fig. 1. A diagram of the proposed generative model.

The Generator network takes in the latent noise vector and the auxiliary data vector. The network consists of five blocks of convolutional layers with upsampling, followed by batch normalisation and ReLU activation function. The final block consists of a traditional convolutional layer with a Tanh activation function.

The Critic network is made up of four blocks of strided convolutional layers with layer normalisation and a LeakyReLU activation function ($\alpha = 0.2$), while the fourth and last block consists of a convolutional layer with a flattening layer and a linear activation function. Each data point of the auxiliary vector is transformed into an array of the same dimension as the data samples (28×100) and the transformed arrays are fed into the Critic during training as channels along with the corresponding data samples, resulting in 8x28 $\times$ 100 inputs.

2.4 Model Training

The model training is based on the Wasserstein GAN with gradient penalty [9]. The Wasserstein GAN was introduced to improve the traditional GAN training with improved training stability [2]. The original paper used weight clipping to enforce the Lipschitz constraint on Critic's model to calculate the Wasserstein distance. A later study proposed using a gradient penalty to enforce the Lipschitz constraint [9]. The loss function is

$$\begin{aligned} L = & - \mathop{\mathbb{E}}_{\mathbf{x} \sim p_{\mathbf{r}}} [D(\mathbf{x})] + \mathop{\mathbb{E}}_{\tilde{\mathbf{x}} \sim p_{\mathbf{g}}} [D(G(\mathbf{z}))] \\ & + \lambda \cdot \mathop{\mathbb{E}}_{\hat{\mathbf{x}} \sim p_{\hat{\mathbf{x}}}} (\|\nabla_{\hat{\mathbf{x}}} D(\hat{\mathbf{x}})\|_2 - 1)^2], \end{aligned} \tag{1}$$

where L represents the training loss of the Wasserstein GAN Critic with gradient penalty, where D denotes the Discriminator (also referred to as Critic), and G denotes the Generator. The first term represents the expectation over the distributions of real samples ($x \sim p_r$). The second term represents the expectation over generated samples ($\tilde{x} \sim p_g$). The third term in the equation corresponds to the gradient penalty term with a scaling factor λ, where $\hat{x}$ is uniformly sampled from $\tilde{x}$ and x. For the Generator, the loss is calculated by maximising the Critic's prediction of the Generator's fake data.

A noise vector of dimension 1×64 and a batch size of 128 were used. As proposed by the authors of [9], during training, the Critic was updated five times ($n_{critic} = 5$) for each iteration of the Generator, along with a scaling factor λ of 10, and an Adam optimiser with parameters $\beta_1 = 0$ and $\beta_2 = 0.9$ to minimise the loss functions [12]. The GAN was trained for 200 epochs with $\alpha = 0.0001$ learning rates for both networks.

2.5 Fine-Tuning

To examine the effects of fine-tuning the network with data from new subjects, the network was fine-tuned using 1%, 2% and 5% of the validation data. With this method, we aim to see how much data is needed for generating realistic subject-specific samples. For this specific database, 1% amounts to 15–20 gait cycles per subject, 2% to 30–40 gait cycles, and 5% to 70–100 gait cycles). A 5-fold cross-validation was used, splitting the data by subjects into training and validation sets. The resulting validation sets contained 4 or 5 subjects. A total of 2000 epochs and learning rates of 0.00005 were used for fine-tuning the models.

For fine-tuning evaluation, all graphs and scores were computed from the data not included in the fine-tuning dataset.

2.6 Evaluation

Comparison of synthetic and real GRF and kinematic coordinate trajectories was performed using Statistical Parameter Mapping (SPM) two-tailed paired t-test ($\alpha = 0.05$), a methodology that uses statistical analyses to quantify and map significant differences or relationships across a set of biomechanical parameters, providing insights into the spatial and temporal patterns of movement-related variations between the two groups [17]. For each subject, the percentage of significant difference between randomly selected 25 real trajectories and the corresponding generated trajectories was computed for all 28 features.

A qualitative analysis was also performed using t-SNE (t-distributed Stochastic Neighbour Embedding) visualisation [13], an unsupervised statistical method for visualising high-dimensional data.

3 Results

A comparison of synthetic and real kinematic/kinetic trajectories for one subject can be seen in Fig. 3. The shaded areas on the images show significant differences (from the paired t-test) between the synthetic and real data. On average, a 9.6% difference was observed from the SPM analysis. For simplicity, only data from the left leg is plotted (17 features of 28).

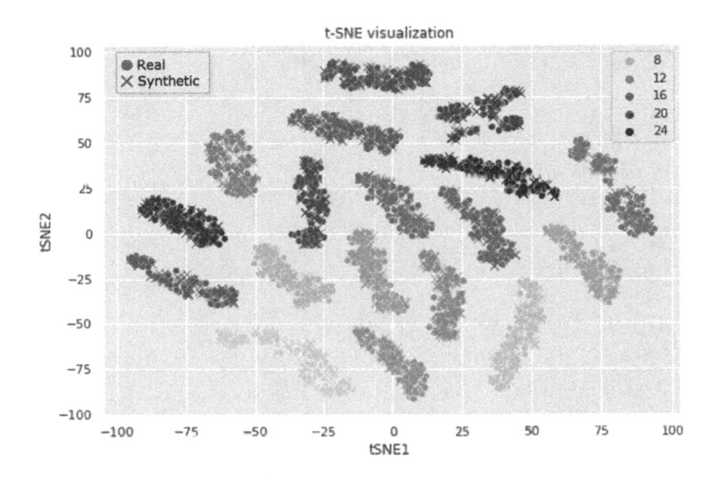

Fig. 2. t-SNE visualisation of the real and synthetic data. Each data point represents a unique gait cycle, with real data as circles, and generated data as crosses. Each color represents a unique subject.

Unsupervised dimensionality reduction visualisation shows that the model is able to generate distinct data for each subject (see Fig. 2).

3.1 Fine-Tuning

In Fig. 4, data is visualised with t-SNE dimensionality reduction. There, it can be seen that the model is unable to generalise to synthesise data from unseen subjects. When fine-tuning the model with just 1% of test data, the model can generate data of similar distribution as the target data. Increasing the size of the fine-tuning set to 2% and 5% further increases the variability of the generated data.

Visual comparison of biomechanical data from one test subject can be seen in Fig. 5 after fine-tuning with 1% of data.

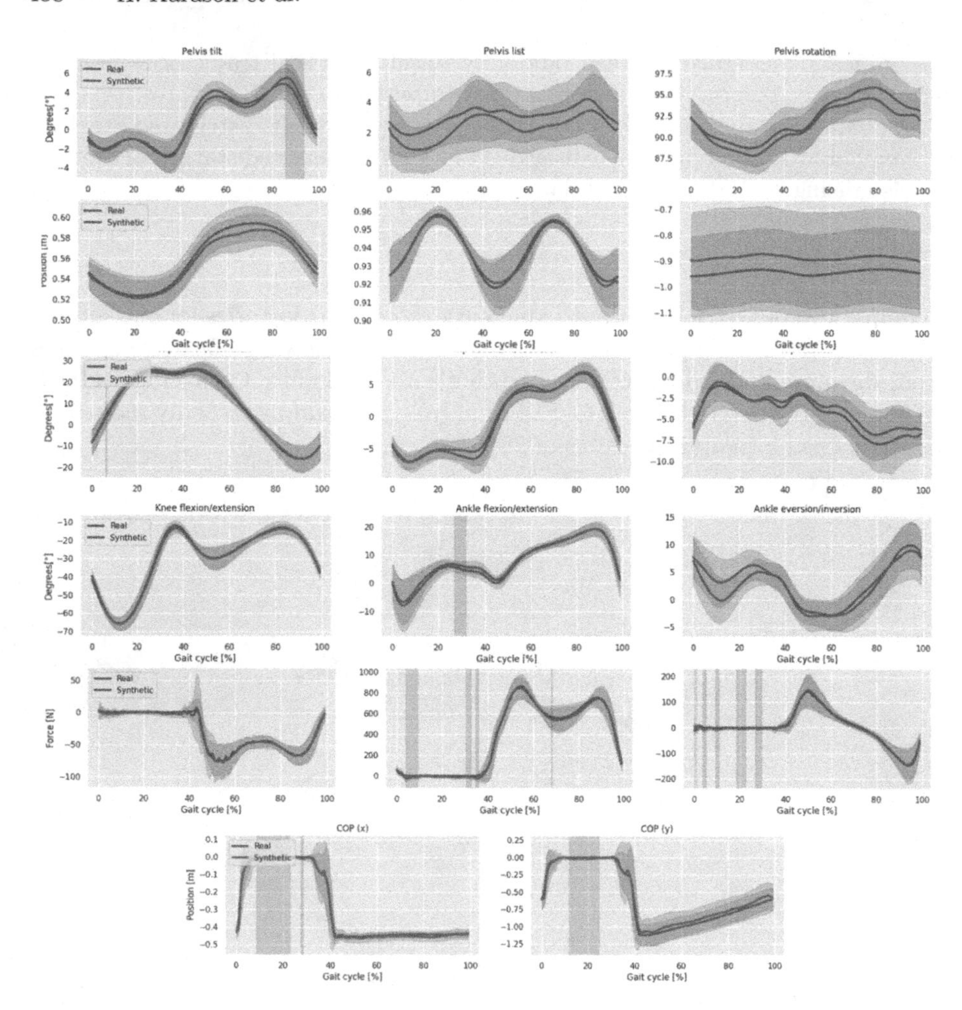

Fig. 3. Comparison of real and generated biomechanics from one subject. Pelvis positional (x-, y-, and z-axis), joint angle coordinates, ground reaction forces, and centres of pressure. The shaded areas show significant differences between the two groups.

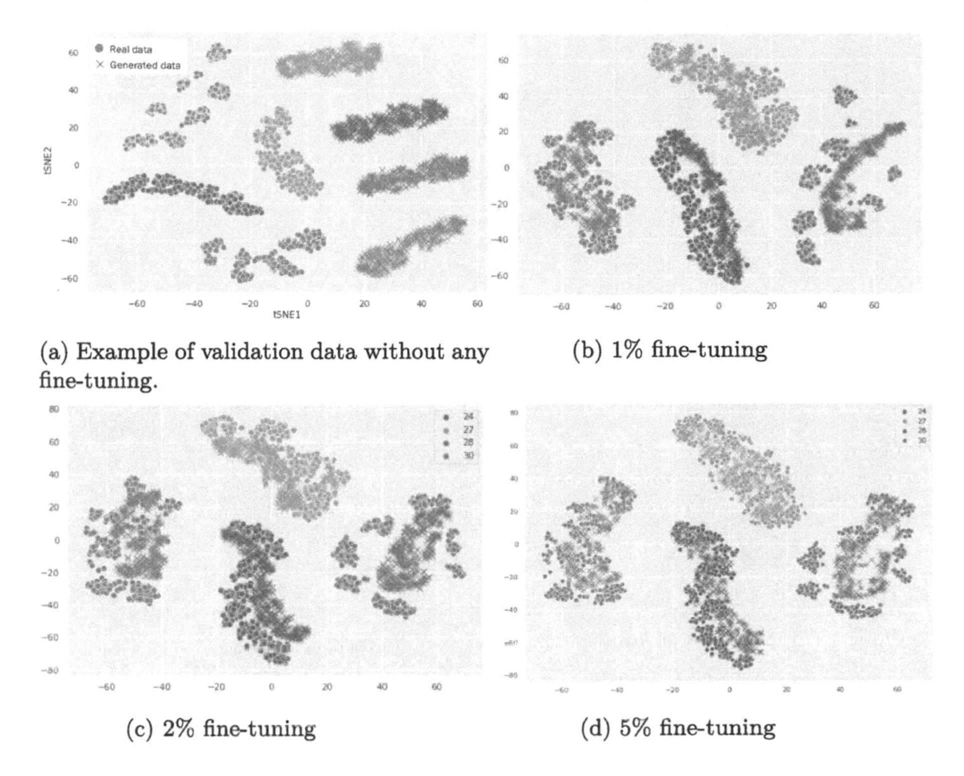

(a) Example of validation data without any fine-tuning.

(b) 1% fine-tuning

(c) 2% fine-tuning

(d) 5% fine-tuning

Fig. 4. t-SNE visual comparison. Example of one fold from the 5-fold cross-validation. Each point on the graph represents a unique gait cycle, and the different colours represent unique subjects, with circles as real gait cycles and crosses as synthetic gait cycles. For simplicity, only data from the left leg are used.

4 Discussion

Here, a GAN network was proposed to synthesise realistic data of human movement and the associated forces, conditioned with anthropometric measures of the subject and gait parameters. However, our preliminary results suggest that our hypothesis, which states that a generative network can be trained to generate realistic movement patterns using only anthropometric measures of the subject, is incorrect. Several limitations of the study may explain these findings. The dataset used in this study consisted of twenty-two subjects, with only seventeen or eighteen subjects used for training the model. It is possible that this sample size was insufficient for the model to learn the relationships between subject measures and movement patterns effectively (Table 1).

On the other hand, an interesting finding of this study is that the pre-trained model can be fine-tuned using a relatively very small amount of data from new subjects (15–20 samples for each subject) to synthesise realistic and diverse data. This implies that the model has the potential to enhance biomechanics datasets and reduce the criticality of collecting large amounts of data from each subject

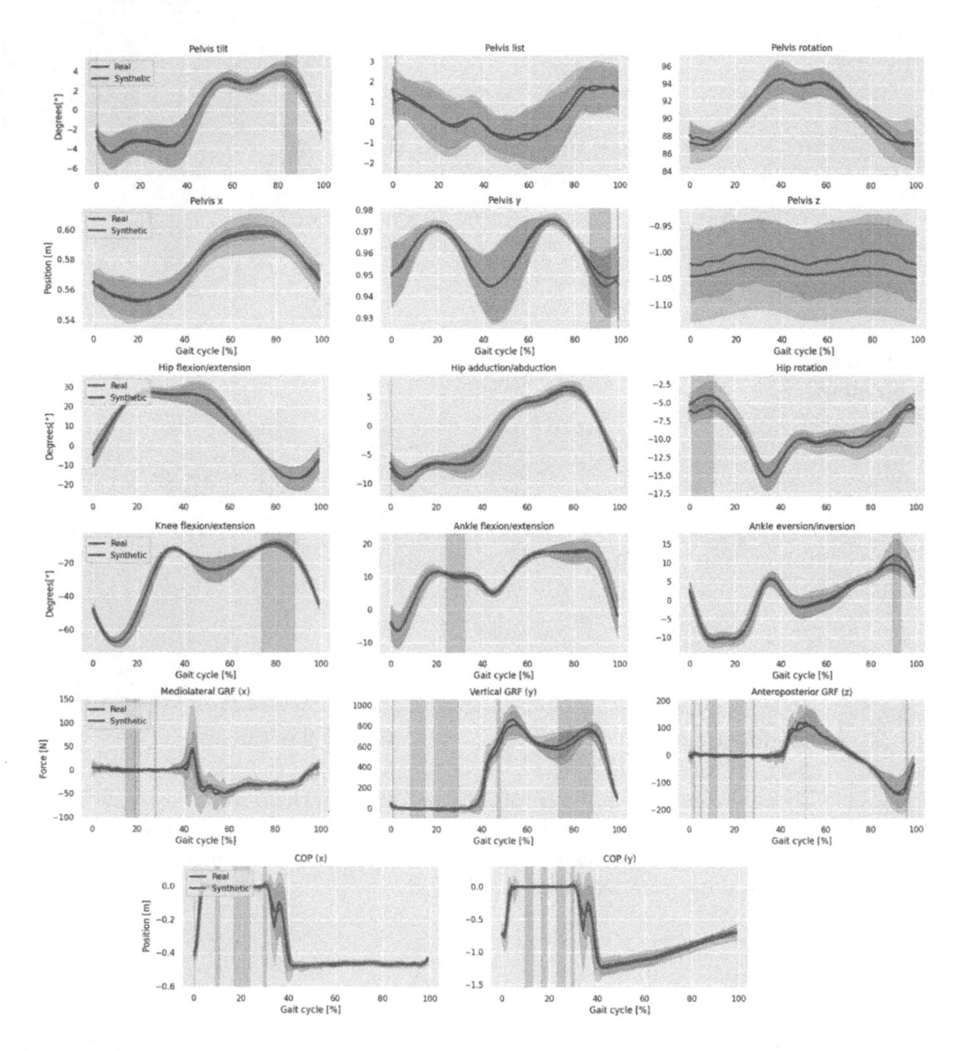

Fig. 5. Comparison of real and generated biomechanics from one subject after fine-tuning with just 1% of data. Pelvis positional (x-, y-, and z-axis), joint angle coordinates, ground reaction forces, and centres of pressure. The shaded areas show significant differences between the two groups.

Table 1. Effects of fine-tuning on the percentage of statistical difference the paired t-test.

Amount of fine-tuning	% of significant difference
0%	87.3
1%	11.29
2%	9.47
5%	9.11

by generating additional samples that capture variations in human movement and forces.

There are several inherent limitations in this study. Firstly, the performance of the model in enhancing a wearable sensor dataset for training a data-driven prediction model was not directly evaluated. Secondly, as mentioned earlier, the number of subjects on which the model was trained could be inadequate for generalising the effects of anthropometric parameters on human movement patterns.

No appropriate metric has been proposed for comparing synthetic and real data in this particular modality. To evaluate the quality of generated images in comparison to real images produced by generative models, the Fréchet inception distance (FID) [10] has been commonly utilised. This score is computed based on the outputs (excluding the fully connected layers) of the Inception v3 model [22], which is trained on the ImageNet database [5]. This approach has been applied to other data modalities like audio [11] and molecules for drug discovery [18], and could possibly be employed for biomechanics data as well.

Future steps will include evaluating the model's performance in enhancing a wearable sensor dataset through data-driven models and comparing the performance to other data augmentation techniques, such as optimal control simulations. Optimisation through hyperparameter tuning should be performed to improve the models' performance further. Increasing the number of subjects in the dataset would provide a more robust understanding of the effects of anthropometric parameters on human movement patterns. Additionally, potential applications and limitations of this generative network have yet to be explored in other domains of biomechanics and human motion analysis.

To our knowledge, this is the first network proposed that generates biomechanical data while allowing for conditioning the generation with anthropometric measures of the subject and reconstructing the gait cycles to the time domain. A model capable of generating synthetic data that represents realistic and physiologically plausible patterns of human movement and the associated forces could be useful for various applications, such as simulations and training ML models. This contribution opens up avenues for further investigation and development in the field of generative models for biomechanics research.

References

1. Alzantot, M., Chakraborty, S., Srivastava, M.B.: Sensegen: a deep learning architecture for synthetic sensor data generation. CoRR abs/1701.08886 (2017). http://arxiv.org/abs/1701.08886
2. Arjovsky, M., Chintala, S., Bottou, L.: Wasserstein gan (2017)

3. Bicer, M., Phillips, A.T., Melis, A., McGregor, A.H., Modenese, L.: Generative deep learning applied to biomechanics: a new augmentation technique for motion capture datasets. J. Biomech. **144**, 111301 (2022). https://doi.org/10.1016/j.jbiomech.2022.111301, https://www.sciencedirect.com/science/article/pii/S0021929022003426
4. Camargo, J., Ramanathan, A., Flanagan, W., Young, A.: A comprehensive, open-source dataset of lower limb biomechanics in multiple conditions of stairs, ramps, and level-ground ambulation and transitions. J. Biomech. **119**, 110320 (2021). https://doi.org/10.1016/j.jbiomech.2021.110320, https://www.sciencedirect.com/science/article/pii/S0021929021001007
5. Deng, J., Dong, W., Socher, R., Li, L.J., Li, K., Fei-Fei, L.: Imagenet: a large-scale hierarchical image database. In: 2009 IEEE Conference on Computer Vision and Pattern Recognition, pp. 248–255 (2009). https://doi.org/10.1109/CVPR.2009.5206848
6. Dorschky, E., Nitschke, M., Martindale, C.F., van den Bogert, A.J., Koelewijn, A.D., Eskofier, B.M.: Cnn-based estimation of sagittal plane walking and running biomechanics from measured and simulated inertial sensor data. Front. Bioeng. Biotechnol. **8** (2020). https://doi.org/10.3389/fbioe.2020.00604, https://www.frontiersin.org/articles/10.3389/fbioe.2020.00604
7. Gholami, M., Napier, C., Menon, C.: Estimating lower extremity running gait kinematics with a single accelerometer: a deep learning approach. Sensors **20**(10) (2020). https://doi.org/10.3390/s20102939, https://www.mdpi.com/1424-8220/20/10/2939
8. Goodfellow, I.J., et al.: Generative adversarial networks (2014)
9. Gulrajani, I., Ahmed, F., Arjovsky, M., Dumoulin, V., Courville, A.: Improved training of wasserstein gans (2017)
10. Heusel, M., Ramsauer, H., Unterthiner, T., Nessler, B., Hochreiter, S.: Gans trained by a two time-scale update rule converge to a local nash equilibrium (2018)
11. Kilgour, K., Zuluaga, M., Roblek, D., Sharifi, M.: Fréchet Audio Distance: A Reference-Free Metric for Evaluating Music Enhancement Algorithms. In: Proc. Interspeech 2019, pp. 2350–2354 (2019). https://doi.org/10.21437/Interspeech.2019-2219
12. Kingma, D.P., Ba, J.: Adam: A method for stochastic optimization (2017)
13. van der Maaten, L., Hinton, G.: Visualizing data using t-SNE. J. Mach. Learn. Res. **9**, 2579–2605 (2008). http://www.jmlr.org/papers/v9/vandermaaten08a.html
14. Mirza, M., Osindero, S.: Conditional generative adversarial nets (2014)
15. Mundt, M., et al.: Estimation of gait mechanics based on simulated and measured imu data using an artificial neural network. Front. Bioeng. Biotechnolo. **8** (2020). https://doi.org/10.3389/fbioe.2020.00041, https://www.frontiersin.org/articles/10.3389/fbioe.2020.00041
16. Norgaard, S., Saeedi, R., Sasani, K., Gebremedhin, A.H.: Synthetic sensor data generation for health applications: a supervised deep learning approach. In: 2018 40th Annual International Conference of the IEEE Engineering in Medicine and Biology Society (EMBC), pp. 1164–1167 (2018)
17. Pataky, T.C.: One-dimensional statistical parametric mapping in python. Comput. Methods Biomech. Biomed. Engin. **15**(3), 295–301 (2012). https://doi.org/10.1080/10255842.2010.527837, https://doi.org/10.1080/10255842.2010.527837, pMID: 21756121
18. Preuer, K., Renz, P., Unterthiner, T., Hochreiter, S., Klambauer, G.: Fréchet chemblnet distance: a metric for generative models for molecules. CoRR abs/1803.09518 (2018), http://arxiv.org/abs/1803.09518

19. Rapp, E., Shin, S., Thomsen, W., Ferber, R., Halilaj, E.: Estimation of kinematics from inertial measurement units using a combined deep learning and optimization framework. J. Biomech. **116**, 110229 (01 2021). https://doi.org/10.1016/j.jbiomech.2021.110229
20. Sharifi Renani, M., Eustace, A., Myers, C., Clary, C.: The use of synthetic imu signals in the training of deep learning models significantly improves the accuracy of joint kinematic predictions. Sensors **21**, 5876 (08 2021). https://doi.org/10.3390/s21175876
21. S.L., D., F.C., A., A.S., A., P., L., C.T., H.A.J., E., G., D.G., T.: Opensim: Open-source software to create and analyze dynamic simulations of movement (2007)
22. Szegedy, C., Vanhoucke, V., Ioffe, S., Shlens, J., Wojna, Z.: Rethinking the inception architecture for computer vision. In: 2016 IEEE Conference on Computer Vision and Pattern Recognition (CVPR), pp. 2818–2826 (2016). https://doi.org/10.1109/CVPR.2016.308
23. Wang, J., Chen, Y., Gu, Y., Xiao, Y., Pan, H.: Sensorygans: an effective generative adversarial framework for sensor-based human activity recognition. In: 2018 International Joint Conference on Neural Networks (IJCNN), pp. 1–8 (2018). https://doi.org/10.1109/IJCNN.2018.8489106

Avatar Reaction to Multimodal Human Behavior

Baptiste Chopin[1], Mohamed Daoudi[2,3](✉), and Angela Bartolo[4,5]

[1] Université Côte D'Azur, Inria, France
[2] Univ. Lille, CNRS, Centrale Lille, Institut Mines-Télécom, UMR 9189 CRIStAL, 59000 Lille, France
[3] IMT Nord Europe, Institut Mines-Télécom, Univ. Lille, Centre for Digital Systems, 59000 Lille, France
mohamed.daoudi@imt-nord-europe.fr
[4] CNRS, UMR 9193-SCALab-Sciences Cognitives et Sciences Affectives, University of Lille-SHS, Villeneuve d'Ascq, 59000 Lille, France
[5] Institut Universitaire de France (Paris), Paris, France

Abstract. In this paper, we propose a virtual agent application. We develop a virtual agent that reacts to gestures and a virtual environment in which it can interact with the user. We capture motion with a Kinect V2 camera, predict the end of the motion and then classify it. The application also features a facial expression recognition module. In addition, to all these modules, we include also OpenAI conversation module. The application can also be used with a virtual reality headset.

Keywords: avatar reaction · human motion prediction · facial expression

1 Introduction

The development of a virtual agent that can interact with users in a virtual environment has potential applications in fields such as gaming, education, and healthcare. Interaction with these virtual agents is provided and has usually been studied, through several modalities: expression recognition and generation [6], action recognition and interaction with objects [5], multi-modal input and output [6]. Some methods can even be used for both virtual agents and robots [8]. However, the challenge of building an interactive virtual agent differs from that of building an interactive robot. Heting et al. [13] discuss the use of virtual agents, specifically avatars, in human-computer interactions. The paper presents an emotionally responsive avatar named Diana, which recognizes human affect and responds with natural facial expressions to improve the user experience in

This work was supported by French government funding managed by the National Research Agency under grants ANR-21-ESRE-0030 (CONTINUUM), by CNRS through the 80-Prime program, and ANR-16-IDEX-0004 ULNE.

G. L. Foresti et al. (Eds.): ICIAP 2023 Workshops, LNCS 14365, pp. 494–505, 2024.
https://doi.org/10.1007/978-3-031-51023-6_41

the interaction. The more recent work focus on facial expression generation in dyadic interactions [11], while our work is interesting by the reaction of an avatar to action.
Daoudi et al. [5] propose a new computational approach to identify human social intention in action by analyzing the trajectories of the human arm. The proposed method for predicting social intention is based on the trajectories of the human arm, which has been used to improve social communication between humans and virtual agents. To our knowledge there exist no virtual agent that is able to react to communicative gesture. Our virtual agent is able to react to communicative gestures using non-intrusive data acquisition devices while having realistic behavior.

2 Overview

We set several constraints to the application we build: the avatar appearance must be realistic, the environment must be believable, the user's motion data must be collected in an non-intrusive manner, the reaction from the agent must be related to the action of the user and realistic, finally the reaction motion must start quickly after the user perform a gesture. To respect the constraints we set for our virtual agent application, we must first be able to react quickly to the user gesture. This implies that, in a short time, we must understand the motion the user is performing and produce a reaction to this motion. In these conditions, waiting for the user to complete the motion is not ideal. Therefore, we choose to start by predicting the end of the user motion based on its start. Then we classify the predicted motion and send the information to the avatar that will react accordingly by selecting the corresponding prerecorded reaction motion. The result of the classification will let the system decide which reaction to perform. However, as we will see in Sect. 3.1, all the gestures we choose indicate discomfort for the user. Depending on the degree of said discomfort, a reaction might not be needed. To take this possibility into account, we add a facial expression recognition module to decide if the virtual agent has to react. Figure 1 shows the overall architecture of the application.

A final issue to address is the need for data. While facial expression databases are plentiful and we do not need data to generate the virtual agent's response, we do need data to train networks to predict and recognize the user's gesture. In summary, the main contributions of our work are:

1. Development of a virtual agent that reacts to gestures and interacts with the user in a virtual environment.
2. A new dataset of 3D gestures of 11 subjects to train our models that are used with the virtual agent, is proposed. Each 3D gesture is recorded by motion capture. The motions were performed by 11 healthy volunteer participants who were asked to perform gestures indicating their mental state.

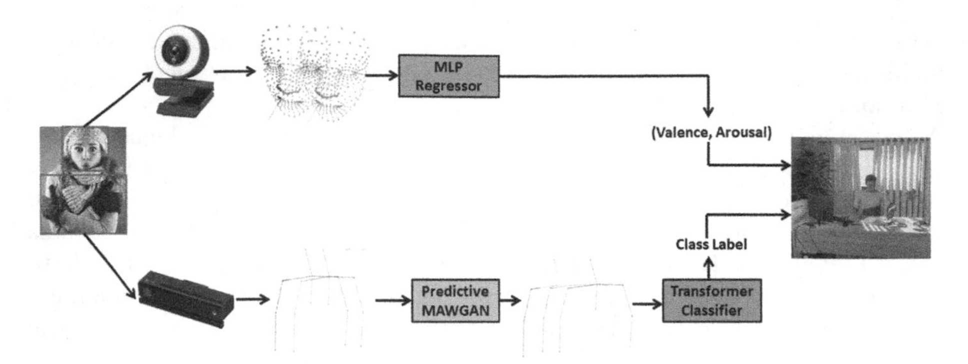

Fig. 1. Overview of the virtual agent application architecture

Table 1. Content of the gesture dataset

Scenario	Motions	number of samples MoCap	number of samples Kinect
I am cold	rub both arms with hands	115	109
I am hot	wave one or both hands near the neck/ mouth	115	115
This is too loud	cover both ears with hands	113	99
This smells bad	pinch nose with one hand or cover mouth and nose with both hands	115	110
It is too bright	cover eyes with one or both hands	111	109
Hello	wave with one or both hands	100	121
Idle	random and meaningless motions	59	65
Total		728	728

3 Proposed Method

3.1 Motion Database

There is currently no database containing the kind of communicative gestures that we are looking for when interacting with the avatar (e.g. "I'm cold", "My ears hurt"...). Most motion databases contain motions that represent action performed alone (walking, kicking...) or in interaction with other persons (exchanging an item, fighting...). In the latter, the interactions require physical contact between the two parties, either directly or through an object. Gesture databases are also available, however, the gestures included are gestures that do not require a reaction from another person. Furthermore, the existing gesture databases that could be used to make another person react are culturally coded, the motion performed might not have the same meaning or no meaning at all for a French person. To get data that we could use in our scenarios we build a database of communicative motions.

We show in Table 1 the different scenarios and motions used in our database as well as a description and the number of samples recorded. Since different motions can be used for each of the scenarios given to the participants, they were also instructed to avoid certain types of motion (*e.g* they were not allowed to pull the collar of their shirt during the "I am hot" scenario). We added an additional class of motions containing random idle motions to train our networks. It is used to recognize when no motion of particular significance is being performed. By doing that, we can avoid the misclassification of random motions as meaningful ones leading to an unwanted reaction from the avatar. The scenarios were selected from situations where people might perform a gesture to indicate their mental state but that can also be performed to elicit a response from another person (e.g. the gesture of "coming here" oblige observers to modify their behavior. In this case, people move toward the executor). Mental state gestures are gestures that communicate a feeling (*e.g.* I am cold). In all our scenarios the motion indicates a discomfort and the avatar must react by helping the user (Fig. 2).

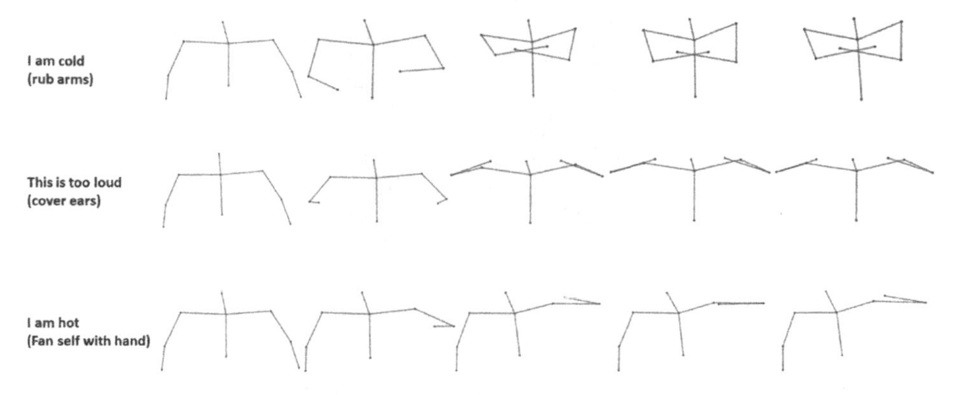

Fig. 2. Examples of motion from the new database (motion capture). We show three different classes. Visualization obtained after motion sequences were cut and resampled to have a duration of 3*s* (see Sect. 3.3)

The motion capture was performed at the Equipex Continuum using the facilities of the Fédération de Recherche "Sciences et Cultures du visuel" at Tourcoing, France. We recruited 11 healthy volunteer participants among the students and staff of the lab team (8 males, 3 females; age between 20–60 yrs). Participants did not receive financial compensation for their participation in the study. Participants were asked to stand in front of the Kinect at a place marked by a cross on the ground. Figure 4 shows the view participants had. Their initial position was then adjusted depending on their height (to allow the capture of all markers by the motion capture system). They were asked to perform the gestures from Table 1. The motions from class "Idle" were only performed by 2 participants as they are random, meaningless, motions. The participants performed each motion at least 10 times. Each participant spent between 30 and 45 min to complete the entire recording session. The motions are captured on two different systems. The first is a high-precision motion capture system with 6 Qualisys motion capture cameras and nine markers on the participant's body: one on the forehead between the eyebrows, one at the base of the neck, and one above the navel, for each arm we have markers on the top of the shoulder, on the elbow and on the outer side of the wrist. Figure 3 shows the position of the markers on a participant. The motion capture system is calibrated to have an average positional error of less than 2 mm. Each sample is recorded for exactly 5 s regardless of the actual duration of the motion at 100 fps.

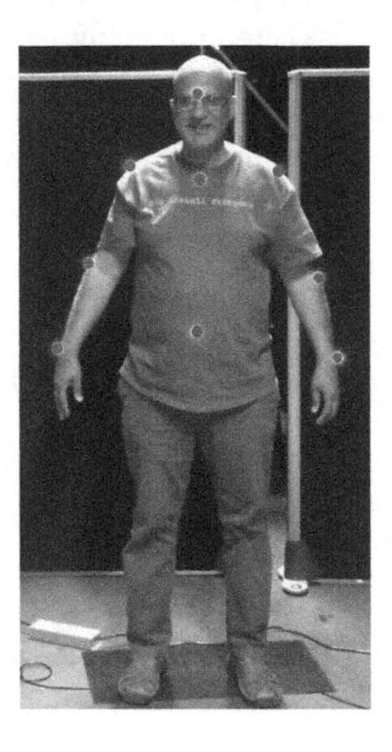

Fig. 3. Motion capture markers position

The second capture system we use consists of a single Kinect V2 camera. We decided to also capture the motion with a Kinect camera since the motion capture system would give us extremely precise data that we will not be able to obtain with the system we have put in place for our application. The Kinect allows us to capture real-world data of the motion present in the motion capture database which will help our models become more resilient to the noise there is in real-world data. With the Kinect, we capture joint positions of the entire body but then only keep the nine joints corresponding to those from the Motion capture. The capture setup is illustrated in Fig. 4. Due to the difficulty of setting up synchronous capture between the Kinect and the motion capture system, the capture from the Kinect is started and stopped by an experimenter following the instruction from the experimenter working with the motion capture system. Due to this, the duration of the Kinect capture varies between 73 and 530 frames with a median of 155 frames and a mean of 157 frames with a framerate of 30 fps which means that on average the motion duration average is close to 5 s. Samples that were too noisy or where joint positions were lost for too long were removed. When a joint position was lost for a short amount of time we estimate its position to be the last known position. All recorded samples were checked by a human curator before being used in the database. This amount to 728 motion samples for the MoCap data and 728 samples for the Kinect data.

Fig. 4. Motion capture and Kinect setup. Point of view of the captured subject.

3.2 Data Capture

To avoid having to wear captors when using the application, we can capture the motion data using depth cameras or RGB cameras. Both allow for the capture of skeletons through different toolkits. For example, OpenPose [3] for RGB camera or the Kinect SDK for the depth cameras. We tried a RGB camera with BlazePose [1], Kinect depth camera, and Intel real sense depth camera with NuiTrack SDK[1]. The result led us to choose the Kinect camera. Blazepose, while good for 2D joint detection, fails to correctly estimate the depth coordinates and causes the recorded data to be noisy. The intel real sense with NuiTrack often fails completely to detect the joint position of the arm and the recorded motion looks nothing like the actual motion. On the other hand, Kinect gives an accurate recording of the motion even if it is quite sensitive to occlusion.

For the face data, we use a RGB camera with facial landmark detection software. Out of the many libraries that exist (OpenFace[2], dlib[3]...) we choose to use Mediapipe FaceMesh[4] which uses a facial detector [2] coupled with a face mesh predictor [9]. We choose to use this method due to the speed and light weight of the model but also because of its accuracy and higher resistance to occlusion. Even if it uses a face mesh instead of proper landmarks, the results on expression recognition are similar to other landmarks detectors.

3.3 Action Prediction and Recognition

Action Prediction. Prediction Module is based on [4]. We have shown that the prediction speed is fast enough to be used in real time which is mandatory

[1] https://nuitrack.com/.
[2] https://github.com/TadasBaltrusaitis/OpenFace.
[3] http://dlib.net/.
[4] https://google.github.io/mediapipe/solutions/face_mesh.

to be used in our application. In [4] we predicted the same number of frames as what we were given as input. Here, however, we take an input of 25 frames ($1s$) to predict 50 frames ($2s$). The Predictive Network is trained on our dataset, where we mix the Kinect and Mocap data. The motion sequences were cut and resampled to have a duration of $3s$.

Table 2. Recognition accuracy on our dataset. Comparison between $1s$ of motion captured by the Kinect, $3s$ of motion captured by the Kinect, and $1s$ of motion captured by the Kinect concatenated to $2s$ of predicted motion

	Capture $1s$	Capture $3s$	Capture $1s$ + Prediction $2s$
Accuracy	47.7%	63.6%	60.2%

Action Recognition. To classify the predicted motions we use a Transformer encoder that uses spatial attention and the skeleton adjacency module presented in [4]. The Transfomer encoder is followed by a simple multi-layer perceptron to classify the motions. Figure 5 shows the architecture of the action recognition network. . We then train the network on $3s$ of motion composed of $1s$ of real data and the $2s$ of motion predicted by the prediction module. We train with predicted data instead of real data because the recognition module will always receive predicted data. This module estimates the class of the concatenation of the recorded motion and the predicted motion. Table 2 shows the accuracy of the network on our dataset. For this experiment, we split the dataset between train and test. Out of the 11 subjects we randomly choose subject 2 and subject 7 to be the test subjects. For class "Idle", since the samples are not connected to a particular subject we randomly take 11 samples from the Kinect data and 11 samples from the Mocap data to create the test set. With this, we have 1192 training samples and 264 test samples. We compare the result of $3s$ of captured data, $1s$ of captured and $1s$ of captured data completed by $2s$ of predicted data. For each modality, we train the network on the corresponding data. As for the prediction network we use both the Mocap and Kinect data. We see that using $3s$ of data the network correctly classifies 63.6% of action. Some actions of our dataset like "Hello" and "I am hot" (see Table 1 for the detailed motions) are similar and often confused which can explain the relatively low score. If we use only $1s$ of data we see a decrease in performance down to 47.7% accuracy. If we use our prediction network to predict $2s$ of motion and combine it with the $1s$ of input data to get a $3s$ sequence we increase the results to 60.2% close to the original 63.6%. This highlights the advantage of using our prediction network. We can are able to classify motion with only one-third of the data nearly as accurately. In practice this allows the avatar to react thrice as fast since the prediction process only takes a few milliseconds.

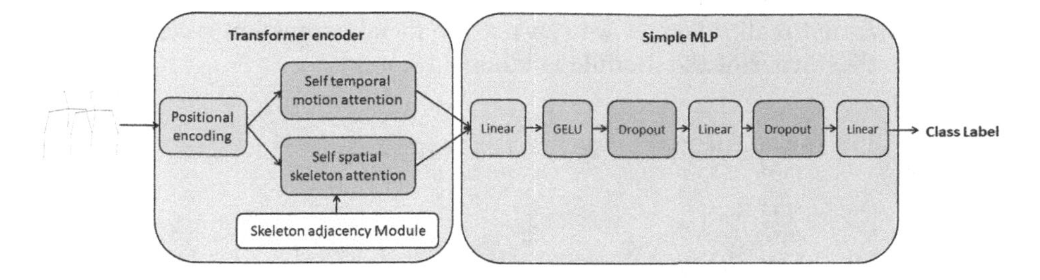

Fig. 5. Architecture of our Interformer-based motion classifier

3.4 Facial Expression Recognition

To avoid reacting to random motion or communicative motion that does not require a reaction from the avatar, we add a facial expression recognition module that estimates the valence and arousal value of the user expression. We use a simple MLP consisting of 4 Dense layers with dropout that we train on the AFEW-VA [10] dataset using landmark data captured with Mediapipe. While the network might seem extremely simple, it is efficient. On the AFEW-VA dataset, with valence and arousal values comprised between −10 and 10, we obtain an error of only 1. State of the art method might perform even better but since our goal is only to set a threshold for discomfort while being fast we do not need as much accuracy as the state-of-the-art methods. We want to prevent the avatar from reacting when the user is in a neutral or positive mood since the motion performed might not indicate discomfort in that case. To do so we ignore motions performed while $valence > 0$ as it correlates with positive feelings. To avoid reacting when the user is in a neutral mood, we decide to only consider motion performed while $valence < -3$ and $|arousal| > 3$. We capture the face of the user with a second camera independent from the Kinect. The Kinect must be far enough to see the entire upper body and can not be simultaneously used to capture a detailed face.

4 Avatar Environment

Our avatar module is a Unity application where our avatar evolves in a scene consisting of a room with furniture as shown in Fig. 7. The room contains a desk behind which the avatar sits, the avatar can interact with several objects in the scene: cigarette, window, curtains, HiFi system, and a keyboard. We choose a simple appearance for the avatar with a tee shirt so that users will not focus on its appearance but rather on its reaction. When reacting to the user gesture the avatar will stand up (except for the "This smells bad" scenario) and perform an action behind the desk (See Table 3 for a complete description of each reaction). This module can also work in virtual reality and the entire room is modelized in unity as seen in Fig. 6. When using the virtual reality version, since the user needs to wear a virtual reality helmet we can not use facial expression recognition. In that case, the output of the module is ignored.

Fig. 6. Aerial view of the scene in the unity builder (ceiling is removed). The entire room is modelized for virtual reality.

Table 3. Avatar reaction to the gestures described in Table 1

Scenario	Motions
I am cold	If the window is open: stand up, close the window, sit down
I am hot	If the window is closed: stand up, open the window, sit down
This is too loud	if the volume is high: stand up, turn the volume down on the hi-fi, sit down system
This smells bad	if the cigarette is lit: put out cigarette
It is too bright	if the curtain is open: stand up, close the curtain, sit down
Hello	Stop idle motion and turn to face and look at the user
Before turning attention to the user	look at the computer screen while typing on the keyboard. Sometimes turns its head to look at the notebook
Attention state	look at the user and sometimes at objects on the desk

The full application combines the aforementioned modules. The motion data is captured with a Kinect V2 camera and the face data with any RGB camera such as a webcam. From the motion data from the Kinect, we keep only the 3D skeleton data that we send to our prediction network every 25 frames. In the prediction network, the data will first be normalized and transformed into The Square Root Velocity (SRVF) [12], [7], [4]. Then we predict the 50 future

Fig. 7. Our avatar in its resting state.

frames of the motion in SRVF format and rebuild the corresponding skeleton sequence. The predicted sequence is concatenated with the sequence recorded by the Kinect for a total length of 75 frames. This concatenated sequence is fed to the action recognition network which outputs a class label. Simultaneously 25 frames of the face data are sent to the facial expression recognition module where we extract the facial landmark from the RGB data with MediaPipe's face mesh. The landmarks are treated by the network which outputs arousal and valence values. Our scenarios are all negative ones, the user is bothered by something and wants the avatar to do something about it. We choose valence and arousal intervals that contain expressions that users may perform in these scenarios. If the values obtained by the network are outside these intervals we consider that the motion performed by the user does not show a discomfort needing a reaction or that it is unrelated to the scenarios. In these cases, we prevent the avatar from reacting. Depending on the output of the motion recognition and of the facial expression recognition network the application will decide if the avatar needs to react and if so start the appropriate motion. If no reaction motion is required because the predicted class is 'Idle' or due to the valence and arousal values, the avatar stays in its 'Attention' state. This cycle is repeated until the application is closed. When using virtual reality users can move the camera with their heads and can move around in the room. However, the position of the avatar remains unchanged and a change in user position too drastic will interfere with the body data capture from the Kinect. Figure. 1 describes the full application.

In addition to the above modules, we have also developed a module for facial expression with VR headset. Indeed, recognizing emotions while wearing a VR headset is a much more complicated task, as the headset hides information. Vive offers a facial tracker that uses sensors to capture facial movements. Facial expression is defined on the basis of 38 characteristic movements, which are assigned a weight according to their presence in the expression. To recognize certain emotions (joy, anger, surprise, sadness), we selected from the 38 movements those that best defined the desired emotion, then set a threshold on the weight of these movements. If the weight associated with an emotion exceeds this threshold, then the user is considered to be expressing that emotion. To be able to converse with the avatar, we used the OpenAI API integrated into Unity, as well as Microsoft Azure's text-to-speech and speech-to-text services. We convert the user's voice to text and send it to the OpenAI model (we use the gpt-3.5-turbo model) using speech-to-text conversion. We then use text-to-speech conversion to make the model's response audible.

5 Conclusions and Future Work

In this paper, we propose a virtual agent application that reacts to gestures and a virtual environment in which it can interact with the user. The application captures motion with a Kinect V2 camera, predicts the end of the motion, and then classifies it. It also features a facial expression recognition module and an OpenAI conversation module. The application can be used with a virtual reality headset. In addition, we propose a new database of communicative motions that were recorded using a Kinect V2 camera. The motions were performed by 11 healthy volunteer participants who were asked to perform gestures indicating their mental state. One limitation of our work is to establish criteria to evaluate the human-avatar interaction.

Acknowledgments. The authors would also like to thank Hugo Pina Borges and Julian Hutin for performing part of the experimentation.

References

1. Bazarevsky, V., Grishchenko, I., Raveendran, K., Zhu, T.L., Zhang, F., Grundmann, M.: Blazepose: On-device real-time body pose tracking. ArXiv abs/2006.10204 (2020)
2. Bazarevsky, V., Kartynnik, Y., Vakunov, A., Raveendran, K., Grundmann, M.: Blazeface: Sub-millisecond neural face detection on mobile gpus. ArXiv abs/1907.05047 (2019)
3. Cao, Z., Hidalgo Martinez, G., Simon, T., Wei, S., Sheikh, Y.A.: Openpose: realtime multi-person 2D pose estimation using part affinity fields. IEEE Trans. Pattern Anal. Mach. Intell. **43**. 172–186 (2019)
4. Chopin, B., Otberdout, N., Daoudi, M., Bartolo, A.: 3-D skeleton-based human motion prediction with manifold-aware GAN. IEEE Trans. Biom. Behav. Identity Sci. **5**(3), 321–333 (2023). https://doi.org/10.1109/TBIOM.2022.3215067
5. Daoudi, M., Coello, Y., Descrosiers, P.A., Ott, L.: A new computational approach to identify human social intention in action. In: 2018 13th IEEE International Conference on Automatic Face & Gesture Recognition (FG 2018), pp. 512–516 (2018)

6. DiPaola, S., Yalçın, O.N.: A multi-layer artificial intelligence and sensing based affective conversational embodied agent p. 2
7. Drira, H., Ben Amor, B., Srivastava, A., Daoudi, M., Slama, R.: 3D face recognition under expressions, occlusions, and pose variations. PAMI **35**(9), 2270–2283 (2013)
8. Hua, M., Shi, F., Nan, Y., Wang, K., Chen, H., Lian, S.: Towards more realistic human-robot conversation: A seq2seq-based body gesture interaction system. In: 2019 IEEE/RSJ International Conference on Intelligent Robots and Systems (IROS), pp. 1393–1400 (2019). https://doi.org/10.1109/IROS40897.2019.8968038
9. Kartynnik, Y., Ablavatski, A., Grishchenko, I., Grundmann, M.: Real-time facial surface geometry from monocular video on mobile gpus. In: CVPR Workshop on Computer Vision for Augmented and Virtual Reality 2019. Long Beach, CA (2019). https://arxiv.org/abs/1907.06724
10. Kossaifi, J., Tzimiropoulos, G., Todorovic, S., Pantic, M.: Afew-va database for valence and arousal estimation in-the-wild. Image Vision Comput. **65** (02 2017). https://doi.org/10.1016/j.imavis.2017.02.001
11. Luo, C., Song, S., Xie, W., Spitale, M., Shen, L., Gunes, H.: Reactface: multiple appropriate facial reaction generation in dyadic interactions. CoRR abs/2305.15748 (2023)
12. Srivastava, A., Klassen, E., Joshi, S.H., Jermyn, I.H.: Shape analysis of elastic curves in Euclidean spaces. PAMI **33**(7), 1415–1428 (2011)
13. Wang, H., Gaddy, V., Beveridge, J.R., Ortega, F.R.: Building an emotionally responsive avatar with dynamic facial expressions in human-computer interactions. Multimodal Technol. Interact. **5**(3) (2021)

Author Index

G. L. Foresti et al. (Eds.): ICIAP 2023 Workshops, LNCS 14365, pp. 507–511, 2024.
https://doi.org/10.1007/978-3-031-51023-6

R

S

T

U

V

Printed in the United States
by Baker & Taylor Publisher Services